W9-CKR-678

PLANT BREEDING SYSTEMS

Second edition

A.J. RICHARDS

Department of Agricultural and Environmental Science
University of Newcastle upon Tyne
UK

CHAPMAN & HALL

London · Weinheim · New York · Tokyo · Melbourne · Madras

Published by Chapman & Hall, 2–6 Boundary Row, London SE1 8HN, UK

Chapman & Hall, 2–6 Boundary Row, London SE1 8HN, UK

Chapman & Hall GmbH, Pappelallee 3, 69469 Weinheim, Germany

Chapman & Hall USA, 115 Fifth Avenue, New York, NY 10003, USA

Chapman & Hall Japan, ITP-Japan, Kyowa Building, 3F, 2-2-1 Hirakawacho, Chiyoda-ku, Tokyo 102, Japan

Chapman & Hall Australia, 102 Dodds Street, South Melbourne, Victoria 3205, Australia

Chapman & Hall India, R. Seshadri, 32 Second Main Road, CIT East, Madras 600 035, India

First edition 1986
Reprinted 1990, 1994
Second edition 1997

Typeset in Palatino by Best-set Typesetter Ltd, Hong Kong
Printed in Great Britain at The University Press, Cambridge

ISBN 0 412 57440 3 (Hb) 0 412 57450 0 (Pb)

A catalogue record for this book is available from the British Library

Library of Congress Catalog Card Number: 96-72119

∞ Printed on permanent acid-free text paper, manufactured in accordance with ANSI/NISO Z39.48-1992 and ANSI/NISO Z39.48-1984 (Permanence of Paper).

Contents

Acknowledgements

I owe a special debt of gratitude to the research (graduate) students and associates who have worked under my supervision over nearly three decades. They have provided me with ideas, discussion, companionship and a stimulating environment, as well as with many interesting results, some of which are used in this book. Those who have been particularly associated with work on breeding systems have been Halijah Ibrahim, Liz Gynn (née Culwick), Mike Mogie, Henry Ford, David Stevens, Guido Braem, the late Chris Haworth, Jane Hughes, Fran Wedderburn, Kath Baker, Valsa Kurian, Hussen Al Wadi, Suleiman Haroun, Juno McKee, Michelle Tremayne, Liz Arnold and Amy Plowman.

Several colleagues from other UK Universities have provided me with constructive criticism and advice over the years, and among these I would particularly like to mention Terry Crawford (who read all of the first edition), Richard Abbott, Clive Stace and Quentin Kay. My friend Brian Waters produced many of the line figures for the first edition, many of which are reproduced here. I have taken the majority of the photographs in this volume, but I am grateful to Michael Proctor for permission to use some of his quite unsurpassed monochromes, and for the cover art work.

My especial thanks go to my wife Sheila for her unfailing support and encouragement for more than 30 years.

Preface

Undoubtedly, Charles Darwin can be considered as the father of plant breeding system studies. In his two books *The Effects of Cross and Self-Fertilisation in the Vegetable Kingdom,* and *The Different Forms of Flowers on Plants of the Same Species,* Darwin pioneered work on self-incompatibility, inbreeding depression, heteromorphy, gynodioecy, pollination biology and gene flow, although he himself had no concept of the gene.

The discovery of plant chromosomes, and the important role played by chromosomal mechanisms in plant evolution dominated plant evolutionary genetics in the first half of the 20th century. Despite the lead taken by Darwin, studies on plant breeding systems proceeded very slowly during this period, although work during the 1920s revealed the genetic basis of self-incompatibility mechanisms. At least two generations ahead of his time, Sewall Wright was providing a sound and sophisticated model basis to the subject, the complexity and relevance of which is still being revealed today.

After the Second World War, a considerable surge of interest in agronomic genetics led to much work on the theory of inbreeding, led by Allard, while Bateman and others published pioneering work on gene flow in crop plants such as fruit trees. At about the same time, Verne Grant was working on the effect of pollinator behaviour and flower syndromes on speciation in North American wildflowers, while Dan 'Incompatibility' Lewis published influential papers on topics as diverse as self-incompatibility, heteromorphy and gynodioecy (he is still active five decades later). Herbert Baker's detailed study on heteromorphic systems lasted for more than two decades, but after he reached California his interests blossomed into several distinct fields.

The genesis of the great explosion of 'modern' work on plant breeding systems can be dated from about 1970. In retrospect, the significant technical advance which triggered this explosion was the use of gel electrophoresis to separate and visualize 'isozymes', the direct products of the genes. Theoretical advances are sterile and non-substantive when unsupported by experimental evidence, but isozymes rapidly revealed the vast influence that breeding systems have on evolutionary processes. Work proceeded rapidly, so that by 1979, major surveys such as that of Hamrick, Linhart and Mitton (1979) had provided a sound experimental basis to the whole subject. (As yet, the DNA itself, which has provided such insights for taxonomy, phylogeny and evolution, has had only a small impact on breeding system studies.)

During the 1970s and 1980s, theoretical advances were led by the hugely influential New Zealander, David Lloyd, and by Deborah and Brian Charlesworth. In consequence we understand much more about the evolutionary routes which breeding systems are likely to have taken, and the importance of resource allocation to components of sexual function in determining the nature of breeding systems. At the same time, zoologists turned to plant pollination as a system to which foraging theory could be usefully applied.

Most recently, primarily in the 1990s, use of sophisticated statistics to undertake paternity analysis in plants has allowed a number of workers such as Diane Marshall, Maureen Stanton and Susan Mazer to investigate complex patterns of mate choice and sexual selection which had hitherto been the province of zoologists. Models for breeding system evolution have increasingly been matched by field data, a field in which Spencer Barrett has led the way. Also, molecular biology is finally catching up with the remarkably intransigent problem of pollen recognition, a field in which Hugh Dickinson and the Nasrallahs have been particularly influential.

In recent years, many more workers than these have had a considerable influence and it is perhaps invidious to name names. As in most biological fields, the growth in the literature, and diversification of research fields have grown exponentially since 1970, so that excellent volumes have recently been produced devoted mostly or entirely to specialist areas such as heterostyly (Barrett, 1992), sexual selection (Wyatt, 1992) and floral biology (Lloyd and Barrett, 1996). A single author such as myself might have attempted a single in-depth coverage of the whole of plant breeding systems in 1986, but this is certainly no longer the case in 1996.

In 1986, I was able to write optimistically that I could perceive trends towards a synthetic approach for plant breeding systems, but diversification has latterly tended to channel studies. Pollination biologists still ignore features of incompatibility, floral timing or flower number, whereas studies on maternal fitness rarely progress beyond seed number. Above all, many theoretical and experimental studies derive complex models and sophisticated results which assume an adaptive basis, which in turn assumes that the attributes that they study are heritable. As I wrote in 1986, evolutionary genetics still underpin all plant breeding system work, but in almost no case is any genetics undertaken. From this point of view, most modern ideas remain to be validated.

In 1986, I found it necessary to promote the study of plant breeding systems as vital to crop breeding, genotype, species and habitat conservation. Since then the word 'Biodiversity' has exploded upon an unsuspecting world. A decade ago, it seemed unlikely that the year 1996 would herald very large UK grants to seed banks, or to new National Botanic Gardens, that the rate of destruction of tropical forest would have declined by more than 80%, or that rare species Threat Ratings or Recovery Schedules would routinely consider genetic population structure, mating system or pollinator(s). Our understanding of the importance of plant breeding systems in the overall evolution and ecology of plant species has improved markedly during the intervening years.

Basic mysteries still remain for the enquiring student of the next generation. The role played by temperature in sex, and how flowers control gynoecial temperatures, is a large subject in its earliest infancy. It seems to me remarkable that large plants such as trees, which bear many thousands of flowers, should still be mostly, and efficiently, outcrossed. How this occurs at the very low density typical of tropical forest trees is still more mysterious. Apomixis is probably much more common than we suspect, especially in tropical forest. We still have no idea whether embryo genotype can influence seed size and seedling success, to what extent there is a paternal input into early offspring fitness, and above all, what is the role played by the paternal contribution to the endosperm. Very little is known about the adaptivity and population genetics of nectar production. And we are still lamentably ignorant about plant recognition processes in incompatibility.

Rather to my surprise, the first edition of *Plant Breeding Systems* has stayed in print for over ten years, for which I am grateful to both publishers (Chapman & Hall have succeeded Allen & Unwin). It seems that the book, despite the production of many symposium volumes and reviews, still has no direct competitor and has filled a need.

However, almost from the day of publication, I was aware of a pressing need for an improved, updated, second edition. We can now see that the study of plant breeding systems was in its infancy when the first edition went to press in 1985. More than 80% of the literature in the area has been published within the last decade. Some topics, for instance multi-allelic incompatibility, heteromorphy, resource allocation studies, and inbreeding genetics have progressed so quickly that passages in the first edition are not only dated, but in parts plain wrong. Also, I was unhappy with some written and illustrative sections of the first edition, in which I felt that the balance between topics could have been improved. Almost unbelievably in a modern context, the first edition was typewritten. I am very grateful to the publishers for scanning that edition, which made the rewrite much easier to wordprocess.

In the present edition, I have retained most of the basic structure of the first edition, but have greatly expanded chapters on sexual theory, and autogamy, while curtailing the discussion on sexual reproduction. After some thought I have completely removed the section on vegetative reproduction. Although I felt it had some interesting points to make, they were made at some length and the subject lacks a direct connection to the main theme of the book. Also, all references to plant groups other than the flowering plants have also been lost. North American readers may have found that too many of the examples used in the first edition were European. In this edition there is perhaps a better balance between the five Continents. Nearly all of the book is completely rewritten, and a majority of the figures and tables are also new. More than 300 (now dated and supplanted) references have been lost from the first edition, but 500 have been added to make a total of more than 900.

In the first edition, I attempted to blend basic information with recent data, and standard theory with informed speculation. Avowedly, it was a book based on natural history rather than mathematics, with very little modelling included. I have always been more interested in what plants do, rather than what an equation suggests they ought to do. The latter teleological approach has tended to pervade the subject in an unhelpful manner. One should never ask why a plant has not done something, as the relevant mutation might not, or could not, have occurred.

In the present volume I have aimed for the same mix, although hopefully in a less discursive and more concentrated fashion, so that rather more, useful, models have been included. This preface has also been modernized.

A. J. Richards
Newcastle upon Tyne

CHAPTER 1

Introduction

A COMPARISON OF THE MATING SYSTEMS OF PLANTS AND ANIMALS

The breeding systems of plants differ from those of animals in three major particulars.

First, most animals, and all vertebrates actively use sensory perceptions and behavioural responses in the choice of their mates. In these ways they influence the genetic make-up and evolutionary patterns of their populations. The scientific study of animal mating systems tends to concentrate on features such as display, song, territoriality, mate-choice, fidelity, clutch size, after-care, dispersal, migration and other characteristics which arise from their behaviour.

Plants can also behave, in the sense that they respond to external stimuli in predictable and adaptive ways. As for animals, this behaviour often involves movement as a part of mating behaviour. Consider for instance how flowers open and shut according to the weather, track the path of the sun, or how plants can grow towards a sunny patch.

However, plants do not have a central nervous system, so that the complexity of their perception and response is severely limited. Plants are essentially passive maters; they cannot actively choose their sexual partners.

Second, being of a relatively simple construction, many plants have a totipotency, so that they can repeat themselves in a modular fashion. A genetical individual (**genet**), formed from a single zygote, may be represented by many physiologically independent individuals (**ramets**). In contrast, this type of asexual reproduction does not occur in complex animals such as vertebrates and arthropods, although it is found in simple animals such as corals and parasites. Unlike complex animals, the sexuality of plants is constantly challenged by asexuality.

Third, all plants have an alternation of generations, so that sexual reproduction is achieved by gametes which are released from a haploid, gametophyte, generation. The significance of this system, missing from animals, is that any harmful recessive mutants which are phenotypically expressed in the gametophyte generation (the pollen tube and the embryo-sac in flowering plants) are screened out by failure or gametophytic competition, and do not persist through to the dominant sporophyte generation (p. 418). In this way, the 'genetic load' carried heterozygously, should be substantially reduced in plants in comparison with that in animals.

These fundamental differences between plants and animals have many consequences, but of these the most obvious concerns the distribution of sexual function. Most complex animals, and all vertebrates, have unisexual (**gonochoristic**) individuals (although some fishes and reptiles can change gender). However, most plants (95% of the flowering plants) have hermaphrodite (cosexual) individuals with both male and female function.

In comparison with hermaphrodites, unisexual mothers have an automatic twofold disadvantage, for about half their offspring will be 'unproductive' males, which do not themselves bear offspring. For animals (and dioecious plants), the unisexual system must have a counterbalancing advantage of at least equal magnitude. It seems probable that this advantage accrues from the processes of sexual selection.

Two antagonistic selective forces operate on the process of sexual fusion. The more gametes that are produced, and the more active that these gametes are, the more zygotes should be produced. However, the more zygotes that are produced for a given amount of maternal resource, the less well each zygote will be resourced, and the less successful each will be. In most organisms, this disruptive selection has encouraged the evolution of **anisogamy**, where gametes are of unequal size and number. Gametes which are small, abundant and motile are by definition male, whereas those which are large, few and sessile are called female.

It follows that the fitness of a male (in terms of the successful zygotes that he fathers) will depend on the number of his gametes, and the success that he has in arranging for his gametes to fertilize egg cells. However, the fitness of a female will depend more on the quality of the zygotes that she mothers, rather than on their quantity. To be successful, fathers should invest in quantity, mothers in quality.

This disparity between male and female function increases with the expense of the mating system, and with the requirement for maternal aftercare of the zygote. In plants and in simple animals, male costs only represent a small fraction of the father's total energy budget. However, most male vertebrates employ very expensive techniques such as territoriality, display and harem-gathering to ensure mating success. Also, for lower plants and most simple animals, female costs are completed by the production of a fertilized zygote. However, most vertebrate mothers (and in some cases, fathers) invest heavily in the aftercare of the zygote which has to learn complex behavioural patterns from parental tutelage, while under parental protection.

For plants and for simple animals, the relatively low cost and sophistication of both male and female function allows both functions to coexist successfully within a single hermaphrodite individual. However, where male and female functions are expensive in terms of the total energy budget, they are likely to maximize their fitness when freed from the conflicting demands and constraints of the other gender. The specialization of gender function, also, selects for unisexuality. We would be surprised if a harem-gathering male also invested heavily in the aftercare of his offspring.

This simple model which describes the selective forces governing gender distribution seems to be borne out by the flowering plants. Most flowering plants do not have particularly expensive or sophisticated gender-specific functions, at least in comparison with vertebrates. However, in cases where particularly expensive strategies are successful (as for the 'big bang' phenology and expensive fruit production of many sparsely distributed tropical forest trees, p. 301), dioecy (unisexuality) has been selected for (Chapter 8, Bawa, 1980).

Lloyd (1992) has viewed any such departure from stable hermaphrodity as a 'gender disability'. He has taken the language and modelling protocols of evolutionary ecology to show how the contrasting forces of kin selection and sib competition can result in an evolutionarily stable strategy (ESS) in plants as well as in animals. He suggests that the gender disability conferred by, for instance, male steriles in a gynodioecious population, can nevertheless benefit their kin (for instance offspring) by reducing levels of offspring competition from that of full sibs to that of half sibs. However, in the majority of (outcrossing hermaphrodite) plants in large populations sib competition is minimized and no strong conflict between gender functions occurs, so that ESS for gender disabilities do not evolve very frequently as they have in animals.

It seems likely that the constant challenge of asexual reproduction in plants and some simple animals also selects for hermaphrodity. Unlike unisexual organisms, hermaphrodites are able to self-fertilize. Clonal organisms may find outcrossing difficult, not least because distances to the next genet may be large. Selfing provides a convenient 'fail-safe' position for organisms with poor reproductive assurance, allowing for an alternative to asexuality which can be damaging in the total absence of sex (Chapter 10). There is a good account of levels of selfing found in hermaphrodite plants and animals in Jarne and Charlesworth (1993).

It seems probable that flowering plants are able to tolerate more selfing than do animals because of their lower 'genetic load' of disadvantageous recessives carried heterozygously. Many of these will have been 'screened out' having been exposed to selection in the haploid pollen tube or embryo-sac.

The flowering plants and the mammals, of all living things, are the only organisms which habitually succour and develop the fertilized zygote internally (although vivipary is occasionally found in other groups of animals). There the similarity ends, for whereas most mammals continue to invest heavily in their behaviourally complex offspring after birth, adult flowering plants do not care for their seedlings. Thus, it is not surprising that unisexuality is found in all mammals, but in only 5% of the flowering plants.

Nevertheless, the internal aftercare of the zygote by the mother does unite the mammals and the flowering plants in one particular. Haig and Westoby (1988) point out that in cases of multiple paternity, a conflict of interest arises between paternal and maternal requirements for the offspring. All the offspring that the mother bears carry her genes. Thus, it is in the interests of the mother that she nurtures equally her relatively many offspring. In this way she will maximize the number of her offspring carrying her genes, within the selective constraints of size at birth. In contrast, it is of evolutionary advantage to the father to ensure that his own offspring are nurtured efficiently by the mother, at the expense of offspring fathered by other individuals. Thus, mechanisms which allow the father some control over the maternal resourcing of embryos should be selected for. In the mammals, this is apparently achieved by tissue of embryonic origin (which contains paternal genes) functioning within the placenta. In the flowering plants, this control potentially derives from the endosperm (Haig and Westoby, 1989) (Chapter 3), a tissue unique to this group, which contains a set of paternal chromosomes and which nurses the embryo. However, there is as yet little evidence of a male control of embryo growth-rates (p. 36).

THE ROLE PLAYED BY FLOWERING PLANT BREEDING SYSTEMS IN THEIR EVOLUTION

Although they cannot choose their mates, plants have, nevertheless, a diverse armoury of methods by which their breeding systems manipulate and control the genetic structure of their populations, and the patterns of their evolution. Like animal behaviour, plant breeding systems are under genetic control and can themselves be selected for. They are rarely fixed and static, but are fluid and respond to selection pressures in an infinite variety of subtle and interrelated ways.

Because breeding systems control genetic variability, and are themselves controlled by components of genetic variability, they are liable to feedback. This can take three forms:

1. Positive feedback; for instance, inbreeding reduces genetic variability, and so should reduce the variability in the breeding system, which may reinforce the inbreeding (Chapter 9);
2. Negative feedback; for instance, agamospermy (Chapter 10) usually arises through the recombination of several different genes or groups

of genes mediated by outcrossing events, often of an extreme kind, such as hybridization. The genetic effect of agamospermy is to restrict or totally halt all genetic recombination and variability.

3. Stabilizing feedback; for instance, the frequency of self-incompatibility alleles in a gametophytic system should be equal for all alleles in a population. This equality is maintained by the frequency-dependent selection inherent in the mating system, and is reproductively the most efficient distribution of *S* alleles (Chapter 6).

Most seed plants have a wide range of reproductive options available for selection. These may arise constantly, through genetic variation, or intermittently, through mutation, in most plant populations. Broadly speaking, these are as follows.

1. Hermaphrodity versus unisexuality; hermaphrodites may be able to self-fertilize whereas unisexuals can only cross-fertilize (Chapter 8). Self-fertilization tends to reduce genetic variability (Chapter 9), whereas cross-pollination maintains genetic variability.
2. Self-pollination versus cross-pollination in self-fertile hermaphrodites (Fig. 1.1). The amount of selfing from within-flower pollination (**autogamy**) depends on the degree of separation that occurs between pollen donation (from anthers) and pollen reception (on to stigmas) in time (**dichogamy**) and space (**herkogamy**) within the flower. The amount of between-flower pollination (**allogamy**) that also results in selfing (**geitonogamy**) rather than crossing (**xenogamy**) depends on patterns of pollen travel between flowers, on the size of the genets (number of ramets and area of ground covered), and on the number of flowers open together on each genet (Figs. 5.5 and 5.9 and Table 5.5).
3. Self-fertilization versus cross-fertilization; successful pollination does not necessarily imply successful fertilization. Many plants reject selfed pollen through a mechanism known as self-incompatibility (Chapters 6

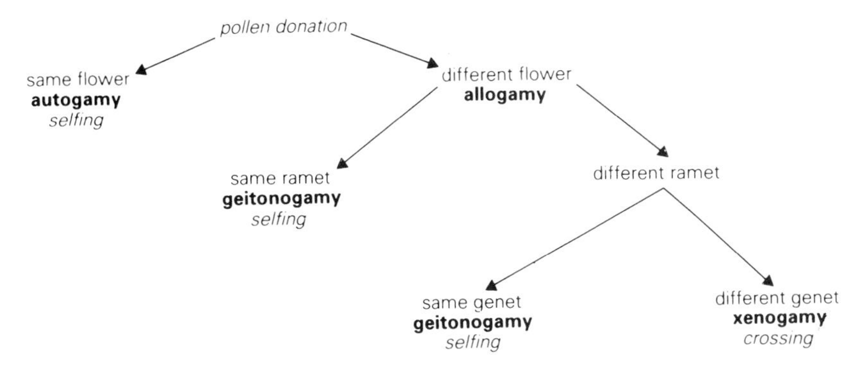

Fig. 1.1 Patterns of pollen transfer within and between flowers and plants.

and 7). Self-incompatibility will lead to more outbreeding and greater genetic variability than will self-compatibility (Chapter 9).

4. Sexuality versus asexuality; there are two main mechanisms by which plants can reproduce asexually. These are vegetative reproduction and by producing seeds without sex (**agamospermy**, Chapter 10). Agamospermy is relatively unusual, but options for greater or lesser amounts of vegetative rather than sexual reproduction are available to most perennial plants. Asexuality engenders little if any new genetic variability, but tends to maintain levels of variability already present.

It is useful to think of the various plant breeding strategies in three dimensions, which can however be more simply represented in two dimensions in the form of a triangle (Richards, 1990d, Fig. 1.2). At the apices of the triangle lie plants with a single strategy.

1. Panmixis. Plants without asexual reproduction which are fully outcrossing (xenogamous), and which randomly mate within an infinitely large population.
2. Self-fertilization. Plants without asexual reproduction which only mate by selfing. These will often be autogamous. However, if only one large self-fertile genet is present in a population, it will be completely selfed, but most matings may be allogamous (geitonogamous).
3. Asexuality. Plants without sexual reproduction. These may be agamospermous, have other special mechanisms of asexual reproduction such as floral proliferation, or they may be sexually sterile, as in many hybrids.

The genetical and evolutionary consequences of each of these mechanisms are entirely different in each case. Panmictic populations will be genetically variable, with a high proportion of heterozygotes present for each polymorphic locus. Such populations will be able to fill many niches within a complex community, and have a potential for ongoing evolution.

Self-fertilizing populations will be genetically invariable, with a very low proportion of heterozygous loci, so that they may lack vigour (p. 384). They will be poorly adapted to fill complex niches. Although they may be the product of rapid evolution resulting from gene fixation, their ongoing evolutionary potential is poor. However, chance outcrossings between differently fixed self-fertilizing populations can give rise to occasional bursts of variability.

Asexual populations will also be genetically invariable, unless a number of diverse asexual 'lines' coexist. However, they may have a very high proportion of loci fixed in the heterozygous phase, so they may be very vigorous, although poorly adapted to complex niches. Their evolutionary potential is, in theory, nil (but see Chapter 10).

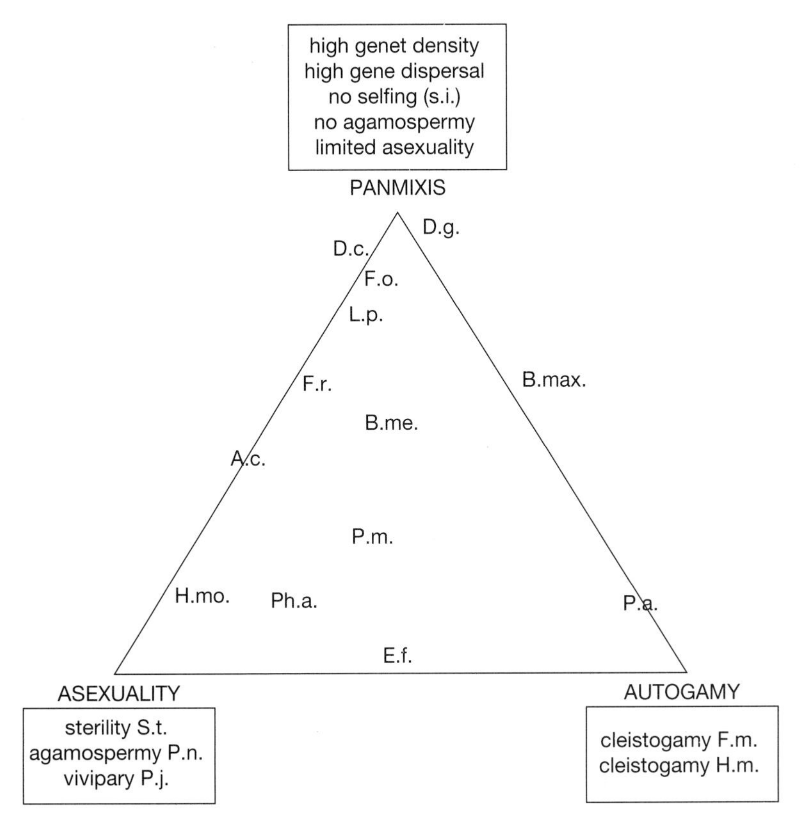

Fig. 1.2 The 'eternal triangle' of breeding system interfaces in grasses. A.c. = *Agrostis capillaris;* B.max. = *Briza maxima;* D.c. = *Deschampsia caespitosa;* D.g. = *Dactylis glomerata;* E.f. = *Elymus farctus;* F.m. = *Festuca microstachys;* F.o. = *Festuca ovina;* F.r. = *Festuca rubra;* H.mo. = *Holcus mollis;* H.m. = *Hordeum murinum;* L.p. = *Lolium perenne;* P.m. = *Panicum maximum;* Ph.a. = *Phalaris arundinacea;* P.a. = *Poa annua;* P.j. = *Poa × jemtlandica;* P.n. = *Poa nervosa;* S.t. = *Spartina × townsendii.* (From Richards, 1990d.)

In practice, most plants have evolved a mixed reproductive strategy (Fig. 1.2). Holsinger (1992) shows that an ESS can develop for mixed mating, because selfing as well as outcrossing can confer fitness benefits onto offspring. Habitual selfers may on occasion outcross, and even obligate outcrossers may have a restricted gene-flow within the population (Chapter 5), so that they are not truly panmictic. In addition, most perennial plants have the capability for at least some vegetative (asexual) reproduction.

The genetic architecture of plant populations, and thus their ecological amplitude and evolutionary potential, will depend on how competitive interfaces between panmixis and selfing on one hand, and panmixis and asexuality on the other, equilibrate. They should do so at thresholds determined by abiotic (density-independent) and biotic (density-dependent) selective constraints (Chapter 2).

It is characteristic of populations with both sexual and asexual reproduction to have highly skewed genet distribution patterns (Fig. 1.3). A very few genets may have proved to be very fit, and predominate in the population, being in some cases thousands of years old, and effectively immortal. Most genets only persist for a short time, and are only represented by one or a few ramets. Such patterns are represented in a wide range of plants with vegetative reproduction, ranging from pasture grasses such as *Festuca rubra* (Harberd, 1961) to trees such as *Tilia cordata* (Pigott, 1969). The genotype depletion curves which result in these skewed distributions are density-dependent, whereas forces which control rates of genet replacement by sexual reproduction are frequency dependent (Chapter 2).

The size and age of a plant resulting from these interactions between sexual and asexual reproduction may also influence its sexual breeding system. Ramets with large numbers of flowers, and genets with large numbers of ramets, are more likely to be geitonogamously selfed than are smaller plants (e.g. Schuster *et al.*, 1993; Harder and Barrett, 1995). If the plant is bee-pollinated, smaller genets with fewer flowers are likely to cross-fertilize more individuals, and over a wider area, than are larger genets with more flowers. This is a function of bee behaviour, for poorly rewarded pollinators will travel further in search of more rewarding 'patches' (Chapter 5). There will be stabilizing selection for genet size. If too small, it will compete poorly with the surrounding

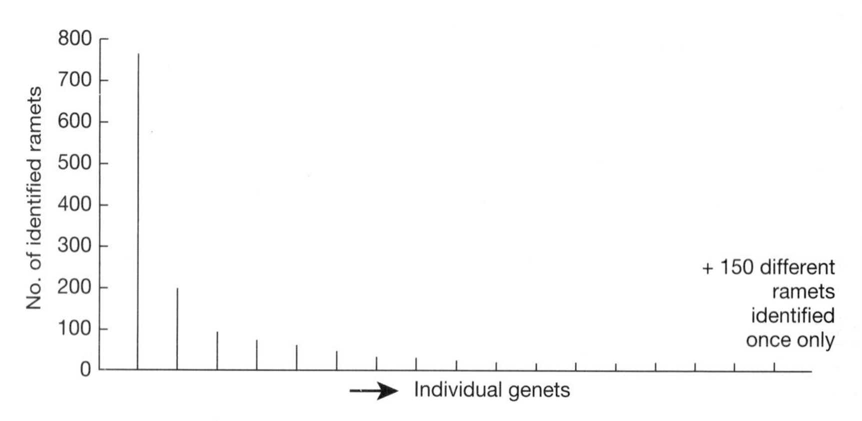

Fig. 1.3 Frequency distribution of numbers of ramets identified as belonging to certain genets in a Scottish population of *Festuca rubra* (after Harberd, 1961). An example of a skewed genotype frequency distribution.

vegetation, but if it is too large, the offspring may suffer a disadvantage from self-fertilization.

For plants with multiseeded fruits, stabilizing selection may also occur for flower number. Individuals with many flowers attract many pollinators, and set many seeds. As a result of competition for resource between seeds, this abundance of seed may cause the seeds to be inefficiently small (Baker *et al.*, 1994; Chapter 9).

GENDER DISTRIBUTION IN FLOWERING PLANTS

Flowering plants are sporophytes which undergo male and female meioses to produce microspores (pollen) and megaspores (embryo-sac initials). In turn, these spores develop into haploid gametophytes (the pollen-tube and the embryo-sac) which produce gametes (generative nuclei and the egg cell).

Therefore, when we describe stamens and pollen as having male function, and the gynoecium as female, these are necessarily shorthands for in reality the sexual function is displaced by a generation. However, once these conventions are understood they are useful, for the parasitic gametophytes are so small, unspecialized and impermanent that they cannot have evolved sexual strategies on their own. Sexual selection in flowering plants is perforce expressed through the sporophyte generation.

It seems very likely that the original flowering plant had hermaphrodite flowers (Taktajhan, 1969), that is flowers bearing both stamens and gynoecia. Not only is this by far the commonest condition in the flowering plants today (p. 10), but it prevails in most of the orders in this class, including those thought to be the least derived from their original founder. Other gender conditions have a scattered distribution systematically, and they rarely predominate within a family or order, suggesting that they are of a relatively recent origin, and are evolutionarily unstable. Also, in unisexual flowers it is usual to find vestiges of the missing gender, so that male flowers may have rudimentary and non-functional gynoecia, whereas female flowers usually have empty stamens, or staminodes.

Three different types of selection pressure may favour gender distributions which depart from the basic pattern of the hermaphrodite flower.

Requirement for outcrossing

Self-compatibility in hermaphrodite flowers may have been originally selected for in conditions of poor reproductive assurance (for instance where pollinators are scarce or unreliable, or where clones are large and few). In such conditions, selfing may become a 'necessary evil'. When reproductive assurance improves, outcrossing may once again become selectively advantageous. The most commonly occurring type of mutation which instantly causes full outcrossing in self-fertile hermaphrodites is male sterility, giving rise to gynodioecy, which may further evolve towards full dioecy (Chapter 8).

Gender resource allocation

As already discussed (p. 2), the resourcing of male fitness attributes may be dissonant with the resourcing of female fitness attributes, as males should select for gamete quantity, and females for zygote quality. The ultimate evolutionary solution is to fully separate gender function onto different genets (dioecy, Chapter 8). Female plants tend to set more fruits than do outcrossing hermaphrodites (Sutherland and Delph, 1984), but of course they do so on only roughly half the number of plants (the other half being male).

However, there exist a number of less extreme gender conditions which help to separate male and female resourcing in time and space (Table 1.1). Some of these, such as gynodioecy and subandroecy, represent conditions on the usual evolutionary pathway from hermaphrodity to dioecy (Chapter 8). Even within an apparently 'normal' hermaphrodite species (the grass *Cynosurus cristatus*), Ennos and Dodson (1987) show that some individuals are genotypically disposed to provide more resource to male

Table 1.1 Some common types of sex distribution within and between flowers, and within and between genets of seed plants. Anthers are designated ♂, and gynoecia are designated ♀. Hermaphrodites are designated ⚥

Name	Distribution of sex organs		Breeding system	Angiosperm species (%)
	Within a flower	Within a plant		
Dioecy	♂ or ♀	♂ or ♀	Xenogamous (outcrossing)	4
Gynodioecy	⚥, ♂ or ♀	⚥ or ♀	Xenogamous, geitonogamous, autogamous	7
Monoecy	♂ or ♀	⚥	Allogamous, some selfing, some crossing	5
Gynomonoecy	⚥ or ♀	⚥	Allogamous and autogamous	3
Hermaphrodity	⚥	⚥	Allogamous and autogamous	72
(other)				9
				100

function, and others more to female function. Webb (1981) suggests that protandry is frequently successful even in small umbellate flowers where it is unlikely to prevent selfing, as it separates male and female function in time, rather than in space, in the hermaphrodite flower. Where such dichogamous conditions operate in a hermaphrodite flower, Geber and Charnov (1986) show that hermaphrodity will usually prove a more efficient reproductive mechanism than dioecy.

Subandroecy, in particular, clearly demonstrates the evolutionary consequences of a disparity between the fitnesses of male and female reproduction. Stable gynodioecy establishes when the offspring of females are fitter than the offspring of hermaphrodites. In this case, it will pay hermaphrodites, which are less fit than females as mothers, to invest more heavily in male function to the detriment of female function. Male-linked mutations which achieve this will be favoured.

The various conditions of monoecy, too, can allow male and female resourcing to be efficiently compartmented (Chapter 8). Monoecy is the condition where only a single gender occurs within a flower, but the plant is hermaphrodite. Sometimes it occurs in conjunction with hermaphrodite flowers, as in gynomonoecy (Table 1.1). Monoecious plants tend to set more fruits than do outcrossing plants with hermaphrodite flowers, presumably because the female flowers, branches or cones are spared from male costs (Sutherland and Delph, 1984).

Monoecious hermaphrodites are relatively rare, particularly among herbaceous plants. Mazer (1992) has advanced the idea that the greater phenotypic constancy in sex expression which is typical of hermaphrodite flowers in comparison with monoecious flowers may often be advantageous when plants are challenged by heterogenous environments.

Monoecy is in fact mostly restricted to two life-forms:

1. Large wind- or water-pollinated plants such as many trees (Fig. 1.4), sedges, water-starworts (*Callitriche*) etc. These plants have inefficient and expensive pollination mechanisms. However, monoecy allows them to compartment male and female effort spatially within a plant, temporally (male flowers usually open earlier, i.e. are **protandrous**), and developmentally (young plants are chiefly male and old plants chiefly female, Chapter 8).
2. Plants with capitulate or umbellate inflorescences, for instance many in the families Asteraceae, Dipsacaceae and Apiaceae. In these cases, a division of labour occurs between flowers, so that those bearing the less expensive gender frequently expend more on secondary sexual characteristics such as showy petals, nectar etc.

Physical interference between male and female function

Attributes of the pollination system, or of fruit dispersal, may select for a separation of gender into different flowers. For instance, in hermaphro-

Fig. 1.4 The monoecious inflorescences of the wind-pollinated alder *Alnus glutinosa*. Female flowers on left, male flowers on right (×1).

dite flowers the large pendulous stamens typical of wind-pollinated flowers can intercept incoming pollen, so that it fails to reach the stigma. Both pollen emission and pollen receipt become more efficient when separated onto different flowers (Fig. 1.4).

Where large fruits have complex means of dispersal, as in the samaras of *Acer* or *Fraxinus*, the wings of the fruit may already be present on the ovary of the flower, potentially interfering with pollen emission from the stamens of a hermaphrodite flower (van der Pijl, 1978).

PRIMARY AND SECONDARY BREEDING SYSTEMS IN THE FLOWERING PLANTS

As the flowering plants evolved into different, short-lived, life styles and colonized marginal environments, self-fertile mutants may have been favoured, so that habitual selfers arose.

It seems to be a feature of the aboriginal gametophytic self-incompatibility (gsi) outcrossing system that selfers cannot backmutate to it. Gsi outcrossing cannot be 'reinvented'. However, new features of other kinds which once again promoted outcrossing may have become advantageous, particularly if conditions favoured cross-pollination. Thus, apart from gsi, it seems likely that all outcrossing mechanisms in the flowering plants are secondary, having evolved from selfing hermaphrodites. These secondary mechanisms are dealt with in more detail in later chapters. They are:

1. Gynodioecy (Chapter 8)
2. Dioecy (Chapter 8)
3. Heteromorphy (Chapter 7)
4. Sporophytic and two-locus incompatibilities (Chapter 6)
5. Monoecy (Chapter 8)
6. Herkogamy and dichogamy (particularly in few-flowered genets (Chapter 5).

The study of plant breeding systems, and their effect on the genetic structure of populations is of basic importance to students of evolution and population genetics. It is also of considerable applied importance. Consider a tomato sandwich. All its constituent parts are made of the products of plant breeding systems, seeds and fruits. The bread is made from the seeds of the inbreeding annual cereal *Triticum aestivum* (wheat). The margarine may be made from the oily seeds of the monoecious, wind-pollinated maize, *Zea mays*, or the insect-pollinated hermaphrodite self-compatible sunflower *Helianthus annuus*.

The tomato itself is the fruit of an insect-pollinated annual, *Lycopersicon esculentum*. In this species, following domestication, selection has occurred for self-compatibility and autogamy from an ancestor which was originally self-incompatible and herkogamous (Simmonds, 1976). Pollen is only released from the poricidal anthers by vibration, so it is necessary to shake or hose the flowers in the absence of insect visits if fruit is to be set.

The pepper (*Piper nigrum*) is a perennial vine, the fruits of which are dried and ground to give the black pepper used to flavour the tomato sandwich. This species was originally dioecious and insect-pollinated, but modern cultivated clones are monoecious, and are often cross-pollinated artificially. Remarkably, not much appears to be known about the pollination biology of this little-studied crop.

Most of the staple crops of the world, wheat, maize, rice, sorghum, millet, beans, bananas, coconut and many others, are fruits or seeds, the yield of which is a direct product of the breeding systems of the plants and their efficiencies. The better we understand these systems, the better shall we all be fed.

CHAPTER 2

Sexual theory in seed plants

FUNDAMENTALS OF SEXUAL REPRODUCTION

Variability

In eukaryotic organisms, such as seed plants, sexual reproduction has three major features that generate variability: recombination, segregation and sexual fusion.

Recombination Recombination constitutes the resorting of alleles of different loci linked on homologous or homoeologous chromosomes, by chiasmata during the heterotypic phase of meiosis (usually meiosis I).

Segregation Segregation is the inclusion into a gametophyte, usually at random, of any one of the four chromatids resulting from a bivalent association of two homologous chromosomes at the heterotypic phase of meiosis. Chromatids of different chromosomes enter gametes randomly with respect to each other, and alleles of heterozygous unlinked loci are randomly resorted with respect to each other between the parent and the gamete.

Segregation is a very powerful mechanism for generating variability. Irrespective of recombination, and assuming all pairs of homologous chromosomes in a diploid are heterozygous at at least one locus, the number of different gamete genotypes generated by a single diploid parent is 2^n, where n is the haploid chromosome number. In a plant where n is 20 (a common and not untypical number), over 10^6 (1 000 000) different gamete genotypes can be created by a single parent. In fact, for most plants, the number of potential gamete genotypes far exceeds the actual number of female gametes produced per parent; even for male gametes, this is likely to happen when n is greater than about 12.

Sexual fusion, or syngamy

Syngamy is the fusion of haploid, or functionally haploid, male and female gametes to form a diploid, or functionally diploid, zygote. As a result of the power of recombination and segregation in creating gametic variation, the number of potential zygote genotypes created even from matings between the same parents, or from selfing, is immense. It can be estimated by $(2^n)^2$. In fact, the rare chance of exactly the same gamete genotypes meeting will render the actual figure slightly lower, and in largely homozygous habitual inbreeders it will be much lower. However, in the example quoted in the section on segregation, the potential number of zygote genotypes from a single diploid mother is about 10^{12}, very much larger than the actual number of offspring produced by a mother. For a sexual outbreeder, it is very unlikely that any two offspring would be genetically identical.

However, it should be realized that species with low chromosome base numbers will create a much lower level of zygotic variability. Thus, where $n = 4$ (a condition not uncommon in higher plants), the potential number of zygote genotypes, ignoring recombination, is only 256. This may be fewer than the number of seeds produced by a mother. Thus, an important consequence of low chromosome base number is a restriction on the release of variability (Chapter 9).

Gene migration

A second important feature of sexual reproduction in nearly all plants is that it allows for gene migration. Gene migration has two components:

1. Gene travel. The motility of pollen and seeds (Chapter 5) allows genes to travel within, and to a lesser extent between, populations. In this way a successful mutation, or linkage group can spread spatially;
2. Gene incorporation. Cross fertilization allows a successful mutation to spread between different mother–offspring lines.

The importance of gene migration is best illustrated by reference to asexual plants, for instance the agamospermous dandelions, *Taraxacum* (Chapter 10). In most *Taraxacum* species, seed is produced without fertilization, and pollen is therefore non-functional. It might be supposed that a selective advantage would be gained by male-sterile plants without pollen (pollen and anthers comprise about 5% of the dry-matter production of reproductive effort in dandelions). Indeed, about 17% of Britain's 200 or so *Taraxacum* agamospecies do not produce pollen. Some species vary in this respect. For instance, *T. rubicundum* usually produces pollen in the south of Europe, but is male sterile in the north, but many populations of this species have male-fertile and male-sterile individuals coexisting. It is evident that continuing mutations produce further possibilities for male sterility. Occasional male sterile individuals are found in otherwise polliniferous species, as I found for a single individual of *T. brachyglossum*.

Mogie (1992) has noted that where asexual dandelions coexist with sexuals, pollen-bearing asexuals will be advantageous, for they may be able to act as fathers to sexuals (whereas sexuals cannot father asexuals).

However, assuming that male sterility is advantageous for most agamospermous *Taraxacum*, the spread of these mutants is very seriously hampered by asexuality. A new mutant gene is strictly limited to the direct line of descent from the parent in which it arose. Its success will entirely depend on the total fitness of the line in which it occurs. It is useful to consider the genome of a seed clone in *Taraxacum* as a single linkage group. Thus, the new mutant becomes a 'hitch-hiker' (p. 23) on that linkage group. The mutant will be subject to the selection pressures on that linkage group as a whole. Its success will therefore be subject to minority-type disadvantage (p. 47). Male-sterile types can only become more common by repeated mutations, and by the relative success of male-sterile lines in contrast to male-fertile ones. It is not surprising to find the scattered and inconsistent pattern of male sterility in *Taraxacum* that we observe.

Yet, if a similarly successful mutant had arisen in a sexual population, it would be rapidly disseminated throughout the population, and would spread to other populations. It would not be limited to one linkage group, but could be associated with all the genotypes produced. By hitch-hiking on successful chromosomes, it might soon become fixed in the population.

DISTRIBUTION OF SEXUAL REPRODUCTION

Sexuality is a striking and pervasive feature of nearly all eukaryotic organisms. Although many plants and animals have alternative asexual modes of reproduction, sexuality is absent from only a relatively very few parthenogenetic animals, agamospermous plants and sterile (usually hybrid) plant clones. In fact, recombination and a kind of sexual fusion substantially predate the eukaryotic cell, and may thus be nearly as old as life itself.

Because all eukaryotic individuals have separate chromosomes, which undergo meiosis prior to gamete formation in a remarkably constant way, the principles of recombination, segregation and fusion are common to all higher organisms. This remarkable uniformity in the highly complicated procedures of sex among all eukaryotic organisms from algae to elephants, and protozoa to pine trees promotes two deductions.

First, meiotic mechanisms, and consequently recombination, segregation and fusion, are primitive to the eukaryote cell. Second, there are evolutionarily vital features of sexual reproduction. As a result, only those lines which have maintained all the complex features of meiosis, and hence of sexual reproduction, have survived to the present day.

This is particularly striking, for some features of sexual reproduction appear to be disadvantageous in comparison with asexuality. In many plants, the need to maintain an inefficient sexual reproduction is being constantly challenged by an ever-present asexual reproduction. In order to have survived so consistently, we must presume that sexuality has some ongoing major advantage over asexuality.

ADVANTAGES OF SEXUAL REPRODUCTION

The outstanding feature of sexual reproduction in outbreeders is the creation of an effectively infinite array of genetic diversity. Thus, it is commonly supposed that the great success of sexuality has resulted from ongoing requirements for evolutionary change. However, genes controlling breeding systems can only evolve in respect to the selection pressures they encounter. If asexuality is favoured at a given moment in evolutionary time, and genes that promote asexuality are present, sexuality will be lost, never again to be recaptured by that new asexual line. A plant cannot develop a breeding system that anticipates its future evolutionary needs.

In recent years, two complementary models have been developed which between them describe the selective forces which have favoured sexual reproduction so consistently over evolutionary history.

The 'Tangled Bank' and 'Red Queen' hypotheses

All natural environments are highly heterogeneous in time and space. Fine-grain environmental heterogeneity is of very great selective importance, as has been shown on many occasions (e.g. McNeilly, 1968; Snaydon, 1970). As the productivity of a habitat, its 'environmental carrying capacity' increases, then so do the number of 'patches' forming distinctive niches available to an organism in that habitat. Intuitively, one might expect that a mother who produces variable offspring would be fitter than one who produces less variable offspring. The variable offspring should be adapted to fill a greater range of niches, so that more of them should be able to survive.

A counter argument suggests that many of the variable offspring might not be very well adapted to any niche in the environment. The successful genotype of the mother, which was by definition well adapted to one niche for her to become a mother at all, would be frittered away by the processes of sexual reproduction (the so-called 'cost of meiosis'). Evolutionarily, it might pay a mother to create all her offspring in her own, successful, image. However, the gross overproduction of offspring typical of most living things allows for the 'wastage' of many maladapted progeny.

In developing a 'Tangled Bank' hypothesis (which using the terminology of Bell, 1982, concerns spatial heterogeneity), Felsenstein (1974, 1988) has concentrated on sibling competition. His models show that it will in fact pay a mother in a productive ('Tangled Bank') environment to produce variable offspring, because in this way competition between siblings is reduced. However, one mother producing variable offspring will only be fitter than a variety of (asexual) mothers producing lines of invariant offspring when she produces offspring which are as variable as those of the whole population. In unproductive or otherwise non-complex habitats, a variety of asexual lines may well prove to be at least as evolutionarily successful as competing sexuals.

In some ways, kin selection is antithetic to the principles of sib competition. However, where lines show degrees of relatedness (as in successive generations of outcrossing half-sibs), this may allow them to evolve ESS breeding systems such as gynodioecy or self-incompatibility which lower individual fitnesses but increase the fitness of the line as a whole (Lloyd, 1992).

Frequency-dependent selection

A central feature of the Tangled-Bank hypothesis is its reliance on frequency-dependent selection. This suggests that a 'rare' genotype will tend to succeed at the expense of a 'common' genotype because it is less likely to find itself in direct competition with its congeners.

Many observations lend support to this generalization, for example in dioecious species where it is found that a 'rare' sex will outperform a 'common' sex in experiments (Chapter 8). A good anecdotal example concerns a newspaper report of the yield of willow (*Salix*) cuttings when grown for electricity generation in Northern Ireland. It was discovered that most clones gave the highest yield when surrounded by other clones, and that there was a direct relationship between the overall yield of the plot and the number of different clones grown therein. In a genetically diverse population, such as that derived from sexually outbreeding mothers, all genotypes are 'rare' genotypes, and perform correspondingly well.

Put another way, 'rare' genotypes occupy a distinctive niche in the environment, and so they suffer less intraspecific competition than do commoner genotypes. They may root more deeply, flower earlier, be less light sensitive, be better at tolerating acidity, or show less seed dormancy than most of their direct competitors.

A classical series of experiments which demonstrated the power of frequency-dependent selection was conducted by Antonovics and Ellstrand (1984) and Ellstrand and Antonovics (1985), using the hay grass *Anthoxanthum odoratum*. These authors examined the yield of single tillers (ramets) produced sexually and asexually in mixed and pure stands at different frequencies and densities. Essentially, they found that a 'stranger' tiller, surrounded by tillers cloned from a single genotype, nearly always performed more strongly than they or their competitors did as single genotype stands. This effect was most marked at high density, and occurred irrespective of whether the 'stranger' was generated sexually or asexually (Fig. 2.1).

A special case of frequency-dependent selection is the so-called 'Red Queen hypothesis', where environmental heterogeneity occurs in time, rather than in space. In Carroll's *Alice Through the Looking Glass*, the Red Queen has to run as fast as possible in order to stay in the same place. Many environmental factors may change with time, most obviously abiotic, density-independent, selectants such as the climate, poisons or pollutants. In such cases, 'rare' genotypes may be favoured during the

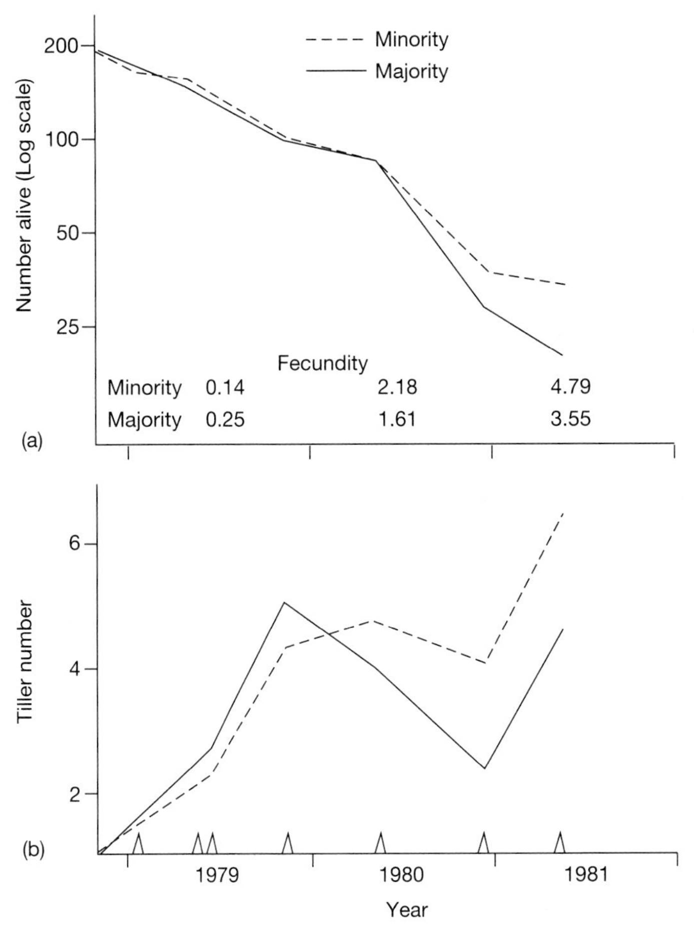

Fig. 2.1 (a) Age-specific survival (number alive), fecundity (number of inflorescences per surviving individual), and (b) size (tiller number) of minority and majority test individuals of the grass *Anthoxanthum odoratum*. (After Antonovics and Ellstrand, 1984.)

environmental shift. Clearly, rats would not have become resistant to 'warfarin' or mosquitoes to DDT if the populations exposed to these toxins had not shown genetic variability for these resistances.

However, the 'Red Queen' more specifically addresses changes to the biotic, density-dependent environment, with respect to interacting organisms such as predators, herbivores, parasites and pathogens. Naturally, the predator/pathogen adapts to its habitat, such as its prey or host, through the sexual process just as readily as does the prey/host itself. As the herbivore adapts to a new food-plant, or the pathogen develops a new strain capable of parasitizing a novel host, so will the food-plant or host be favoured if the sexual process yields genotypes which can avoid or are resistant to this new threat from grazing or infection.

To take one example, the elms (*Ulmus*) of southern and central England have shown almost no resistance to the aggressive strain of elm disease (*Ophiostoma ulmi*). These southern elms are represented by relatively few sterile clones which were propagated vegetatively. In the north of England and in Scotland, only the sexual wych elm (*Ulmus glabra*) occurs, and here a proportion of the trees has usually survived in most populations. It seems likely that in these variable populations, some plants by chance show a measure of resistance to this pathogen.

In an interesting piece of work, Schmitt and Antonovics (1986) show that aphid infection from an inoculum spreads less rapidly through *Anthoxanthum* populations when they are planted so that relatively unrelated seedlings form neighbours, than through monogenetic stands (Fig. 2.2).

Density-dependent selection

Selection pressures which arise from intraspecific competition increase with ramet density, as is made clear by the competition experiments of Antonovics and Ellstrand, reported above. Thus, 'Tangled Bank' and biotically driven 'Red Queen' frequency-dependent selection for sexual variability should also be density-dependent.

This density-dependence has major ecological implications, for instance with respect to seral colonization within communities. I have reviewed this topic elsewhere in greater detail (Richards, 1990d). For plants such as perennial grasses which are also capable of asexual, vegetative reproduction, sexual reproduction becomes a rare event once seedlings are unable to penetrate the closed community. Typically, the density of genets increases rapidly during an initial establishment phase, and then decreases as a few well-adapted genets persist and co-dominate the community (e.g. Law, Bradshaw and Putwain, 1977; McNeilly and Roose, 1984; Gray, 1987). This pattern of the density of genets increasing, and then decreasing as the density of ramets increases during the colonization of a community

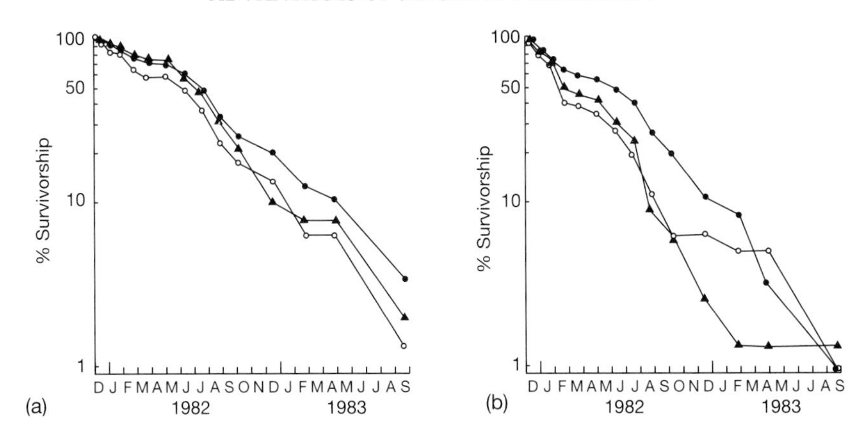

Fig. 2.2 Effect of neighbour treatment on survivorship percentage in *Anthoxanthum odoratum*. (a) Plants free from aphid attack. There were no significant differences between the 'no neighbours' and 'unrelated neighbours' treatments (Gehan–Wilcoxon test, $X^2 = 2.96$, d.f. = 1, $P > 0.08$) and no significant differences between the 'sibling neighbours' and 'unrelated neighbours' treatments (Gehan–Wilcoxon test, $X^2 = 0.02$, d.f. = 1, $P > 0.8$). (b) Plants attacked by aphids. Survivorship was significantly lower for plants without neighbours than for those with unrelated neighbours (Gehan–Wilcoxon test, $X^2 = 6.44$, d.f. = 1 $P < 0.02$). Plants with sibling neighbours had significantly lower survivorship than those with unrelated neighbours (Gehan–Wilcoxon test, $X^2 = 3.91$, d.f. = 1, $P < 0.05$). Open circles, no neighbours; closed triangles, sibling neighbours, closed circles, unrelated neighbours. (From Schmitt and Antonovics, 1986.)

has been described by Grime (1973) in his 'hump-backed' model. An example is given in Fig. 2.3.

Doubtless, for plants in communities where little potential for asexual reproduction is expressed, such as rain-forests, frequency-dependent se-

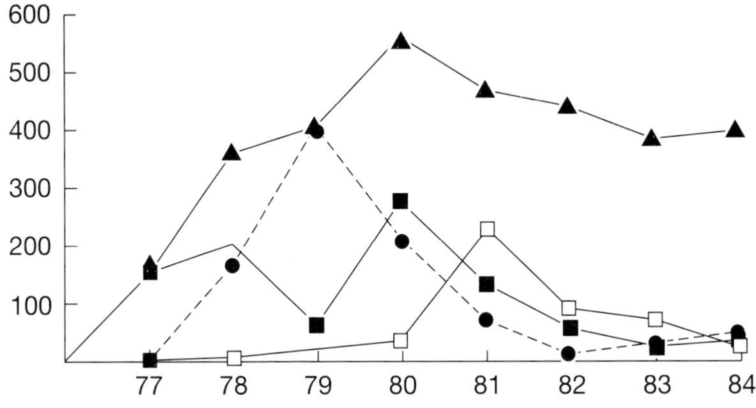

Fig. 2.3 Changes in a population of *Agrostis curtisii* colonizing an area of Hartland Moor burnt in August 1976. ▲, Numbers of plants; ■, numbers of new recruits; □, numbers of deaths; ●, numbers of inflorescences divided by 10. (After Gray, 1987.)

lection for rare genotypes increases directly with increasing density. However, in many such communities, much of the competition tends to be interspecific, not intraspecific.

Linkage disequilibrium and the 'Hill–Robertson effect'

With regard to the ever-present advantages of sexuality, there is a second school of thought, persuasively argued in Maynard Smith (1978), concerning linkage disequilibria. These occur when the frequency of combination of alleles at two linked loci *A*/*a* and *B*/*b* on chromosomes, or gametes, differs from that predicted by the product of the allele frequencies. Linkage disequilibrium is likely to be increased by:

1. Close linkage of *A*/*a* and *B*/*b*;
2. Low rates of recombination (chiasma formation);
3. Selection for particular allelic combinations;
4. Small population size, or selfing.

Some forms of linkage disequilibrium are strongly favoured by selection, and may be maintained by cytological features, e.g. inversions or reciprocal translocations. In the heterostyly supergene of *Primula* (Chapter 7), most populations have maintained linkage disequilibrium to the extent that recombinants between the *S* supergene and the *s* supergene rarely occur (they are often homostyles), and when they do occur they usually appear to be outselected. The linkage groups *GAP* (thrum) and *gap* (pin) are common, and although recombinants such as *Gap* and *gAP* do occur, they rarely become established. This is an adaptive linkage disequilibrium, a linkage group or supergene. In this case, the *S* alleles at different loci are coadapted, as are the *s* alleles. If the linked loci were not coadapted, linkage disequilibrium might be harmful.

Consider a situation in which the linked loci *A*/*a* and *B*/*b* have the following fitnesses (*W*) in environments Ā and B̄:

environment Ā	*A* 1.2	*B* 0.9
	a 1.0	*b* 1.0
environment B̄	*A* 1.2	*B* 1.1
	a 1.0	*b* 1.0

Then, in environment Ā, the preferred chromosome linkage group would be *Ab*, assuming simply additive fitnesses. On migration to environment B̄ (or a change in the same locality from Ā conditions to B̄ conditions), the preferred linkage group would be *AB*. However, in any of the situations listed above which favour linkage disequilibrium, linkage equilibrium would not occur quickly enough to favour *AB* recombinants. Thus, the chromosome *Ab* might predominate to fixation although *AB* was more successful; particularly if, as would probably be the case, allele *B* was originally scarce. Thus, the disadvantageous allele *b* would 'hitch-hike' to success in its new environment $\bar{B}$ by being closely linked to the successful allele *A*. A specific case is argued elsewhere (p. 205), where disadvantageous genes might persist in populations when tightly linked to incompatibility *S* alleles which are rendered scarce, and thus advantageous, by this disequilibrium (O'Donnell and Lawrence, 1984).

In such a way, disadvantageous genes will accumulate in linkage groups without recombination. If the whole genome is effectively one linkage group, as in an asexual *Taraxacum* (p. 15), the whole genome will accumulate disadvantageous mutants which arise in that line by chance ('Muller's ratchet'), with no means of shedding them by recombination.

This combination of fixed linkage disequilibrium and mutant accumulation which is typical of asexuals has been termed the 'Hill–Robertson Effect' by Felsenstein (1974). According to Maynard Smith it is the most potent current selectant for sexuality. In panmictic populations, this effect will be minimized by large haploid chromosome numbers, high rates of chiasma formation, outbreeding and large effective population sizes. These factors will encourage linkage equilibrium by maximizing recombination.

It is important to note that although recombination in this sense refers to the breakage of linkage groups through chiasmata, it will be rendered effective by the other elements of sexuality, the random segregation of the recombined chromosomes, and the random fusion of the resultant gametes.

A further point of fundamental importance is the role played by sex in not only breaking, but also in creating linkage groups. The processes of sexual reproduction will help advantageously coadapted genes to come together in tight linkage on the same chromosome. Thus sex can both promote useful linkage, and also dissipate damaging linkage.

I thoroughly agree with Maynard Smith that the negation of the Hill–Robertson Effect is a most vital component of the function of sexuality, but I believe very strongly that it has no meaning, except when viewed in the context of the highly heterogeneous and unpredictable environment.

This is elegantly demonstrated by a computer-generated model illustrated by Maynard Smith (1978, p. 104) (Table 2.1). In this simulation of the effect of sibling competition, the proportion of sexual to asexual individuals resulting after ten generations is given, where the initial frequency was 0.5. There are five heterozygous loci with one allele dominant, each allele fit, or not fit, and five paired features of the environment. There are thus 2^5 (i.e. 32) different environments, and 32 different phenotypes, and each phenotype will be the fittest for one environment. Asexuality will only predominate where sibling and total competition (*RN*) is low, and the reproductive advantage of asexuality (*K*) is high (encircled values in Table 2.1). With realistic levels of sibling and non-sibling competition, and

Table 2.1 Results of simulating a model of sibling competition. Proportion of sexual individuals after ten generations (initial value, 0.5) (from Maynard Smith, 1978)

Population size	Intensity of selection: Between families	Within families	Total	Advantage of asexual reproduction (K)					
L	*R*	*N*	*RN*	1.0	1.2	1.4	1.6	1.8	2.0
400	20	1	20	0.48	–	–	–	–	–
400	1	20	20	0.51	–	–	–	–	–
200	2	4	8	0.97	0.69	0.29	0.09	0.10	0
200	4	4	16	1.0	0.92	0.79	0.37	0.38	0.14
200	6	6	36	–	–	0.92	0.95	0.56	0.54
400	6	8	48	–	–	–	–	0.78	0.64

asexual advantage, sexuality is more likely to win, because more of the offspring produced sexually will be better adapted to the different environmental niches.

In a similar experiment, Maynard Smith (1978, p. 106) shows that an allele for high recombination will always be favoured over an allele for low recombination when reproductive advantages are equal. Thus, even in the unrealistically low levels of environmental heterogeneity allowed by computer models, sexual reproduction will succeed over asexual reproduction. In most cases, this will be true even for unisexual species with an automatic disadvantage, compared with asexuals and selfing hermaphrodites, as high as 1:2 (Chapter 8).

DISADVANTAGES OF SEXUAL REPRODUCTION

Maynard Smith (1971, 1978) has stated that in a unisexual, or dioecious species, sexuals will suffer a twofold disadvantage against asexuals, because roughly half the offspring they produce are male. In contrast to asexuals which produce only females, half the reproductive effort of unisexuals is wasted by producing unproductive males. This is indeed a powerful argument in favour of some compensatory selective advantage for sexual reproduction in groups such as the vertebrates, in which unisexuality dominates (p. 2).

For 95% of the organisms that are the subject of this book, this argument is much less persuasive, because they are hermaphrodite. However, for hermaphrodites, there may still be disadvantages inherent in the sexual process, compared to asexuals.

1. Sexual mothers spend maternal resource on offspring (seeds) which are likely to include some unfit individuals (the 'cost of sex', or 'the cost of meiosis'). The offspring of an asexual will have the same genotype as

their mother, and should therefore be just as fit as their mother in her particular niche.

2. A newly arising asexual mother may still be potentially a sexual father (Mogie, 1992). The asexual will be able to donate asexual genes to sexuals, but will not receive any sexual genes itself. This is disadvantageous to coexisting sexuals.
3. Male-sterile asexuals will save on male reproductive effort in comparison with sexual hermaphrodites.
4. Outbreeding sexuals may be less reproductively efficient than asexuals in conditions of limited cross pollination, such as very harsh environments.

GENDER RESOURCE ALLOCATION AND POLLEN/OVULE RATIOS

Model systems anticipate that resources should be partitioned between genders according to the contribution to the next generation made per individual of that gender (if dioecious) or per unit of reproductive resource of that gender if hermaphrodite. The reciprocal of the amount of gender resource required to parent a single offspring can be regarded as the fitness of that gender for a hermaphrodite. In recent years, distinctions are often drawn between the morphological gender of a plant, and its 'functional gender'. The functional gender concerns the proportion of total gender resource spent which is successfully devoted to a single gender function by an individual plant (e.g. Lloyd, 1980a).

In Fig. 2.4, 'fitness curves' are illustrated (derived from Charnov, Maynard Smith and Bull, 1976 and Charnov, 1984). In these, the fitness of the individual genders are scaled on the X and Y axes. In a dioecious (unisexual) population of stable size, the mean fitness of each gender should be two, that is, each individual should donate to the next generation on average two offspring. For an hermaphrodite in a stable population, the mean fitness of each parent should be one.

For a hermaphrodite, male fitness may differ from female fitness within a single individual. Consequently, resource expenditure on male and female function may differ considerably. Consider a monoecious species which bears 20 flowers, and male and female flowers are equally expensive to produce. If it produces 10 male flowers and 10 female flowers, male and female functions are equally fit, and if each plant donates one offspring to the next generation, the male fitness W_m and the female fitness W_f will both be one. However, if each plant donates one offspring to the next generation, but 15 male flowers and 5 female flowers are produced, male fitness will be $1 \times 5/10 = 0.5$ and female fitness will be $1 \times 15/10 = 1.5$ compared with the population average (Fig. 2.4). The intersect of these relative fitnesses is on the straight line for the fitness of the hermaphrodite

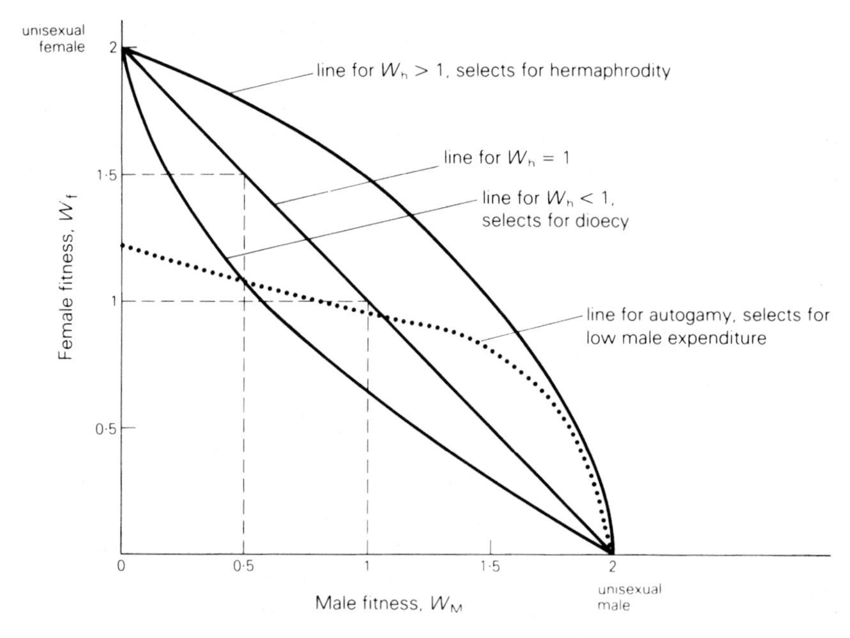

Fig. 2.4 Fitness curves for sexual function in hermaphrodite and dioecious plants. The fitness of each sex, in terms of contribution to the offspring relative to reproductive effort, are scaled on the *x* and *y* axes. See text for further details. (After Charnov, 1984.)

plant $W_h = 1$. If the intersects of the relative fitnesses are to the right of the $W_h = 1$ line, W_h is greater than one, and hermaphrodity will be favoured (the W_h line will be convex). If, however, the intersects of the relative fitness are to the left of the $W_h = 1$ line, W_h will be less than one, and dicliny will be favoured (Chapter 8). That is, the gender fitnesses when isolated are greater than when they are combined in a hermaphrodite. Such conditions might arise in a hermaphrodite if, for instance, a heavy female reproductive load resulted in insufficient allocation of resource to male function for total fitness W_h to reach one (there might be too few male flowers in a monoecious species for pollination to be effective).

Such a differential compartmentation of gender resource to male function or to female function may even occur between individuals in a single hermaphrodite population, as for the grass *Cynosurus cristatus* (Ennos and Dodson, 1987).

Some simple predictions of gender resource allocation fitness curves (e.g. Charnov, 1982) are that

1. selfers have greater levels of male fitness than do outcrossers, and thus it will pay a selfer to allocate less resource to male function;
2. cross-pollinated plants with inefficient pollination mechanisms (e.g. wind-pollinated, or polytropic species) will have lower male fitness

than will efficiently pollinated species, and thus should devote correspondingly more resource to male function.

In a review on this topic, Goldman and Willson (1986) point out that such assessments of gender resource allocation tend not to fit these predictions very well, and some of these failures may be attributed to difficulties encountered in deciding what should actually be measured.

Pollen/ovule ratios

The less expensive a successful contribution by a gender to reproductive fitness is, the greater will be the total fitness of that gender. Trade-off theory predicts that it will pay a hermaphrodite plant to reduce expenditure on that gender in order to maximize expenditure on the more expensive gender (however, such trades-off between gender are not always found, as in Devlin, 1989 and Mazer, 1992). Expenditure on the more expensive gender should be increased to a threshold point beyond which the fitness gain curve flattens off. It is typical of an autogamous plant with high levels of selfing for male fitness to be optimized by a low expenditure compared with female expenditure (Chapter 9). That is , for all ovules to be self-fertilized, the flower need only produce a small amount of pollen (dotted line on Fig. 2.4).

A study of reproductive resource allocations compared between four species of *Primula* section Aleuritia (Mazer and Hultgaard, 1993) provides a good example of reproductive strategies and trades-off (Fig. 2.5). One of these species, *P. farinosa*, is a diploid outcrossing heterostyle from relatively southerly and lowland localities in Scandinavia and Britain (see Table 7.12, Chapter 7). The other three species examined, *P. scotica*, *P. scandinavica* and *P. stricta*, are largely selfing homostyle polyploids from more northerly or upland localities.

The selfing homostyles produce only between 10% and 20% of the number of pollen grains per flower than does *P. farinosa*. Perhaps as a result of the transference of resource that this saving allows, the homostyles produce nearly twice the number of ovules per flower compared with the heterostyles, and the pollen/ovule ratios of homostyles are much lower than for heterostyles. Some resource transference may have also occurred for corolla characteristics. The homostyles tend to have flowers of a smaller diameter which are less likely to attract pollinators (saving of a secondary male resource), but have longer corolla tubes which may serve to protect the ovary from predation, desiccation or cold (investment into a secondary female resource).

One interpretation of these evolutionary trends is that the male contribution to a fertilized seed is relatively much cheaper than is the female contribution in the homostyle selfers, compared with the heterostyle outcrossers. Consequently, male function is much fitter than female func-

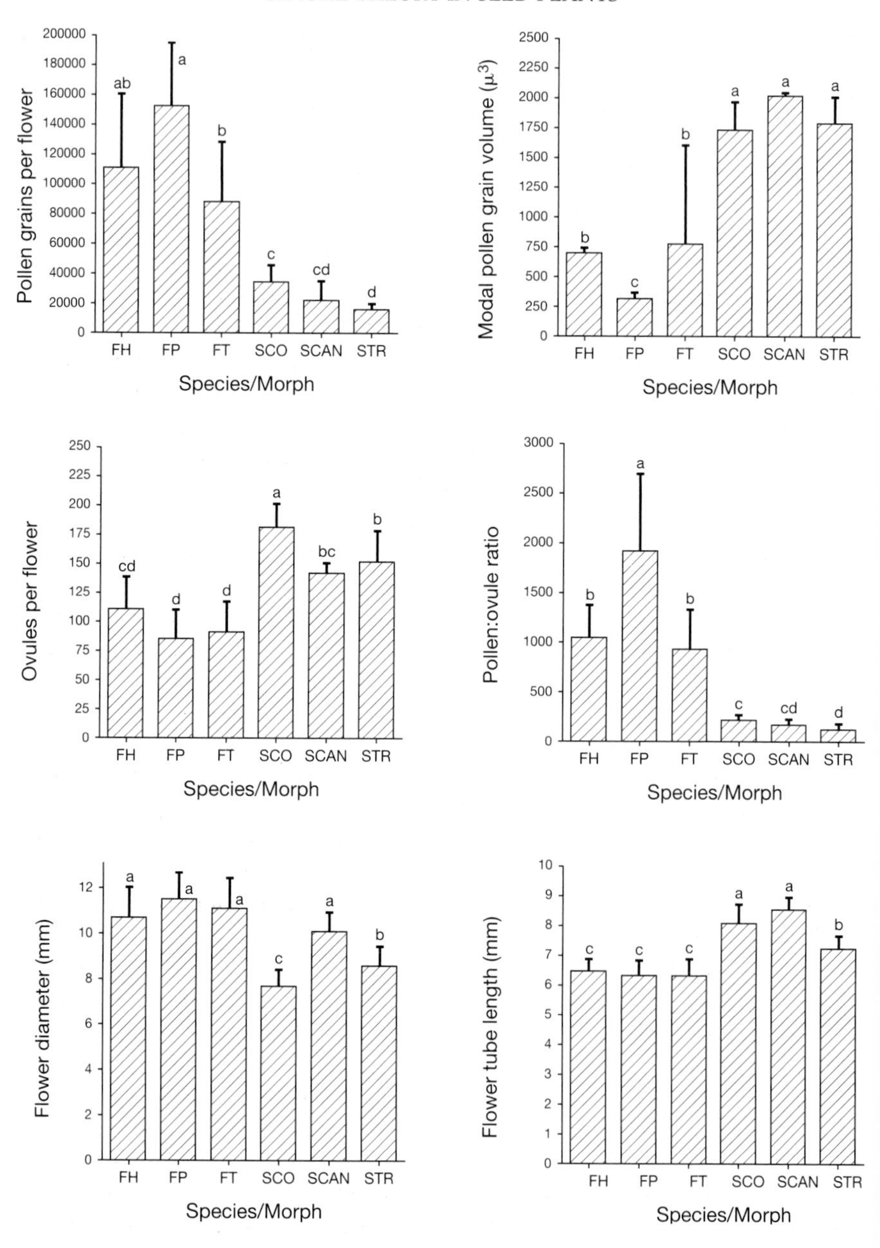

Fig. 2.5 Mean values of floral traits in each species or morph. Error bars are one standard deviation. These mean values represent the means of the genets contributed by a taxon or morph, not the means of all available flowers. Species/morph abbreviations are as follows: FH, *Primula farinosa* (homostyles); FP, *P. farinosa* (pin morph); FT, *P. farinosa* (thrum morph); SCO, *P. scotica*; SCAN, *P. scandinavica*; STR, *P. stricta*. (After Mazer and Hultgaard, 1993.)

tion in the selfers as seeds are much more expensive to produce than are pollen grains. It will pay the plant to concentrate reproductive resource into the less fit and more expensive female function.

Intraspecific variation for reproductive resource allocation also occurs. In an exhaustive study of the wild radish *Raphanus sativus*, Mazer (1992) shows that female resource allocation decreases sharply with increasing density of plants, whereas male resource allocation scarcely changes (Fig. 2.6). Sparsely planted individuals maximize their reproductive performance by investing largely in fruits and seeds, whereas densely planted individuals largely invest in stamens and pollen. Mazer suggests that where the variance of pollen dispersal radius is more than double that of seed dispersal radius, it will pay poorly resourced plants to invest in pollen which may disperse genes into patches which are better resourced. (To validate this idea, it is necessary to demonstrate the corollary, that poorly resourced plants with a high seed dispersal/pollen dispersal variance invest more in female attributes when stressed.) Unusually for this type of study, Mazer recognizes that it is necessary to show that high levels of heritability exist for the capability for plasticity (in female resource allocation) as well as for genetic variability in the trait itself (male resource allocation) if natural selection for gender allocation is to operate. However, her data suggest that many of these attributes are not highly heritable.

Reproductive resource allocation may have wider implications with respect to the breeding system of plants, influencing, for instance, pollination systems. Cruden and Miller-Ward (1981) recorded stigmatic area, pollen-grain surface area, pollen volume and pollen-bearing area on pollinators as well as pollen/ovule ratio for 19 North American flowering plants (Table 2.2). Rather surprisingly perhaps, inefficient systems with high pollen/ovule ratios favour the provision of highly localized stigmatic areas which will in turn tend to promote specialized, oligophilic pollination systems (Chapter 4, p. 76). Thus, there are intrinsic as well as extrinsic features of an outcrossing system which tend to favour specialized pollination.

Female resource allocation

In the flowering plants, fitness is usually equated with the number of seeds produced by a genet. However, fitness concerns the replacement of one reproducing individual with another. Seeds and seedlings vary considerably in their survivorships and there is a large body of evidence (summarized in Baker, Richards and Tremayne, 1994) that relatively large seeds almost always survive and grow between than do smaller ones.

A noteworthy exception is described for *Thymelaea hirsuta* by El-Keblawy *et al.* (1996), in which seeds which have been resourced by the

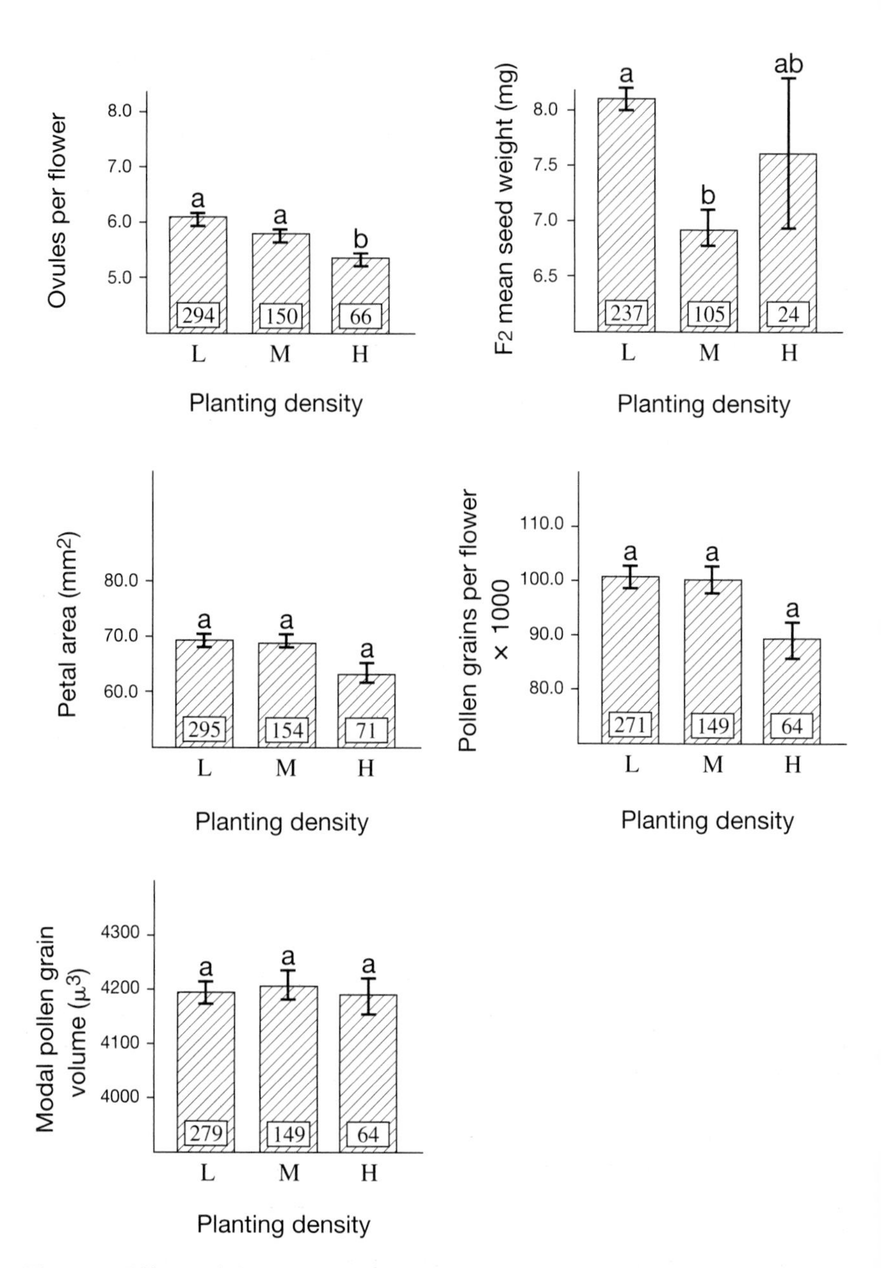

Fig. 2.6 Effects of density on mean phenotype of floral traits in wild radish in each of three planting densities. Bars indicate mean phenotype values; lines indicate the standard error of the mean for each trait. Sample sizes are given at the base of each bar. L, low density; M, medium density; H, high density. With each graph, shared letters indicate that mean phenotype values do not differ significantly between densities, as determined by Tukey's test following a two-way ANOVA (detecting the effects of block and density on mean phenotype). (After Mazer, 1992.)

Table 2.2 Differences in pollen:ovule ratio, pollen surface area, pollen volume and number and stigmatic area: pollen-bearing areas in 19 flowering plants (from Cruden and Miller-Ward, 1981)

Characteristic studied	High values	Low values
Pollen:ovule ratio	in outbreeders	in selfers
Stigmatic area: pollen-bearing area	for low pollen:ovule ratio	for high pollen:ovule ratio
Pollen surface area	for high stigmatic area: pollen-bearing area	for low stigmatic area: pollen-bearing area
Pollen volume	for large pollen number	for small pollen number
Pollen number	for large ovule number	for small ovule number

mother in the absence of male load for the longest time give rise to the fittest seedlings, but these seeds are in fact no larger than those which are less fit.

For multiseeded fruits, the limitation of available resource almost invariably results in an inverse relationship occurring between seed number and average seed size. It follows that, paradoxically, mothers which produce fewer seeds may in fact be fitter than mothers which produce more seeds, because their fewer, larger seeds stand a much better chance of producing reproductive offspring than do their smaller competitors.

Competition for female resource can theoretically occur at several levels:

1. between flowering branches
2. between flowers
3. (between follicles)
4. between seeds

Using *Primula farinosa*, Baker, Richards and Tremayne (1994) showed that seed number tended to increase with flower number per scape, and that average seed weight correspondingly decreased (Fig. 2.7) up to 11 flowers/scape. We suppose that multiflowered scapes may be better resourced, and may also attract more flower visitors, thus allowing them to set more seed. For scapes with numbers of capsules in excess of 11, between-flower competition for resource became evident, so that for these seed number decreased, and average seed weight increased once more (Fig. 2.8). Thus, competition for resource occurs at levels (2) and (4) in this species. As seedling fitness can be shown to be strongly influenced by seed size in *P. farinosa*, we suppose that a disruptive selection for flower number per scape may occur. Scapes with less than nine or more than 12 flowers will tend to set fewer but larger seeds. (In practice, a stabilizing selection for scapes with between five and eight flowers may be more common, as plants with more than 12 flowers per scape are unusual.)

Interestingly, however, there are few if any examples of female resource competition occurring between single-seeded fruits. For instance, Baker,

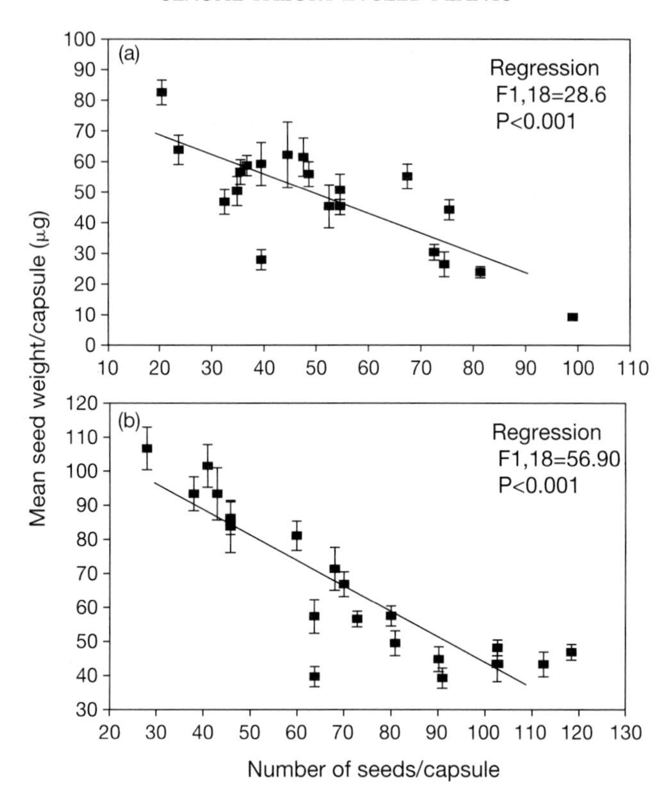

Fig. 2.7 Relationships between the number of seeds/capsule and the mean seed weight/capsule (±2 × SE) for 20 capsules taken from 10 plants of *Primula farinosa* sampled from two subpopulations (Sunbiggin 1 (a) and Sunbiggin 2 (b)). (After Baker, Richards and Tremayne, 1994.)

Richards and Tremayne (1994) show that in the single-seeded *Armeria maritima*, fruit weight increases consistently with flower number per capitulum (Fig. 2.9), presumably because heads with large flower numbers are better resourced and the fruits do not compete. These circumstances should directionally select for large flower number, and may have influenced the evolution of the complex capitulum in families with single-seeded fruits such as the Asteraceae and Apiaceae. When such seed size-driven directional selection is caused by pollinator choice for large inflorescences, it becomes a form of sexual selection.

SEXUAL SELECTION

Sexual selection, where male competition for female choice leads to directional selection for features which may advertise male fitness, but could in

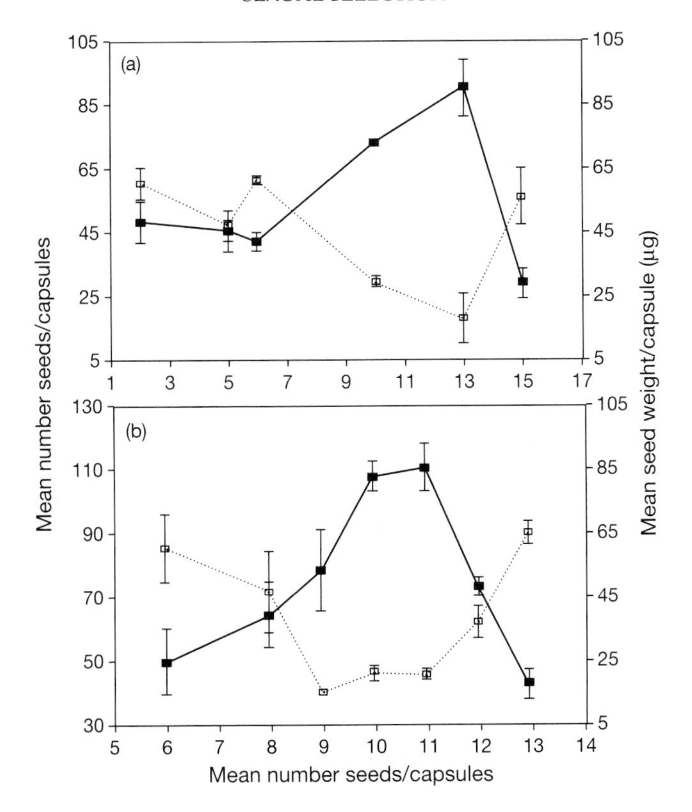

Fig. 2.8 Relationships between the number of capsules/scape and the mean number of seeds/capsule ($n = 20$ capsules) (■), and the mean seed weight/capsule ($n = 20$ seeds) (□) of samples from two subpopulations of *Primula farinosa* (Sunbiggin 1 (a) and Sunbiggin 2 (b)). Values recorded $\pm 2 \times$ SE. (After Baker, Richards and Tremayne, 1994.)

themselves be disadvantageous, was first identified by Darwin (1871). As the female function of flowering plants is essentially sessile, and plants cannot actively choose mates, it might be considered that the concept of sexual selection does not apply to plants.

Selection for female choice

In a review of this topic, Queller (1987) suggests that whereas maternal parents cannot choose the male parents to their offspring directly, they might well be able to select between gametophytes (pollen tubes) derived from different males, or between embryos (or even endosperms) fathered by different males. Such behaviour would be selected for if it enhanced the quality of the zygotes the mother was carrying, at no extra cost.

Typically, investigations into female choice have compared the off-

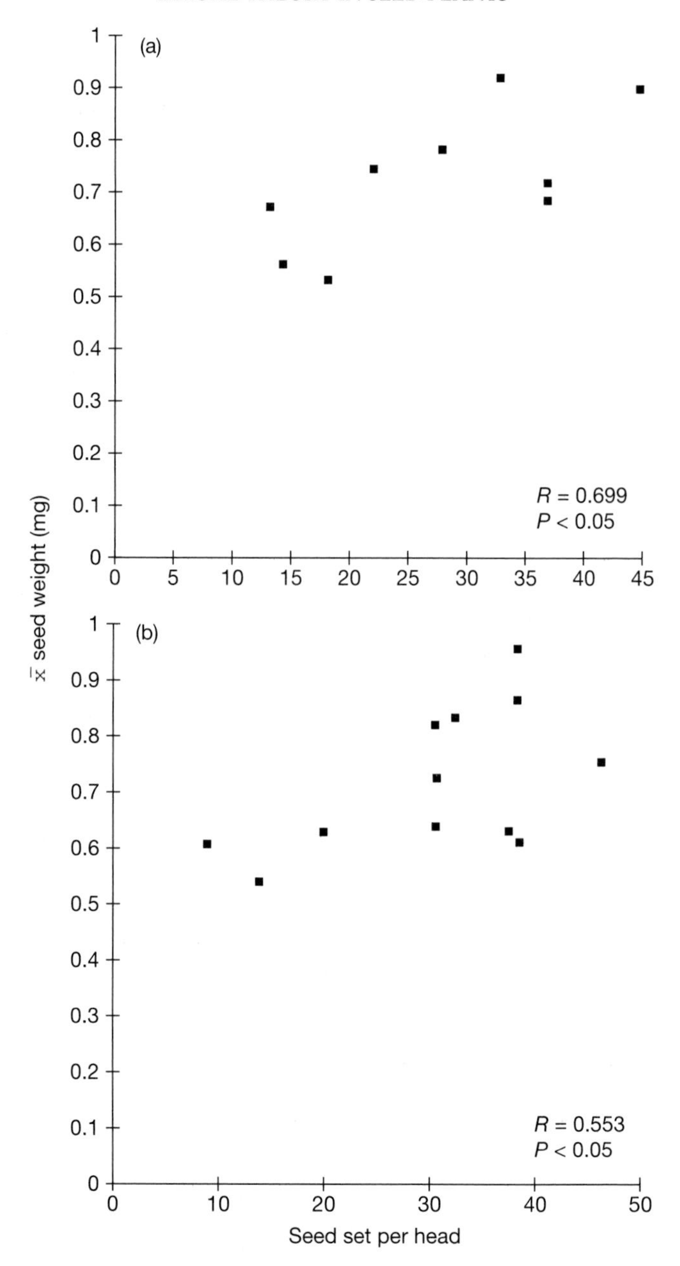

Fig. 2.9 Relationships between numbers of seed set per head and the mean weight of seed set on those heads, for field samples of *Armeria maritima* growing at Underbank. (a) Cob mating type mothers; (b) papillate mating type mothers. (After Baker, Richards and Tremayne, 1994.)

spring of pollinations made at different positions on the stigma (Mulcahy and Mulcahy, 1987). The assumption is that gametophytic competition is maximized for distally placed pollinations, as pollen tubes have further to grow than for proximally placed pollinations. In general, seedling vigour has proved to be greater for the offspring of distally placed pollen, suggesting that genes which are involved in gametophytic vigour also influence sporophytic vigour. Purrington (1993) provides a clear example of this for *Silene latifolia* (Table 2.3).

Most obviously, such female sexual selection occurs through self-incompatibility (s-i) (Chapter 6). Where outcrossed pollen supply is adequate for fertilization, Harder and Barrett (1995) point out that female function should select for s-i, as females avoid incestuous matings by rejecting their own pollen and some pollen from close relatives; however s-i is not favourable to male function as the majority of pollen is selfed and thus rejected by the s-i system.

Through late-acting self-incompatibility, a limited maternal selection for paternity can also sometimes occur through ovule abortion (Seavey and Bawa, 1986). However, there seems to be little if any evidence that a mechanism exists whereby mothers can select between cross-compatible pollen-tubes from different fathers. To demonstrate such sexual selection, it would be necessary to show that genetic variation between mothers occurs for male choice. In one of the few studies of this kind, Marshall and Folsom (1992) were unable to demonstrate such variation.

When zygotes are competing for a limited female resource, the mother should control seed and fruit number, and might be able to do so in a way that is selective between fathers. However, as is seen in the previous section, it may well prove that seed size rather than paternity choice is the main selective parameter which determines zygote fitness.

Roach and Wulff (1987) have reviewed the literature concerning the maternal control of seedling fitness and concluded that the maternal phe-

Table 2.3 Differences in numbers of seeds set in *Silene latifolia* after pollination at the tip and at the base of the stigma (after Purrington, 1993)

	Treatment			
	High nutrient		Low nutrient	
	Base pol.	Tip pol.	Base pol.	Tip pol.
Seed production per capsule	341 ± 32	258 ± 61	368 ± 20	265 ± 50
% Flowered	92 ± 2	93 ± 2	87 ± 3	89 ± 5
% Female	61 ± 3	65 ± 4	60 ± 3	55 ± 3
Emergence time (days)	7.6 ± 0.2	6.9 ± 0.1	8.4 ± 0.2	8.1 ± 0.2
Age at flowering (days)	33.4 ± 0.4	33.5 ± 0.3	33.6 ± 0.3	32.9 ± 0.4

notype, the cytoplasmic genetic component of the embryo, and the maternal nuclear component of the endosperm all interact.

The crucial question, which remains to be answered fully, is whether the paternal components of the genotype of the embryo and endosperm can influence seed size. As yet, the evidence, summarized in Baker, Richards and Tremayne (1994), suggests that seed size is mostly or entirely under maternal control. Mothers are most likely to retain large ovules at abortion, but it is not at all clear that mothers can select for embryonic characteristics (and so, in part, for paternal characteristics) in this way. It is also far from certain that intrinsic vigour attributes of the embryo (for instance those from heterosis) can override extrinsic effects resulting from a large seed size with respect to seedling performance (although heterotic effects are certainly expressed at a later stage in zygote development, Chapter 9). Work in the African orchid *Disa* shows that seed resulting from cross-pollinations is about twice as heavy as that resulting from selfs (Johnson, 1994), but this seems to be a lone example.

Theoretically, it is possible that female mate choice could be controlled by a manipulation of the pollinator system so that male partners were restricted to those which flowered at a particular time, or possessed a particular flower colour or flower height. In such cases, it would be necessary to show that those characteristics in the male and female which encouraged such subgrouping were closely linked or otherwise associated with features which promoted male fitness. As far as I am aware, no such association is known.

Thus, it is far from clear that sexual selection for female choice ever occurs in the flowering plants, except with respect to near-relative mating.

Selection for male competition

Paternity analysis has shown that when a flower is pollinated by several different fathers, genotypes of the offspring do not always fully reflect the potential paternity (Bookman, 1984; Bertin, 1986; Marshall and Ellstrand, 1986; Marshall and Folsom, 1992). Such non-random mating patterns create the opportunity for the evolutionary development of male sexual selection. Several studies show that pollen grains which reach a stigma first usually out-compete later arrivals, and so early-flowering individuals may be so favoured, despite concomitant disadvantages (e.g. Epperson and Clegg, 1987, for *Ipomoea purpurea*).

Marshall and Ellstrand (1986) and others have also shown that multiply-sired offspring tend on average to be fitter than are full-sib offspring. Such results have been interpreted as a demonstration of female mate choice. However, Marshall and Folsom (1992) found no significant interactions between pollen donor and maternal plant effects.

In the absence of an explanation as to how mothers are able to select between zygotes sired by different fathers, it is in fact simplest to assume that these results merely reflect differences in competitive ability between the pollen of different fathers (reviewed by Mulcahy, Mulcahy and Searcy, 1992). As approximately 60% of the genes expressed in the sporophyte are also expressed in the male gametophyte (Willing, Bashe and Mascarenhas, 1988; Ottaviano and Mulcahy, 1989), there is a hint that fitness attributes of the (haploid) pollen tube could persist into the (diploid) zygote it fathers. If such heritable attributes for male fitness are differentially expressed in both the gametophyte and sporophyte generations, the mechanism exists to generate a powerful selection for male success.

In fact, alternation of the generations in the flowering plants can be viewed as a mechanism which has the potential to screen out disadvantageous recessive mutants, i.e. the 'genetic load' (p. 383). Such genes are not protected from selection by heterozygosity in the haploid pollen tube (or the embryo-sac on the female side), so that those mutants whose low fitness phenotypes are expressed in the gametophyte generation should not persist into the sporophyte generation.

Using results from paternity analysis, this subject has been reviewed by Snow and Lewis (1993). These authors do not always demonstrate that clear-cut relationships occur between sporophytic and gametophytic success. Plants which potentially appear to be the most successful fathers do not necessarily sire the most offspring. For instance Meagher (1991) found no relationship between male reproductive success and plant size or flower number in *Chamaelirium luteum*, whereas Devlin, Clegg and Ellstrand (1992) show an asymptotic relationship between flower number and male reproductive success in *Raphanus sativus*, so that the most floriferous plants were not always the most successful fathers.

However, Galen (1992) working with the sky-pilot *Polemonium viscosum* showed that levels of bee visitation to plants are associated with amounts of pollen removed from flowers, with pollen deposited on stigmas of other plants, and with the number of mates reached in the population. She assumes that bee visitation is associated with male expenditure and display. Likewise, in plants with pollinia such as *Asclepias exaltata*, flower number and rates of pollinium removal both provided excellent estimates of male reproductive success, and male reproductive success and female reproductive success were highly correlated (Broyles and Wyatt, 1990, 1991). Such plants (including the Orchidaceae) with an 'all or nothing' pollination mechanism often provide no reward to the pollinator. Male and female reproductive successes are largely determined by the strength of signals sent to the limited number of flower visitors.

The detailed study by Harder and Barrett (1995) suggests that Galen (1992) may not always be correct when she suggests that the successful export of pollen to other individuals should be strongly associated with floral display. These authors point out that a fine floral display might well attract more insect visitors, but it will also usually be associated with the production of more flowers in an inflorescence. Consequently, the chance of geitonogamous self-pollination increases with display, so that there is more 'pollen discounting' (wastage) from many-flowered inflorescences than from few-flowered inflorescences (Fig. 2.10).

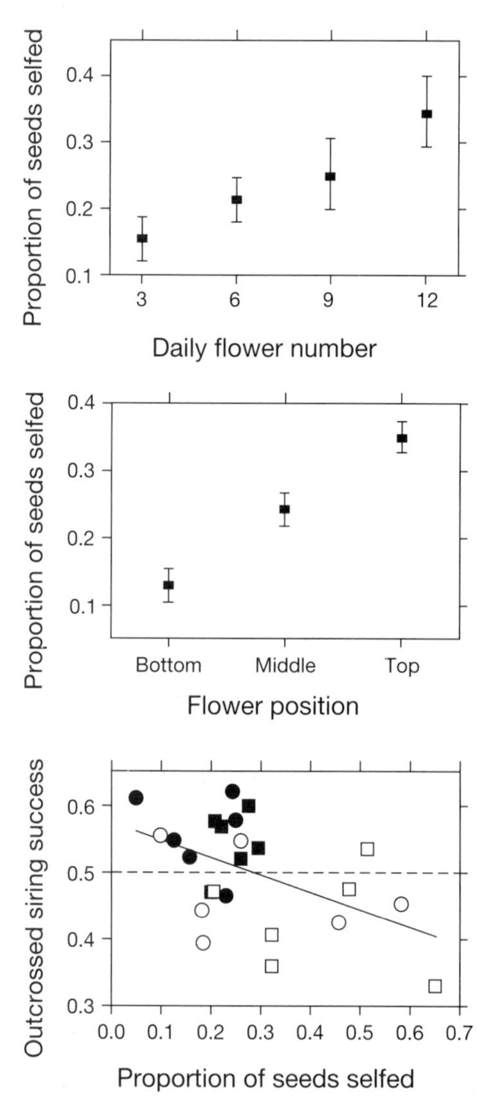

Fig. 2.10 Effect of floral display and flower position on levels of selfing and outcrossing in *Eichhornia paniculata*. ●, three flowers; ■, six flowers; ○, nine flowers; □, 12 flowers per inflorescence. (After Harder and Barrett, 1995.)

Flower number and flower size, major components in floral display, should respond to stabilizing male sexual selection for successful pollen export to other individuals, which will be maximized at an equilibrium point between the contrasting constraints of rates of pollinator visitation and rates of geitonogamous pollen discounting.

In less specialized plants such as *Chamaelirium luteum* (Meagher, 1991) or *Raphanus sativus* (Devlin and Ellstrand, 1990), male and female reproductive success is poorly correlated within individuals. For instance, when ovules within a linear fruit mature sequentially, as in the Brassicaceae, genotypic differences in pollen-tube vigour (and hence in the timing of fertilization) may favour multiple paternity during a single pollination event. Centrally placed ovules are fathered first, by the most competitive of the fast-growing pollen tubes. Apical and basally placed ovules are fathered later by less vigorous tubes. Marshall and Ellstrand (1986) found superior fruit weights after many-fathered pollinations compared with those for single-fathered pollinations in radish (*Raphanus raphanistrum*). Marshall and Folsom (1992) show that this effect probably results from interactions between pollen tubes from different fathers at the stigma which delay fertilization and may cause effective competition to occur at all ovular positions.

In the special cases of dimorphic, trimorphic (Chapter 7), gynodioecious or dioecious species (Chapter 8), individual male fitnesses are dependent on the frequency of the morph which they express (Barrett, Kohn and Cruzan, 1992), but such frequency-dependent effects on male fitnesses are unlikely to occur in monomorphic species.

Sexual selection for female attributes should select for zygote quality, whereas selection for male attributes should select for the competitive ability of gametes (p. 2). In the flowering plants, large components of this competitive ability may be provided by flower number, and by the ability of the inflorescence to attract flower visitors through display and reward. Various methods of estimating potential male fitness are reviewed by Stanton *et al.* (1992). Thus, unlike female sexual selection, male sexual selection in flowering plants can resemble that for animals, in that the attributes which reflect the fitness of the male parent as a whole may also cause the plant to be particularly fit as a male parent by producing more pollen, and by being a more attractive target to pollinators. Naturally, this correspondence between general fitness and male fitness will depend on the plant transferring much of its resource into floral production. That is, directional male sexual selection in zoophilous flowering plants should result in 'r' strategies, where much of the available resource is devoted to flowering, and consequently the plant is monocarpic (fails to perennate).

In a few examples, for instance the spectacular monocarpic saxifrage of the Pyrenees, *Saxifraga longifolia*, or the century plant *Agave americana*, such directional selection seems to have proceeded to this logical but rather ludicrous extreme. These could be thought of as the plant equivalents of the antlers of the (extinct) giant Irish Elk or of the peacock's tail. That selection for male attributes seems to have rarely proceeded so far in the flowering plants can be explained in three ways.

1. In most flowering plants, new ramets are produced both sexually and asexually. In most circumstances, the advantage gained by fathering new genets will scarcely match the disadvantage of not producing new asexual ramets.
2. Highly successful, large and showy individuals are not necessarily the most successful fathers. As prime targets to predators and pathogens, they may be the most likely to be eaten or diseased, and they may be the most liable to damage by wind or rain. Also, they may be so successful at attracting flower visitors that the visitors become loath to travel elsewhere, or at least very far, so that ironically, they are rendered poor fathers to other genets by their very success at attracting visitors (Harder and Barrett, 1995) (Chapter 5, p. 171). Most plants demonstrate stabilizing selection for size and vigour.
3. As over 90% of the flowering plants have hermaphrodite flowers, the dissonant requirements of male and female sexual selection will also tend to stabilize expenditure on flower display.

Thus, Pyke (1982b) showed that increases in conspicuousness of flowers and in potential reward to the pollinating birds of large inflorescences on tall stems in the waratah (*Telopea speciosissima*) may provide directional sexual selection for much larger inflorescences than the fruit-carrying capacity warrants. Fruit-set is therefore always low, even after efficient pollination (Fig. 2.11). Wyatt (1981) similarly demonstrated that in five tropical Leguminous shrubs, seed-set per plant is increased by producing more flowers rather than by increasing the seed-set per fruit. The generality of this principle has been demonstrated by the wide-ranging studies by Sutherland and Delph (1984) and Sutherland (1986). These authors show that the relative fecundity of a flower is normally lower for outcrossers with hermaphrodite flowers than it is for plants with unisexual flowers (Table 2.4). Presumably, hermaphrodite flowers which are produced in excess according to the requirements of male sexual selection have a greater tendency to jettison excess female load postfertilization. However, where male and female functions are separated onto different flowers (monoecy) or different plants (dioecy), female flowers are spared male costs and can thus afford to carry a higher proportion of fertilized fruits.

Bawa (1980) has suggested that selection in tropical forest trees for large, energy-rich animal-dispersed fruits has encouraged the separation

Fig. 2.11 Infructescence of *Banksia quercifolia* (Proteaceae) near Sydney, Australia. Only about 1% of flowers have set fruit; the regular distribution of set fruits on the infructescence suggests that pollination was adequate, and that fruit set is limited by resources available to female function. Fruits remain closed for years until burnt by bush fires, when seeds are released, as here (×0.25).

Table 2.4 Comparison of breeding-system-specific percentage fruit-set patterns. Statistical analysis was based on arcsine-transformed data. $\bar{X}$ and s values in this table have been back-transformed to percentages (after Sutherland and Delph, 1984)

Breeding system	$\bar{X}$	s	N	t		
	% fruit-set					
Hermaphrodites	42.1	19.2	316			
SI hermaphrodites	22.1	13.6	187	12.40	4.24	
SC hermaphrodites	72.5	12.5	129	$P < 0.001$	$P < 0.001$	
Monoecious and dioecious	61.7	14.1	129		4.03	0.10
Monoecious	53.8	10.5	80		$P < 0.001$	$P > 0.90$
Dioecious	73.8	17.6	49			

SI = self-incompatible; SC = self-compatible.

of male and female function, leading to dioecy. It would certainly seem that the dissonant requirements of male and female fitness might in such extreme cases encourage the separation of sexual function onto discrete genets. However, both dioecy (less than 5% of species) and monoecy, where male and female function is separated onto different flowers within the same hermaphrodite individual, are scarce in the flowering plants. This suggests that the differential requirements of sexual selection in flowering plants are rarely of a magnitude such as to overcome the inherent disadvantages of separating male and female function into separate flowers or individuals (pp. 303–10).

SEXUAL SYSTEMS AND GENOTYPE FREQUENCIES

Theoretical panmixis requires a sexual population of infinite size and random distribution, in which all individuals are equally as likely to exchange genes with each other and all are equally viable. Such constraints are considerable, and will be true for very few populations of plants. As is discussed below and in Chapter 9, plant population sizes are often very small, often indeed much smaller than is immediately apparent. As Chapter 5 shows, pollen flow is often far from random, and in Chapter 9 it becomes clear that many plants are far from being wholly outbreeding. Further, the distribution of plant populations is rarely random, but shows infinitely variable gradations of clumping, regular spacing and density, from dense groups to sub-isolated islands. However, mathematically, many populations may approximate to panmictic ideals sufficiently closely to give genotype frequencies that do not statistically depart from those predicted. It is when predicted frequencies vary from those observed, and particularly when heterozygote frequencies are lower, or higher, than those expected, that one may suspect that non-panmictic breeding systems are in operation.

As is well known, panmictic genotype frequencies are predicted by the formula independently conceived by G. H. Hardy, the Professor of Mathematics at the University of Cambridge and friend of the geneticist, R. C. Punnett, and by the German medic Weinberg, in 1908 (Punnett, 1950). This principle, equation, law or equilibrium (for it is called all four) is the basic model of population genetics from which all others are derived, or which they assume, and is named after its inventors 'the Hardy–Weinberg Law'. It is not the function of this book to derive, or even expound, this law to any depth, for this is ably achieved by many primers of population genetics. Rather, I need to briefly recount its application at a simple level in breeding system theory, to which it is also basic.

The Hardy–Weinberg Law refers to genotype frequencies of individuals in a panmictic population for a locus polymorphic for two alleles which we will call A and a. It requires the determination of allele frequencies $p(A)$ and $q(a)$, where $p + q = 1$. These frequencies are readily determined when there is incomplete dominance and the heterozygote Aa is phenotypically distinct from either homozygote AA or aa. Most usually, this is achieved by examining isozyme banding using gel electrophoresis. In these cases:

$$p = (2AA + Aa)/2n$$

and $q = 1 - p$, where n is the sample size. Where there is dominance, it will be necessary to assume Hardy–Weinberg genotype frequencies to establish allele frequencies, and if the object is to investigate departures from Hardy–Weinberg ratios, circularity is involved. The frequency of the homozygous recessive $aa = q^2$ (as will be seen below), from which q, and hence p, can be calculated, if panmictic assumptions are made.

The law predicts that genotype (and hence in the absence of dominance, phenotype) frequencies will occur at:

$$p^2(AA) + 2pq(Aa) + q^2(aa) = 1$$

This equation thus presents two very important, although perhaps rather obvious, conclusions:

1. Genotype (and phenotype) frequencies are reliant on allele frequencies, and nothing else, in a panmictic population in which A and a are neutral with respect to each other.
2. Allele, and hence genotype, frequencies will not vary significantly from one generation to another in the absence of differential selection, migration or mutation, in panmictic conditions.

To these, we may add another conclusion which is perhaps more important to the present topic:

3. Significant excesses, or deficiencies, in the frequency of heterozygotes from those predicted by $2pq$ will most probably occur as a result of non-panmixis, for instance by disassortative mating.

However, heterozygote excess may also occur as a result of heterozygote advantage, or by genetic duplication (for instance in an allopolyploid), or by asexual fixation of advantageous heterozygotes, for instance by agamospermy. Differential selection of A as against a phenotypes may cause excesses or deficiencies for homozygote frequencies p^2 and q^2, but should not seriously bias $2pq$.

Non-panmictic deficiencies in heterozygote frequency will result from (a) assortative mating; (b) selfing; (c) small populations; and (d) sibling or near-relative mating.

Before we go on to examine those interesting abnormalities of breeding systems which lead to departures from expected heterozygote frequencies, it is worth briefly examining modifications of the formula for multi-allelic loci, and for polysomic loci.

Multi-allelic loci

Hardy–Weinberg genotype ratios can be easily expanded to account for loci with more than two alleles, thus for three alleles:

$$p^2(AA)+2pq(Aa)+2pr(Aa')+q^2(aa)+2qr(aa')+r^2(a'a')=1$$

where r is the frequency of the third allele a' (and so on).

Polysomic loci

These will occur in polyploids, where autopolyploidy, or repetition of loci in different, but homoeologous chromosomes, leads to a locus being present more than twice, and thus polysomic inheritance occurs. This leads to quite complicated frequencies; thus even at a tetrasomic locus:

always assuming that you can estimate allele frequencies, which is likely to be very difficult.

Disassortative mating

Disassortative mating, in which mating between two phenotypic expressions of different alleles at the same locus occurs more frequently than expected, is not common, even among animals. However, there are many examples of frequency-dependent disassortative mating among insects such as *Drosophila* and certain Lepidoptera, where distinctive phenotypes are particularly sought after as mating partners when rare in a population.

The only examples that I am aware of in plants that could be construed as disassortative mating are where the population is formed of two distinctive phenotypes which mate with each other, but cannot mate among themselves. Dioecy (unisexuality) is the most obvious example (Chapter 8), and heteromorphy (Chapter 7) also falls into this category. In these, loci linked to the chromosome on which control of the sexual dimorphism lies will be much more heterozygous in the heterogametic phase (*XY* in the case of dioecy) than predicted by Hardy–Weinberg equilibria, which is merely another way of describing the effects of sex linkage.

Assortative mating

Assortative mating is a common phenomenon in both animals and plants, and is very important as a vital trigger to the inception of sympatric speciation. Clearly, if gene flow is restricted between two or more phenotypes in the populations, this population subdivision may allow genecological differentiation of the two subdemes which may in due course lead to speciation. Assortative mating in outbreeding plants can be due to one of two causes:

1. temporal separation, due to two phenotypes in a population which have different flowering times;
2. ethological separation, due to two phenotypes in a population which attract different animal visitors, or create different spatial patterns of pollen donation and reception.

There are numerous examples of both classes in the literature, and ethologically induced assortative mating is further discussed in Chapter 5, especially with regard to flower-colour polymorphisms.

Assortative mating in otherwise outbred plants will lead to an excess of homozygotes at those loci that control the assorting phenotypes, and at other loci linked to them. However, it will not lead to general homozygosis.

Selfing

A special case of assortative mating is selfing, which is very common in plants, and occurs in nearly all self-compatible hermaphrodite seed plants, which may be a quarter or more of all species. In highly autogamous species it is the dominant breeding system (Chapter 9). When genes for selfing and for outcrossing coexist within a population, selfers have an automatic transmission advantage over outcrossers, because selfers provide both the male and female parents to their offspring, but outcrossers can only control the female parentage to their offspring.

Selfing will result in an excess of homozygotes, and obligate selfing may lead to total homozygosis, except perhaps at a few loci with marked heterozygote advantage. During an initial bout of selfing, offspring commonly experience a diminution of vigour (inbreeding depression) which is attributable both to the homozygous condition of deleterious recessives, and to the loss of heterozygous advantage (p. 385, Chapter 9). However, habitual selfers consist of largely homozygous populations which have become 'purged' of deleterious recessives (Charlesworth and Charlesworth, 1987, 1990). The lack of genetic variability associated with high levels of homozygosis may present evolutionary obstacles to this kind of breeding system; high levels of selfing are usually restricted to opportunist annuals for which fitness is maximized by reproductive efficiency.

Small populations

Theoretical panmixis requires infinitely large populations of randomly interbreeding individuals. Very small populations tend to show two characteristic features:

1. inbreeding (the mating of related individuals), resulting in the loss of genetic variability. Fewer loci are polymorphic in populations and heterozygous in individuals than is the case for larger populations;
2. the tendency for alleles to fix within small populations at random, or even somewhat against selective forces ('genetic drift').

Inbreeding

The rate of the loss of genetic variability varies with $1/2N$, where N is the actual population size of interbreeding individuals. Thus, if panmictic principles otherwise hold, where 43% of the genetic variability in a population is lost in a population with a constant size of ten individuals after ten generations, if the population is of 100 individuals, the corresponding loss of variability will be only 5%.

Of course, population size is rarely constant from one generation to another, and the effective population size Ne can be estimated by

where $N1$, $N2$ and $N3$ are the actual population sizes in succeeding generations and n is the number of generations (see also p. 187 for a discussion on neighbourhoods).

When population sizes crash due to some catastrophe, the population then passes through a 'bottleneck' before regaining its former size again. In this case, the average population sizes N required to restore the genetic variability in the population to the condition found before the crash (Ne) after t generations can be estimated by

where N^* is the population size at its lowest point. To take one example, where $Ne = 100$, and $N^* = 25$, it will require a population size N of 1650 to restore variability after ten generations; 2400 to restore it after five generations.

Genetic drift

The effect of small population size on genetic structure can also be examined by considering the sampling error on allele frequencies in populations of different sizes. We assume that these populations have been derived from a single founder at generation t with allele frequencies of p and of $(1 - p) = q$ and have a size (number of interbreeding individuals) of N. For any generation T, the frequency of an allele will be distributed with a mean value of p (assuming no selection) and variance of:

$$pq\left[1-(1-1/2N)^{T-1}\right]$$

The variance exhibited between populations will be a function of N. If N is relatively large, variance between populations will be low. If N is very small, the variance in allele distribution between populations will be high.

This chance or stochastic fluctuation between populations, and for one population between generations, in allele distribution is known as genetic drift, or the Sewall Wright effect after the American geneticist who derived its theory. Naturally, genetic drift is also dependent on the rate that new alleles enter the population by mutation or migration, and Wright (1931) showed that genetic drift would only operate when $1/4Ne$ was substantially greater than the product of mutation rate, the migration rate, and the selection coefficient for or against the mutating allele. Where migration rates (p. 185) are greater than one immigrant individual into a population (irrespective of its size) every second generation, Wright suggested that gene drift effects would be negated.

Ellstrand and Elam (1993) show that in small populations of 32 species the estimated number (Nm) of immigrant individuals per generation varied from 0 to 40 and this variation was representative of that in plant populations as a whole. They consider that Nm is usually the most important parameter with respect to the genetic structure of any small population.

The effect of genetic drift in small populations will be that chance fluctuations may cause a polymorphic allele to become extinct. Such genetic fixation will result in evolution in the absence of differential selection (I have assumed the alleles are selectively neutral with respect to one another).

Earlier generations of British population geneticists, for example R. A. Fisher and E. B. Ford, did not consider genetic drift to be an important feature of evolution. Their studies of insect populations suggested that small populations (for instance less than 100) rarely, if ever, occurred. In fact, some isolated animal populations (for instance many large mammals and birds) are sometimes that small. However, plant populations are very frequently extremely small. Small population sizes in plants can be caused not only by local rarity, but also by clonality. Many apparently huge plant populations may in fact be composed of only a few genets or even of only one.

Using data from the California Department of Fish and Game RAREFIND database of endangered species, Ellstrand and Elam (1993) examine the population size of 2993 populations assigned to 559 plant taxa. Of these, they find that the modal population size (22% of cases) is for populations of between 11 and 50 individuals, and that 52.5% of the populations sampled have population sizes of less than 100 individuals (Fig. 2.12). For 484 British populations of a single genus, *Carex*, I have analysed the data of R. W. David, and find that exactly two-thirds of populations have less than 100 individuals (Richards, 1986).

In most groups of plants, small populations of less than 100 individuals are not the exception, but the norm. As yet, there has been very little work on wild populations of plants to assess whether the genetic effects experienced by such small populations are deleterious. For many rare species, where populations are habitually small, populations may have been 'purged' of deleterious recessives, so that little inbreeding depression is experienced. Karron (1989) examined a number of widespread and localized species of *Astragalus* for symptoms of inbreeding depression, but in only one case did he suspect that localized populations were exhibiting harmful effects. However, Byers and Meagher (1992) suggest that self-incompatible species with small populations are likely to suffer from a shortage of *s* alleles which will hinder their successful reproduction (Chapter 6). This is clearly exemplified by species such as the reed *Phragmites australis* where large populations can be composed of few genotypes which never set seed (McKee and Richards, 1996).

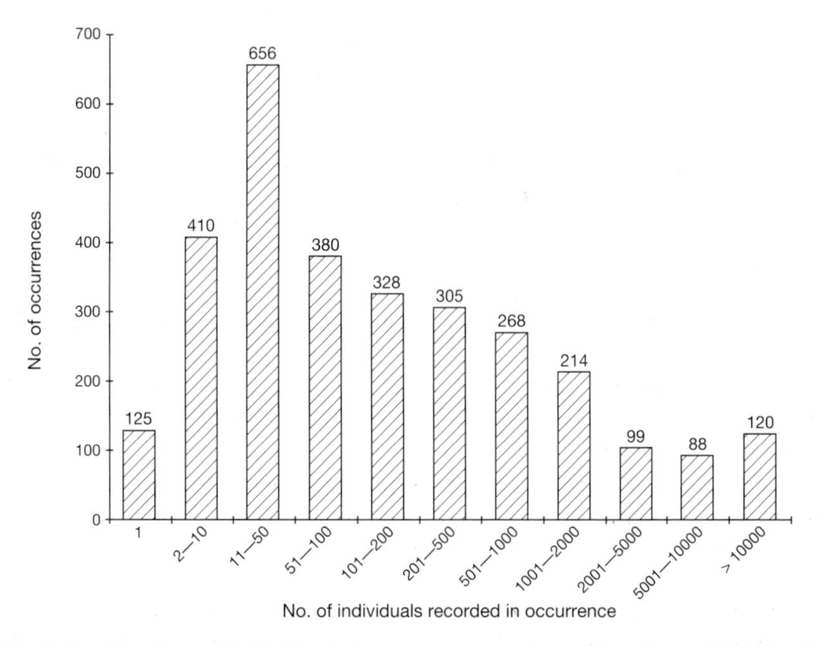

Fig. 2.12 Number of individuals in occurrences of sensitive flora of California. (Each occurrence was assumed to constitute a single population.) See text for more information. (After Ellstrand and Elam, 1993.)

An interesting potential interaction between small population size and the breeding system arises out of suggestions by O'Donnell and Lawrence (1984). They note that the frequency-dependent advantage of rare self-incompatibility *S* alleles will increase as the number of *S* alleles in the population decreases. In very small populations of gsi species with few *S* alleles, *S* alleles which are in linkage disequilibrium with disadvantageous alleles of other genes may be advantaged to such an extent that the linked disadvantageous alleles may persist through 'hitchhiking'.

The results of genetic drift in small populations are not necessarily harmful, and in some cases may even result in the fixation of characteristics which become successfully adaptive at a later stage (Moore and Lewis, 1965).

Ellstrand and Elam (1993) suggest that harmful effects resulting from small population size are in fact most likely to result from outbreeding depression (p. 187). This phrase is used to describe the condition where the products of distant matings are less fit than the offspring of near-neighbour matings. This phenomenon presumably results from differential adaptations within populations such that the products of matings between localized ecotypes are maladapted in comparison to the offspring of crosses within the ecotypes. Waser (1993) reviews 25 cases and finds evidence for outbreeding depression in more than 70% of these. The effects of outbreeding depression are most obviously expressed when a rare species hybridizes with a more abundant and aggressive congener, giving rise to maladapted and frequently sterile hybrids. Clearly, the likelihood of outbreeding depression becomes greatest when populations are very small.

CHAPTER 3

Sexual reproduction in flowering plants

Seed plants such as the Gymnosperms and Angiosperms have acquired two major advantages over the Pteridophytes: they have dispensed with the need for water in order to achieve sexual reproduction; and they protect and disperse the resultant zygote in a seed. Yet only 700 species of Gymnosperm survive today, in contrast with about 300 000 species of Angiosperm. The Angiosperms have adopted a much wider range of growth forms, and have been able to inhabit a much wider range of ecological niches than have either the Pteridophytes or the Gymnosperms. Also, they have been able to adopt a much wider range of means of achieving sexual reproduction in terms of spore travel (pollination), control of mating partner both before and after pollination and fertilization, and seed protection and dispersal.

This variety of mating systems, so characteristic of the flowering plants, has been largely responsible for the structural variety of this group, and hence its ecological success. This structural variety has been achieved in two ways:

1. by maximizing the opportunities for the reproductive isolation of evolutionary lines;
2. by utilizing manifold features of the environment, very frequently other organisms, to transport pollen and seeds to specific sites. This has greatly multiplied the potential number of species-specific niches within an environment.

These features have arisen as a result of the development of the gynoecium, which I consider to have been the single most significant advance in the evolution of land plants.

THE SIGNIFICANCE OF THE GYNOECIUM (PISTIL)

The word 'angiosperm' translates as 'enclosed seed'. Just as the seed plants enclosed the process of sexual fusion and protected the zygote by allowing the male and female gametophytes to remain within the megasporangium (ovule = seed), so the Angiosperms have further enclosed the ovule within a modified sporangium-bearing leaf (megasporophyll), which is termed the carpel. The gynoecium (pistil) consists of the total megasporophylls of a flower and their contents: the carpels, whether free or fused into an ovary, and their ovules, styles and stigmas (Fig. 3.1). Thus, the sexual process has been progressively enclosed in two sets of concentric containers. The second enclosure, of the ovules in the carpel, has had the following major consequences.

First, as a result of the enclosure of the micropilar area of the ovule, it became necessary for a pollen reception area to develop outside the carpel (the stigma). The stigma provided pollen reception in optimal positions. In particular, it would have removed the potential antagonism between the functions of ovule protection (by the megasporophylls) and pollen reception which appears to be a feature of many modern Gymnosperms.

Second, the development of the carpel freed some of the foliar organs of the floral axis from the functions of ovule protection. These organs evolved other functions such as the attraction of animals (tepals, petals), reward of animals (nectaries), or protection of the floral axis from damage and desiccation (tepals, sepals, etc.). In this way the Angiosperm flower evolved. The carpel also allowed the functions of pollen issue and reception to be combined within one floral axis; in other words, the flower could be hermaphrodite, the usual and probably the ancestral condition in the Angiosperms (Chapter 4).

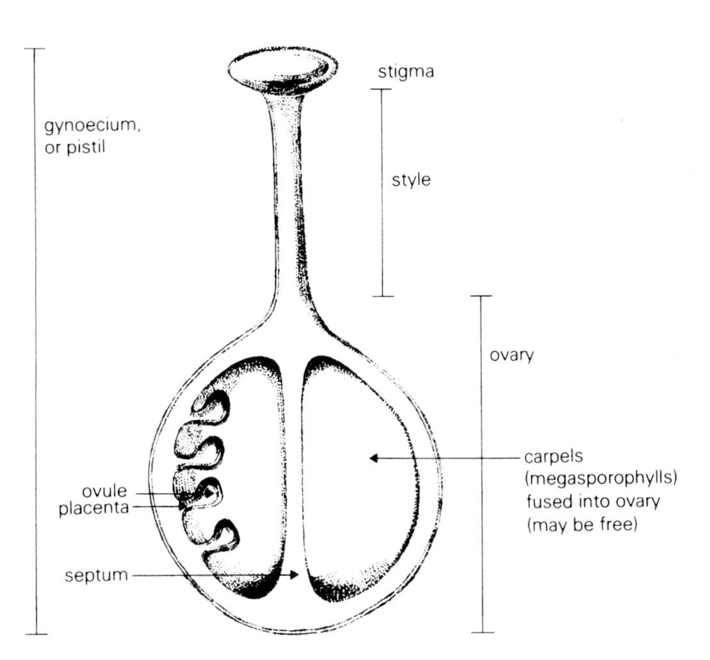

Fig. 3.1 Conventionalized diagram of a gynoecium in an Angiosperm in the family Cruciferae (Brassicaceae).

Hermaphrodity allowed a greater diversity of breeding systems to evolve, including self-pollination. Animal pollination is greatly encouraged by the hermaphrodite condition, for all flowers produce pollen which many animals collect. Although some animal pollination may predate the Angiosperms, the development of the flower which attracts, receives and rewards visitors, is unique to the Angiosperms.

Third, the continued enclosure of the ovule (seed) after fertilization allowed great diversification to develop in the processes of protection, transportation and release of the seed, and in the structure of the seed itself. In the Angiosperms, ovaries (fruits) have developed colours, odours and food to encourage ingestion by animals; hooks, glue and external secretions to encourage external transport by animals; wings and hairs for transport by wind; bladders and tough exteriors for water transport; and many other devices. Neither could the minute, windborne seeds of the orchids and other families successfully reach maturity without external protection during their development. Much of the major niche diversification exhibited by the Angiosperms is a product of their highly specialized fruits, or of highly specialized seeds that develop in fruits. One may mention epiphytes, partial or obligate saprophytes, partial or obligate parasites, hydrophytes, halophytes and maritime psammophytes as being among those Angiosperm life styles which are absent, or virtually so, from the Gymnosperms, and which rely heavily on a specialized fruit or seed.

STAMENS, POLLEN AND THE MALE GAMETOPHYTE

Most Angiosperm stamens are composed of a narrow stalk (filament), homologous with the microsporophyll of some Pteridophytes, and an anther, which is a microsporangium (Fig. 3.2). The anther is usually composed of two segments, joined longitudinally by the connective, which each contain two pollen sacs (locules), although these may be fused within each segment. Within each locule, an archesporial cell divides mitotically to yield many pollen grain mother cells (PMC), which undergo meiosis to give haploid tetrads of microspores. In most cases, these separate and develop as pollen grains within the anther. They are surrounded by a single layer of endothelial cells, the tapetum, which are secretory nurse cells and are irregularly and highly polyploid. The mature anther most often dehisces longitudinally, towards the inside (introrse) or the outside (extrorse) of the flower, but may dehisce by apical or basal pores. In a few families, notably the Orchidaceae and Asclepiadaceae, there is a single indehiscent anther, the locules of which have formed two (or one) pollinia containing tetrads; the pollinia are dispersed whole.

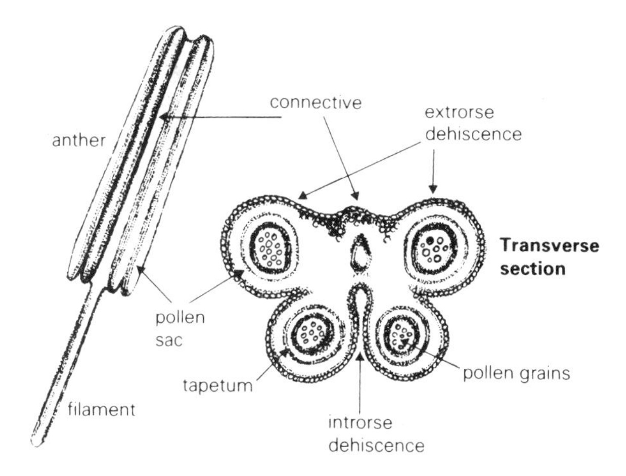

Fig. 3.2 Conventionalized diagram of a stamen in an Angiosperm.

Stamens vary from one to hundreds per flower and are borne laterally around the floral axis, proximally to (below) the gynoecium, but distally to (above) the corolla and calyx. Although most often borne free on the receptacle, the filaments and even the anthers may at times be fused to the petals, or more rarely to the nectaries or gynoecium. In the latter case, the anthers may be apparently fused to the gynoecium, often very close to the stigma, forming a gynostegium. This is once again typical of the unrelated dicotyledonous Asclepiadaceae, and monocotyledonous Orchidaceae, a clear case of parallel evolution (Chapter 4).

The pollen grain, a haploid microspore, is not dissimilar in structure to Gymnosperm pollen, or even Pteridophyte spores, but is consistently diagnosed by the columnar structure of the outer wall (exine), between the continuous inner layer (nexine) and the perforated outer layer (tectum) (Hickey and Doyle, 1977). The exine is constructed primarily of a lipoprotein, sporopollenin, whereas the inner wall (intine) is made of cellulose (Heslop-Harrison, 1975a). The exine may be variously decorated with spines, furrows, etc., and usually has from one to four pores, which may or may not be associated with furrows (Fig. 3.3). The characteristic size, shape and decoration of a pollen grain persists into the fossil record.

Within the grain, one or two mitoses have taken place to yield two or three haploid nuclei. The timing of the first pollen grain mitosis varies widely, occurring rapidly in those species with fast pollen grain development (e.g. most tropical species). However, in boreal plants in which floral initiation and meiosis occur the previous autumn (for instance many species of *Rhododendron*), pollen grains may over-winter in a uninucleate phase. The pollen grain mitosis is usually unsynchronized, except

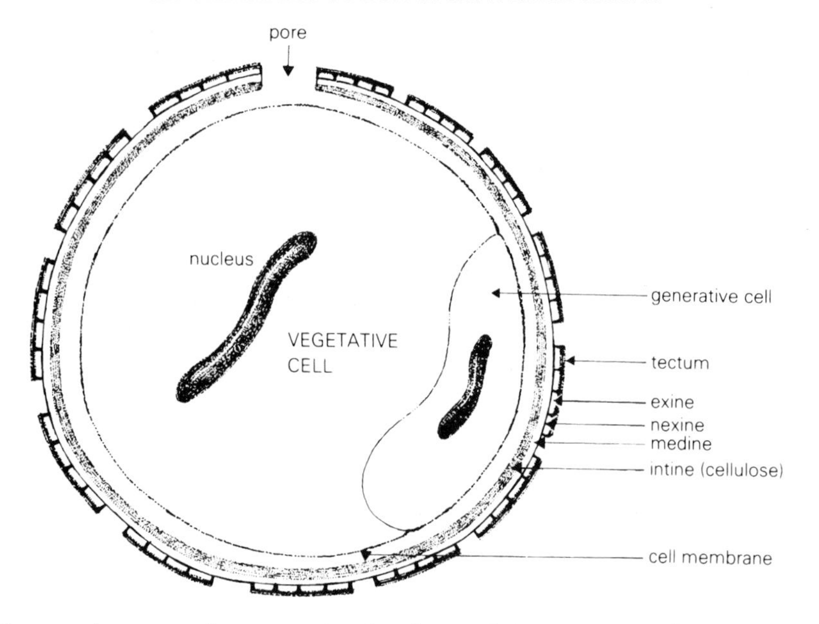

Fig. 3.3 Anatomy of a conventionalized typical Angiosperm pollen grain. The gametophyte may contain two or three cells.

when the pollen grains are in intimate contact within pollinia, as in the orchids. The mitosis is abnormal, being asymmetrical, the pole next to the cell wall being blunt and the free pole acute in shape. After the first division, two unequally sized cells, without a cellulose partition, arise. The larger central cell (vegetative), and the smaller peripheral cell (generative) together form the highly reduced male gametophyte. The nucleus of the vegetative cell frequently assumes curious forms, sometimes becoming very thin and filamentous. The generative cell loosens itself from its peripheral position latterly, and often becomes spindle shaped. It can divide into the two male gametes at an early stage, or this second mitosis may wait until after the germination of the pollen grain, although in the latter case it may have entered prophase before germination. In some cases, the second mitosis occurs after the pollen grain alights on the stigma, but before germination, during the hydration phase (Maheshwari 1949). Thus some pollen grains are binucleate, with a vegetative and a generative nucleus, whereas others are trinucleate, with a vegetative nucleus and two male gametes (sperm cells). There is a strong correlation between these conditions, the nature of the cuticle of the stigmatic papillae (wet or dry), and the control of incompatibility, as follows (see Chapters 6 and 7):

Pollen grain	Cuticle of stigmatic papilla	Control of incompatibility
binucleate	'wet'	gametophytic
trinucleate	'dry'	sporophytic

The second mitosis giving rise to the male gametes (sperm cells) is also usually abnormal in configuration, apparently because of limitations in space caused by the constriction of the generative cell or pollen tube. In wheat (*Triticum*), sperm cells show the following four developmental stages (Chu and Hu, 1981):

1. Naked cell stage in which the sperm cell is round and naked, enclosed only by a discontinuous plasma membrane;
2. Walled cell stage in which the sperm cell is completely surrounded by membranes of its own and of the vegetative cell; the space is filled with callose;
3. Stage of cytoplasm increase during which the size of the cell and the number of organelles increase and the cell changes in shape, becoming more elongated;
4. Mature sperm in which the cytoplasm and organelles are concentrated at one end to form a tail-like structure.

Usually, both male gametes (sperm cells) are indistinguishable in shape and size, but Russell (1981) describes how those of *Plumbago*, which abnormally lacks synergids in the mature embryo-sac, has very distinct sperm cells. One only has a long slender projection which is associated with the vegetative nucleus; this sperm, which eventually fuses with the primary endosperm nucleus, contains the greater proportion of mitochondria. The other, which fuses with the egg cell to form the embryo, receives nearly all the plastids. There are a number of other cases cited in Maheshwari (1949) in which the sperm cell fusing with the egg cell is smaller than that which helps to form the endosperm.

POLLEN GERMINATION

When mature pollen is released from the anther, it normally has a very short life. In some grasses this may be as short as 30 min, and even in insect-pollinated species with sticky pollen it rarely exceeds one day, although much longer periods of natural viability have been recorded in some fruit trees and Gymnosperms (Maheshwari and Rangaswamy, 1965). Pollen which has been freeze-dried under a partial vacuum and stored at temperatures of about –50°C and less than 50% relative humidity may survive for much longer periods, certainly in excess of ten years. Heslop-Harrison (1979a,b) and Shivanna and Heslop-Harrison (1981) have related pollen viability to the condition of the plasmalemma of the vegetative cell. If this is able to reassume a continuous lamellar structure of lipid bilayer on rehydration, its osmotic properties are recovered, and the turgescence which is a vital preliminary to pollen germination can be attained. However, if the partially disordered plasmalemma encountered in the dry grain is further dissociated by heat or desiccation, it will become unable to reorder, and thus to act osmotically, and the grain will be non-viable. The state of the plasmalemma of the vegetative cell, which clearly plays a key role in the germination of the pollen grain, is readily assessed by the use of the dye fluorescein diacetate, which is cleaved by esterase activity to fluorescein and accumulates within the intact membrane in a viable cell.

The only essential requirements for the germination of many types of pollen grain seem to be water and oxygen. Pollen may germinate on the moist surface or a corolla (Maheshwari, 1949) or in a water-filled corolla tube (Eisikowitch and Woodell, 1975), and I have seen selfed pollen of a *Primula* germinate in nectaries. However, pollen tubes that germinate in water usually burst from turgor pressure. A solution of the correct osmotic potential is necessary for pollen-tube survival, and a carbohydrate source is required for continued pollen-tube growth. Rather unexpectedly, boron is necessary for the successful germination of pollen in many species. Boron is deficient in pollen, but occurs at comparatively high concentrations in stigmas and styles. It has been shown to aid sugar uptake, and to be involved in pectin synthesis in the developing pollen tubes, perhaps by its coenzymic function in the synthesis of the D-galacturonosyl units of pectin (Stanley and Loewus, 1964).

Small quantities of calcium are also necessary for successful pollen germination and pollen tube growth. The requirement seems to be complex, calcium being involved in pectate synthesis, in the suppression of inhibitory cations, and in osmotic regulation (Brewbaker and Kwack, 1963). The population effect, or mass stimulation effect, observed when many pollen grains germinate much better than do isolated grains *in vitro*, is also apparently due to calcium borne on the outside of the pollen, although other substances such as organic acids may also be implicated (Maheshwari and Rangaswamy, 1965). This mass stimulation effect may provide the rationale underlying the use of mixtures of old and new date palm pollen by Arab cultivators (where fresh pollen is scarce).

Calcium gradients are also unambiguously implicated in the chemotropism of the pollen tubes within the gynoecium in some, but not all, plants. A very wide range of substances has been shown to have a stimulatory effect on pollen-tube growth *in vitro*, ranging from amino acids to auxins, and purines to sugars. Few, however, have been implicated in tropism, although it is often easy to show experimentally that stylar fragments exert an (unidentified) chemotropic effect.

Stigmas are usually receptive to pollen over a relatively short period while the flower is open, which is generally marked by the turgidity of the stigmatic papillae, which often secrete a sugary solution, and is concluded by the necrosis of these cells. Legitimate pollen arriving on a ripe stigma will adhere to it. However, it is often possible for fertilization to be achieved by artificial pollination of bud stigmas, particularly if the stigmatic surface is damaged.

There are two main types of stigmas recognized, termed 'wet' and 'dry'. These can be differentiated as follows:

	Stigmatic exudate	Papilla cuticle	Pollen hydration	Pollen entry	Incompatibility
wet	present	gappy	external	intercellular	gametophytic
dry	absent	continuous	internal	intracellular	sporophytic

A wet stigma produces a stigmatic exudate containing free sugars, lipids, phenolic compounds and traces of enzymes. This exudate appears to perform four distinct functions:

1. Attaching the pollen to the stigma surface;
2. Hydrating the pollen via the discontinuous exine to the absorbent medine, the pectin- and protein-containing layer situated between the exine and the intine;
3. Providing a suitable medium for the initial growth of the pollen tube;
4. Preventing the stigmatic papillae with their discontinuous cuticles from drying out.

In contrast, a dry stigma, typical of most sporophytically controlled incompatibility systems, has no stigmatic exudate. The pollen adheres by virtue of its external cover of lipoprotein 'pollenkitt', or 'tryphine' which is secreted by the tapetum on to the developing grain. Pollen grains break down the cuticle of the stigma papilla enzymically. They are then able to absorb water from the turgid cells of the stigma papillae and become hydrated.

POLLEN TUBE GROWTH

On germination, after hydration, pollen tubes emerge through the pores of the pollen grain. In most plants, only one tube is produced per grain, but in a few cases (e.g. the Malvaceae), grains produce more than one tube (polysiphonous), although only one contains the pollen-tube nuclei. The other tubes may have haustorial or absorbative functions. Pollen tubes (microgametophytes) have a very simple structure, being composed of a pectin sheath and three cells lacking cellulose walls, and are filled largely with water and solutes. Organelles concentrate largely at the apex, which is rich in endoplasmic reticulum and ribosomes. The main amino acid in pollen grains is proline, which is rapidly metabolized and converted into glutamic acid in pollen tubes. RNA is synthesized by the developing pollen tube, and both RNA and ribosomes are released into interstylar tissue by the tube apex. Pectinase and β-1,4, glucanase are associated with the pollen tube, and probably mediate the plasticity of the growing tube apex, as well as assisting entry into the stigma.

In gametophytically controlled systems, with a wet stigma, pollen tubes grow down the outside of the stigma papilla, and penetrate the stigmatic tissues at the base of the papillae by dissolving the cell-wall middle layer of pectate. In systems with dry stigmas, such as the Brassicaceae, the pollen tubes penetrate the cuticle (of lipoprotein), and the thin intermediate pectic layer, and grow in between the cellulose lamellae of the cell wall of the papilla. Thus both cutinase (esterase) and pectinase activity by the pollen tube apex are required for successful stigmatic penetration in sporophytic systems.

Having reached the style, the pollen tube continues to grow intercellularly in the spaces between the elongated stylar transmitting tissue; these spaces are also pectic in nature. There are two main types of style. In open styles, typical of the monocotyledons, a central stylar canal is present, lined with a glandular canal that functions as transmitting tissue. In closed styles, as in most dicotyledons, the centre of the style is filled with transmitting tissue, between which the pollen tubes grow towards the ovules.

In most plants, growth of the pollen tube lasts between 12 and 48 h from pollen germination to fertilization. In exceptional cases, as for instance *Taraxacum kok-saghyz*, the Russian rubber dandelion, fertilization occurs between 15 and 45 min after pollination. In contrast, in some trees such as oaks (*Quercus*), hazel (*Corylus*), alder (*Alnus*), *Garrya*, *Carya*, *Hamamelis* and *Ostraya*, pollen tube growth may be arrested for weeks or months, with pollination taking place in the winter or early spring, but fertilization occurring in the summer. In *Quercus velutina*, fertilization is achieved only in the season after pollination, 13 months later. Similar situations prevail in some orchids, where the ovules only develop after pollination has occurred.

The distance travelled by the pollen tube varies widely, being less than 1 mm in some species with short or no styles and small ovaries, and as much as 20 cm or more in species with exceptionally long styles, which are frequently bird or bat pollinated. It is abundantly clear that the pollen tube is self-sustaining on resources of water, sugars and amino acids provided by the style, and that pollen size or pollen resource is not related to tube growth. (However, Plitmann and Levin (1983) showed that there is a close relationship between pollen size, pollen tube diameter, stigma papilla width and style length among a number of species in the family Polemoniaceae.) Growth rates also vary, and readings from 1.75 mm/h to 35 mm/h have been obtained in different species. Growth rates are temperature dependent, and are usually optimal between 25° and 30°C, being totally inhibited above 40° and below 10°C. In cold climates, temperatures may thus be limiting to successful fertilization, and various features which aid heat conservation, or promote 'solar furnaces' (Chapter 4) may be favoured.

As the growth of the pollen tube progresses, the vegetative cell, which had originally assumed an apical position, becomes diffuse and is left behind, often disappearing entirely by the time of fertilization. In contrast, the two male gametes or sperm cells, which arose by the second pollen grain mitosis from the generative cell and were initially proximal, actively move into an apical position. Some controversy still surrounds the mode of movement of these cells. By this time they are often elongated with a tail-like structure, and despite the absence of a well-marked flagellum or cilia, may still in some sense 'swim'. However, active cytoplasmic streaming is also observed, and their transport may be passive.

Proximally, the mature pollen tube is characteristically filled with plugs of the complex carbohydrate 'callose' (β-1,3, glucan). This fluoresces under ultraviolet radiation when stained with, for instance, aniline blue and this is the basis of microscopic techniques for tracing the path of the pollen tube. Callose is a substance that can be a conspicuous feature of several different aspects of Angiosperm pollination and fertilization, and appears to occur principally as a 'wounding response', not only to physical damage, but also as a result of chemical recognition or antagonism. Thus on the style prior to germination, callose appears outside the pollen grain in sporophytically controlled species, and inside the pollen grain in gametophytically controlled species, and it may be especially marked in incompatible reactions. It also occurs within the stigma papilla in incompatible reactions under sporophytic control and may even play a role in causing the incompatibility through blockage. In incompatible reactions under gametophytic control, it becomes increasingly evident around the progressing pollen tube before it halts in the style, although it is not apparently directly concerned with the cessation of growth. Thus its occurrence in the normal pollen tubes of compatible pollinations is something of an enigma.

On arrival at the ovary (carpel), the pollen tubes travel round intercellular spaces on the inside of the ovary wall until they reach an ovule placenta, up which they travel (Fig. 3.4). At some stage during the

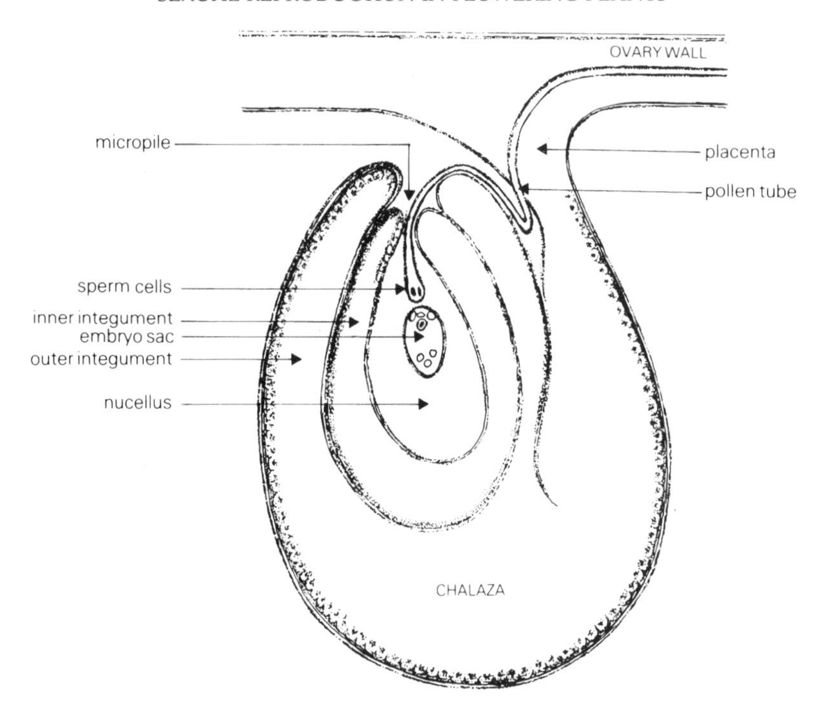

Fig. 3.4 Diagram of the path of the pollen tube in the ovary and ovule of an Angiosperm.

progress of the tube from the ovary wall to the ovule, it emerges from intercellular spaces into free space, where it travels in extracellular secretion to the micropile of the ovule (porogamy). More rarely, the nucellus, or even the embryo-sac, protrude from the micropile, so that micropilar penetration is unnecessary. Occasionally, the chalazal end of the ovule is penetrated. Once one pollen tube has entered a micropile, growth of other tubes is stopped or redirected, and polyspermy caused by the entry of more than one tube into an ovule is rare, although it has been reported. At this stage, tubes occasionally branch, sometimes spectacularly so, branches lacking nuclei entering the integuments and the nucellus. It has been suggested that such anastomization may be functional, allowing metabolite transfer from the nucellus and integuments to the embryo sac via the pollen tube. In any case, the tube transverses the nucellus to the polar end of the embryo sac, again in intercellular spaces in the ovule. The nucellar cells between which it passes frequently appear different from the remainder.

On arrival at the embryo sac the tube usually broadens before sperm release, and in some cases bifurcates in the directions of the egg cell and the primary endosperm nucleus, one sperm travelling into each branch.

In its later stages, the direction of travel of the pollen tube is remarkably changeable, 90° turns occurring at the style/ovary junction, the ovary/placental junction, at the micropile, and at the embryo-sac (Fig. 3.4). Strong chemotropic forces are clearly implicated, but these have not been satisfactorily identified. Promising results involving calcium gradients have been at least partially negated by the discovery that calcium levels in stylar transfer tissue are lower than in surrounding tissue.

DOUBLE FUSION

In the embryo-sac (female gametophyte), four unwalled cells lie at the polar (micropilar) end. Of the two synergids, the larger lies against the exterior of the embryo-sac, with the egg cell in close conjunction. The other synergid is close by, and the dikaryotic primary endosperm nucleus is distal to the egg cell (Fig. 3.5). The proximal (larger) synergid has produced outgrowths (filiform apparatus), and after pollination, but

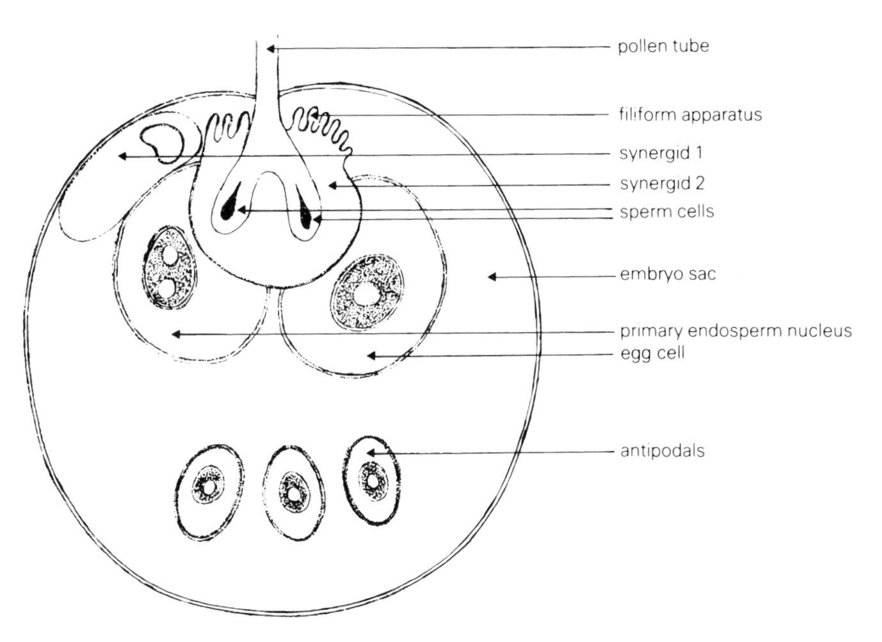

Fig. 3.5 Diagrammatic representation of the process of double fertilization in an Angiosperm embryo-sac. The pollen tube has entered the embryo-sac by means of one of the two synergids that now engulfs it. The pollen tube has bifurcated, with one sperm cell in each branch, and the synergid leads one pollen tube branch and one sperm cell towards the egg cell and the other towards the primary endosperm nucleus (fused polar nuclei). The nuclear membranes of the primary endosperm nucleus and the egg cell will shortly break down, admitting the degenerating synergid and the respective sperm cell.

before the arrival of the pollen tube at the embryo sac, it starts to degenerate. In the absence of pollination, similar effects can be produced by the application of gibberellic acid to ovules *in vitro*, and it is supposed that synergid degeneration is triggered by gibberellic acid carried on pollen grains. The vacuole of the synergid has high levels of calcium which are released on degeneration, and probably act as a close-range chemotrophic attractant to the pollen tube. This penetrates the synergid via the filiform apparatus, and ruptures, releasing the two sperm cells, starch grains and other organelles into the degenerating synergid. The synergid remains envelop the egg cell and the primary endosperm nucleus, carrying the two sperm cells to lie in conjunction with them.

At this point, double fusion takes place, a phenomenon that is unique to the Angiosperms among all living things. One haploid sperm cell (male gamete) fuses with the haploid egg cell to form the diploid zygote, which will form the embryo in the developing seed, and, ultimately, the new sporophyte generation. The second haploid sperm cell fuses with the primary endosperm nucleus. This has arisen from the fusion of two haploid cells of the embryo sac (female gametophyte), and is at this stage either diploid, or dikaryotic. In either case, the result is a triploid primary endosperm cell, which develops into the endosperm.

Fusion occurs through the evagination of the female egg cell and enclosure of the sperm cell, apparently including its cytoplasmic contents. Once enclosed, the sperm cell membrane breaks down to release the nucleus, which lies against the female nucleus. The intervening nuclear membranes then disintegrate, but the respective nucleoli remain distinct, two in the zygote and three in the primary endosperm cell.

The complex mechanism of sexual fusion in the Angiosperms, just described, is remarkably constant throughout its many diverse orders, and has led many authors to suggest that the Angiosperms have a single origin, i.e. that they are monophyletic. Exceptions to the general pattern can be found. Thus in some plants, the synergids are absent. In *Plumbago*, the egg cell itself acquires a filiform apparatus and functions as a synergid, the synergid phase having been lost (Johri, 1981). In other cases, both synergids are involved in sperm transport, often one sperm in each, or the synergids may survive but are not involved in sperm transport. In rare cases, multiple sperms may be produced by the pollen tube and multiple eggs by the embryo-sac (Favre-Duchatre, 1974). In this last case, zygotes may also divide to produce multiple pro-embryos. In certain families with highly specialized seeds, there is no endosperm, and thus double fertilization is not required. Thus in the Orchidaceae, there may be conventional double fusion, but no further development of the endosperm occurs; the process may progress as far as double fusion, but no fusion with the primary endosperm nucleus occurs; or the embryo-sac may be incompletely developed, with four, five or six cells instead of eight (Savina,

1974). In these latter cases, some or all of the antipodals may be missing, and the diploid primary endosperm nucleus is not formed. These are merely a few of the many variants observed to the sexual process in Angiosperms, but it is generally considered that they are secondary, later developments which do not counter the monophylesis argument.

THE OVULE AND FEMALE GAMETOPHYTE

Being heterosporous, Angiosperms produce two types of sporangium on two types of sporophyll, giving rise to two types of spore, and eventually to two quite different gametophytes, one male and the other female:

	Sporophyll	Sporangium	Spore	Gametophyte	Gamete
male	stamen	anther	pollen grain	pollen tube	sperm cell
female	carpel	ovule	megaspore	embryo sac	egg cell

At each of these stages, the various male or female cells or organs are homologous with each other, but are very different in structure and function. As in all seed plants, the female gametophyte generation has become extremely reduced, and is parasitic on the sporophyte megasporophyll (carpel) and megasporangium (ovule).

In primitive Angiosperms, many ovules are borne in each carpel (part ovary), parietally and orthotropously, being attached to the ovary wall by a placenta at the chalazal end, distally to the micropile (Fig. 3.6). The ovule is composed of three layers; starting from the exterior (Fig. 3.4) these are:

outer integument	protective	form the seed coat (testa)
inner integument	protective	
nucellus	nutritive	disappears after fertilization

The nucellus differentiates to form an archesporium with a central archespore which enlarges to form an embryo-sac mother cell (EMC) which undergoes female meiosis. The resultant four megaspores form a row along the ovule axis, and the three at the micropilar end dwindle and eventually disappear. The fourth chalazal megaspore enlarges and undergoes three mitoses to give eight haploid cells. These differentiate to form the mature embryo-sac (female gametophyte) (see Fig. 3.5).

The embryo sac is thus enclosed within the innermost tissue of the ovule, the nucellus; it lies centrally or towards the micropilar end and

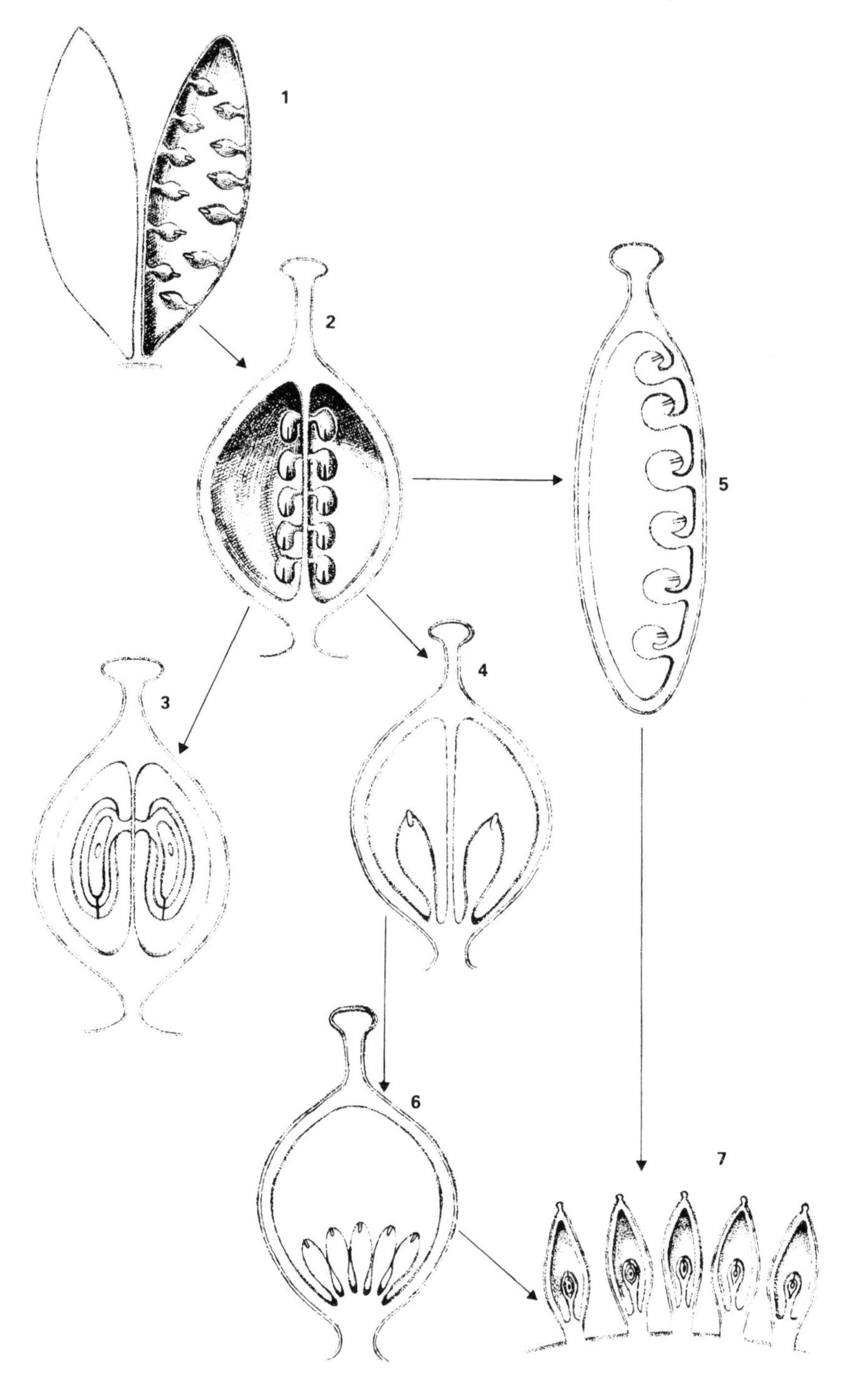
1
2
5
3
4
6
7

usually takes up about half the nucellar volume. The embryo sac is readily distinguished from the parenchymatous, secretory nucellus, the cells of which bear cellulose walls, because it forms a rather empty looking coenocytic sac with a pectic wall and unwalled internal cells. These comprise, at the chalazal end, three antipodal cells of uniform size. At the micropilar end, as already discussed, lie two synergids and the egg cell. One synergid lies at the apex, in close conjunction with the egg cell, and the other, which may be somewhat smaller, is nearby. Distal to these cells, but still at the micropilar end, are the two remaining cells which fuse, usually prior to pollination, although sometimes remaining as a dikaryon. This is known as the primary endosperm nucleus (fused polar nuclei).

The fate of the eight cells of the embryo-sac after double fusion can be summarized as follows.

1. The antipodals disintegrate, or undergo endomitoses, to form a block of antipodal tissue of unknown function in the seed; or they occasionally connect to the zygote by branches of pollen tube, apparently feeding it (rare).
2. The primary endosperm nucleus forms triploid endosperm which acts as nurse tissue (protection and feeding) to the embryo in the seed, thus displacing the ovular (parent sporophyte) nucellus in nurse functions; the nucellus degenerates in the developing seed.
3. The egg cell forms the zygote, which develops into the embryo in the seed, and thus into the new sporophyte generation.
4. The synergids, one or both, start to disintegrate before accepting the pollen tube and sperms; if only one is involved in sperm transport, the other also soon disappears.

There are exceptions to this structure and function of the embryo-sac, the more important of which have been covered earlier in this chapter. After double fusion, no less than four independent and genetically different generations coexist in the seed:

Fig. 3.6 Evolutionary trends in Angiosperm ovaries and ovules. 1, Free multilocular carpels with parietal placentation; 2, fused multilocular carpels with axile placentation; 3, fused unilocular carpels with axile placentation; 4, fused unilocular carpels with basal placentation; 5, single multilocular carpel (derived from fused carpels) with axile placentation; 6, single multilocular carpel (derived from fused carpels) with basal placentation; 7, single unilocular carpels with basal placentation. A trend is observed in carpel fusion followed by loss of carpel walls (septae); in the reduction of number of ovules per carpel and per ovary to one, and from axile placentation to basal placentation. 1 is typical of Magnoliaceae, Paeoniaceae etc., 2 is typical of Cruciferae (Brassicaceae), 3 is found in Rutaceae etc., 4 is found in some Primulaceae, 5 is typical of the Leguminosae (Fabaceae), 6 is found in Papaveraceae etc. and 7 in Compositae (Asteraceae).

1. the parent sporophyte (diploid) – integuments of the ovule, forming the testa;
2. The female gametophyte (haploid) – the embryo sac wall and antipodals;
3. the offspring sporophyte (diploid) – the embryo;
4. the endosperm (triploid) – the product of an independent fusion between a sperm cell and diploid primary endosperm nucleus.

The relationship between the endosperm and the embryo is of great interest. The male genome that enters the embryo should be exactly the same as that entering the endosperm, for the two sperm cells are genetically identical, resulting from the mitosis of a single spore (meiotic product). Similarly, the egg cell should be genetically identical to both of the polar nuclei. Thus the embryo should contain exactly the same nuclear, but not necessarily cytoplasmic, genes as the endosperm. Therefore, there is only one important nuclear difference between the embryo and the endosperm. Whereas the embryo is diploid, with one genome from each parent, the endosperm is triploid, with two maternal contributions to one paternal.

Although it may be assumed that those genes which are independently expressed in the endosperm and embryo have co-evolved harmoniously within a population, this may not be the case for hybrids between demes that have evolved different rates of seed development. As the embryo and endosperm receive different gene doses from the male and female parents, each may develop at different rates in the hybrid seed, resulting in embryo starvation, or in the smothering of the embryo by the endosperm. Either can lead to the death of the seed, or in poor seed viability. Such a case has been described by Woodell (1960a,b) for hybrid *Primulas*, where misalignments between the male interests and female interests in the aftercare of the seed (p. 4) have resulted in a seed incompatibility, particularly for the crosses between the cowslip *P. veris* and the oxlip *P. elatior*.

The independence of the embryo and the endosperm after double fusion is well illustrated by the work of Batygina (1974) on several species of cereals. In these, the fusion of the male gamete with the primary endosperm nucleus is completed in less than 2 h, and is followed by a dormancy of less than 1 h. In contrast, the fusion of the other sperm cell with the egg cell lasts 4–5 h, and is followed by a dormancy of about 18 h, at which stage the endosperm may have four or eight cells.

The stimulus involved in the initiation of development in the embryo and endosperm has received some attention, and often seems to be dependent on the process of cytoplasmic fusion of the gametes. In this respect, observations on the aberrant phenomenon of hemigamy are of interest (Solnetzeva, 1974). Here, cytoplasmic fusion of the gametes occurs, but nuclear fusion is delayed. Within the fused cytoplasm nuclear division of the sperm cell and the egg cell proceed independently, the sperm cells forming some haploid tissue. In most of these instances, the egg cell also goes on to form an embryo parthenogenetically, but this can only happen after cytoplasmic fusion with the sperm cell.

In other agamosperms (Chapter 10), pollination but not fertilization is required for embryony (embryo development) to proceed, and this can be substituted experimentally by an application of auxin to the stigma, auxin being carried externally on pollen (pseudogamy). In all pseudogamous species, fusion of a sperm cell with the polar nuclei and conventional endosperm development proceed as in a sexual plant. For multiseeded fruits, such an auxin-stimulation system can confer reproductive efficiency. In the sexual *Passiflora vitifolia*, the fruits of which bear several hundred seeds, Snow (1982) shows that at least 25 pollen grains must be deposited on the stigma for fruit to set. In this way, the mother is spared the effort of forming expensive fruits which contain few seeds.

Embryony does not universally depend on an auxin message to the stigma, as is demonstrated by the experimental technique of placental pollination *in vitro* (Rangaswamy and Shivanna, 1972). In fact, successful fertilization and embryony can be achieved experimentally by the application of germinated or ungerminated pollen to a variety of maternal tissues, from the cut style to the ovule micropile. In all cases except non-pseudogamous agamosperms, some hormonal message from the pollen grain to the egg cell does seem to be required for embryony to proceed however. There is also some evidence that development of the endosperm precedes that of the embryo, and that embryony rarely proceeds without prior division of the endosperm, except for non-endospermous seeds.

CHAPTER 4

Floral diversity and pollination

An outstanding feature of the Angiosperms is the amazing diversity in form and colour that has been adopted by the inflorescence, sufficient to inspire great art, fuel a major industry and serve as a solace for suffering mankind.

Yet the flower is merely a sex organ, and never has any function except to promote reproduction by seed, usually sexually. The beautiful, weird, sinister, astounding forms that flowers have acquired are strictly pragmatic, and have encouraged the ecological diversification, and dominance, of the flowering plants, as was argued earlier (Chapter 3).

This book does not have the space to catalogue all the manifold adaptations of the inflorescence to pollination by animals or abiotic forces, and to fruit dispersal. In recent years this has perhaps been most fully achieved, and certainly has been best illustrated, by Proctor and Yeo (1973). An interesting, but briefer, account is by Faegri and van der Pijl (1979). Of the earlier accounts, Knuth (1906–1909) and Muller (1883) are outstanding. In the present chapter, a brief survey of the adaptive radiation of floral organs in relation to pollination systems is attempted, with some fuller accounts of exceptionally interesting or complex syndromes. I shall reserve for the next chapter (Chapter 5) an account of the control exercised by the inflorescence on patterns of gene exchange, a theme which is more central to this book.

PATTERNS OF FLORAL EVOLUTION

It is generally, although not universally, accepted that it is most useful to consider the flower of the genus *Magnolia* as the starting point for the adaptive radiation of floral diversity (Taktajhan, 1969; Crepet and Friis, 1987). In all features *Magnolia* exhibits what are thought to be primitive character states, and in this it may not be too distant in structure from the flower of a hypothetical archae-angiosperm. Thus, it is instructive to examine the structure and function of a *Magnolia* flower in detail (Fig. 4.1).

Flowers are very large (up to 30 cm in diameter in *M. grandiflora*) and are borne individually on trees or large shrubs. They may flower sporadically over much of the season. Floral organs are borne spirally on an elongated floral axis (receptacle) in a manner reminiscent of a Gymnosperm cone. Proximally (on the outside) are borne the tepals (together these form the

Fig. 4.1 The primitive flower of *Magnolia grandiflora* (×0.5).

perianth), undifferentiated into petals and sepals. These are many (i.e. more than six which is the largest number of a type of floral organ which is usually constant), large, thick, with leaf-like veining, usually white (occasionally pink or purple, e.g. *M. campbellii*), and are unscented, lacking markings, hairs or nectaries.

Distal to (inside) the tepals are borne the stamens (androecium), also spirally. The stamens are also very numerous, and although smaller and narrower, also share the leaf-like structure of the tepal, the pollen sacs being borne in linear rows on the stamen surface. There is no clear differentiation into anther and filament; rather the organ as a whole is very recognizably a sporophyll. Anther dehiscence is longitudinal and extrorse. Pollen grains are rather large, with a single pore.

Distal to the stamens, and apically (inside), are borne the carpels. These are quite free from one another, are relatively large, usually many, and they bear several large orthotropous ovules parietally. There is no style, the large, somewhat linear, stigmatic area being borne directly onto the carpel, usually subapically. In some species, the carpel is not fully closed. The fruits form dry follicles, which dehisce longitudinally to release large, hard and thick-walled seeds which may lie for months or even years before the testa rots sufficiently to allow germination. There is no specialized mode of transport of the seed or fruit, although they may be eaten by birds and monkeys.

Observation of the flowers of modern Magnolias shows that a wide variety of insects, chiefly composed of the insect orders Coleoptera and Diptera, visits the large white bowls. Many are apparently attracted by 'rendezvous' stimuli. That is, the flower is a pleasant resort for an insect, being sheltered and warmer than the ambient temperature. Most *Magnolia* flowers will probably act as a solar furnace (p. 106). However, many beetles (Coleoptera) and primitive flies with biting mouth parts (Diptera) actively feed on the pollen (Fig. 4.2), which is a food source rich in protein and fat, although with little if any 'instant energy' in the form of soluble carbohydrate.

Of the winged insects, the Coleoptera and Diptera were among the earliest orders to appear in the fossil record, being conspicuous by the Carboniferous period. At this early stage, long before the origin of the Angiosperms, spore feeding may have evolved in these groups of insects, using the cones of Pteridophytes and Gymnosperms.

Those groups of insects that evolved tubular mouthparts to suck up nectar (Hymenoptera, Lepidoptera and Syrphidae) appear much later in the fossil record, in the mid-Cretaceous, coincidentally with the main diversification of the Angiosperms. It is axiomatic that these groups co-evolved with nectar-producing flowers symbiotically. Nevertheless, it is possible that nectar production predated the Angiosperms, and may have

Fig. 4.2 A muscid fly feeding on pollen in the 'solar furnace' primitive flower of the Mount Cook lily *Ranunculus lyallii* (×0.6).

occurred in the Pteridosperms or the Gnetopsida (Crepet and Friis, 1987). Nectar production occurs in modern *Ephedra* (Bino, Devente and Meeuse, 1984), and entomophily is a significant source of pollination in this dioecious genus (Meeuse *et al.*, 1990).

However, present-day plants such as *Magnolia*, which show an apparently primitive state in every floral character, are only visited by the earliest types of insect, for pollen feeding alone. *Magnolia* has no other means of reproduction, so it is not surprising that the flowers are very large, and a great deal of expensive pollen is produced, giving an inefficiently high pollen:ovule ratio. However the ovules are very large and both male and female expenditures per fertile seed will be very high.

Magnolia flowers are homogamous, or weakly protandrous, but they do not often set viable seed as isolated individuals in collections. Most species are self-incompatible, and reference to other relatively primitive families of plants (Ranunculaceae, Papaveraceae, Theaceae) renders it very likely that they have a gametophytic incompatibility (Chapter 6), as did many early flowering plants (Whitehouse, 1950).

Using *Magnolia* as a baseline, I shall now consider developments that have occurred with respect to diversification of pollination mechanisms.

DEVELOPMENT OF PERIANTH SHAPE

Three major structural changes occurred in the development of the perianth.

1. Differentiation of the calyx (sepals, small and green) from the corolla (petals, usually larger and white or coloured). Sepals and petals undertake different functions. The calyx protects the flower bud, and in hypogynous flowers the gynoecium, and in some instances forms nectaries, attractant organs or organs for fruit dispersal; whereas the corolla attracts, receives, shelters and succours the pollinator.
2. Fusion of the calyx and corolla to form bells, trumpets, tubes and traps; although the free (unfused) perianths of more primitive plants may form such shapes, they are more liable to have their protected food sources (pollen, nectar) robbed by pollinators that do not visit the flower legitimately (Fig. 4.3).
3. Loss of radial symmetry (actinomorphy), and adoption of bilateral symmetry (zygomorphy) thus forming 'gullet' and 'flag' flowers (see below). Such flowers are usually characterized by a landing platform and dorsal, often 'sprung' stamens and styles; these are typical of 'bee flowers', but may also be found in other syndromes (Figs. 4.4 and 4.5).

Fig. 4.3 A worker hive bee (*Apis mellifera*) 'robbing' the nectar from the base of the tube of the bluebell (*Hyacinthoides non-scripta*). Photo by M. C. F. Proctor (×1.5).

Zoophily (animal pollination)

Faegri and van der Pijl (1979) have characterized six (or seven) main perianth shapes, and these are listed here with their main types of flower visitor (Table 4.1). It can be seen that beetles and simple flies tend to be associated with dish- or bowl-shaped flowers, whereas birds, butterflies or moths are associated with trumpet-, brush- or tube-shaped flowers. Bees are the most catholic visitors, being commonly associated with all flower types except narrow tubes, and even in this class, long-tongued bees may often be effective flower visitors. In most ecosystems, bees are the most dominant and most diverse flower visitors. Broadly speaking, it is possible to associate various visitor types with various flower shapes, and to suggest that those in each category near the top of the list are the least specialised, and those at the bottom of the list are the most specialised (Kevan 1984).

Nevertheless, for nearly a century there has been a tendency for the literature to overgeneralize with respect to the syndromes of flowers

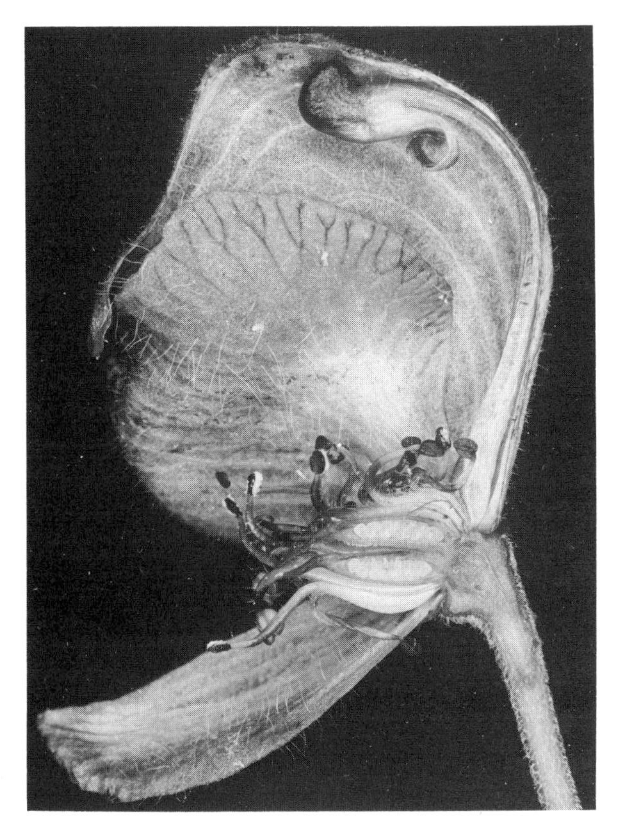

Fig. 4.4 Section of the zygomorphic bee flower of *Aconitum napellus* (Ranunculaceae), showing the large inflated posterior tepal containing two large tubular nectaries in the roof. *Bombus* bees land on the lower tepals and climb upside down (sternotribically) into the roof, passing the stamens and stigmas as they do so. Photo by M. C. F. Proctor (×1.5).

associated with particular visitors. Certainly, some flowers can only be effectively visited by certain visitors (*Arum*, flies; most papilionid flowers, bees; figs, *Blastophaga*; *Yucca*, *Tegeticula* etc). However, in most cases such narrow specificities exist only in certain conditions, where competing flowers, and competing pollinators, are diverse. By comparing floral morphologies, and insect pollinating classes for several Mediterranean floras, Herrera (1996) has quantified such failures in correspondence between pollinators and flower classes. For instance, Herrera infers that floral tube depth may be adaptive to constraints other than those imposed by the tongue length of visitors, including protection against nectar evaporation. Even where the nature of the reward, or the deceit, may be closely tailored to the morphology or behaviour of a single visitor, acci-

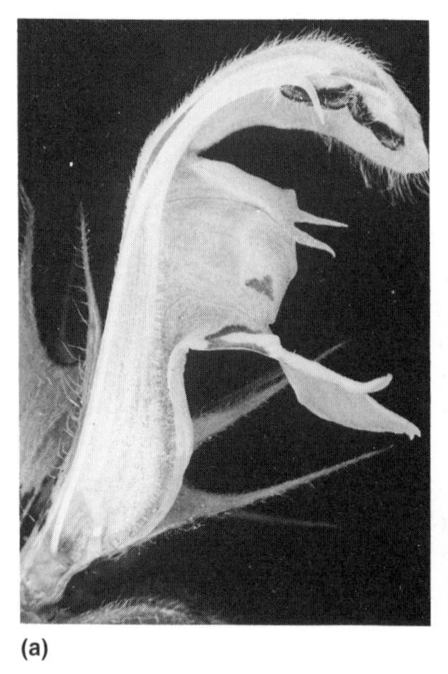

(a) (b)

Fig. 4.5 Section of the zygomorphic bee flower of *Lamium album* (Labiatae) (a) which is nototribically visited by *Bombus* (b). The forked stigma and the four stamens, two of which are shorter than the other two, each bearing divergent locules, are borne under the hood and pollination takes place via the bee's thorax. Photos by M. C. F. Proctor (a ×3; b ×2).

Table 4.1 'Harmonic' relations between pollinators and perianth shape and colour (greatly simplified, adapted from Faegri and van der Pijl, 1979)

Structural blossom class[a]		Pollinator class	Colour preference (human visual spectrum, HVS)
dish or bowl	(unf, a)	beetles	brownish, or dull
bell	(unf or f, a)	flies	
			white or cream
		syrphids	
gullet	(f, z)		
		bees	yellow
flag	(unf or f, z)		
		bats	
trumpet	(f, a or z)	moths	blue or purple
brush	(o, a or z)	butterflies	orange or red
		birds	
tube	(f, a or z)		green

[a] Abbreviations: unf, corolla unfused; f, corolla fused; a, corolla actinomorphic; z, corolla zygomorphic.

dental pollination by other visitors may occur. Thus for many 'bird flowers', only birds may be able to reach the nectar, yet stamens and styles are far exerted, so that the casual attraction of bees, for instance, may also effect pollination. Flowers in gardens are frequently pollinated by species (very often the honey-bee) far removed in structure and behaviour from their native pollinator, and the introduction of either pollinators or flowers into foreign areas may break down pollination specificity. Thus many Australian *Banksia* (bird flowers) are efficiently cross-pollinated by introduced honey-bees, and in Australia, African flame-trees (*Erythrina*) are pollinated by native birds such as honey-eaters, wattle-birds and even drongos. At times, any of the flower types listed in Table 4.1 may be visited by any of the pollinator types, and stable oligophilic relationships may evolve between the most unexpected pairs of species, as for instance, a beetle (primitive) and the orchid *Listera ovata* (in a very specialized group of plants). Broad assumptions about coadaptation are thus only valid at a single time and in a single place, and all the so-called 'adaptive syndromes' (Table 4.2) have many exceptions.

Oligophily and polyphily

Nevertheless, it is true to say that some flowers (and flower types) tend to be oligophilic, having very few species of visitors (often only one), and this tends to be more true of flowers with relatively specialized shapes (flags, trumpets, tubes) than those with unsophisticated shapes (cup, bell, brush). The latter may receive many different types of visitor (polyphilic).

Visitors to oligophilic flowers are likely to be obligate flower visitors, specializing on one or a few types of flower (oligotropic). Thus, pollen donated from a flower is likely to be received by a flower of the same

Table 4.2 The major classes of pollination syndrome

beetles	cantharophily	entomophily	zoophily
flies	myophily, sapromyophily, bee-flies, syrphids	entomophily	zoophily
bees	melittophily	entomophily	zoophily
butterflies	psychophily	entomophily	zoophily
moths	phalaenophily	entomophily	zoophily
birds	ornithophily		zoophily
bats	chirepterophily		zoophily
also: small mammals, monkeys, slugs, ants, thrips, water skaters, others?			zoophily
wind	anemophily		
water	hydrophily		

species. As a result, less energy is required for pollen production per ovule fertilized, and more energy is available for the production of other floral attributes (e.g. more complex shapes, gynoecia that are larger or more numerous, or more flowers).

Whether it is more advantageous to be oligophilic or polyphilic will depend on ecological constraints. Oligophily is likely to be advantageous in ecosystems in which specialist pollinators are numerous and diverse, and where many species of animal-pollinated plants flower synchronously (Table 4.3). In some cases, for instance mediterranean phrygana, pollinator diversity may be extremely high, as for instance in a 20 ha plot near Athens, Greece, where Petanidou and Ellis (1993) recorded no less than 666 species of flower visitor.

In contrast, visitors to polyphilic flowers may not be obligate flower visitors, and in any case they may visit many different types of non-specialist flower (polytropic). These generalist visitors have been called 'the insect riff-raff'; small beetles, flies, thrips, bugs, etc. Thus in polyphilic flowers, pollen donated by a flower is much less likely to be received by a flower of the same species. More energy is required for the production of pollen per ovule fertilized. Many of these plants tend to be self-fertile, and most pollination will be selfed, geitonogamous or autogamous (Chapter 9). Flowers will tend to be numerous, often in large inflorescences, with unsophisticated shapes and simple pollen presentation and receipt. In such systems, accurate cross-pollination is sacrificed for any pollination at all (but with the residual chance of cross-pollination).

These polyphilic systems are likely to be favoured in time/space niches where potential specialist pollinators are few (e.g. early or late in the season, or in cold locations).

In a study of Greek Lamiaceae, Petanidou and Vokou (1993) show that early flowering species are polyphilic, but as a guild are in fact visited by far fewer species of pollinator than is the guild of mid-summer species. There is a tendency for early flowering species to have larger, more showy, more rewarding flowers which flower over a longer period of time

Table 4.3 Patterns of flower visiting, and ecological constraints, on simple and complex flowers

Flower type	Flower visiting	Insect visiting	Diversity of potential visitors	Number of other synchronous species of flower	Pollination efficiency (reciprocal of pollen:ovule ratio)	Breeding system of flower
Simple (bowl, bell, brush)	polyphilic	polytropic	low	small	low	mostly selfed
Complex (gullet, flag, trumpet, tube, trap)	oligophilic	oligotropic	high	large	high	mostly crossed

than do mid-summer flowers. Of the 201 insect flower visitors studied, 22% are highly oligotropic, relying totally on flower visits to this family of plants, and most visit only one flower species. Nearly all of these are midsummer visitors. In contrast, most early season visitors are polytropic.

Although generalist, polyphilic flowers tend to be favoured when the diversity of specialist pollinators is low, it is remarkable how many species of the 'insect riff-raff' can be recorded within a generalist flower. For *Saxifraga hirculus* in Switzerland, Warncke *et al.* (1993) record no less than 76 species of flower visitor, whereas Barrett and Helenurm (1987) identified 103 insect species visiting the flowers of 12 forest herbs, mostly generalists, in New Brunswick forests. For three species of Mediterranean *Cistus*, with showy but generalist flowers, Bosch (1991) records representatives of no less than 31 families of insects visiting flowers.

Kochmer and Handel (1986) undertook a statistical study of nearly 4000 species of zoophilous flowering plant from Japan and Carolina. In general, flowering times are bimodal and are strongly influenced by taxonomic relationships. They suggest that evolution into early flowering, generalist guilds and midsummer specialist guilds may have occurred at a relatively early stage in the phylogenetic development of flowering plant families. Another floral strategy which may be adaptive to pollination efficiency and specialization is the longevity of individual flowers (Ashman and Schoen, 1996). Longevity implies increased costs per flower, and hence will militate against the production of many flowers. Long flower-life is negatively correlated with visitation rate by pollinators, and hence with reproductive assurance. It should pay polyphilic flowers from temporal or ecological environments poorly supported by pollinator guilds to be long-lived. In practice we find that many winter flowers, and alpine flowers, tend to be individually long-lived.

Although most generalist or specialist zoophilous flowers can be classified in shape according to the types listed in Table 4.1, many defy such classification. Some flowers of great complexity have extraordinarily complex methods of pollination, and these lend much of the glamour to the 'story-telling' aspect of so-called 'pollination ecology', which has tended to dominate texts on pollination biology, and latterly, television programmes. Some other flower shapes will be catalogued briefly here.

Spurs

In a variety of families from the 'primitive' Ranunculaceae (*Aquilegia* (Fig. 4.6), *Delphinium*) to the Fumariaceae (*Corydalis*), Balsaminaceae (*Impatiens*), Scrophulariaceae (*Diascia, Linaria*), Lentiburiaceae (*Pinguicula, Utricularia*), Valerianaceae (*Kentranthus*), Violaceae (*Viola*), Tropaeolaceae (*Tropaeolum*), and the very specialized monocotyledonous Orchidaceae (*Platanthera, Gymnadenia*, and many others), corollas (and/or more rarely

Fig. 4.6 Nectar-containing petal spurs in *Aquilegia longissima* (Ranunculaceae), a North American columbine visited by Lepidoptera which reach the nectar by means of a long proboscis (×1.5).

calyces) form spurs. These are broad and rounded to long and very narrow outgrowths near the base of the corolla, which usually contain nectar. Their function is twofold: to render the nectar reward unavailable to all except a few structurally suited flower visitors; and to increase the distance from the reward to the anthers and stigma (pollen presentation and receipt) and/or the landing platform or hovering position. Thus, visitors that can 'work' spurred flowers will have long, narrow mouthparts, and include Lepidoptera (Fig. 4.7), longtongued bees, and some birds. Spurs are very prone to illegitimate visiting by 'nectar-robbing' bees such as *Bombus mastrucatus* which bite a hole in the spur wall.

Constricted flowers

Some flowers are so constructed that a visitor can only work the flower by exerting considerable force, or even by damaging the flower. Visitors are limited to those with accurate structural specifications, and the strength to open the flower. Adey (1982) has recorded the ability of 45 species of bee to 'trip' 27 species of the tribe Genistinae, relatives of brooms and gorses. ('Tripping' is the depression of the keel and the forcing open of the wings of a papilionid flag flower, thus releasing the spring-loaded stamens and style from a ventral position in the keel onto the pollinator (Fig. 4.8).) Using numerical methods she grouped the bees into five classes based on body weight, and the flowers into seven classes based on structural and size characters. There is a strong relationship between bee types and flower types with respect to tripping ability. The papilionid ('pea') type flower is a familiar example of constricted entry in a flag flower, but many

Fig. 4.7 Lulworth skipper butterfly (*Thymelicus acteon*) visiting the tubed and short-spurred flowers of red valerian *Kentranthus ruber* which it is probing for nectar. Photo by M. C. F. Proctor (×2).

Fig. 4.8 Tripped flowers of the broom, *Cytisus scoparius*. The flowers have been visited by bees which have depressed the boat-shaped keel, so releasing the single style and the stamens, which hit the bee and then recoil, having been held under tension (×0.5).

mechanisms that constrict entry have been adopted. The corolla may be constricted at the apex (*Campanula zoysii*); the apices of the corolla lobes may even be fused, allowing only forced lateral entry (*Physoplexis*, Fig. 4.9); the corolla tube may be filled with a plug of hairs allowing entry only to the hard, narrow mouthparts of some syrphids (*Primula primulina*); and there are many other such examples.

Spring-release flowers

Often associated with constrictions are spring-loaded mechanisms whereby the stamens and style(s) are forcibly released from a hidden location onto the pollinator as the flower is entered, often showering the pollinator with pollen. This has already been briefly described for papilionid flowers in the preceding section. In these pea-flowers, pollination is sternotribic, that is the underside of the visitor is dusted with

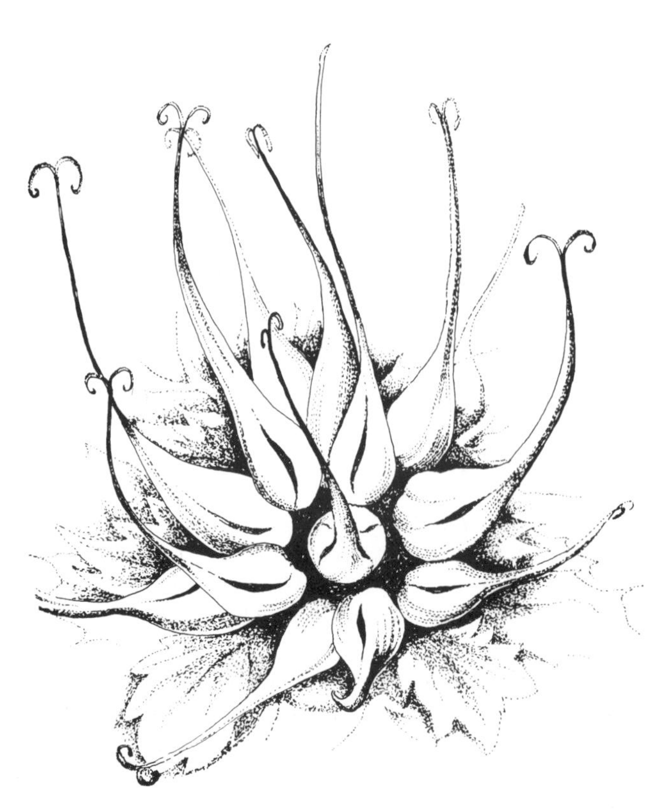

Fig. 4.9 The petal lobes in *Physoplexis comosa* (Campanulaceae), the devil's claw of the European Alps, are fused apically, allowing forced lateral entry by bees and syrphid flies which achieve pollination in the struggle to reach basally presented nectar (×2).

pollen. In many other constricted flowers, for instance many Lamiaceae and Scrophulariaceae (the white dead-nettle, *Lamium album* is a familiar example; Fig. 4.5) pollination is onto the back of the pollinator (nototribic). In a series of comparative observations made over many years on the many North American species of *Pedicularis* (Lousewort) Macior (1982) shows adaptation to both forms of pollen presentation in the same genus (Fig. 4.10).

In spring-release flowers, pollen is often released explosively (as in the broom, *Cytisus scoparius*; see Fig. 4.8), even to the extent of upsetting and discouraging the flower visitor, but other forms of pollen release or presentation, such as that resulting from vibration, or a 'rocker' movement of the stamen (*Salvia glutinosa*; Fig. 4.11) are also well known.

Trap flowers

Many kinds of trap flowers are recorded. Some, as in the Araceae, are modifications of whole inflorescences, and these are also dealt with later (p. 112). The incentive for visitors to enter trap flowers is very frequently sapromyophilous, the flower producing a decaying smell, and being coloured reddish, brownish or purplish in a manner reminiscent of dead flesh. This is also true for *Aristolochia*, and for some orchids, for instance Australian birdwing orchids (*Pterostylis*) and *Paphilopedium*.

In other cases, visitors may be attracted by sweet scent and by what are to the human eye bizarre shapes and colours. In the orchid *Coryanthes*, euglossine *Eulaema* bees attempt to collect scent droplets from the rim of the trap, but become intoxicated and fall in, only to escape eventually through a narrow exit passage, effecting pollination as they do so. (It has been suggested that narcotic nectar may be especially attractive to some flower visitors, as in the sphingid *Manduca quinquemaculata* visiting *Datura meteloides* (Grant and Grant, 1983).)

In the trap flowers of tropical water-lilies, *Cyclocephala* beetles seem to be attracted by scent alone. The water-lily flowers heat up (in the same way as do aroids, p. 112), releasing volatiles during their first evening. The giant annual *Victoria amazonica* smells of 'a mixture of butterscotch and pineapple' (Prance and Arias, 1975) attracting *C. hardyi*, whereas *Nymphaea rudgeana* and *N. blanda* smell of 'aniseed' and 'sweet and fruity, with an admixture like xylene' respectively, and attract *C. castanea* (Cramer, Meeuse and Teunissen, 1975). In each case, the protogynous flower closes at dawn, trapping the beetles. It reopens the following evening, by which time the anthers have dehisced, and the beetles make their escape carrying pollen. (In some other *Nymphaea* species, some pollen-carrying visitors may be killed by the flower (p. 83).)

In the Lady's slipper orchids (*Cypripedium*), so rare in Europe now but still very common in North America, attraction into the trap is provided by nectar. This is a exception to the general rule that trap flowers operate

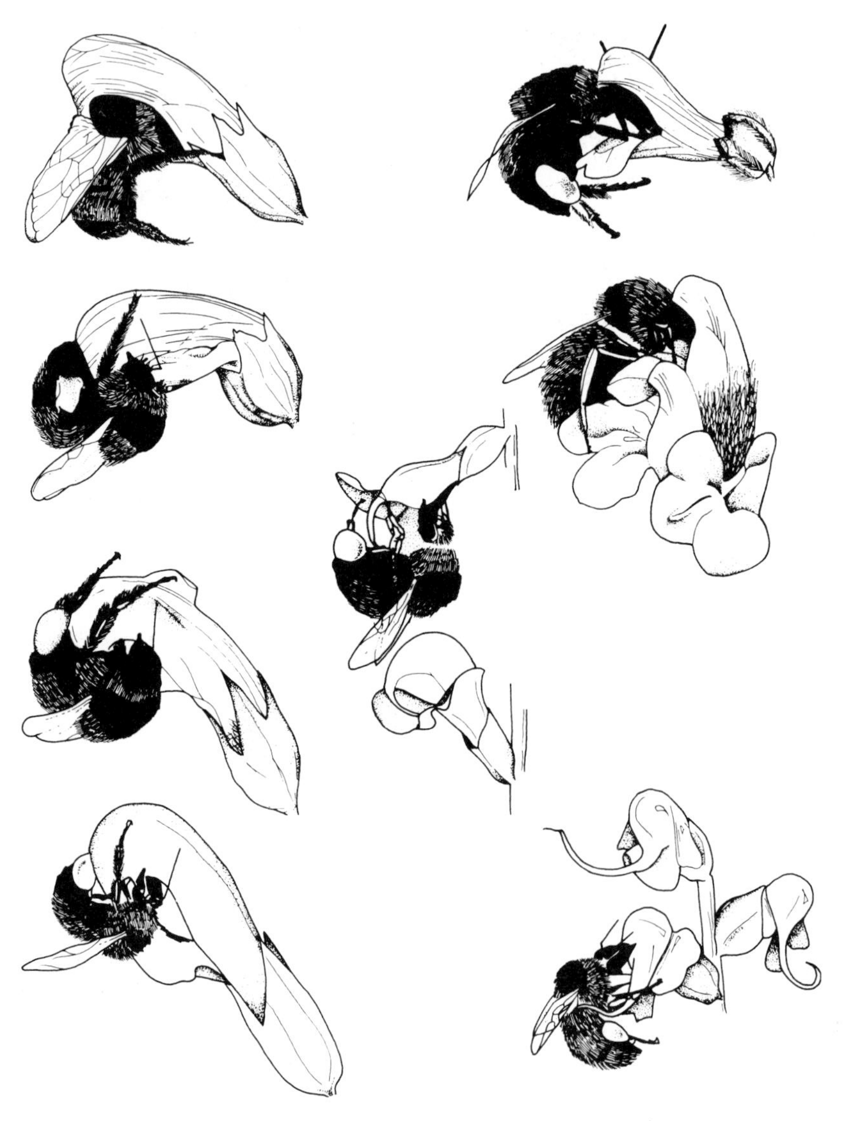

Fig. 4.10 Sternotribic (below) and nototribic (above) pollen presented to *Bombus* species by the flowers of eight species of North American *Pedicularis* (Scrophulariaceae). Drawing reproduced by kind permission of L. W. Macior (all ×4).

by deceit, not offering a reward. Deceit is not surprising, for an insect is unlikely to 'major' on a trap flower, in which it may be forced to spend hours at a time, as a food source. However, deceit also renders the mechanism vulnerable to density-dependent selection. If deceit flowers are com-

Fig. 4.11 *Salvia glutinosa* being visited by *Bombus agrorum*. In the sages, each of the two stamens has an elongated connective which is pressed by the visiting bee, rocking the fertile anther lobe on to its back. Photo by M. C. F. Proctor (×3).

mon in the environment, more so than their model, their strategy is likely to be less successful. In practice, deceit flowers rarely dominate a habitat and usually occur very sparsely.

Trap flowers usually form a pouch with a restricted, slippery entrance, often reinforced by unidirectional hairs. Exits may be separate, as in trap orchids, in which the visitor is forced to squeeze past the stigma and the pollinia (in that order) in order to escape.

In *Aristolochia* and the Araceae (see Fig. 4.12) the entrance and the exit are the same, so a female phase (pollen receipt) always precedes a male phase (pollen issue) (protogyny). The relaxation of stiff unidirectional hairs mediates the operation of such a 'time-trap'. Modifications of the corolla wall into greasy, wax-secreting areas and into hairs enables traps to work. However, in their simplest forms, trap flowers may be merely formed by imbricated whorls of many stiff, inward-bending waxy tepals, which the insect can enter easily, but from which it cannot escape until the petals reflex during the later male phase (*Calycanthus*). In extreme cases, early visitors during the female phase may be killed, either by drowning or by a toxin secreted into the stigmatic liquid (*Nymphaea*). As the male phase progresses, the stamens cover the toxic stigma, and visitors are able to escape, carrying pollen. In this primitive flower, pollen may adhere better to the stigma if the pollinator carrying it dies on the stigma, and thus pollen germination is enhanced.

Complexity in flower outline

Most flower shapes present the visitor with a relatively simple outline, and such outlines seem to attract most visitors. However, it has frequently

Fig. 4.12 The inflorescence of *Arum italicum*. Within the trap (left) which has been dissected open are the female flowers, male flowers and hairs (reading from left to right). During the initial female phase the hairs act as a unidirectional valve, allowing flies to enter the trap, but not to leave it. In the later male phase, the hairs collapse, releasing the flies which escape carrying pollen. The end of the inflorescence axis forms a spadix which heats up through respiratory activity to release foul-smelling skatoles which attract the flies. Photo by G. Chaytor (×0.5).

been noted that flowers that are primarily visited by day-flying Lepidoptera tend to have more complex outlines, as for many Caryophyllaceae where each of the five petals is itself deeply divided. Herrera (1993) tested the hypothesis that flowers with highly dissected outlines are more attractive to the hummingbird hawkmoth, *Macroglossum stellatarum* than are those with relatively entire outlines. The endemic Spanish shrubby violet, *Viola cazorlensis*, varies considerably for corolla outline, and it is visited almost entirely by this moth. Herrera was able to show that individuals with dissected flower outlines set more capsules which contained more seed than did those with more entire outlines, suggesting that they were more likely to attract pollinators.

Flowers with food bodies

In most flowers, reward is provided solely by pollen and/or nectar. However, in some, nutritious food bodies may be produced on the perianth, stamens or staminodes. Such white, granular food bodies are the main reward for the visiting beetle *Colopterus* in *Calycanthus* (see above), and may also be provided by the labellum of certain orchids (*Maxillaria* and *Eria*). In other cases, oil bodies may be produced by special organs (elaiophores), which are collected for their broods by solitary anthophorid bees. This phenomenon seems largely restricted to South America, but is recorded for several different families of plants. It is well known in many species of *Calceolaria* (Scrophulariaceae).

Quite apart from true nectaries, which are located within the flower, usually on the receptacle, ovary or petal base (p. 115), many plants also produce extrafloral nectaries, usually as stalked glands which secrete a sugar solution. These may be produced on many organs, and may sometimes have physiological or protective functions unconnected with floral biology. Often they are located on the outside of the perianth or calyx (Fig. 4.13). Their primary function may be to encourage ants to take up positions on the outside of the flower, thus discouraging flower-robbing by the pollinator at the base of the perianth (in constricted flowers such as *Thunbergia grandiflora*, visited by *Xylocopa* bees). At the same time, extrafloral nectaries may also distract ants from visiting the flowers themselves, as they may sterilize pollen (p. 143). Often, extrafloral nectaries secrete before the flower opens, and in these cases (e.g. three Australian *Solanum* species, Anderson and Symon, 1985) ants may serve to protect the flower buds from predation.

Fig. 4.13 Extrafloral nectaries on the calyx and pedicles of *Arbutus canariensis* (×1).

Sexual mimic flowers

In these, the flower has come to resemble an individual of the pollinator species, and at least in the European 'insect orchids', *Ophrys* (Fig. 4.14) produces pheromone scents and tactile responses that also mimic those of the female insect (Bergstrom, 1978). For the 30 or so species of *Ophrys*, most are visited by males of different species of solitary bee (in the genera *Andrena*, *Eucera* and others) and the wasp *Campsoscolia*. These hatch several weeks before the females, setting up territories, and they are very responsive to sexual signals. Usually, only one species of bee or wasp visits any one species of *Ophrys*, although four or five species of *Ophrys* commonly coexist (Table 4.4). In many cases, male bees are seen to at-

Fig. 4.14 The 'pseudocopulation' flower of the insect-mimic sawfly orchid *Ophrys tenthredinifera* of the Mediterranean. Males of the solitary wasp *Eucera nigrilabris* mistake the flower for a female and attempt to mate with it, thus promoting cross-pollination. The flowers produce a pheromone-like scent.

Table 4.4 List of *Ophrys* species which are pollinated by sexually excited aculeate Hymenopteran males (after Bergstrom, 1978)

Ophrys species/form	Pollinators observed
Fuciflorae	
O. scolopax	*Eucera tuberculata, E. longicornis*, etc. EI
O. fuciflora	*Eucera* spp. EI
O. apifera	*Eucera* spp. EI
O. tenthredinifera	*E. nigrilabris*, etc. EI
Bombyliflorae	
O. bombyliflora	*E. oraniensis*, etc. EII
Araneiferae	
O. sphecodes-atrata	*Andrena nigroaenea*, etc.
O. sphecodes-sphecodes	*A. nigroaenea*, etc. *Colletes cunicularius*
O. sphecodes-litigiosa	*Andrena* spp.
O. sphecodes-provincialis	*C. cunicularius, C. infuscatus*
Arachnitiformes	
O. 'arachnitiformis' (three populations in southern France)	*Andrena* spp., *C. infuscatus*
Fusci-Luteae	
O. fusca-fusca	*Andrena* spp. AII, III, *A. flavipes, Anthophora balearica*
O. fusca-iricolor	*Andrena* spp., *Anthophora* spp.
O. lutea	*Andrena* spp. AI, III (*Chlorandrena, A. ovatula*)
Ophrys	
O. speculum	*Campsoscolia ciliata*
O. insectifera	*Argogorytes mystaceus, A. fargei*

tempt to copulate with the flowers (pseudocopulation), although they rarely ejaculate. Such visits are often brief, but I have observed a species of *Andrena* stay on a single flower of *O. helenae* for at least three hours, making copulatory movements throughout. It had pollinia on its head.

Just as remarkable as the visual mimicry in *Ophrys* is the pheromone mimicry, and Bergstrom has shown that flowers are rarely visited in the absence of this scent. Both the pollinators and flowers produce a wide range of volatile compounds, but some of these are common to both the insect and the flower (citronellol, geraniol) (Bergstrom, 1978).

Sexual attraction is recorded for several other orchid genera, notably in Australia (*Caladenia, Drakaea*) and intermediate conditions between conventional attraction and sexual attraction are well demonstrated (Stoutamire, 1983), illustrating that such bizarre syndromes can arise through Darwinian selection. In extreme cases, thynnid wasps, which abduct females, attempt to carry off mimic labella which are suspended on a long balanced lever. After depressing the labellum, it swings back upwards, throwing the wasp against the column and achieving pollination with remarkable accuracy. It is not correct to consider this as a form of pseudocopulation, however, as copulation seems never to be attempted.

Insect mimicry can also appeal to territorial instincts. The delicate brown and yellow striped hanging flowers of neotropical *Oncidium* (Orchidaceae) give rise to aggressive behaviour in the territorial males of *Centris* bees, which try to fight the flowers, thus pollinating them.

Anemophily (wind pollination)

Not all the extreme forms of floral adaptations are directed towards animal visitors. The wind is extremely important as an agent for pollen travel (anemophily), and many large families of plants, many of which habitually dominate communities, are wind pollinated. The disadvantage of wind as a pollen dispersant is its randomness. Whereas an animal may accurately transport a high proportion of the very little pollen produced large distances to a tiny stigmatic target, wind-transported pollen has no accuracy whatsoever. It is, therefore, not surprising to find this mechanism prevailing among species-poor (temperate) communities, where a few species of grass, sedge, rush or tree co-dominate so that legitimate mating partners are readily reached (Regal, 1982; Whitehead, 1983). In species-rich communities, with low levels of individual ecological dominance, biotic pollen dispersal predominates (thus alpine grassland, Mediterranean maquis and tropical forest are full of pretty flowers).

Wind pollination is most likely to be favoured when:

1. individuals of a dominant species are not widely spaced;
2. unambiguous environmental stimuli can orchestrate a closely coordinated flowering time;
3. pollen release can occur when wind-speed and turbulence are adequate and the probability of rain is relatively low;
4. foliage (e.g. in deciduous trees) is unlikely to filter out pollen.

Thus, there are two main advantages of wind pollination. First, it is independent of weather, relatively speaking, so some wind-pollinated trees bloom very early in the year not only to avoid foliage filtering overmuch pollen, but also to avoid excessive stigmatic contamination with foreign pollen. Second, despite expenditures of energy on massive quantities of pollen, and to a lesser extent on large feathery styles, other floral parts are reduced to a minimum. Thus, while rudimentary perianths persist in the Juncaceae (rushes), they have totally disappeared from most wind-pollinated trees and from the Poaceae (grasses) and Cyperaceae (sedges), where they are replaced by derivatives of the bracts and bracteoles.

It is a mistake to consider anemophily and zoophily as being non-overlapping categories however. Work by Stelleman (e.g. 1979) and A. D. J. Meeuse (e.g. 1978) has clearly shown that in *Plantago* and *Salix*, respectively, apparently wind-pollinated species may nevertheless receive appreciable quantities of animal-dispersed pollen, carried by pollinators as sophisticated as syrphid flies and bees (or even birds, see p. 97) respectively (Fig. 4.15).

No doubt many other flower types (for instance *Chenopodium* or *Urtica*) undergo both wind and animal pollination. Such 'ambophily' (Stelleman, 1984) is reported for a wide range of apparently anemophilous species, whereas several Cyperaceae (*Scirpus maritimus*, *Schoenoplectus* spp., *Carex hirta*) and the bur-reed, *Sparganium erectum* are visited by Syrphidae according to Leereveld (1984). Conversely, the apparently zoophilous sea squill, *Urginea maritima*, which flowers in the Mediterranean autumn attracts few insects despite its massive floral display, and much (mostly

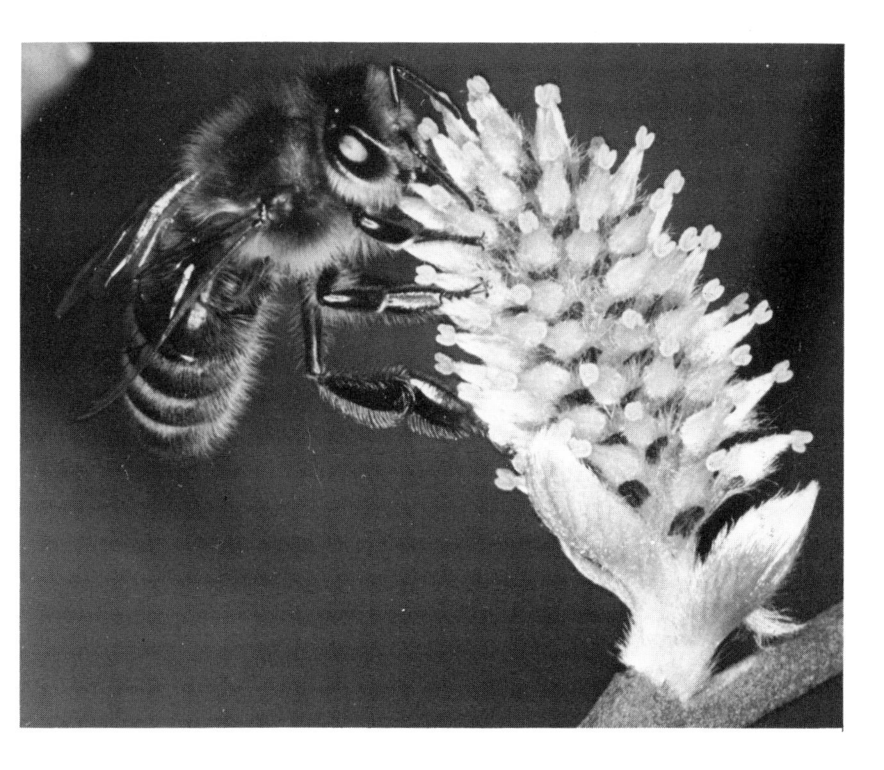

Fig. 4.15 Hive bee, *Apis mellifera*, visiting a female inflorescence of the sallow, *Salix cinerea*. Although the willow flowers lack petals and appear at first sight to be wind pollinated, they have nectaries and are visited by bees. Photo by M. C. F. Proctor (×2).

self-) pollination is wind-mediated (Dafni and Dukas, 1986). For another very attractive autumnal species of the Mediterranean, Petanidou and Vokou (1990) show that *Cyclamen graecum* is initially insect-pollinated, but the older flowers can be pollinated by wind.

With respect to pollination, perhaps the most misunderstood family of plants has been the palms (Arecaceae). Because the huge, monoecious inflorescences lack petals, they have often been considered to be wind-pollinated. However, many species are scented, some have nectar, and nearly all are visited by insects which feed on the copiously produced pollen (Henderson, 1986). In one case the reliance of the oil palm *Elaeis guineensis* on its weevil *Elaeidobius kamerunicus* was such that in its absence, the oil palm industry of Malaysia largely failed until the pollinator was introduced.

In *Thalictrum*, a remarkable diversity of mechanisms are found, and whereas large-petalled, nectar-bearing species such as *T. tuberosum* are typical insect-visited bowl flowers, others such as *T. minus* and *T. alpinum* almost lack perianths, having large hanging stamens and are presumably mostly pollinated by wind. Here is a clear case of the evolution of anemophily within a genus.

Hydrophily (water pollination)

Most water plants produce aerial inflorescences, and the flowers may be sophisticated, involving traps (*Nymphaea*), heteromorphy (*Hottonia*), or spurs (*Utricularia*). Occasionally, inflorescences are produced floating on the meniscus of the water surface, as in the South African water hyacinth (*Aponogeton*), or *Stratiotes* and *Hydrocharis*, which may be pollinated by water-skaters. However, in a few cases, flowers function under water These may be entirely cleistogamous when submerged (*Subularia*, Chapter 9). Alternatively, pollen transport by water may occur (hydrophily). In epihydrophily, pollen grains, or in the famous case of *Vallisneria*, whole male flowers, are released to float to the water surface, where they encounter female flowers with stigmas at the meniscus. As has been pointed out by Faegri and van der Pijl (1979) and Cox and Knox (1986), these cases are unusual, in that pollination takes place in two dimensions only, leading to a greater efficiency in pollen use. Other examples are *Ruppia, Lepilaena, Callitriche, Hydrilla, Elodea* and the sea-grasses *Neptunia, Halophila, Halodule* and *Aeschyomene*.

In hyphydrophily, pollination takes place under water, in three dimensions (e.g. *Najas, Ceratophyllum* and *Zostera*). The case of the sea-grass *Zostera* is especially interesting in that the tendency to a loss of exine exhibited in most cases of hydrophily reaches an extreme, and a pollen tube-like male gametophyte is dispersed through the sea, wrapping itself around any likely object. In almost all cases of hydrophily, perianths are extremely reduced, being small and green or absent. In *Vallisneria*, however, considerable modifications have occurred to the tepals of the male flowers, which rise to the surface by virtue of aerenchyma in the tepals, which are at first tightly closed but immediately open on the surface of the water to form a floating platform for the tiny male flower.

DEVELOPMENT OF PERIANTH COLOUR, SCENT AND NECTAR

Apart from shape, many other features of the perianth have shown a remarkable diversification from the Magnoliid baseline, and these are also concerned with animal attraction. They can be divided into colour attractants, scent (odour) attractants and nectar reward.

The behaviour of a sophisticated flower visitor such as a bee, butterfly or bird with respect to an inflorescence is complex. The behaviour of bees, and especially honey-bees (*Apis*), is best known (Kevan and Baker, 1983). Long-range (primary) attraction is generally by colour; bee vision is highly responsive to colour, but resolves poorly. Thus a suitably coloured object such as an article of clothing, or a motor car will frequently attract bees which then fly off at a distance of a metre or so. Occasionally, a strongly scented plant can also act as a long-range attractant and bees will react in experiments to hidden, scented flowers. However, most strongly scented flowers primarily attract other pollinators, especially moths (pale, night-scented flowers often with long tubes), or flies (brown or purple disgustingly scented flowers). Bats, which in common with moths fly principally at night, also usually visit scented flowers.

Once within a metre of a flower, a bee will respond to flower shape and pattern, which it is now able to resolve. Identification of the shape of an individual pollination unit (e.g. a flower or inflorescence) will allow the bee to orientate correctly, and colour guides on the flower will reinforce this orientation. A high proportion of bee flowers have differential markings which are sometimes invisible to the human eye but can be discerned by the use of ultraviolet-sensitive film. These guide marks characteristically lead from the landing site to the reward (Figs. 4.16 and 4.17). Such nectar-guides are unusual in flowers primarily visited by other classes of pollinators.

On alighting on the flower or inflorescence, the bee will disregard visual and olfactory cues, and its behaviour is then determined principally by touch. Thus, hairs, rugosities, callouses, etc., which aid adhesion of the bee to the flower, may also act as tactile guides. In a flower with a 'bearded' lip (Fig. 4.16) (as in many Lamiaceae) the bee will walk up the beard into the flower towards the reward. Such features may also act as constrictants (see p. 78) and aid in pollen issue and receipt. The stamens may also be bearded at the base (Fig. 4.17).

Fig. 4.16 Brownish spots in the bearded throat of the gullet of the monkey flower *Mimulus guttatus* act as nectar guides to visiting bees (×2).

Fig. 4.17 *Rhododendron macabeanum* provides a diversity of colour markers to the insect visitor; a purple basal blotch to mark the basal nectar, a red stigma and purple anthers which provide a pollen reward.

How long the bee remains at a flower or an inflorescence will largely depend on how rewarding the bee finds the visit (see Chapter 5, p. 148).

Complex behavioural responses such as these in bees render it easy to oversimplify the role played by any one factor in attraction.

Colour attraction

Flower colours depend on the interplay between reflectance spectra of floral organs such as petals, and the ability of the flower visitor to perceive parts of these spectra. The spectral composition of energy reflected from floral organs depends on both the structure of the epidermal and subepidermal layers of the organ, and on the pigments contained therein (Kay, Dauoud and Stirton, 1981).

Two main types of pigment are responsible for the selective absorption of some wavelengths of light, resulting in the selective reflectance of other wavelengths. Pigments soluble in cell-sap (vacuoles) are flavonoids. These are classified into anthocyanins and anthoxanthins. The main pigments occurring, and their main reflectance colours to the human eye are:

anthocyanins	pelargonidin	scarlet
	cyanidin	red, magenta
	delphinidin	purple, blue
anthoxanthins	quercetagetin	ivory
	isoliquiritigenin	yellow

Anthocyanins and anthoxanthins can occur together within the same cell (copigmentation) and can interact to reflect, for instance, vivid blues.

Non-soluble pigments occur on plastids and are involved in the photosynthetic process. As well as the green-reflecting chlorophylls, these include the orange, red or brown reflecting carotenes, and the yellow reflecting xanthophylls. Various combinations of soluble and non-soluble pigments can occur in either or both of the epidermal and subepidermal petal layers, resulting in overlay, masking and pattern effects.

The structure of epidermal and subepidermal cells and their interlying air spaces can also strongly influence reflectance spectra through the physical reflection and refraction of incident light (for instance, petals lacking pigments can nevertheless appear to be a matt white). Petal colour as perceived by humans results from a very complex interaction between the absorption and reflectance characteristics of a variety of physical and chemical factors.

However, petal colours have evolved to attract not humans, but various flower visitors which have very different spectral sensitivities from ourselves. Thus, flower colours and patterns which are invisible to humans will be perceived by flower visitors, and vice versa.

Relatively little is known about the receptivity of many types of pollinator to flower colour. The broad correspondence of certain flower colours to certain classes of pollinator is well known (for instance Table 4.1), but is little understood. Most flower-visiting bats are nocturnal and probably have no visual clues, thus the sombre colours of most bat flowers may help to prevent predation or visitation of these flowers by other animals (Fig. 4.18). This may also be true of flowers visited by other

Fig. 4.18 The large terminal male flower of the bananas and plantains (*Musa*) is coloured an inconspicuous brownish-purple and is visited by bats (×0.5).

Fig. 4.19 The silver-Y moth, *Autographa gamma*, visiting the pale-coloured night-scented flowers of the honeysuckle, *Lonicera periclymenum*. Note extrafloral nectaries on the perianth tube. Photo by M. C. F. Proctor (×1).

pollinators which work principally by scent (e.g. slugs, ants and flies). In contrast, many moths visit flowers at night by hovering and, in the absence of close-range tactile clues, they need to employ acute night vision for an accurate approach (Fig. 4.19). Thus, most moth flowers are pale or white, reflecting the small amount of available light (Fig. 4.20).

Bird pollination

Perhaps the most widely remarked association between flower colour and pollinator class is the affinity of birds for orange, red, or less commonly green, flowers. It is most striking that this association has clearly originated on several different occasions. Thus, the main avian flower visitors in the Americas are hummingbirds (Trochilidae), and they do not occur outside that vast continent. Many important families of plants (for instance the Cactaceae) are largely or entirely American, and have many bird-pollinated species, with scarlet flowers. Yet, in South Africa the entirely Old World sunbirds (Nectarinidae) similarly visit scarlet flowers of endemic genera such as *Kniphofia* and *Haemanthus*.

Similar associations are found in Australia (honey-eaters, Meliphagidae and lorikeets, Trichoglossidae with many endemic scarlet-flowered genera of the Myrtaceae, Proteaceae, Epacridaceae, Goodeniaceae, Rutaceae, Loranthaceae, even Lentiburiaceae and many more). In New Zealand the endemic bellbird associates with the endemic *Metasideros*, and in Hawaii

Fig. 4.20 The white, night-scented tube flower of the moth-visited *Rhododendron hertzogii* from south-east Asia. The tubes are up to 6 cm in length and the basal nectar can only be reached by moths with long probosces. Compare the bee-pollinated *Rh. macabeanum* (Fig. 4.17) (×0.6).

the endemic honey-creepers, Drepanididae, so many of which are now sadly extinct, associate with endemic scarlet lobelias etc. (Plate 1).

The reason for this colour association is not hard to find. Birds have poor colour vision, and although they receive a spectral range similar to that of humans, they do not distinguish well between many yellows, blues and purples. Only at the long-wave limit of their vision do they distinguish colours well, and thus they are very responsive to red. As it happens, most insects receive a range of wavelengths of light shorter than that perceived by vertebrates, and they do not see red well; scarlet flowers in a garden bed are not visited by bees as was observed by Darwin (1876) for *Lobelia fulgens*.

Thus an ecological niche, for instance for the flower pigment pelargonidin, is left vacant by insects, and is readily filled by birds.

In some cases, it is clear that an evolutionarily transitional status between bee pollination and bird pollination still persists. Macior (1986) shows that the scarlet corollas of the Californian *Pedicularis densiflora* are attractive to Anna and rufous hummingbirds, but the purple calyces and bracts of this plant tend to attract bees, chiefly *Bombus edwardsii*. Younger flowers, which have shorter tubes are bee-visited, but the longer-tubed old flowers can only be successfully visited by birds. One race of this species, *P. densiflora* v. *aurantiaca* is only visited by birds, and has presumably evolved recently into an obligate bird flower.

Of course, birds do not only see red flowers, and not all bird flowers are red. Thus, many of the Australian bird flowers are yellow or white, as is the case for most *Eucalyptus*, and many flowers habitually visited by birds in Hawaii, South Africa and other areas are not red. Yellow forms of *Crocus* are savagely attacked in British gardens by sparrows (*Passer domesticus*), although purple crocuses are generally left unharmed. Although the cause of this aggression seems to be mysterious, I have observed that yellow-flowered plants may set better seed after such an attack.

Kay (1985a) notes that the catkins of willows (*Salix* species) are visited in the UK by blue tits (*Parus caeruleus*) and *Salix* pollen can be recovered from the birds' heads when they visit female catkins, so that cross-pollination can presumably occur by this means. It appears that the birds are drinking nectar from the yellow catkins.

Generally, bird pollination is regarded as a phenomenon which is restricted to relatively non-seasonal tropical and subtropical floras where specialist flower-birds can visit bird flowers at most times of year. However, in north America, where flower-visiting by humming-birds is an important strategy even at alpine levels, and as far north as subarctic Canada, humming-birds migrate to tropical sites in winter.

Nevertheless, many birds, like blue tits, are facultative flower feeders, and it is perhaps surprising that virtually no bird-flowers have evolved in western Eurasia, from whence many small birds migrate to Africa, where they feed on flowers, in winter. Recently, it has been shown that the western Asiatic *Fritillaria imperialis* is effectively visited by blue tits in British gardens, and it seems likely that related scarlet lilies such as *Lilium pomponium* and *L. chalcedonicum* from southern France and Greece are also visited by birds.

Human and insect visual spectra

In contrast to birds, bees generally visit flowers with colours in the middle of the human visual spectrum (HVS), that is to say yellow to blue, or with mixed and wide spectral reflectances including some of these wavelengths (purple, pink, white). It has long been known that bees, at any rate, do not receive the same spectral range as vertebrates, but this has given rise to some confused thinking on the subject. However, Kevan (1978) gives a splendid account of the relationship between the HVS and the 'insect' visual spectrum (IVS).

Human colour vision extends from violet (380 nm) to deep red (780 nm), and bee vision from 300 nm to 700 nm. Thus, as already discussed, most HVS red tones are invisible to a bee, which ignores red flowers, but not to birds, whereas wavelengths of reflected light in the ultraviolet (300–380 nm) range which are visible to bees, cannot be seen by humans. Equally importantly, the peak sensitivities to colour vision also vary between vertebrates (humans) and insects such as bees and butterflies; Table 4.5). Thus, HVS sensitivity reaches peaks at 436 nm (blue), 546 nm (green) and 700 nm (red), and those of insects are at 360 nm, 440 nm and 588 nm. Kevan's arguments rest on two important premises. First, all perceived colours are best interpreted as mixtures in varying energies of the three sensitivity peaks (blue, green and red in HVS); they are thus best interpreted on a trichromatic colour scale (Figs. 4.21 and 4.22) with these three pure colours each represented at an apex of a triangle; equal emissions of each sensitivity peak wavelength are perceived as white (and this is in fact Kevan's definition of white), which is thus represented centrally on the diagrams in Figs. 4.21 and 4.22.

Second, it is reasonable to equate the three sensitivity peaks in IVS with those in HVS in terms of perception (although this is not by its nature susceptible of proof), as follows:

	HVS	IVS
red	700 nm	588 nm
green	546 nm	440 nm
blue	136 nm	360 nm

Table 4.5 Some equivalences between trichromatic colours in the human visual spectrum (HVS) and the insect visual spectrum (IVS) (after Kevan, 1978). Peak sensitivities are given in parentheses

HVS (380–780 nm)	IVS (300–700 nm)
red (700 nm)	
orange	
yellow	purple to red (588 nm)
green (546 nm)	mauve to red
white	yellow to white
blue (436 nm)	green (440 nm)
violet–purple	green
	blue (360 nm)
	violet–purple

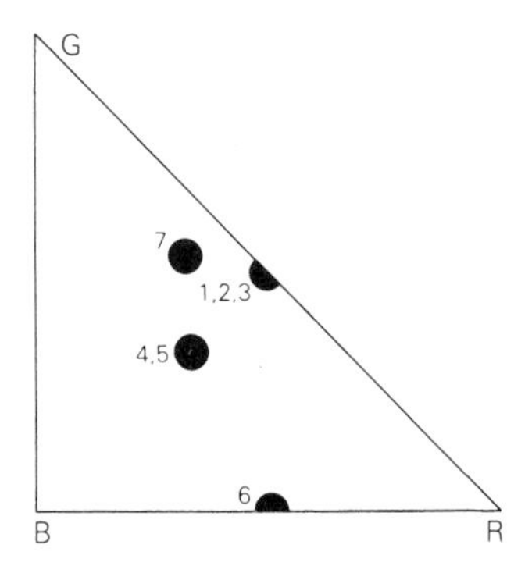

Fig. 4.21 Trichromatic plots for the HVS colour triangle. G, B and R are monospectral green, blue and red, respectively. Points 1 to 7 correspond to those in Fig. 4.22; 1, 2 and 3 are yellow to HVS, 4 and 5 are white, 6 is purple and 7 is greenish yellow. (After Kevan, 1978.)

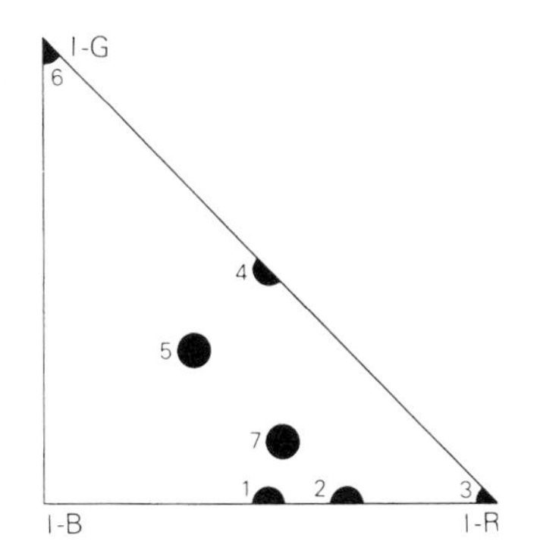

Fig. 4.22 Trichromatic plots in the IVS. I-G, I-B and I-R refer to the monospectral peaks of maximum sensitivity to insects (Table 4.5). Points 1 to 7 correspond to those in Fig. 4.21. 1 is insect purple, 2 is insect red–purple, 3 is insect red (all yellow in HVS), 4 is insect yellow and 5 is insect white (both white in HVS), 6 is insect green (HVS purple) and 7 is insect mauve (greenish-yellow in HVS). (After Kevan, 1978.)

As a result, to take a case of almost exact correspondence between HVS and IVS sensitivity peaks, human blue (436 nm) equals insect green (440 nm). But most colours perceived from reflecting surfaces are complex mixtures of the three peak sensitivities, and must be interpreted from trichromatic diagrams. Because peak sensitivities differ so much between human and insect perception, a single emitted source will take a very different position on HVS and IVS trichromatic diagrams (Figs. 4.21 and 4.22). In the seven colours represented here, only one (labelled 5) is by chance perceived the same by humans and insects (white). Thus, assuming the same colour names for their respective sensitivity peaks, some kind of 'translation' from human colours into insect colours can be achieved (Table 4.5).

From this argument stem three conclusions.

1. There will be colour contrasts within the flower, and between the flower and its background, which will be visible to insects, but not to humans. 'Invisible' nectar guides which reflect in the ultraviolet region, and are thus visible to insects, but invisible to humans (except by the use of ultraviolet-sensitive camera film) are a case in point. However, it is also worth reflecting that yellow flowers on a green background of foliage could be much less visible to bees than to humans. Galen and Kevan (1980) suggest that flower colour differences between shaded (pale blue) and alpine (purple) populations of *Polemonium viscosum* may relate to the visibility of flowers to insects against different backgrounds.
2. Some flowers, and other floral organs, which are inconspicuously coloured to the human eye, might be more attractive to an insect (in the ultraviolet region).
3. Differences in flower colour within a flower or between flowers which cannot be perceived, or are poorly received by the human eye, may be much more vivid to insects. A good case is described by Kay (1982) for the yellow/white polymorphism of *Raphanus raphanistrum*, in which the yellow flowers visited by long-tongued bees and butterflies are insect purple, whereas the white flowers visited by *Bombus pascuorum* are insect cream.

It is important to emphasize the role played by spectral sensitivity. Thus bees do in fact perceive some red light, right up to 700 nm in wavelength, but their sensitivity at this wavelength is low, and they are not attracted by it.

The range of flower colours perceived by the human eye is large, and very nearly encompasses all the colours we can enjoy. When massed in a herbaceous border, or alpine meadow, such a range of colours can indeed make a fine spectacle. Yet Kevan has pointed out that the range of colours available to insects within a habitat is much greater than it is to the human eye. After all, the plants in the wild have adapted to pollinator perception, not to human perception, and in many areas the pollinators solely consist of insects. It is no accident that the most dazzling flower colours to humans often come from vertebrate-flowers from tropical or subtropical areas, nor that the flamboyant colours of many of the birds of these areas (e.g. hummingbirds, sunbirds, lorikeets and parrots) are in fact cryptic. The birds are less visible while feeding and so less vulnerable if they are coloured like the flowers they visit.

Kevan has plotted the trichromatic receptivities in the HVS and the IVS of the colour reflectances of 53 species of Canadian weed (Figs. 4.23 and

4.24). It is clear that insects perceive a much wider range of colours among these flowers than we do. Adey (1982) has devised a technique by which insect colour vision can be conceptualized in human terms, using Kevan's theories. Monochrome photographs were taken of flowers using separately the following filters:

Kodak filter no.	Transmission range (nm)	Human colour	Insect colour
18A	300–400	ultraviolet	blue
47	400–600	blue	green
61	500–600	green	red
29	600–750	red	–

These filters were used in conjunction with neutral density filters, and photographs taken against a uniformly reflecting 'grey scale' so that adjustments in density and contrast can be made (for the ultraviolet transmitting filter, the flash-gun output was adjusted). The three negatives obtained could then be sequentially printed on to colour-sensitive paper through colour filters appropriate to the IVS:

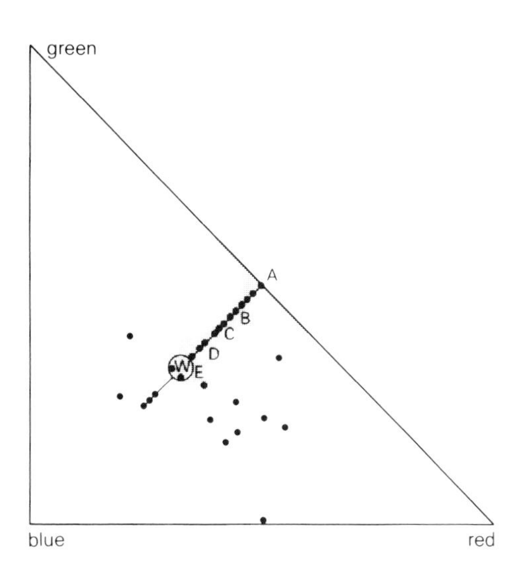

Fig. 4.23 Trichromatic plots (as in Fig. 4.21) for 53 species of Canadian weed in the HVS. Most species appear white or yellow to the human eye. W is the equal energy white point; A, B, C, D and E are points with multiple observations. (After Kevan, 1978.)

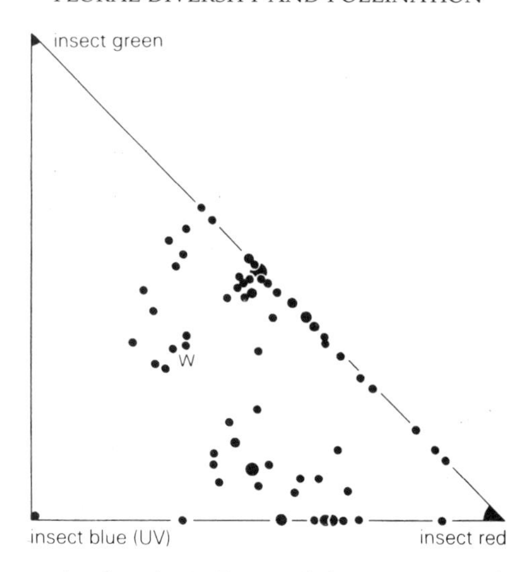

Fig. 4.24 Trichromatic plots (as in Fig. 4.22) for 53 species of Canadian weed in the IVS. There is a much greater variety of colour to insect vision than to human vision (Fig. 4.23). W is the equal energy white point for insects. (After Kevan, 1978.)

Transmission range of negative filter (nm)	Insect colour	Printing filter
300–400	blue	47
400–500	green	61
500–600	red	29

The resulting prints then contained colour mixes representative of the IVS conceptualized in HVS terms.

Petal guides

Petal colours are not necessarily constant, and changes in petal colour are often adaptive, or apparently so. Thus in the bee-visited flowers of the Boraginaceae, there is frequently a marked change between the colour of the buds and recently opened flowers, and those that are donating or are receptive to pollen. This change is often from long wavelength reddish colours, not easily detected by bees, to short wavelength bluish colours to which bees respond (Muller, 1883). It is a feature of many anthocyanins that they can change the wavelength of their reflected light markedly in response to minor physiological changes, for instance in pH value. Floral guidemarks may often change colour as well, frequently as the flower ages, and the stigma cease to be receptive. Thus in *Androsace villosa* (Fig.

4.25), fresh flowers have a yellow annulus at the mouth of the tube, which turns red on ageing. This apparently occurs through an increase in, or change in the nature of, anthocyanins in this tissue, thereby masking the carotene pigment, and rendering the guide invisible to insects. Similarly, the yellow spots on the whitish petals of the horse chestnut, *Aesculus hippocastanum* turn red after the stigmas wither. In the prophet flower, *Arnebia echioides* (Fig. 4.26), the fine black petal spots (supposedly representing the results of handling by Mohammed) fade with time and older flowers do not therefore bear nectar guides. In such cases, the persistent corolla may protect the developing fruit after fertilization, but the plant avoids unproductive, and perhaps discouraging, visits by the pollinator to overmature flowers. *Ranunculus glacialis* is frequently quoted as a flower that starts white, but turns red after pollination. This is a much less certain case, for some populations are entirely reddish from anthesis, others stay white, and there is no evidence that those that change do so in response to pollination. As in other cases quoted, it is much more likely to be a response to ageing of the flower.

Mimicry

Petal colour may have many other functions, some of which have been surveyed by Kevan (1978). Thus, petals are frequently involved in deceit,

Fig. 4.25 The flowers of the European alpine *Androsace villosa* (Primulaceae) have a yellow annulus which acts as a guide to visiting syphids, but which turns red and becomes invisible as the flower ages.

Fig. 4.26 In the prophet flower *Arnebia echioides* (Boraginaceae), the black petal spots on the yellow flowers fade with age and make the flower less attractive to insects (×0.5).

as has been documented above (sapromyophily, pseudocopulation, etc.). Also, flowers may show colour mimicry, not only to insects, but also to other species of flower (Dafni, 1986). Clearly, it may be advantageous for a plant that is a minority in a patch to display colour signals to pollinators very similar to those displayed by the dominant zoophilous flower in the patch, particularly if pollinators are scarce or low in diversity. The advantage of this Batesian mimicry will be increased where the potential mimic is self-compatible, but is reliant on insect visits for within-flower pollination, or geitonogamy.

A good example of colour mimicry is found in two colour forms of the South African orchid *Disa ferruginea*. These closely mimic *Kniphofia uvaria* (orange) and *Tritoniopsis tritacea* (red), respectively (Johnson, 1994) (Fig. 4.27). Studies showed that the mimics agreed well with four postulates for flower/flower mimicry, as follows.

1. The mimic only occurred with the model, and flowered at the same time.
2. The mimic closely resembled the model in the height, size, patterning and reflectance spectra of the flowers.
3. The mimic flowered at a much lower density than did the model (not usually exceeding 10%).

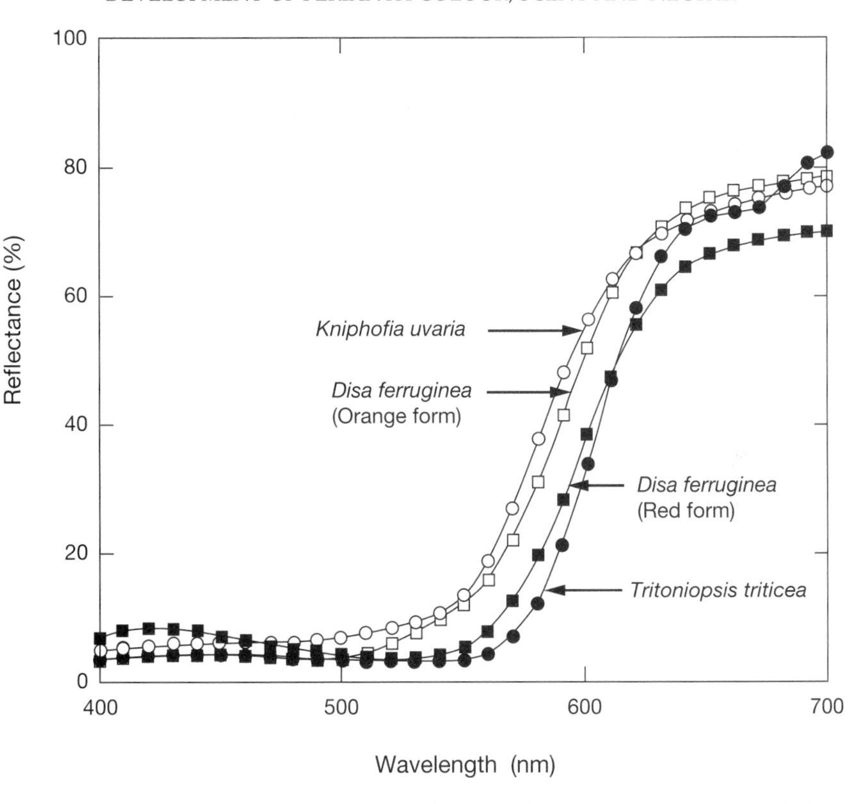

Fig. 4.27 Spectral reflectance curves of flowers of the two colour forms of *Disa ferruginea* and their putative models.

4. The mimic was more reproductively fit when the model was present than in its absence, and was also reproductively fitter than related non-mimic species.

Johnson points out that as mimicry only succeeds at low densities, mimic flowers may suffer from poor reproductive assurance. In the case of *Disa*, sticky pollinia, sticky stigmatic cavities, and a single lepidopteran pollinator all serve to increase reproductive efficiency at low density. With their accurate mechanisms for pollen presentation and receipt, and 'all or nothing' pollination mechanisms, orchids are particularly preadapted to be successful mimics at low density.

Other possible examples of Batesian mimicry for flower colour and shape are quoted by Proctor and Yeo (1973), for instance between *Euphrasia micrantha* (an inconspicuous annual eyebright) and *Calluna vulgaris* (heather) on British heather moors. Vokou, Petanidou and Bellos

(1990) have suggested that the self-incompatible gesneriaceous palaeoendemic on Mt. Olympus, *Jankaea heldreichii*, which offers no nectar reward to visiting bees, mimics nearby chasmophilous *Campanulas*.

Dafni (1986) suggests that, in addition to such apparent examples of Batesian mimicry, Müllerian mimicry may occur between co-dominant species, particularly if they have unspecialized flowers. Thus, Proctor (1978) suggests that the very similar shape, colour and size of the unrelated flowers of *Ranunculus* (buttercups), *Potentilla* (cinquefoils) and *Helianthemum* (rock roses) in temperate and alpine meadows in many parts of the world may be coadaptive. Müllerian mimicry among coexisting flowers has been reviewed by Kay (1987b).

Müllerian mimicry may also occur between common and rare species. Several authors have noted the striking resemblance between the totally unrelated flowers of species of the common figworts, *Scrophularia* and those of the orchid *Epipactis helleborine* and its close relatives. In each case, numerous small greenish flowers borne in racemes of similar stature display a central, purple, cup-shaped nectary. In both cases, this is extremely attractive to wasps, mostly *Vespa germanica* in the UK, and these form virtually the only flower visitors to these plants. Wasps rarely visit flowers otherwise, and these form the only clear examples of wasp flowers in the European flora. Presumably, the chief attraction to wasps lies in the distinctive, fruity, acetone-like scent possessed by the flowers of both genera (and maybe the flavour of the nectar) which also appears to discourage visits by other potential pollinators such as bees.

In a survey of the uv reflecting patterns of flowers in the Fabaceae, Kay (1987a) shows that unrelated but coexisting species tend to show similar flower patterns, particularly when they share a single visitor. However, related species are more likely to show different uv reflectance patterns, features which may have led to their reproductive isolation during speciation. Kay tentatively suggests that the tendency to find striking flower colour variation in the IVS, but none in the HVS may also serve to protect flowers from vertebrate predators when the guild contains at least some inedible models (Batesian mimicry). As yet, such suggestions seem not to have been the subject of experimental investigations, and this would seem to be a fruitful field for further research.

Thermoregulation

By absorbing, trapping or reflecting heat energy, flower pigments may not only act as an attractant, but may provide a reward to small poikilothermic insects. The study of the role played by the colour and architecture of flowers in their own thermoregulation, and in the thermoregulation of visiting insects is as yet in its infancy. So far, such studies have largely been confined to heliotropic, parabolic solar furnace flowers of the arctic (Kevan, 1972a,b, 1975, 1989).

Recently, we have shown (unpublished) that the closed flowers of *Crocus* also thermoregulate, but by a glasshouse effect. Incoming light energy is transmitted through the tepals and some of this radiation is dissipated as heat energy which is then trapped within the flower. This can result in internal flower temperatures of as much as 7°C above ambient (Fig. 4.28). Most crocus flowers open when the internal temperature is at about 12°C, so that on a colder, bright day, they may still be caused to open by this glasshouse effect. Once open, the flowers then act as solar furnaces, so a second physical principle takes over the maintenance of relatively high internal temperatures.

When closed, white and yellow crocuses attain lower internal temperatures than do those with purple flowers, as less incoming energy is absorbed at the tepal surface in the former. White and yellow flowers may, however, reflect more heat when open. It is possible that a disruptive selection operates which for different reasons favours both pale flowers, which are more attractive to microfauna when open, and dark flowers, which are more likely to open at cold temperatures, thus encouraging larger pollinators. Possibly, this may explain why so many early spring flowers of the Mediterranean have striking pale/dark flower colour polymorphisms.

Colour-mediated internal thermoregulation may also benefit the sexual process of the flower itself. Jewell, McKee and Richards (1994) show that stylar temperatures of *Lotus corniculatus* are, in sunshine, higher within individuals with dark, anthocyanin-pigmented keels than they are within plants with yellow keels (Fig. 4.29). As dark-keeled plants tend to inhabit cooler microsites, it seems possible that this particular polymorphism is maintained by a combination of microsite temperature heterogeneity and optimal stylar temperatures for sexual efficiency. The keel is hidden by the wings from visiting pollinators, and in general, Jones *et al.* (1986) find that visitors do not distinguish between flowers with different keel colours.

Kevan (1990) shows that silky hairs of *Salix* (willow) catkins absorb and trap heat, but the long-lived pistillate flowers heat up more than do the ephemeral male catkins. For these arctic species, Kevan suggests that floral heating may encourage the successful completion of sexual fusion and embryo development within the gynoecium.

Odour attraction

Despite the advent of the highly sophisticated and sensitive techniques of gas chromatography, relatively little is known of the chemistry of flower scent. Perhaps half of all animal-visited flowers appear scented to the human nose, and unlike many other features we have been discussing, floral odour may be primitive in the Angiosperms. Some primitive beetle

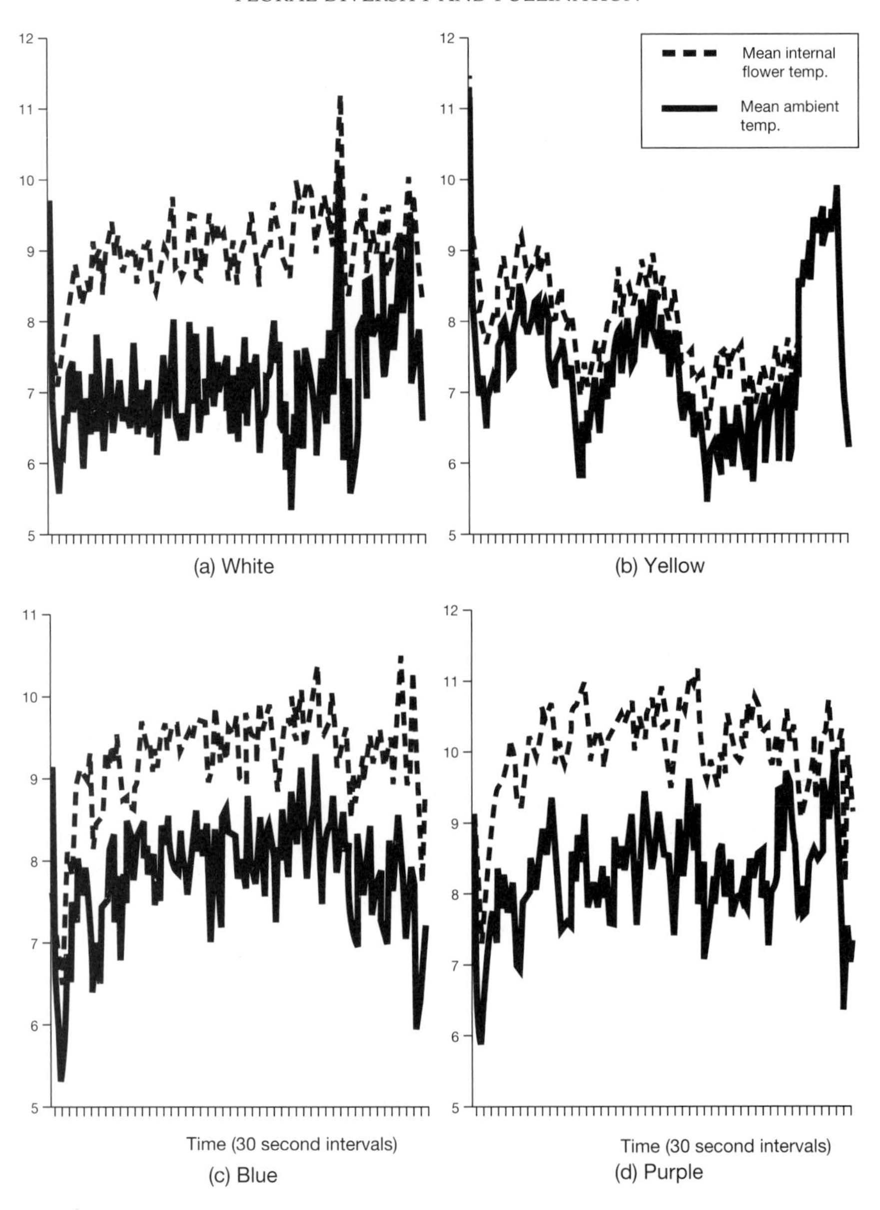

Fig. 4.28 Comparisons of means of mean internal flower and ambient temperatures for different coloured flowers of *Crocus*.

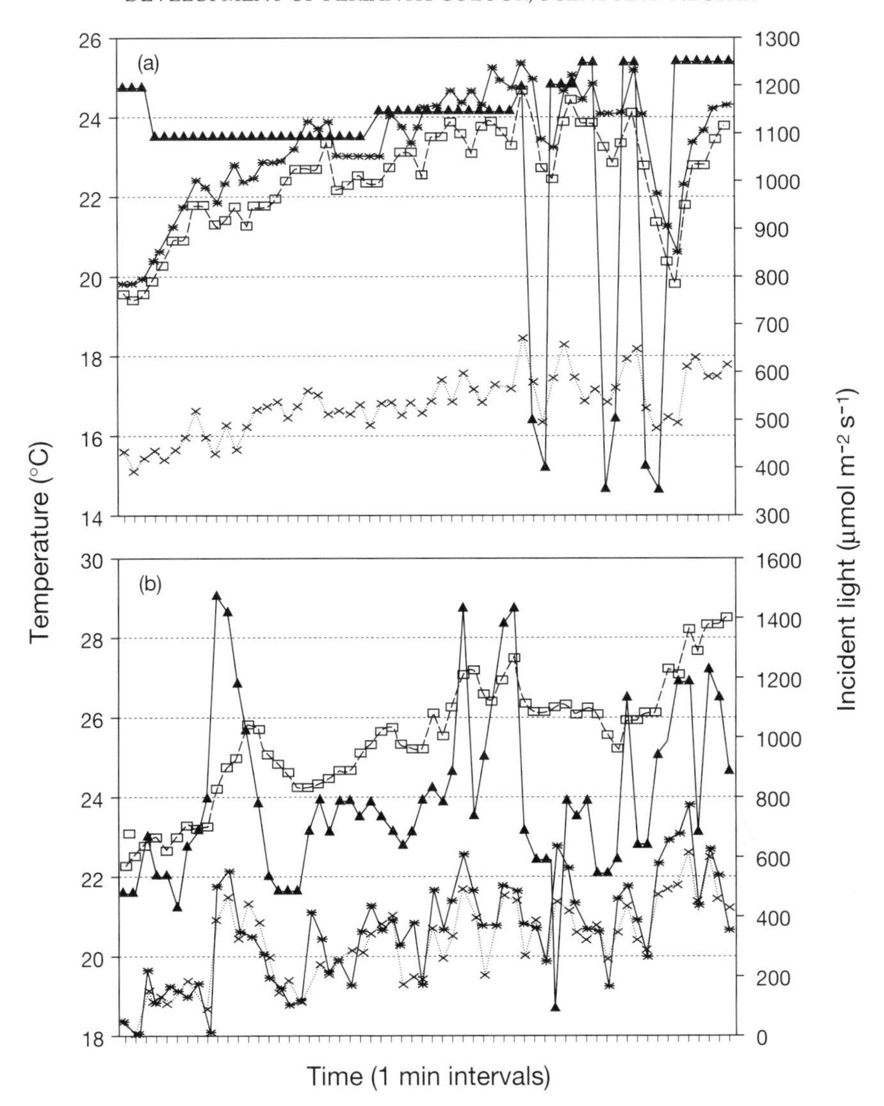

Fig. 4.29 Graphical representations of simultaneous individual temperature readings made within the keels of light-keeled (crosses) and dark-keeled (squares) flowers of *Lotus corniculatus*. The heavy continuous line represents ambient, and the triangles show the incident light readings. (a) Readings were made in sunny, almost cloudless conditions; (b) readings were made in bright conditions with some hazy sunshine. (After Jewell *et al.*, 1994.)

flowers such as *Drimys* and *Nymphaea* etc. are strongly scented, especially at night, and odours with various functions are found in living representatives of still earlier groups such as the cycads, and certain Marchantiales (liverworts).

However, diversification in the nature of flower odour must have occurred during the adaptive radiation of the flowering plants, to give rise to odours as distinctive as the sweet scent of honeysuckle (*Lonicera*, a moth flower; see Fig. 4.19), the foetid stench of various aroids (sapromyophilous), the pheromone (sex hormone) imitating scent of the pseudocopulatory *Ophrys* (Fig. 4.14), or the 'cabbagy' scent of many bat flowers (*Cobaea*). Variation in odour (to the human nose) can occur within a species. Galen and Kevan (1980, 1983) record 'sweet' and 'skunky' odours within populations of *Polemonium viscosum*, and report that *Bombus* bees preferentially visit flowers of this species with sweet odours.

Flower odours are usually produced by the petals or tepals, from the epidermis. They are most usually produced by volatile oils, and, in these cases, the scent-producing areas can be stained by immersing flowers in 0.1% aqueous solutions of neutral red at room temperature for 2–10 h (Vogel, 1962; Adey, 1982). Scent-producing areas in the Leguminosae are known as 'Duftmale', and Adey has shown that a wide variety of Duftmale patterns are produced by different species in tribe Genistinae, and that these are taxonomically distinctive (Fig. 4.30). Often, Duftmale patterns correspond to general morphology, thus they may be limited to the wings or the standard (but never to the keel), or they may sometimes correspond to guidemarks on the flower.

As has already been described, odour is most often employed as a primary long-range attractant to the flower. Experiments by Brantjes (1976a,b, 1978) on various moths showed that they can respond to scent alone, and in some cases can show limited orientation to scent, although in others, scent merely elicits a feeding response. In practice, moths respond to stimuli of scent and vision simultaneously. The range at which scent can attract a pollinator is little known, but scents may be apparent to bees at a concentration of only one-hundredth of their apparency to man.

However, scent may have other functions. Bees carry scent back to the hive, where the scent of a productive source is detected by other workers and is used, together with other evidence from 'bee dances', to trace the productive patch. Some tropical euglossine bees collect perfume from a variety of types of flower (*Catasetum* and other orchids, Araceae such as *Anthurium*, *Gloxinia* and others). By scraping up the odorous cells, the bees appear to become drugged, but later they fill their leg slits with the odour and perform apparently territorial flights (Vogel, 1966). Whether the purpose of this behaviour is really territorial or sexual is not clear, but it may

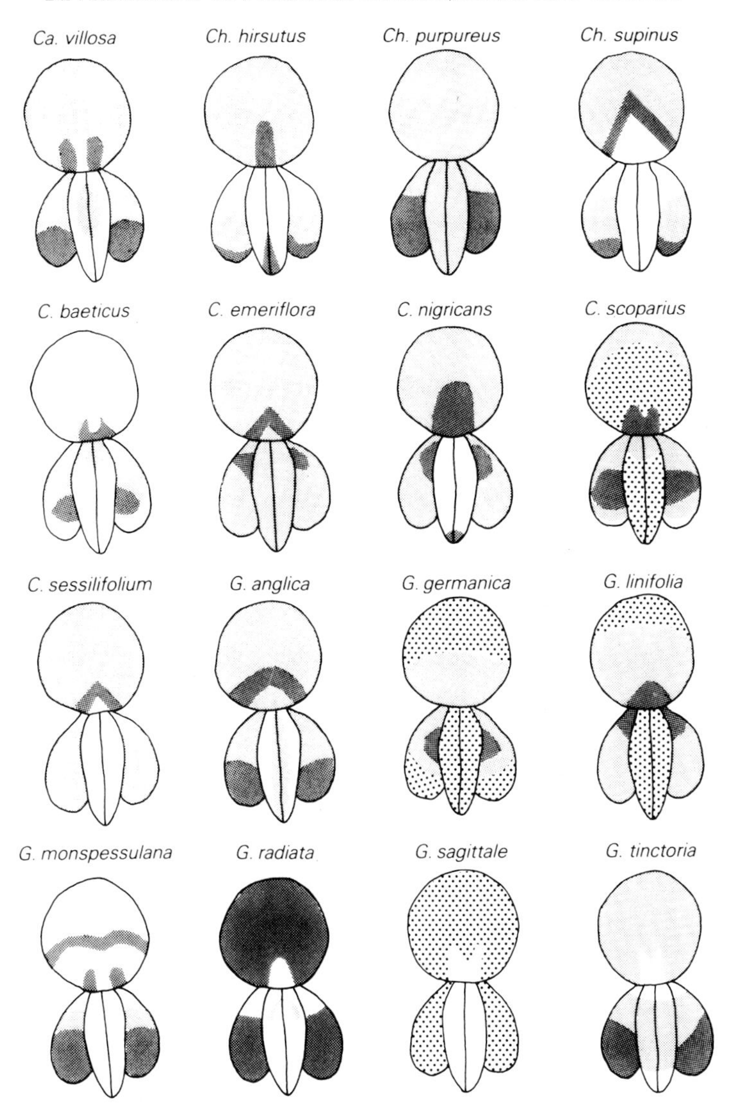

Fig. 4.30 Representations of 'Duftmale' patterns on the flowers of various species in the tribe Genistinae, family Leguminosae (Fabaceae). Duftmale secreting areas produce scents as volatile oils; heavily secreting areas are dark, moderately secreting areas are lightly shaded and lightly secreting areas are stippled. *Ca*, *Calicotome*; *Ch*, *Chamaecytisus*; *C*, *Cytisus*; *G*, *Genista*. Drawing reproduced by courtesy of Margaret Adey.

provide novel stimuli for flower visiting, freed from the constraints of energy budgets. Thus, scent-collecting bees may be encouraged to make much longer species-specific flights between widely dispersed flowers in tropical forests ('trap-lining'; Williams and Dodson, 1972) than would be the case if they were merely collecting nectar.

In sapromyophilous flowers, scent may be limited to narrow hanging or erect appendages to the flower ('osmophores') (Vogel, 1962), which may act as primary alighting sites for attracted flies (*Arisaema*, some *Caladenias* (Fig. 4.31), *Tacca*, etc.). More often, the foul scents emitted by sapromyophilous flowers, such as aroids (*Arum*, *Dracunculus*) are produced by the spadix (a modified end to the inflorescence axis) (Fig. 4.32). Amines, ammonia and skatoles are volatilized by a tremendous respiratory effort, in which the spadix may heat up to 22°C above ambient

Fig. 4.31 The Australian orchid *Caladenia filamentosa* has very long, narrow perianth segments, the dark distal areas of which are 'osmophores' and produce an unpleasant scent that attracts flies (×1.0).

Fig. 4.32 The sapromyophilous inflorescence of the dragon arum, *Dracunculus vulgaris*, of the Mediterranean. This stands one metre high; the spadix projects from the spathe and emits a foul odour, thus attracting oil beetles.

temperatures, rendering it hot to the touch (if the stench permits the experimenter to get so close) (B. D. J. Meeuse 1978). Such vile smells are usually preludes to a trap mechanism, but other aroids (*Amorphophallus*) have no trap, yet flies are so attracted that they may spend several days in close proximity to the inflorescence.

Although Kullenberg (1956) has suggested that the pheromone-mimicking odours of *Ophrys* (Fig. 4.14), which are faintly discernible to the human nose, may in fact be the pheromones themselves, Bergstrom (1978) shows that only a few of the volatiles are shared by the orchid and the pollinator. Flower odours are involved in other deceit mechanisms, as in *Arum conophalloides* which exclusively attracts blood-sucking midges (females only), although it does not trap them.

Of the more attractive odours (to the human nose), it is not easy to classify types of odour to types of pollinator. Moth flowers (Fig. 4.20) are usually very strongly and sweetly scented, principally by night. Bee flowers are poorly scented, but simple bowl flowers attracting beetles, syrphids and small flies (Fig. 4.2) are often sweetly scented. Wasp flowers may have a fruity scent, as do some bat flowers (Fig. 4.18), although others have disagreeable cabbagy scents. Bird flowers usually have no scent at all, and birds do not respond to flower odour (this negative evidence serves to suggest that most or all flower odours play a positive role in pollinator attraction, and odour production expends enough energy to render its loss evolutionarily successful where it has no function). As yet we know little about the chemistry of odours with respect to different attraction syndromes but it is likely to be complex. An analysis of human responses (25 individuals) to 12 different varieties of *Narcissus* which were hidden from the subjects, suggested to me that considerable variation occurred not only with respect to odour detection (this is well documented in humans, for instance in the ability to smell *Freesia*), but also in the *Narcissus*. A study by Adey (1982) on the Genistinae produced similar results.

Nectar reward

The evolution of nectar feeding

We have seen that many primitive flower types are visited only for pollen, and that most primitive flower visitors cannot utilize nectar. However, specialized flower visitors (bees, wasps, butterflies, moths, birds and bats) depend crucially on the production of nectar by many flower types.

The first nectar feeders probably utilized stigmatic secretions. Often stigmas, including those of many 'primitive' flowers, are 'wet' (p. 57), producing a nectar-like secretion which adheres, hydrates and succours the pollen. Where stigmatic surfaces are large, which once again is often found in 'primitive' flowers, stigmatic secretions can be substantial. In such cases, these secretions may have the dual function of receiving the pollen, and rewarding the flower visitor. In the dioecious tropical fruit trees, *Garcinia*, pollination is undertaken by *Trigona* bees which feed exclusively from stigmatic nectar which is also produced by male flowers (Fig. 4.33) (Richards, 1990b).

Sophisticated flowers which receive pollinia within stigmatic cavities may also produce significant quantities of nectar in these cavities which is utilized by visitors, as in *Asclepias syriaca* (Kevan, Eisikowitch and Rathwell, 1989). This is also true for many orchids. In the genera *Orchis* and *Dactylorhiza* many species produce no nectar in their spurs, and they are frequently thought to be Batesian mimics of nectariferous models (reviewed in Dafni, 1984, 1986; Fritz and Nilsson, 1996). However, Dafni

Fig. 4.33 Worker of the sweat bee *Trigona laevipes* visiting the stigma of the female flowers of the Malaysian fruit tree *Garcinia hombroniana* for the purpose of feeding off stigmatic exudate, a primitive form of nectar.

and Woodell (1986) show that visitors to the spotted orchid *D. fuchsii* feed on stigmatic nectar, and we have also shown that this occurs in *D. purpurella*. In the latter species there may also be an automimicry, in that in some populations individuals with nectariferous spurs and non-nectar-bearing spurs coexist (p. 121).

Once the habit of stigmatic nectar feeding evolved in primitive flower visitors, it would pay flowers to produce nectar secretions elsewhere, thus protecting the stigmatic fluid. In this way, specialized localized nectaries would have evolved, and in time, guilds of flower visitors would have evolved which specialized on drinking floral nectar.

Nectar is essentially a phloem secretion, in which sucrose usually predominates, with fructose and glucose also being present, and maltose and mannose (which may be toxic) on occasions. It also contains amino acids and lipids. It is secreted by localized densely packed groups of specialized cells (nectaries). These are most usually located on petals or sepals, often at the base, or are hidden in a flap (*Ranunculus*; see Fig. 4.2), pouch (*Impatiens*), spur (*Aquilegia*; see Fig. 4.6) or tube (*Plumbago*), although the nectaries themselves may form special tubular organs derived from tepals

(*Helleborus*, *Trollius*), or may be located on the receptacle (many Rosaceae etc.). Some flower spurs merely act as a vessel to contain the nectar which is produced proximally (*Viola*).

In bird flowers and bat flowers, nectar production is often copious, and many millilitres may be shaken out of flowers. In many flowers nectar may be difficult to locate, however, and a simple colorimetric test for sugar (for instance with diabetic test papers, e.g. 'Diastix') is often useful. Once the nectary has been located, volumes of liquid nectar can be measured using microcapillary tubes, and the concentration of sugar estimated using a refractometer. In this way, the energetic reward to a visitor provided by a flower can be quantified in calories, and the diurnal output, and rhythm, of a flower producing nectar can be investigated (Corbett, 1978; Best and Bierzychudek, 1982). As monosaccharides have half the refractive index of disaccharides, it is possible to measure directly the nectar strength in sucrose equivalents without conversion factors; the calorific equivalent of sugar is 3.7 cal/mg (15.5 J/mg).

A good deal of work has been published on the strength and composition of nectar (e.g. Percival, 1961, 1965; Baker and Baker 1973a,b, 1975, 1977) but as Faegri and van der Pijl (1979) point out, much work has concerned itself only with the volume of nectar, and not its calorific value, which may be more significant to visitors. Maximum calorific rewards per flower per day have been observed as:

Sinapis alba	400 (cal)	1673 (J)	Corbett (1978)
Echium vulgare	100 (cal)	418 (J)	Corbett (1978)
Digitalis purpurea	50 (cal)	209 (J)	Best and Bierzychudek (1982)
Rubus fruticosus	4.5 (cal)	19 (J)	Yeboah Gyan and Woodell (1987)
Prunus spinosa	1.1 (cal)	5 (J)	Yeboah Gyan and Woodell (1987)
Crataegus monogyna	1.1 (cal)	5 (J)	Yeboah Gyan and Woodell (1987)
Arctostaphylos otayensis	1.5 (cal)	6.3 (J)	Heinrich and Raven (1972)

Undoubtedly, rewards may be higher than this for vertebrate-visited flowers. Hummingbirds can assimilate more than 60 cal (250 J) while hovering, and some bird and bat flowers may contain in excess of 1 kcal (4200 J) of sugar equivalent.

However, sugar content per flower varies greatly within most species. Many variables act on the amount of sugar available, including the position of flower, the age of flower, time of day, weather, the vigour of the plant, and whether pollinator visits have occurred. If nectar volumes and concentrations are to be measured, it is most important that the flowers are protected from visits by pollinators for a period of 24 h before the measurement is made. Maximum rewards vary very much between species. In the foxglove, *Digitalis purpurea* (see Fig. 5.1) nectar content per flower increases with the age of the flower, and may indeed be greatest after the corolla has fallen (Best and Bierzychudek, 1982; Arrow and Richards, unpublished). This species secretes nectar continuously and at a fairly even rate throughout the life of the flower. However, in the orchid *Dactylorhiza ericetorum*, most nectar is found in the spur of the first flower to open, on its first morning, and the volume decreases throughout the life of the individual flower (even in the absence of visits), and is less in subsequent flowers. On the whole, large spikes of this orchid bearing many flowers secrete more nectar per flower than do less well resourced plants (Fleming and Richards, unpublished). In *D. purpurea*, too, nectar production seems to be resource-limited, at least in stressed, low density populations. In contrast, vigorous, large plants show no relationship between flower number and nectar production, although considerable, apparently innate differences in nectar production are found between individuals (Parry, unpublished).

Often, it appears that the pattern of nectar secretion within and between flowers has become adapted to control the behaviour of visitors (Chapter 8). In dioecious species, females often secrete more nectar than do males, perhaps to compensate for the absence of pollen as food, and the poorer flower display of females. In *Silene dioica*, female flowers secrete about twice as much nectar as do male flowers (Kay *et al.*, 1984; Gillon and Richards, unpublished).

In most flowers, the concentration of nectar in sucrose equivalents lies between 25 and 75% (Percival, 1961; Yeboah Gyan and Woodell, 1987). Harder (1986) suggests that the nectar concentration which would normally maximize energy returns to a bee for a unit of foraging effort is about 60%, although this figure would be lower (nectar more dilute) in long-tubed flowers.

Corbett (1978) showed that in *Echium vulgare*, the normal daily fluctuation in concentration was between 20 and 58%, whereas in *Sinapis alba* it was between 20 and almost 100%. Both species secreted the daily supply of nectar early in the morning, and again in the evening; this seems to be a common pattern of secretion in many species. Sugar concentration was highly linked with flower temperature, which reached a peak for *Echium* at 13.00 hours, and for *Sinapis* at 17.00 hours on experimental days. Yeboah Gyan and Woodell (1987), working with three British woody members of the Rosaceae, also showed that nectar volumes drop and concentrations increase as the day proceeds.

Examination by Percival (1965) of the range of nectar concentrations in flowers with different types of visitor, shows that nectar concentration can be highly adaptive in limiting pollinator visits (Table 4.6). Insects are probably not responsive to sugar concentrations below 10%; Vansell, Watkins and Bishop (1942) showed that only when orange blossom (*Citrus*) nectar was concentrated by daytime heat to 30% was it attractive to

Table 4.6 Sugar concentrations in the nectar of flowers of various pollination classes (Corbett, 1978, after Percival, 1965)

Pollinator	Concentration of sugar (%)	Number of species of plant
Moth	8–18	2
Bat	14–16	2
Bird	13–40	7
Butterfly	21–48	2
Bee	10–74	24

bees; undoubtedly proboscis drinkers such as butterflies and moths will require more dilute nectar than this, as will 'lappers', such as bats, and suckers, such as many birds. At concentrations above 70%, nectar becomes very viscid, and at high concentrations it is crystalline. It will then become more suitable to feeding by insects with relatively unmodified mouthparts such as flies, and some beetles and bees; nectar of unspecialized umbel flowers may always be in this state. Corbett (1978) suggests that the pollinator suite of *Sinapis* changes through the day, long-tongued bees and butterflies being replaced by flies, and this broadens the chances of successful pollination, while limiting the pollinators at a given moment. This apparently successful strategy may be a very common one, and may explain why many flowers have a 'limb and tube' shape; the salver-shaped limb concentrates heat, and allows nectar to become more concentrated during the day; the tube limits visits earlier in the day to specialized long-tongued feeders. Dafni and Werker (1982) show that the autumn bulb *Sternbergia clusiana* diversifies its visitors by providing basally produced nectar which is taken by *Apis* bees, and nectariferous hairs at the perianth apices which are visited by syrphids. Two whorls of stamens are produced, each relevant to the feeding patterns of one of the two pollinator classes. Much more work is needed on the role played by nectar in providing 'time-niches' and 'space-niches' for the pollinator suite.

For butterflies, larger moths and some birds, nectar provides the total food input (hence the presence, or at any rate the function, of trace amounts of nitrogenous substances), but for bats, other birds and bees it acts as a liquid fuel. It may also act as a source of water; thus bees may visit more dilute nectar sources on hot days. Many pollen-collecting bees will collect pollen on one species, and fuel with nectar on another. Thus workers of *Bombus lucorum* in Cumbria, UK, on a June day in 1983 (ambient temperature 21°C) were pollen collecting exclusively from *Cerastium fontanum*, which has no nectar, and drinking exclusively from *Trifolium repens*. For many different individual bees, the ratio of pollen visits to nectar visits was about 10:1. On a colder day, the proportion of nectar visits will increase. Bees rarely fly at under 10°C, below which temperature the thoracic–ambient temperature differential is such that any feeding flights will have a negative energy budget.

Naturally, spring flowers tend to have higher nectar concentrations; some autumn flowers (*Solidago* (Fig. 4.34), *Eupatorium, Campanula thyrsoides*) have dense heads so that pollen-collecting bees can crawl, thus saving energy.

Nectar robbing

Nectar robbing may also save energy for the flower visitor. I have observed the smaller-bodied (alien) honey-bee (*Apis*) visiting the flowers of the native *Iris pseudacorus* and the introduced *I. versicolor* legitimately in marshes on the shores of Windermere, UK. However, the larger bodied

Fig. 4.34 Flower of *Solidago virgaurea*. Stigma branches initially fused at the apex and bearing pollen; the branches part latterly, exposing the receptive inner surface of the stigmas, and allow geitonogamous self-pollination to occur. Photo by M. C. F. Proctor (×6.0).

Bombus lucorum visited *I. pseudacorus* legitimately, but had difficulty with the smaller-bored flowers of *I. versicolor*, which it robbed by levering the base of the false tube. Legitimate visits lasted an average 25 s, but illegitimate visits only 5 s.

A few species of flower visitors are habitual nectar thieves; thus *Bombus mastrucatus* usually bites holes at the base of corollas which it could readily visit legitimately (Fig. 4.3). Macior (1966) has described how the queens of *Bombus affinis* pierce the spurs of *Aquilegia* (bird and sphingid flowers) for nectar, while collecting pollen (and probably achieving pollination) legitimately. Examination of any population of, for instance, red clover, *Trifolium pratense*, will demonstrate that many flowers with perforated bases have been robbed. Such behaviour is bound to affect gene flow by rendering robbed flowers unattractive to legitimate visitors.

Although many flowers have guide marks on petals which lead towards nectaries, nectaries themselves are usually inconspicuous, but they can be visible and glistening, especially in some beetle flowers. However, the glistening substance itself is often not nectar. *Parnassia* is among nectarless flowers that produce dummy nectaries (Fig. 4.35).

Fig. 4.35 The flower of *Parnassia palustris* has five staminodes which bear about 10 filiform processes with shiny terminal knobs each. These have been regarded as dummy nectaries which attract the pollinator (here the hover fly *Neoascia podagrica*) which on landing is able to discover the concealed true nectaries. Photo by M. C. F. Proctor (×3).

Nectar as food

In general, despite great variation in the production of nectar even within a single flower, the chemical composition, concentration, calorific value and position of nectar produced by a flower will tend to be adaptive with respect to its habitual visitors. Larger visitors will receive greater energetic rewards, but may have to expend more energy in acquiring them (Heinrich, 1975). Total flower feeders will receive a more balanced nectar diet (more amino acids and lipids) (Heinrich 1979a,b) than mixed feeders. Sucking feeders will receive more dilute nectar, in greater volumes, from narrower receptacles, at greater cost than dabbing feeders. Thus butterflies may expect to receive relatively large volumes of dilute nectar with amino acids and spend long periods of time at each flower, in comparison with short-tongued bees, or flies, which will receive small volumes of concentrated pure-sugar nectar on short visits. Baker and Baker (1983) and Kevan and Baker (1983) give average concentrations of amino acids in nectar of flowers with different visitor syndromes, and show that sapromyophilous carrion fly-visited flowers have about 12 times the concentration of amino acid of any other flower type. They also list occurrences of individual amino acids, and show that alanine is the most widespread amino acid in nectar, but that 13 amino acids occur in more than half the sample of 395 species.

By such strategies, the plant tends to narrow the range of visitors to those to which it is best adapted for pollination and to those most suitable to the time/space niche in which it flowers.

Variation in nectar production: automimicry

Because nectar volumes and concentrations are so variable within a plant, and because they also seem to vary in response to the resourcing of the plant, studies of the genetic variability of nectar production between individuals seem rarely to have been undertaken (Lopez-Portillo, Eguiarte and Montana, 1993). In *Ipomopsis aggregata*, Pleasants (1981) shows that significantly different levels of nectar secretion occur between individuals, but Lanza *et al.* (1995) failed to show such differences in populations of *Impatiens capensis*. Only in *Trifolium pratense* has it been demonstrated that differences are in fact heritable and can therefore respond to natural selection (Hawkins, 1971). In 1995, co-workers and I found that only about 15% of individuals of the orchid *Dactylorhiza purpurella* (in a genus said to be nectarless) consistently lacked nectar. We also noted variation in the average level of nectar produced in the foxglove, *Digitalis purpurea*.

For such variability, one would expect to find an evolutionarily stable proportion of individuals in a population producing little if any nectar. These individuals should be fitter, being able to redirect resource to male and/or female function. The frequency of such 'nectarless' individuals should be governed by frequency-dependent selection. At low frequency, nectarless plants should be visited as commonly as are nectariferous individuals. As nectarless plants become more common, pollinators would be more likely to detect their mimicry, and the nectarless gene would be rendered less fit.

However, when the species itself occurs at low density, and resembles a common model of another species, deceit mechanisms may succeed to the point at which nectar production is lost for the entire species (p. 114). In these cases, which are particularly common in the Orchidaceae, display in terms of flower size, flower number and length of flowering display are strongly selected for in *Anacamptis* and in two *Orchis* species (Fritz and Nilsson, 1996) (we have found the same for two British *Dactylorhiza* species). Interestingly, plants at low density are more effectively visited than those at high density, just as mimicry theory would predict.

These phenomena should prove to be fertile fields for future studies.

DEVELOPMENT OF THE ANDROECIUM AND GYNOECIUM

The androecium (stamens)

Pollen not only forms an attractant and reward to flower visitors, but the dispersal of pollen from stamens to stigmas is in fact the reason why flowers attract visitors at all. Clearly, there is a potential antagonism between these two functions. Primitive 'pollen flowers' which lack nectar tend to overcome this antagonism expensively, by producing far more pollen than is solely needed for fertilization of all the ovules. In practice however, many primitive flower visitors ('insect riff-raff') indulge in rather inaccurate flower-visiting procedures ('mess and soil pollination') and the niceties of efficient pollen collection and grooming are beyond them. Indeed, many of these unsophisticated microfauna visit flowers for reasons other than those of pollen-eating ('rendezvous attraction').

However, bees are quite a different matter, and bees that visit efficient flowers with low pollen:ovule ratios for the purposes of pollen collection may transfer very little pollen to other flowers, as they groom themselves very efficiently. Various floral techniques have evolved to overcome this problem, for instance the sequential opening of anthers which is very common in long-flowering winter flowers such as *Viburnum farreri* and *Jasminum nudiflorum*. Pollen 'packaging' and 'dispensing' schedules such as these will be very important as determinants of male function (p. 37). They should determine crucially the proportion of pollen produced which is removed by pollinators from anthers (estimate of male fitness, p. 25), and the number of different visitors which remove this pollen. The latter statistic is considered by Lloyd (1984) as the single most important determinant in male mating success (fitness). The whole topic of pollen packaging and dispersal syndromes is reviewed by Harder and Thomson (1989).

A division of labour between stamens is also often observed. This division of labour is most often achieved between pollen-bearing anthers which are cryptic (as in *Commelina coelestis*), and conspicuous staminodes, which lack much pollen, or by conspicuous tufts of hairs on the filaments or connective which mimic productive stamens. Alternatively, two types of stamen may occur, one of which is conspicuous and extruded and is designed to reward the visitor, while the other more cryptic set is ignored by the visitor which brushes against them while feeding (*Lagerstroemia indica* and many Scrophulariaceae, e.g. *Verbascum*). In extreme cases, 'feeding stamens' may provide no pollen as such, but provide another type of food reward to the visitor (*Calycanthus occidentalis*, *Tibouchinia*). Vogel (1978) provides a useful review of such changes in anther function.

Dichogamy may also cause functional antagonism in strongly protandrous pollen flowers. During the later female phase of the flower, anthers will have shed all their pollen, and may no longer be attractive to insects. In response, the familiar African violet, *Saintpaulia* maintains attractive yellow anthers throughout the female phase, long after the pollen has been shed, thus attracting pollinators by deceit. Pollen flowers are almost never protogynous, as they would not be attractive to pollinators during the initial female phase.

Anther size varies greatly between different species, but most of this variation probably responds to constraints on the number of pollen grains produced. The topics of male fitness and pollen/ovule ratios are more fully discussed in Chapter 2. Pollen-reward flowers which lack nectar and are monoecious or dioecious present a particular problem. Schemske, Agren and Le Corff (1996) have shown that female *Begonia* flowers, which present no reward, nevertheless receive adequate visit rates to optimize seed production, although they are smaller than male flowers. In this case, female flower numbers should be lower than male flower numbers if the deceit is likely to be successful.

In primitive pollen flowers, pollen is usually dry, and is often released simultaneously from many anthers, extrorsely, by the feeding action of the visitor. In more specialized nectar flowers, pollen is more often sticky and adheres to the visitor as it brushes past the anthers; in these cases pollen dehiscence is more often introrse, towards the centre of the flower, thus mediating pollinator contact. Anther dehiscence may take place sequentially over a period of days, or even weeks, thus enhancing male fitness (p. 26). In some flowers with corolla tubes constricted distally to the anthers, anther dehiscence is apical and poricidal, onto the head of the long-tongued visitor (Ericaceae, *Solanum*). Pollen release from such anthers (and some others such as *Rubus*) is mediated by vibration from the buzzing or hovering visitor. Such vibration-released pollen is dry. Dry pollen may also be released explosively, as in the sprung-trap flowers of many Leguminosae, a puff of released pollen being clearly visible as the striving visitor suddenly releases the stamens from the keel (see Fig. 4.8).

In brush, trumpet and some tube flowers adapted to hovering visitors (hummingbirds, sphingids, some bees and syrphids), stamens are long-extruded from the flower on slender, but strong, filaments (Figs. 4.19, 4.20 and Plate 2). In contrast, gullet and some bell flowers usually have included stamens (Figs. 4.4 and 4.5), and in these the filaments are frequently fused to the wall of the corolla, so that the anthers appear to arise from the corolla itself. In narrow bell flowers (Campanulaceae, Ericaceae, Dipsacaceae), the filaments may be fused to each other to form a staminal tube with an apical whorl of anthers between which the pollinator forces itself, or its mouthparts. This system reaches its evolutionary climax in the Asteraceae with its heads of reduced tube florets each containing a staminal tube surrounding the single style (Figs. 4.34 and 4.36).

Staminal filaments may also become fused to the gynoecium, so that the stamens are closely attached to the ovary or style, often adjacent to the stigma. Thus, potential antagonisms between the separate functions of pollen release and pollen reception have been overcome by combining both functions on the same organ (gynostegium); the style replaces the filament as the anther-presenting organ. In many Asteraceae there is a further step, in that the style and stigma actually present the pollen to the visitor. Initial self-pollination is avoided by protandry (the stigma lobes do not open and recoil to achieve self-pollination until a later stage; Fig. 4.36).

In the gynostegial Asclepiadaceae and Orchidaceae, both dichogamy and herkogamy prevail. For instance, in many orchids pollinium release occurs before the stigmatic cavity is receptive (protandry), although this is not true of orchids with a trap mechanism (e.g. *Cypripedium*, *Coryanthes*). In addition, the stigmatic cavity is usually hidden with respect to the pollinia (herkogamy) in such a way that within-flower pollination is impossible. Only in rare cases can within-flower pollination occur in this vast group. This becomes possible either through the loss of the intervening viscidium in the primitive *Epipactis* (Fig. 4.37), or through the possession of unusually long caudicles to the pollinia which droop onto the stigma (*Ophrys apifera*; Fig. 4.38) (Chapter 9). In any case, herkogamy prevails, despite the gynostegium. That is, although the anthers are borne on the column (= modified style), they are spatially separated from the stigma.

Fig. 4.36 Disk-florets of the golden rod, *Solidago virgaurea*, family Compositae (Asteraceae). Each capitulum bears many reduced flowers (florets) which are tube shaped and have five small lobes, although the marginal ray-florets (Fig. 4.34) form a strap-shaped ligule. The five stamens in each floret are fused into a tube, through which the style with closed pairs of stigma arms passes. The outside of the stigma arms forms the pollen-presenting organ. Later the stigma arms are recoiled and self-pollination can occur. Photo by M. C. F. Proctor.

The Orchidaceae are indeed a remarkable case which demands further description. Replacement of the filament by the style as the anther-presenter has allowed a considerable reduction and specialization to occur in the anther. In the primitive Lady's slipper orchids (*Cypripedium*) and their relatives, each flower has two anthers that still release particulate pollen. However, in most orchids a single anther remains which forms two, and in some cases only one, pollinia from its two locules. These pollinia have sticky stalks (caudicles) which may be joined together, but are more usually dispersed separately. These caudicles have been formed from an ancestral connective. The filament is entirely missing, and the pollinia merely consist of indehiscent membranous sacs full of blocks of pollen nuclei (massulae) still in the tetrad state, and without pollen grain walls. Each pollinium can contain between 500 and 5000

Fig. 4.37 (a) Flower of the outbreeding *Epipactis helleborine*, a common European orchid. The dark shiny hypochile contains nectar and attracts wasps (*Vespa* spp.). Immediately above this is a white knob, the rostellum, bearing a sticky viscidium on which rest the two pollinia which drop out of the anther. Within-flower pollination cannot take place, at least in the absence of a wasp visit, and the pollinia are removed by means of the sticky viscidium. (b) Flower of a selfing *Epipactis* in which the viscidium is absent and the rostellum rapidly withers, allowing the pollinia to fall directly onto the stigma and self-pollinate.

nuclei, and this becomes the pollination unit, being dispersed on the pollinator whole.

Pollinia may vary remarkably in form, even in a single tribe such as the Oncidineae (Fig. 4.39). Dressler (1980) gives a splendid account of pollinium form and function. Pollinium form may depend on the shape and behaviour of the pollinator. As many as 18 different positions for normal pollinium attachment on an insect have been enumerated; these vary from a butterfly's proboscis (it may not be able to recoil the proboscis as a result) to the underside of the abdomen of a bee. Pollinium shape will depend on the position of its attachment on the pollinator, but also on the position and shape of the pollinium-receiving stigmatic cavity on the column, to which it is coadapted. In many orchids, as Darwin (1862) observed for *Orchis mascula*, the pollinia attach to the bee in an outward-pointing posture (Fig. 4.40), so that when the bee visits other flowers in the same inflorescence, or even colony, it cannot pollinate other flowers. Only when the bee undertakes a lengthy flight does the caudicle dry out, and

Fig. 4.38 In the European bee orchid, *Ophrys apifera*, pollinia drop from the anther on long sticky caudicles, allowing automatic self-pollination in the absence of an insect visit. The anther is within the hood at the top of the flower, and the caudicles and pollinia can be seen below this. Most *Ophrys* species can only be pollinated by pseudocopulatory solitary wasps (see Fig. 4.14) (×2).

the pollinium bends forward to the direction in which the bee is flying. It is then in a position to pollinate subsequently visited flowers.

All orchids are relatively specialized flowers. The invention of the gynostegium has encouraged the evolution of the pollinium, and this results in an 'all or nothing' approach to pollination. Roughly equal numbers of ovules occur in an ovary as pollen nuclei occur in a pollinium. Thus, on the relatively rare occasions that a successful pollination takes place, very large numbers of seeds are fertilized, but these are, by constraints of space, very small and reduced.

Although these seeds can travel long distances by wind (at least for terrestrial orchids; sheltered epiphytic orchids show much more limited patterns of distribution and more local endemism), the safe sites for seed

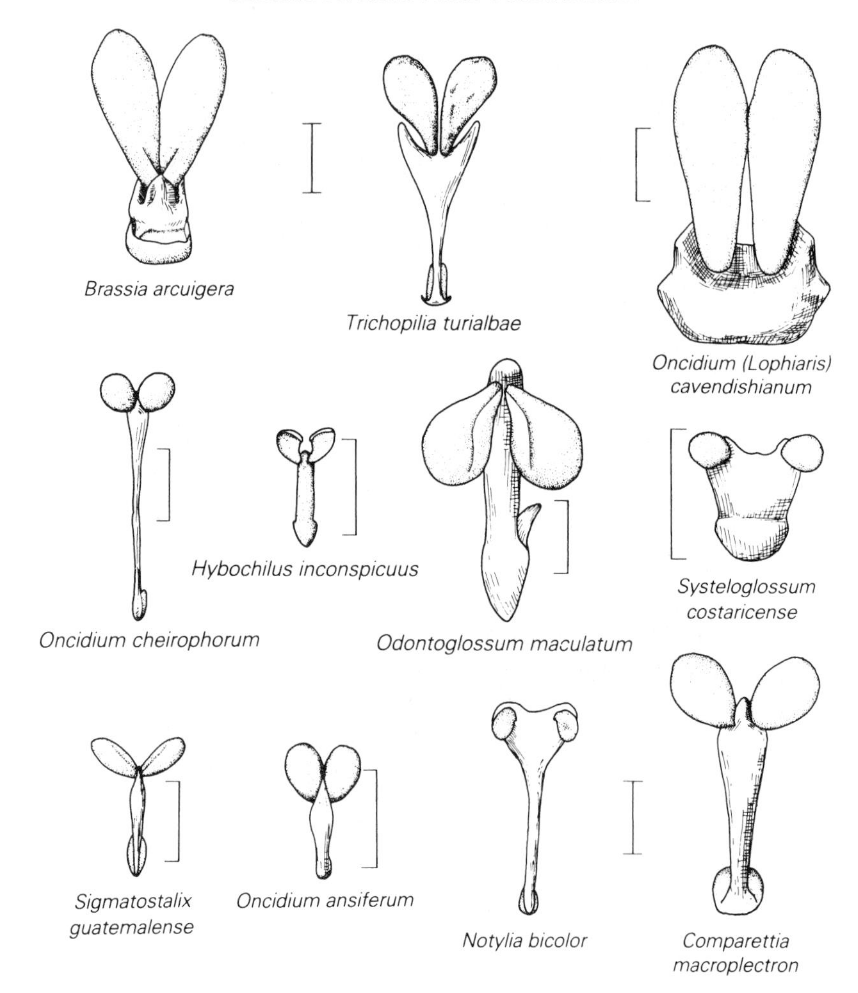

Fig. 4.39 Diversity of pollinium and anther shape among members of the orchid tribe Oncidineae. Scale bar = 7 mm. Reproduced by kind permission of R. L. Dressler.

establishment are few, and most need to form a symbiotic association with a fungus before development can occur.

The gynostegium also requires very accurate patterns of behaviour by the pollinator if pollen donation and pollen receipt are to be effective. Target areas are very small, although they may be increased by columnar wings in the small-flowered, pendulous, wind-blown, *Centris*-attacked *Oncidium* (Dodson and Frymire 1961). Thus orchids tend to have very specialized, complex, pollinator-specific flowers to aid this accuracy.

Curiously, relatively few orchids offer rewards such as nectar; their complex modes of attraction usually involve deceit, as is the case in many

Fig. 4.40 Pollinia removed from flower of the European common spotted orchid, *Dactylorhiza fuchsii*, are initially pointing in an outward direction and rarely achieve pollination. Only after the caudicle dries out, often after an insect flight, does the pollinium bend forward into a position in which pollination is likely. This mechanism will discourage geitonogamy and encourage xenogamy. Photo by M. C. F. Proctor (×3).

highly oligophilic (pollinator-specific) species (p. 76). For some reason, orchid pollinia seem not to be eaten by flower visitors, so orchid flowers are not attractive to primitive pollen-eating flower visitors. By inventing the pollinium, orchids have avoided the antagonism between pollen reward and pollen dispersal.

There is an interesting parallel in the dicotyledonous Asclepiadaceae, which has also evolved a gynostegium, and this provides a test for the theory of the role played by the gynostegium in allowing the evolution of the other remarkable features of the orchids. Despite the very distant relationship of the two families, many features are indeed common to both. Both have pollinia, which can be remarkably similar in form in the two families, and both have groups, which have been divided into subfamilies, with one or two pollinia, respectively. Both have dichogamous and herkogamous stigmatic cavities, and both produce, in most cases, many small seeds, although those in the orchids are more numerous and more reduced. The Asclepiadaceae do not, as far as it is known, have a symbiotic requirement for seedling establishment, although this is a matter that may deserve further investigation.

Both families also tend to have specialized pollination mechanisms, although one such is unique to the Asclepiadaceae (tribe Ceropegieae, including the succulent stapeliads) in which the legs or proboscis of the visitor are trapped in elastic anther appendages. The struggle for release leads to pollinium attachment. However, although the variety of form of the plant (from lianes such as *Hoya* to succulents such as *Stapelia*) and of the flower found in the asclepiads almost rivals that of the orchids, the range of pollinators and pollination mechanisms is much more limited. Most asclepiads attract flies by deceit (sapromyophily), although they use a variety of mechanisms to do so. It is likely that the regular, actinomorphic flowers of the asclepiads preclude the attraction of many more sophisticated visitors such as bees. Such visitors prefer landing platforms as provided by zygomorphic flowers like the orchids. It is interesting, however, that deceit plays a major role in the pollination syndromes of both families.

Pollen as food

Pollen forms a complete food for many groups of insects, for instance bee broods when predigested by worker bees. It contains proteins (16–30% by weight); carbohydrates, usually in the form of starch, and lipids. In general, larger pollen which is often wind-dispersed is starch-rich, whereas smaller pollen which is usually insect-dispersed is lipid-rich (Baker and Baker, 1983; Petanidou and Vokou, 1990). Weight for weight, lipid-rich pollen tends to have a higher energetic content than does starch-rich pollen. Pollen usually has a higher energy investment per gram of organic tissue than do other plant parts (Siafaca, Adamandiadou and Margaris, 1980), although this is not the case for the wind-pollinated gymnosperm *Cupressus sempervirens*.

Although the honey industry has inspired a good deal of work on the energetics of nectar feeding (pp. 116, 149), there is remarkably little published information on the energetic content of pollen. Colin and Jones (1980) and Petanidou and Vokou (1990) present the two main studies, both on mediterranean ecosystems, in California and Greece, respectively. Such ecosystems typically have high diversities and densities of pollen-eating flower visitors.

Petanidou and Vokou (1990) show that animal-dispersed pollen has a higher calorific content than does wind-dispersed pollen, and is more often lipid-rich. They also provide results which suggest that plants with the most rewarding pollen tend to be visited by a wider range of pollinators (polyphilic) than are those with less rewarding pollen. However, they do not find average differences in the energetic content of pollen between taxonomic groups, or between plants flowering at different times of year. Nevertheless, their results suggest that plants adapt the energetic and chemical content of their pollen to their pollination syndrome. Colin and Jones (1980) do not find such differences, but their study only included a few insect-visited species.

Wind- and water-pollinated species

Naturally, the requirements for pollen release and dispersal differ significantly between zoophilous and abiotic forms of pollination. The stamens of wind-pollinated flowers are typically long-exserted, having long, slender filaments which may be pendulous (Fig. 4.41). They have large anthers containing much dry, rather smoothly ornamented pollen, which dehisce

Fig. 4.41 The common sedge, *Carex nigra*. This has monoecious inflorescences and is wind pollinated. There are three male spikes (top), the uppermost of which has shed pollen, and the lower two have yet to do so. The two female spikes have receptive stigmas, so this species is homogamous. Photo by G. Chaytor (×1.5).

readily, especially in warm, humid conditions (which therefore result in the greatest suffering to those with pollen allergies).

Pollination by water has already been discussed in this chapter; hydrophily has required the evolution of some distinctive male characteristics, such as anthers that will dehisce when submerged, floating pollen, pollen without walls, and even floating male flowers.

The gynoecium (carpels or ovary)

The gynoecium shows a variety of adaptations to forms of pollination as does the androecium, but in addition gynoecial adaptations may be directed towards fruit and seed dispersal. The very different functions of pollen reception and fruit dispersal may in some instances be apparently antagonistic, as has been discussed in van der Pijl (1978), or they may be coadaptive. Once again using the orchids as an example, the very large number of highly reduced ovules has clearly originated in response to the evolution of the pollinium. Correspondingly, the very small wind-borne seeds have allowed the orchids to acquire several distinctive ecological functions (epiphyty, saprophyty) as well as an unusual capability for dispersal.

The adaptive radiation of fruit and seed morphology in the Angiosperms scarcely merits a full discussion in this book. More relevant is variation in the position of the ovary in the Angiosperm flower and the evolution and diversificiation of the style, and of the stigma. Although the primitive, Magnoliid condition is for the carpels to be free, most Angiosperms (outside the Magnoliales and some Ranales) have carpels fused into an ovary. This development probably accompanied the evolution of a style, and of localized stigmata. Flowers with free carpels, as in *Magnolia* (see Fig. 4.1), *Paeonia* or *Helleborus*, generally have a diffuse stigma and a very short style, or no style, on each carpel. Although multistyled fused ovaries do occur, as in the Caryophyllaceae, they are not widespread, and are typical only of unspecialized flowers. Carpellary fusion was most usually followed by stylar fusion leading to localized stigmata typical of specialized flowers.

An evolutionary trend is also discernible in the position of the ovary with respect to the remainder of the flower. In the Magnoliales, Ranales, Nympheales, Liliales and other putatively primitive groups which tend to have unspecialized flowers (as well as anemophilous groups such as the Cyperales, Poales and Fagales), the ovary is superior (hypogynous) with the stamens and perianth inserted at its base. Such a syndrome may have some benefits to the plant, especially with respect to gynoecial temperatures within solar-furnace bowl flowers in cold climates. However, a superior ovary may well be more liable to predation, being obviously presented, and it may interfere with the function of complex specialized flowers with tubes or traps. (However, the predominantly gullet-flowered, bee-pollinated Labiatae and Scrophulariaceae, and the flag-flowered Leguminosae have superior ovaries.) Nevertheless, it is significant that a parallel trend from superior ovaries to inferior ovaries occurs in both the monocotyledons (e.g. Liliaceae to Orchidaceae) and the dicotyledons (e.g. Ranunculaceae to Asteraceae), and in both classes this trend tends to be associated with increasing specialization in flower form.

Stigma number often reflects phylogenetic constraints, usually depending on the number of fused carpels, that is the number of segments in the ovary. Thus, the Geraniaceae have five stigmas, the Liliaceae have three stigmas, the Cruciferae have two stigmas (or sometimes one), and the Poaceae have a single stigma. Sometimes stigmas are more numerous than are carpels, as in the Orchidaceae which have two stigmatic cavities, but a single ovary. In families with single-seeded fruits (achenes), the Cyperaceae have three or two stigmas, and the Asteraceae have two stigmas. In other cases (e.g. Ericaceae), the number of stigmas (one) may be fewer than the number of segments in the ovary (five).

Stigmas and styles show rather less development in form than do stamens with respect to pollination biology. In anemophilous plants, stigmas are typically exerted, long and feathery (Fig. 4.41), often being pinnate in shape. In entomophilous flowers they may vary very much in size, in relation to the rest of the flower, tending to be larger in unspecialized flowers. Flowers with efficient pollination will tend to have sparse pollen and small stigmas. The nature of the stigmatic papillae will depend on the incompatibility system, being 'wet' or 'dry' (Chapter 2). Even in very complex and specialized inflorescences such as the traps of *Arum* or *Aristolochia*, the brood flowers of *Yucca* or *Silene*, or the enclosed inflorescences of *Ficus*, the form of the stigma does not differ a great deal. Only in the gynostegial flowers of the Orchidaceae and Asclepiadaceae does it change radically to form a cavity in the column (style) to receive the pollinium. Unlike free pollen, the pollinium is too large to lodge successfully on an exposed stigma.

Dulberger, Smith and Bawa (1994) describe caesalpinaceous shrubs in the genera *Cassia*, *Senna* and *Chamaecrista* as having stigmas with a conventional appearence save for a small apically placed cavity within which most of the papillae are hidden. It is suggested that in some species at least, a stigmatic 'drop' secretion is hidden within the cavity, to be extruded only by the vibration (at a certain, bee-specific frequency) of the visitor's wings. When the visitor disappears, the drop, together with newly received pollen, draws back into the cavity again. It is suggested that such a mechanism may have evolved as a device to protect pollen on stigmas from being washed off by rain in the wet tropics.

In most cases, the pollen is captured by the stigmatic papillae themselves. However, Nyman (1993) shows that in *Campanula* the style possesses special pollen-sized cavities, derived from hairs. These collect the pollen from bees which delve deeply in the flower in search of nectar. Later, the recoiling stigmas scavenge pollen from their own style. Similar mechanisms may well operate in the family Asteraceae.

Although style and stigma structure is fairly uniform, the position of styles and stigmas may be highly adaptive with respect to animal pollination. Flowers receiving large, hovering visitors with long tongues (e.g. hawk-moths and hummingbirds) usually have long-exserted styles with smallish capitate stigmas; the styles may be up to 20 cm long (*Datura*, *Hibiscus*). Other bird flowers and flowers pollinated by small mammals such as bats, mice, small marsupials and lemurs, are usually brush flowers with large numbers of stamens, the filaments of which may, together with the tough, single styles, form the main and attractive part of the flower. This is particularly prevalent in Australian Myrtaceae (*Eucalyptus*, *Callistemon*) (Fig. 4.42) and Proteaceae (*Banksia*, *Grevillea*, *Hakea*). A parallel development occurs in the Mimosoideae, notably in *Acacia*, although these are more often visited by bees.

In sprung flowers, as has already been discussed, a rather conventional style is held in tension to hit the visitor when entrance to the flower is forced (Fabaceae, some Fumariaceae). In many more complex trap, gullet, trumpet, flag and tube flowers, the style and stigma are unconventionally positioned (herkogamy), thus maximizing pollen receipt from other flowers, rather than from the same flower.

In *Iris*, styles become petaloid. In the related, but apparently more primitive, *Crocus* the tepals are poorly differentiated, forming a bowl, and there is a tripartite stigma, which may be brilliantly coloured, for instance orange or red, as an attractant (Fig. 4.43). In *Crocus banaticus*, however, the

Fig. 4.42 Inflorescences of the Australian gum *Eucalyptus calophylla* (Myrtaceae). Stamens are many and are coloured yellow, forming the main attractant to pollinators (×0.5).

(a)

(b)

Fig. 4.43 (a) *Crocus chrysanthus* (Iridaceae) illustrating a simple early spring bowl flower with undifferentiated tepals and a tripartite yellow stigma. (b) The more specialized bee-visited *Iris warleyensis* has six perianth segments differentiated into three falls and three narrow erect flags. The styles are petaloid and form the upper half of a gullet-shaped pollination unit; the fall forms the lower half. This flower type has evolved from the simple flower of *Crocus*.

inner tepals are narrow and erect, forming a flag, as are the inner tepals of *Iris*. This tepal differentiation removes the unspecialized bowl shape of the flower. The flag-tepals in *Iris* render each of the three facets of the flower effectively zygomorphic to the pollinator, but do not localize the approach of the pollinator sufficiently for efficient operation as a zygomorphic flower. This problem has been solved brilliantly and uniquely by the enlargement and broadening of the style into a petaloid form, to become the upper half of a tube, the landing platform being formed by the lower, larger tepal (fall) (Fig. 4.43). The flag tepal retains its function as a long-distance attractant, although in some species it has become secondarily reduced. Thus, the six-tepalled flower produces three pollination units (gullets) made of the fall and the style. Each contains a stamen and a stigma, the latter being on the inner side of the top (style) of each tube as a small flap. The visitor generally visits each of these units in turn.

A very remarkable stigma is found in a familiar and very aberrant member of the Liliaceae, *Aspidistra*, in which the flower, rarely seen in cultivation, produces a stigma which uniquely acts as the attractant and the reward. The flowers of *Aspidistra* are produced at the end of underground stems, away from the leaves, at ground level. The six brownish tepals form a bell, the mouth of which is entirely blocked by a very large, fleshy, disc-like stigma (Fig. 4.44), beneath which are hidden the stamens. The stigma is attractive to slugs and isopods (Kato, 1995), which partly eat the stigma, thus gaining access to the stamens below. When the slugs emerge from the flower they will pollinate the remains of the stigma on that or other flowers. A similar mechanism is also found in the quite unrelated genus *Asarum*. Curiously, this remarkable pollination mechanism, which seems to have no parallel, has escaped mention in most standard texts on pollination biology.

DEVELOPMENT OF THE INFLORESCENCE

In *Magnolia*, and other supposedly primitive woody Angiosperms, flowers are borne singly. With the evolution of the herbaceous habit in the Ranales came a tendency for flowers to be aggregated into inflorescences, probably originally as a panicle, as in *Ranunculus aconitifolius*, and later as a raceme, developed from the panicle, as in *Delphinium* and *Aconitum*. Aggregation of the inflorescence had the following consequences with respect to pollinator attraction and gene flow:

1. The primary attractant (colour, scent) could increase in size, whereas the number of pollination units (i.e. flowers) could be increased within a single attraction unit.
2. A greater reward per pollinator flight could be provided, thus tending

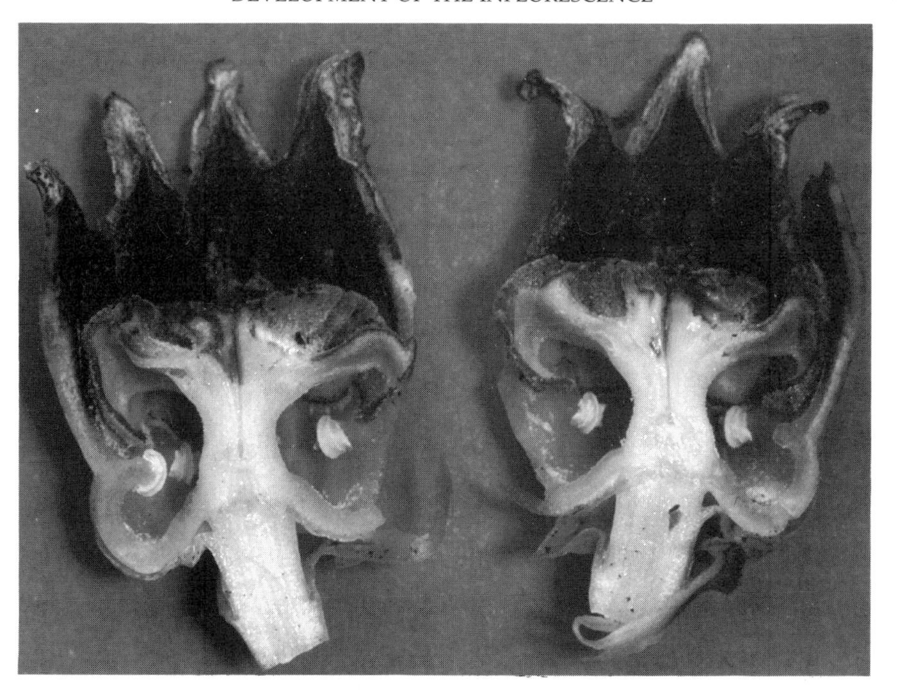

Fig. 4.44 Section of the flower of the tropical *Aspidistra lurida* (Liliaceae). Flowers are inconspicuous, brownish in colour and borne on the ground. A large fleshy stigma which blocks the perianth is attractive to slugs which eat it, revealing the hidden stamens. Pollen sticks to the slug and, if part of the stigma is left uneaten, pollination takes place (×2).

to make specialization by the pollinator on one plant species more worthwhile, and encouraging oligophily, oligotropy and pollination efficiency for both the pollinator and the plant.

3. By separating receptivity of the different flowers in the inflorescence in time, the inflorescence could act as an attraction unit for a longer period; coupled with dichogamy, sequential receptivity of flowers would encourage a mixture of allogamy and geitonogamy, allowing both some outbreeding and acceptable levels of seed-set to occur.
4. Division of labour could evolve between different flowers in the inflorescence with respect to:
   (a) attraction (the large, sterile marginal flowers in the umbel of *Viburnum opulus* or *Hydrangea villosa* (Fig. 4.45) or the terminal sterile florets in *Muscari spreitzenhoferi* (Fig. 4.46) for instance;
   (b) sexual function (the female ray-florets contrasted with the hermaphrodite disk-florets in Asteraceae such as *Senecio* or *Doronicum*; Fig. 4.47) Andromonoecy and gynomonoecy are common in complex inflorescences (p. 337); or

Fig. 4.45 Inflorescences of *Hydrangea villosa* showing division of labour within an umbel. Marginal florets are large, sterile and showy; central florets are relatively inconspicuous, but produce nectar and are fertile (×0.25).

   (c) reward (the nectarless marginal florets of *Scabiosa* in comparison with the nectar-bearing central florets);
   (d) timing (the sequential opening of the flowers in most aggregate inflorescences).

5. Other organs in the floral axis apart from flowers could be used as attractants, or more rarely as rewards or defence mechanisms. Most flowers are subtended by bracts or, in the case of some complex multiple inflorescences, the inflorescence is subtended by a bract, and the flowers by bracteoles. Such bracts may have a wide diversity of function some of which are listed in Table 4.7.

Stems may also act as agents of attraction or reward, by being brilliantly coloured (*Polygonum persicaria*), or by bearing extrafloral nectaries (e.g. *Arbutus*; Fig. 4.13). Even leaves can act as reproductive agents, especially

Fig. 4.46 In the grape hyacinth, *Muscari spreitzenhoferi* from Crete, the upper flowers are sterile but showy, acting as an attractant.

in cases where bracts perform other specialized functions; thus in the familiar house plant 'poinsettia' (*Euphorbia pulcherrima*) the upper leaves are brilliantly red, presumably acting as a primary attractant, while the bracts act as landing platforms, and house the independent nectaries, in the curious division of labour between the floral units typical of that genus.

Leaves can also aid in the dispersal of fruits, as in the grass *Sporobolus*, and other desert species which form balls of fruit which are blown along the ground by the wind. The extraordinary fibrous balls of the maritime grass, *Posidonia* discovered on many Mediterranean shores (for instance Mallorca) may also aid fruit dispersal, either by sea or, when stranded, along beaches by wind.

In some cases, flowers may move with respect to the inflorescence in a functional way. Some species of the epiphytic orchid genus *Oncidium*

Fig. 4.47 Inconspicuous hermaphrodite tubular disc florets, and conspicuous female marginal ligulate ray florets dissected from the head of the Oxford ragwort, *Senecio squalidus*, illustrating division of labour within the capitulum of a member of the Compositae (Asteraceae). See also Fig. 4.34. Photo by M. C. F. Proctor (×5).

have small, solitary, pendant, yellow and brown spotted flowers which are thought to mimic territorial males of *Centris* bees, thus provoking aggression (Dodson and Frymire, 1961) and pollination. G. J. Braem (personal communication) has observed in *O. henekenii* in the Dominican Republic that flowers move from a patent, apparent position to a reflexed, non-apparent position only after receiving pollinia. They do not so move after a visit during which pollinia are donated. Such non-apparency of pollinated flowers, often mediated by a colour change, may well be functional in limiting pollinator visits to receptive flowers. In a frequency-dependent function based on deceits such as this, a limitation of apparency would be particularly valuable.

SPECIAL CASES OF POLLINATION SYNDROME

Brood flowers

One interesting class of pollination syndrome has been scarcely mentioned until this point, that of the total symbiotic relationship seen in brood flowers.

A simple, non-obligate brood-flower syndrome has been recounted by Brantjes (1978) for moths in the subfamily Hadeninae, which lay eggs, usually singly, in the ovaries of flowers from which they have fed on nectar. The moth thus benefits at both the adult and larval stages (the larvae eat some but not all of the developing ovules), and the flower is pollinated and produces some seeds. This occurs in various members of the Liliaceae, Amaryllidaceae and Caryophyllaceae. In the familiar white campion, *Silene latifolia*, the moth *Hadena bicruris* visits flowers of both sexes, but only oviposits on female flowers, which it is apparently able to

Table 4.7 Some functions of bracts in the reproduction of flowering plants

Species	Nature of bract(s)	Function of bract(s)
Ajuga pyramidalis, *A. orientalis*	Large, showy, reddish or purplish	Act as primary attractant to bees, flowers have become small and unshowy
Saxifraga grisebachii	Showy, red, covered with glandular hairs	Act as primary attractant and reward to small flies, etc.
Dracunculus vulgaris, *Arum muscivorus*, etc.	A spathe, enclosing inflorescence, mottled dark purple	Primary attractant to flies by production of foul scent, secondary attractant by provision of visual guide marks, formation of trap by shape and provision of slippery surface
Lamium maculatum, *Coleus* spp.	Large, green with white blotches	Provision of guide marks
Freycinetia insignis	Coloured, fleshy, sugary	Attractant and reward to fruit-eating bats
Eryngium spp.	Stiff, spiny, sometimes coloured bluish or yellowish (Fig. 4.48)	Protection of the inflorescence from grazing; primary attractant
Compositae	Form an involucre around the capitulum, sometimes spiny	Protection of young flowers from extremes of climate and from grazing
Tilia spp.	Adnate to the peduncle, forming a wing	Assists dispersal of the fruit by wind, using 'propeller' action
Cornus spp.	Four white, yellow or red bracts closely subtend reduced inflorescences (Fig. 4.49)	Form secondary bowl-shaped 'solar furnace' type floral unit

differentiate by a quantitative scent difference. In several unpublished studies with students, I have found that the frequency of oviposition by *H. bicruris* on red campion, *Silene dioica*, is dependent on both the size and density of campion plants. Most fruits are predated (and flowers pollinated) when the campion patch is biggest and densest, and the tallest plants, and those with most flowers, suffer the heaviest predation.

A more complex syndrome is observed in American *Yucca*, which is visited by a genus of moth (*Tegeticula*, formerly known as *Pronuba*), with only one species, *T. yuccasella*, pollinating all the eastern species. The female moth gathers pollen which is shaped into a sticky ball and thrust into the stigmatic tube of another flower; at the same time the moth lays one egg in each cell of the ovary. By carefully ensuring pollination, the

Fig. 4.48 The prickly blue bracts of *Eryngium alpinum* (Umbelliferae) protect the inflorescence from grazing (Table 4.7) (×0.33).

Fig. 4.49 The white bracts of the pan-arctic dwarf cornel, *Chamaepericlymenum* (*Cornus*) *suecica* make an inflorescence of small flowers attractive to insects (Table 4.7) (×1).

moth provides ovules for its larvae. One ovule in each *Yucca* carpel next to the egg grows abnormally large and feeds the larva; the remainder develop normally.

The ultimate brood inflorescence is *Ficus*, the figs, a very large genus of trees and shrubs, many of which are ecologically very important in the tropics. The *Ficus* mechanism is too complex to be dealt with in detail here, and the reader is referred to McLean and Ivimey-Cook (1956), Ramirez (1969), Galil and Eisikowitch (1969) or Storey (1975). The essence of the story is that the inflorescence is a closed sphere which may contain male, female and neuter flowers. The edible fig, *Ficus carica*, is gynodioecious and may have three types of inflorescence. One, produced by the hermaphrodite in the winter, produces neuter and a few male flowers ('caprifig'). Female chalcid wasps of the genus *Blastophaga* penetrate the sphere and lay eggs in neuter flowers. The resulting offspring hatch in the spring, the male wasps fertilize the females, and then die. The female wasps escape, and enter a second class of inflorescence, produced in the spring, which has female and neuter flowers. The female flowers are fertilized with pollen carried from the male inflorescence, and the wasps lay eggs in the neuter flowers. The next generation of female fertilized wasps seek out a third type of inflorescence with only neuter flowers where the annual cycle is completed. Although some primitive fig populations ('Smyrna figs') require cross-pollination for fruit to set, and thus must contain some hermaphrodite 'caprifigs', most cultivated figs ('common type') set fruit without pollination (parthenocarpy), and are grown without males unless male caprifigs are required for breeding purposes. Thus, much fig culture is able to persist in the absence of the *Blastophaga* pollinator.

Ant pollination

Most classes of pollinators have been mentioned, however briefly, in the foregoing chapter. However, ants form a special case. Ants are among the most numerous and widespread animals on earth, and they forage widely, not least on inflorescences. Thus, even though most ants do not fly, it might be supposed that they would form an important class of pollinators.

Beattie (reviews in 1985, 1991) has made a special study of ants as pollinators. In fact, very few plants seem to have become adapted to pollination by ants. Indeed, many produce extrafloral nectaries which appear to distract ants from visiting the flowers themselves (p. 85). Beattie has shown that many ants produce an antibiotic, known as myrmicacin, most frequently from metapleural glands. This antibiotic, which seems to protect the ant from attack by fungal spores, also has the effect of preventing pollen (spore) germination. In a few cases, for instance the Australian orchids *Microtis parviflora* and *Leporella fimbriata*, flowers are only visited by ants which do not appear to be secreting antibiotic. However, it is noteworthy that in orchids such as these, many grains in the pollinium would not in any case come into direct contact with the antibiotic.

Altogether, Beattie (1982) lists some 12 species of plant which are thought to be primarily pollinated by ants. Many are dwarf, with inconspicuous flowers borne at ground level, which however produce a concentrated, lipid- and sterol-rich nectar. It is not yet clear how many of these avoid the sterilizing effect of the ant antibiotic.

In contrast, ants are prolific and important dispersers of plant seeds, many of which are specifically adapated to ant carry (Chapter 5).

Other classes of pollinator

Natural history films enjoy showing some of the more arcane pollination systems, and the recent (1995) showing of David Attenborough's 'The secret life of plants' (BBC) has given examples of several of these.

For instance black lemurs are the only Madagascan animal strong enough to part the basal bracts, allowing it to feed on the flowers, of the magnificent native palm *Ravenala madagascariensis*. A South African *Protea* has hidden inflorescences near to the ground, and these are visited solely by a local mouse. This system has an interesting parallel in western Australia where the inflorescences of the proteaceous *Banksia repens*, also borne on the ground, are visited solely by a mouse-like marsupial, the dibbler.

Attenborough even shows the giant gecko pollinating the flowers of the New Zealand tree with red brush-flowers, *Metasideros robusta*, although these are in fact usually bird-flowers.

Slugs and water-skaters as pollinators have been mentioned already in this book, but for me the most remarkable story I have encountered recently comes from my own north England county. Bowey (1995) has frequently observed that moorhens, a water bird the size of a small chicken, climb the stems of reed-mace (*Typha latifolia*) to feed off the pollen. In doing so, they cover themselves with pollen, and undoubtedly cause cross-pollination in this plant, which must presumably have adapted primarily to an anemophilous pollination system.

CONCLUSION: POLLINATION AND THE ECOSYSTEM

With respect to pollination ecology, habitats are heterogeneous for many variables, for instance:

- temperature means and ranges at various times of year;
- rainfall and sunshine hours at various times of year;
- average height of vegetation;
- herbivory and flower predation (Hendrix, 1988);
- diversity, density and pattern of other plant species (Weiner, 1988);

- number of species of potential flower visitors;
- number of individuals of flower visitors.

For any one species of flowering plant, a suite of coadaptive flower characteristics has evolved to fit multidimensional niches within these variables (Waser, 1983; Waller, 1988). Where variation within a species exists, for instance in flower colour or nectar production, it should be maintained by balanced selection forces within an evolutionarily stable strategy (ESS).

Each strategy coexists with very different levels of reproductive efficiency. A small annual on an anthill in a closed grassland is likely to flower early, be dwarf, have unspecialized, mostly autogamous flowers, and have large seeds with poor dispersal. This syndrome will favour an 'r' type (Chapter 2) reproductive efficiency, and more than 50% of its total energy budget may be expended on its single reproductive effort.

In contrast, a neighbouring perennial orchid in closed grassland may flower in mid-summer, be taller, attract and reward only one species of pollinator, which is scarce in the habitat, and which alone can effectively mediate pollination. Thus, this orchid may rarely set seed. However, when seeds are set, they are very small and produced in very large quantities. Such a species may spend much less than 10% of its total annual energy budget on seed reproduction. It has a 'K' type reproductive efficiency, in which much more of its energy is spent on perennation and vegetative multiplication.

How many different types of reproductive strategy can coexist in a habitat will, to a great extent, depend on the total biomass productivity of the habitat. Thus, on an Arctic tundra, the flowering season is short and time niches few; exposure to wind and cold renders all plants dwarf, so height niches are almost absent; pollinators are species-poor and few in number, so most flower types are generalist. Conversely, in the very productive tropics, flowering can occur over 12 months of the year, trees may exceed 80 m in height, the variety and the total quantity of potential flower visitors is vast. Clearly, there are very many more reproductive niches in the tropical forest, and it is no accident that tropical habitats harbour many of the more spectacular and arcane pollination syndromes. However, an absence of time-niche signals, and the filtering effect of surrounding vegetation render wind-pollination unusual in the tropical forest.

Unfortunately, we have very little data to support such energetically based models, and, in particular, our knowledge of the allocation of energy budgets in different plants, and in different habitats, is almost nil. Also, we have very few studies that compare different habitat types with respect to their pollination syndromes. That of Moldenke (1975) examines oligophily to polyphily in plants, and oligotropy to polytropy in pollinators in a wide range of Californian habitats over a considerable

Table 4.8 Pollination syndromes and the rôle of pollination in speciation

Pollination syndrome	Specialist successful family	Rôle of pollination in speciation	Number of ovules per flower
Anemophily	Gramineae	None	1
Generalist zoophily	Compositae	Low	1
Specialist zoophily	Orchidaceae	High	many

altitudinal transect. He found that the spectrum of flower visiting pattern and pollinator receipt pattern varied rather little over a wide range of habitats. For instance, most visitors were moderately oligotropic, visiting from two to five plant species, and extreme monotropy (one host only) and polytropy (more than ten hosts) were surprisingly rare. Extreme monotropy and extreme polytropy were surprisingly more frequent in high-altitude tundra vegetation.

In two south American studies involving altitudinal transects, Cruden (1972) and Arroyo, Primack and Armesto (1982) show changes in pollinator guilds with increasing altitude. In tropical regions, birds become more important pollinators than do bees at high altitude, perhaps because they are less inhibited by cloudy or rainy conditions. Further south, in the temperate zone of Chile, bees once again become less important as pollinators in the alpine zone, but here they are replaced by butterflies and flies. These flower visitors are spared the constraints and energetic loads involved in brood feeding, and are more likely to wander into the alpine zone.

Proctor (1978) divides plants into three broad types of pollination syndrome which he supposes evolved at an early stage in Angiosperm evolution, and between which types plants have been changing ever since. However, he points out that in some important plant families, adaptations have apparently reached a dead end, precluding further between-type evolution. This is of no apparent disadvantage, as almost every habitat will provide different niches (reproductive strategy ESS) covering between them all three of Proctor's classes (Table 4.8).

Proctor's field survey is made of a number of phytosociological communities on the limestone region of the Burren, western Ireland. Like Moldenke, he finds a wide and not dissimilar spectrum of pollination types is typical of most communities: that is, each bears a diversity of reproductive strategy ESS, and it is typical of no community of plants to specialize in only one reproductive strategy, or pollination syndrome. It is interesting that the more productive, more stable communities have slightly higher proportions of specialist flowers; this might well be predicted as they will have more reproductive niches available.

CHAPTER 5

Pollination biology and gene flow

INTRODUCTION

In the last chapter, no attempt was made to describe how animals mediate patterns of pollen flow within and between plants, nor how efficient various pollination mechanisms are in the dispersal of pollen and in achieving seed-set. The behaviour of pollinators strongly influences pollen flow, and the ways that plants have adapted to these behavioural characteristics play an overriding role in gene dispersal, and the genetic structure of plant populations.

In contrast, abiotic pollination by wind and water is relatively unresponsive to subtle adaptation, and patterns of gene flow in wind-pollinated species depend to a great extent on chance. Some consideration will, however, be given to the distances over which pollen can disperse by wind.

In order to understand the coadaptation of zoophilous plants to pollinators with respect to gene flow, it is first necessary to give an account of pollinator behaviour and foraging strategies of flower visitors. Subsequent sections of this chapter will examine pollen travel within and between plants, the efficiency of pollen travel, neighbourhood, interruptions to pollen flow and the role of ethological isolation in plant speciation.

FORAGING THEORY

The foraging strategy of pollinators was originally modelled by Pyke (1978a,b,c, 1979, 1980a,b, 1982a), Pyke, Pulliam and Charnov (1977), Krebs (1978) and Waddington and Heinrich (1981) and is based on food-gathering and predator–prey models devised by Royama (1971), Lawton (1973), Schoener (1969) and others. There is a useful review by Waddington (1983). Flower visitors gather food at the first trophic level, which is the most energetically efficient. Data on the energetic costs involved in flower foraging have been published by Wolf and Hainsworth (1971), Wolf, Hainsworth and Stiles (1972), Stiles (1975), Heinrich (1972a,b, 1975, 1979b), Waddington (1981), Waddington and Heinrich

(1981), Waddington, Allen and Heinrich (1981), Best and Bierzychudek (1982), Zimmerman (1979) and Bertsch (1984) among others.

At the most simple level, foraging theory predicts that animals will visit flowers in the most energetically efficient manner. There is, however, a massive assumption built into these predictions. The evolution of behaviour is no different from any other type of evolution in that natural selection can only operate on inherited variation. Undoubtedly, heritable variation for behavioural traits does exist in bees and other pollinators. However, it can never be assumed that this variation will always have allowed a pollinator to evolve a sophisticated series of behavioural traits 'fine-tuned' to a maximum efficiency of energy gathering in an infinitely variable environment. To do so is teleological in the extreme.

Energy costs in foraging will be composed of:

1. energy expenditure per distance travelled;
2. mean distance travelled per flower visit;
3. energy expended in winning a unit of reward from a flower (greater for viscous nectar in long-tubed flowers);
4. energy expenditure in maintaining body temperature, a function of the difference in temperature between ambient and body temperature, of wind speed, and of travel speed;
5. energy costs incurred from the carrying and excretion of extra water incurred from the gathering and metabolism of nectar (Bertsch, 1984).

Energy gains or savings are considered as:

1. energy reward per flower visited (from nectar for suckers, and from pollen or food bodies for chewers);
2. energy savings for small flower visitors made in sheltered warm flowers with respect to heat loss, reduced need to seek mates, evade predators etc.
3. energy savings made by the taking of concentrated nectar so that less water is absorbed.

Brood feeders (bees, birds) may also gain reward for non-feeding relatives, which is taken back to the hive or nest.

FORAGING STRATEGIES

Social bees

Foraging efficiency can be expressed in terms of profit, as a ratio of calories gained to calories expended. It can be predicted that flower visitors will cease to visit a patch of flowers when other more profitable sources exist nearby, and will cease to fly at all when all available sources are no longer profitable, as on cold days early or late in the season.

Some important flower visitors, such as colonial bees, 'learn' the position and characters of a rewarding patch, using information transmitted by experienced members of the hive or nest. Scent collection, and bee dances teach young foraging bees which 'major' (a North American word for graduate) on currently rewarding patches (Heinrich, 1976, 1979c). Waddington (1985) describes in detail how bee dances accurately communicate levels of energetic reward available from artificial flowers.

Worker hive bees are probably only able to remajor onto different sources on a few occasions, which may limit their foraging efficiency. Having majored on a flower type, bees increase their efficiency (rate of feeding) on that type with experience (Heinrich, 1976, 1979c). Within a hive, different individual workers may have majored on different flower types, and thus relearning may involve other workers that have majored on other patches. It is expected that all patches worked from a given hive should yield similar levels of net energy gain to all workers. Patches that drop below this gain threshold due to overworking or cessation of flowering should no longer be worked.

Quality and quantity of reward

As has been suggested by Waddington and Heinrich (1981), colonial bees with workers have a flower-visiting strategy composed of the following actions.

1. A novice bee will sample a number of different flower types, in addition to receiving messages about local rewarding flower patches.
2. From this information, it will 'major' on a rewarding flower type, which will be the most rewarding one encountered in terms of energy received/energy spent. This flower will be common, clumped, in full flower, with good pollen and nectar production, and will be conspicuous and accessible to bees. It may not be the most potentially rewarding species present, which may already be heavily worked by other individuals, thus decreasing its attractiveness.
3. The bee's behaviour pattern with respect to flower visiting will then depend on rewards subsequently received, as detailed in Table 5.1.

Bee behaviour is likely to be important in determining patterns of pollen flow, and hence of gene flow (Zimmerman, 1979; Schmitt, 1980; Waddington, Allen and Heinrich, 1981; Thomson, 1982).

To take one example in detail, Waddington (1981) has investigated the behavioural responses of *Bombus americanus* to differences in nectar volume in wild populations of *Delphinium virescens* (Table 5.2). He showed that nectar volume per flower decreases from the bottom flower of the raceme upwards. This is a common pattern in racemose flowers which is also observed in the foxglove, *Digitalis purpurea* (Best and Bierzychudek, 1982). Visitors generally visit the bottom-most flower first, moving up-

Table 5.1 Foraging strategies amongst workers of colonial bees

Components of foraging strategy	Strategy on major flower			
	Rewards high	Rewards intermittent	One reward poor or absent	Rewards poor
1. Number of flower species visited per trip, and frequency and distribution of visits to each	One species (oligotropic)	One (major) species to few (minor) species, majoring on one	Two to few species, regular alternation between pollen flowers and nectar flowers	Remajor to new species
2. Distance of hive or nest to foraging patch	Near to distant	More distant	Nearer	–
3. Number of inflorescences visited per foraging trip	Few	More	More	Many
4. Distance of flights between inflorescences	Short	Longer	Longer	Long
5. Directionality of flights between inflorescences	Random	Approaching 180°	Variable	Approaching 180°
6. Time spent at a flower	Short	Variable	Short	Long
7. Proportion of flowers in inflorescence visited	High	Lower	High or low	Low
8. Direction of movement within inflorescence	Unidirectional from most rewarding flower, not skipping	More skipping	More skipping	Random
9. Proportion of time spent on inflorescence on feeding	Low	Higher	Variable	High

Table 5.2 Behaviour of bees with respect to nectar volume of the bottom-most flower of a raceme in *Delphinium* and *Digitalis*

	Nectar volume	
Characteristic studied	Large	Small
Quantity of pollen collected per flower visit	Large	Small
Number of flowers visited per inflorescence	Many	Few
Distance travelled between inflorescences	Short	Long

wards, and leaving the spike when rewards per flower drop below the mean flower reward for the population. This is usually at a point above where the protandrous flowers in the raceme are in male phase rather than female phase, so that this visiting pattern maximizes the requirements of both female function (pollen receipt on bottom flowers) and male function (pollen export from upper flowers).

Racemes with large nectar volumes in the bottom-most flowers also have relatively large volumes in other flowers, and there is a strong correlation between the nectar volume of a flower and the amount of available pollen per flower. This probably depends on the age of the flower, the number of visits it has received, and on the physiological vigour of the plant.

Ironically, bee behaviour suggests that fit, well-resourced flowers bearing much nectar and pollen are more likely to be selfed, or to suffer near-relative matings than are poorly resourced flowers.

We can infer from Table 5.2 that inflorescences producing much nectar will receive some geitonogamous pollination due to pollen carryover between flowers of the same inflorescence. At the same time, substantial quantities of pollen will be carried between inflorescences, but most will only travel short distances to neighbouring inflorescences which may be related to the pollen donor. However, Harder and Barrett (1995) point out that the protogynous raceme is a system which is favoured by male sexual selection, as it minimizes geitonogamous 'pollen discounting' for hermaphrodite flowers.

In contrast, inflorescences producing relatively little nectar will donate and receive relatively little pollen. However, geitonogamy will be less common and on average outcrossed pollen will travel further than from flowers with high nectar production.

It is very possible that there is an ESS threshold for the optimal production of nectar and pollen per flower. Flowers which produce too little nectar or pollen would receive too few insect visits, and would donate too little pollen to be successful, whereas those which produce much pollen and nectar would be disadvantageously selfed and suffer near-relative matings. This hypothesis is dependent on variation in nectar and pollen production being heritable (further discussed on p. 121 and 129).

Similarly, ESS thresholds for inflorescence size and flower number may result in a stabilizing selection for those attributes too. Large plants with many flowers may receive many visits from pollinators, but they may also be more susceptible to predation (as we have found for seed predation in *Silene dioica* p. 141) and to mechanical damage. Working with the water hyacinth, *Eichhornia paniculata*, Harder and Barrett (1996) show firstly that *Bombus* bees prefer to visit inflorescences which bear more flowers, and secondly that they visit more flowers on multiflowered inflorescences than on few-flowered inflorescences. These results demonstrate clearly the stabilizing penalties imposed by flower number in racemose inflorescences. Too few flowers, and reproductive assurance suffers; too many and levels of outcrossing suffer. Similar results are reported for *Aconitum*, *Delphinium* and, interestingly, for the non-racemous, single-flowered cushion New Zealand forgetmenot, *Myosotis colensoi* (Snow *et al.*, 1996).

In both *Delphinium virescens* and *Digitalis purpurea*, flowers are protandrous, and open sequentially from the bottom upwards (Fig. 5.1). Thus, the lower flowers are functionally female and the upper ones functionally male. Best and Bierzychudek (1982) predicted that increasing nectar production during the life of a flower should ensure that a bee moves upwards on an inflorescence, and will not usually leave it until it reaches a flower in the male phase. In this way, geitonogamy should be minimized, and pollen will usually be carried to the first flower of the next raceme visited, thus promoting cross-pollination (*Digitalis*, at least, is self-compatible, so that 'fail-safe' geitonogamy mediated between flowers of the same inflorescence will help seed-set in poor flying conditions, or for isolated plants).

Such detailed field studies can be complemented by laboratory work, as Waddington, Allen and Heinrich (1981) have done for *Bombus edwardsii*, using artificial flowers with rewards of sugar solutions of known volume and concentration. This study reached the following conclusions.

1. Better rewards encourage more rapid flower visitation, with more flowers visited, but less time spent at each.
2. Using artificial flower colours as signals, more reliable rewards are preferred to less reliable rewards, even if the total reward from the two types is equal. More reliable rewards are considered to reinforce a behaviour pattern more frequently.

Directionality of flight

Directionality of between-inflorescence flights is an important component of gene travel. Mogford (1974), working with the marsh thistle, *Cirsium*

Fig. 5.1 The racemose inflorescence of the common foxglove, *Digitalis purpurea*. Flowers open sequentially upwards and are protandrous. Usually about 10 are open together, the lower in a female phase and the upper in a male phase. The calorific value of nectar reward increases downwards, so bees visit lower flowers first and work upwards. In doing so, they pass from flowers in a female phase to flowers in a male phase, and so maximize cross-pollination between inflorescences in this self-compatible biennial (×0.5).

palustre, distinguished between sequential and overall directionality. Woodell (1978), who examined the direction of bee flights on salt-marsh populations of thrift (*Armeria maritima*) and sea-lavender (*Limonium vulgare*), considered that a strong sequential directionality in flights was merely a symptom of overall directionality imposed on the bee by climatic and topographical features such as the almost unidirectional winds in the coastal locality in which he worked (Scolt Head Island, Norfolk, UK). When the wind is strong bees prefer to fly, and especially to land, into the wind. Overall directionality might also be imposed by the topography of a patch, or habitat (e.g. road verges, wood margins and cliff edges), by the position of the hive or nest with respect to the patch, or even by the position of the sun.

In contrast, Pyke (1978a) has considered sequential directionality to be a direct product of foraging strategy. He examined sequential flight directions of *Bombus flavifrons* working *Aconitum columbianum*, and *B. appositus* working *Delphinium nelsonii*, and in both cases he found a unimodal sequential directionality, with modal peaks very close to 0° (=180° from the direction of arrival) in each instance. He suggests that such patterns optimize foraging success, as they minimize the chance of a bee revisiting an inflorescence. In this work, overall directionality and environmental constraints were not considered.

Zimmerman (1979), also working with *Bombus flavifrons*, but visiting Jacob's ladder, *Polemonium foliosissimum*, found no directionality in sequential inflorescence visiting, only totally random movements. This he ascribes to the very different pattern of flower presentation in this species. The mean number of flowers per plant was 90, and the mean number of plants per patch 193. Only an average of 3.3% of flowers per patch are visited on each foraging journey, and thus the chance of a bee encountering a flower it has already worked is low. Because the species grows in a highly aggregated manner, directional foraging is inefficient: the visitor would rapidly leave the patch.

Clearly, both sequential unidirectionality (optimal foraging) and overall directionality (environmental constraints) may occur, and in some localities both may operate together. Thus the bee may forage unidirectionally and the environment may determine in which direction this is.

Majoring patterns with time and patch 'switching'

Initial choice of a patch on which to 'major' will involve consideration of the profitability of a patch (modelled by Real, 1983). Patches are only likely to be adopted as others become less profitable, so they may receive few visits early in their flowering period, but will remain popular right up to the time that flowering ceases. Thomson (1982) examined patterns of flower visitation to entomophilous flowers in sub-alpine meadows in the

Rocky Mountains of Colorado, and described five patterns of visitation to a patch with time. The most frequent pattern showed low visitation early in the flowering of a patch, but high visitation right up to the end of flowering. He found no consistent relationship between the degree of overlap in flowering time of species with similar pollinators and visit frequency.

For the monoecious cucumber, *Cucumis sativus*, Handel and Mishkin (1985) showed that, although pollinator visits remain high to the end of the flowering season, fruit-set declines considerably for late-season flowers. This was attributed to competition for resource by earlier set fruits causing fruit-set failure in later flowers. In such a case, it would pay a plant to devote less resource to late season flowers.

For dominant species, an overlap in flowering time may generate an inefficient competition between species which reduces levels of visitation in each. Selection pressures will encourage the evolution of different flowering times among co-occurring dominants. This has also been modelled by Real (1983). Nevertheless, for minority species in the habitat, an overlap in flowering time can enhance the attractiveness of different species with a similar appearence through Müllerian mimicry (p. 106), allowing bees to major on cohorts of coadapted species. However, Thomson (1982) suggests that the evolution of such overlaps may be constrained by a reduced seed-set resulting from stigmatic sites being blocked by ineffective pollen from other species.

Macior (1983), working in sub-alpine North American communities in Montana, throws an interesting light on the effect of the majoring habit of bees on pollination efficiency. He concludes that the more frequent a bee species is in a habitat, the more different species of flower that species of bee will visit. That is, an abundant bee species will need to major on a greater diversity of patch types if it is to exploit the community successfully. Nevertheless, oligophilic flowers, which receive visits from only one or a very few insect species, benefit from just as many visits in total as polyphilic (generalist) flowers that receive visits from many different species. A consequence of the majoring habit is that floral specialization does not disadvantage any plant species.

Within a habitat, there should be an ESS threshold in specialization of insect-flowers between high-visit low-accuracy generalizers and low-visit high-accuracy specializers. In a pollinator-rich community, both types of flower may be able to coexist, as Macior has shown (p. 146).

Flight height

Flight height is one of the many features of foraging behaviour that enable a bee to remain faithful to its majoring patch. Foraging bees fly at a remarkably constant height. A species of plant may adopt an inflorescence height different from those of coexisting plants with similar syndromes. Waddington (1979) shows that species in Colorado meadows that share pollinators are less alike in inflorescence height than those that do not share pollinators. Once again an ESS threshold between too little and too much specialization may operate in spatial niches such as this. The greater the number of pollinators available in the habitat, the greater variety of generalist and specialist types that are likely to be able to coexist (e.g. as different inflorescence heights).

Butterflies and moths (Lepidoptera)

Generalist flowers may receive visits from other classes of pollinators as well as bees. Schmitt (1980) demonstrates the different flight patterns of foraging bees and butterflies, and the very different patterns of pollen flow that result, using three North American ragworts, *Senecio amplectens*, *S. crassulus* and *S. interrigimus* (Fig. 5.2). These are visited by both long-tongued bees and butterflies of many species. To quote Schmitt:

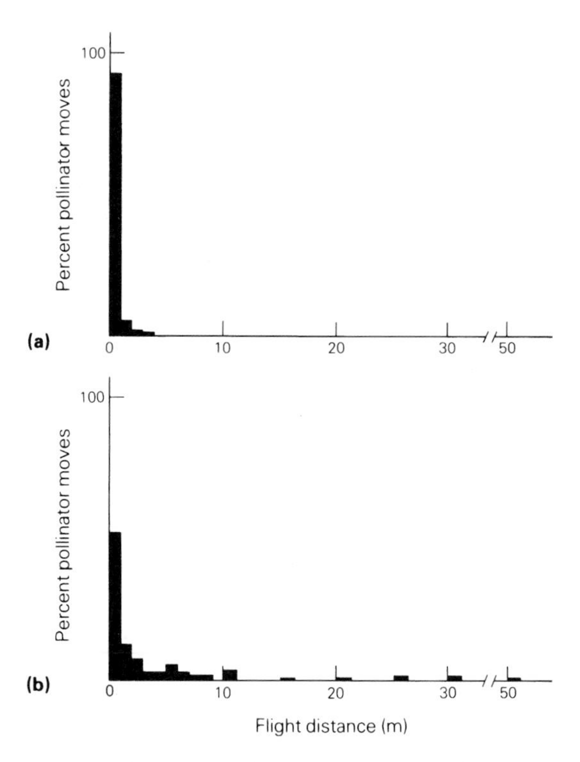

Fig. 5.2 Flight distance between inflorescences of three North American species of *Senecio* made by visiting bees (a) and butterflies (b). (After Schmitt, 1980.)

> the individual fitness of a butterfly depends not only on nectar uptake, but also on mating success, and for females, oviposition success. In a patchy environment, butterfly fitness may be maximized by flying longer interplant distances so as to maximize the probability of encounters with mates, or with larval foodplants. Levin and Kerster (1969a,b) have shown that a strong relationship exists between plant spacing means and bee flight distance means. Beattie (1976) found a similar relationship for solitary bees, but not for lepidopterans, observed foraging on *Viola*. One can therefore expect differences in foraging behaviour on the same flower resource between bumblebees and butterflies that could be reflected in plant neighbourhood characteristics.

Schmitt goes on to show that owing to the irregular behaviour patterns of butterflies, effective population sizes, expressed as neighbourhood sizes, differ by huge factors of between 50 and 500, depending on whether the visitor is a bee or a butterfly. The large variance in between visit distance characteristic of butterflies is due to the fact that butterflies thermoregulate by basking, whereas bees and syrphids require the expenditure of energy for thermal control. Butterflies are less dependent than bees on regular fuelling stops at flowers.

Anyone who, like me, has chased foraging butterflies for purposes of capture or photography, will indeed testify to the maddeningly indecisive, long-distance and essentially unidirectional flight patterns that they show (Baker, 1969). Perhaps for this reason, butterflies have been little studied as foragers and pollen dispersants. Yet they are undoubtedly very important visitors to the flowers of late summer with inflorescence heads of many narrow-tubed flowers (e.g. thistles and knapweeds, *Cirsium*, *Carduus*, *Carlina*, *Centaurea*, scabious, *Scabiosa* and *Knautia*, valerians, *Valeriana* and *Kentranthus*). Although these flower types are also visited by bees, syrphids and beetles, I estimated that ten times as many butterflies as bumblebees were visiting these flowers in a meadow in Auvergne, France in August 1983.

In an intriguing study, Olesen and Warncke (1989) compare the effect of syrphids and the day-flying moth *Zygaena trifolii* in effecting pollen travel in *Saxifraga hirculus*. Here they show that moths also carry pollen further than do syrphids, but for all pollinators, visit patterns were rather aimless. For instance, flight directions were random, and on average travelled as far as the eighth nearest neighbour, although flower revisitations (11% of all visits) were also common. The authors point out that these relatively non-specialized visitors focus less on foraging, but undertake diverse activities such as preening, mate-searching and predator avoidance which are foreign to social bees.

Relatively little is also known about foraging strategies of night-flying moths. Brantjes (1978) recounts how slow, aimless flights are replaced by much more urgent 'seeking flights' ('Nahrungflug') on the detection of odour. The moth then orientates along a scent gradient upwind until it makes visual contact with an inflorescence, from which it feeds. Flights between inflorescences are therefore likely to be relatively distant and random in direction.

In common with other poikilothermic ('cold-blooded') foragers, moths are highly influenced by the ambient temperature, and as they fly at night they will more frequently encounter low temperatures than do day-flying congeners. Cruden *et al.* (1976) have shown that in Mexico and the western USA hawk-moth (sphingid) activity is restricted by temperatures below 15°C at sundown (Fig. 5.3). These authors demonstrated that the proportion of flowers visited by moths varies strikingly with altitude (Fig. 5.4), and that at high cold altitudes seed-set is impaired by poor visit rates. They suggest that the primary determinant for altitudinal limit of moth-visited flowers is moth activity.

At high altitudes, sphingid flowers such as *Aquilegia pubescens* and *Castilleia sessiliflora* in the Californian Sierra Nevada are day-flowering. They are visited by the hawk-moth, *Hyles lineata*, at elevations of over 3500 m. It is suggested that this syndrome has evolved in the absence of humming-birds which would otherwise compete for tube flowers during the day. Being homoiothermic ('warm-blooded'), birds are normally more efficient flower visitors in the uncertain climates of high altitudes.

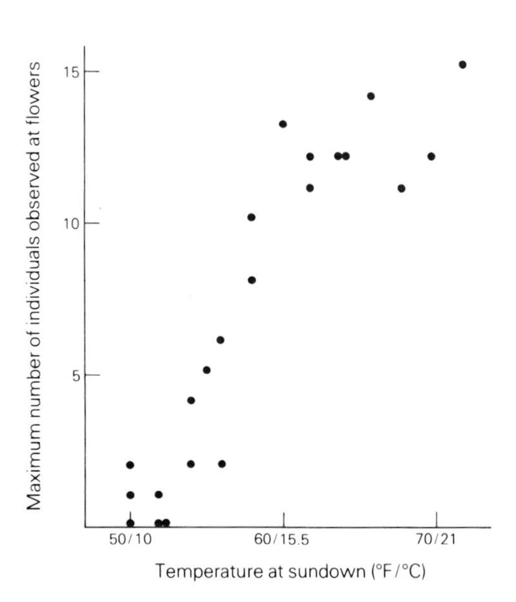

Fig. 5.3 Maximum numbers of hawk-moths (sphingids) foraging at flowers of *Oenothera caespitosa* in the western USA at dusk at different temperatures. (After Cruden *et al.*, 1976.)

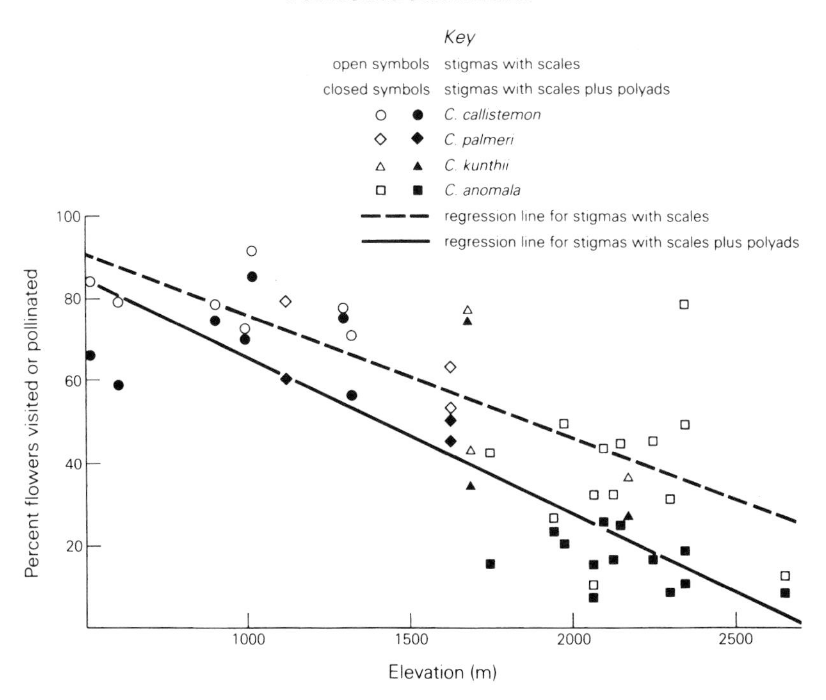

Fig. 5.4 Proportion of flowers visited by moths on plants of various species of *Calliandra* (Leguminosae) at different altitudes in Mexico. (After Cruden *et al.*, 1976.)

Birds

In common with long-tongued bumblebees and lepidopterans, birds prefer to feed from long-tubed flowers (Chapter 4). A longer proboscis, or beak, renders nectar extraction from a single tubed or spurred flower more efficient, but also lengthens the time interval between probes of different flowers (Hainsworth and Wolf, 1972; Inouye, 1980; Pyke, 1981, 1982a). This relationship has been substantiated for *Bombus* foraging on *Delphinium* and *Aconitum*, and for humming-birds feeding on artificial flowers. Field data on the time taken at flowers by humming-birds with different beak lengths is contradictory (Wolf, Hainsworth and Stiles, 1972; Stiles, 1975). Nevertheless, it is likely that long-beaked birds would be best served by specializing on long-tubed flowers, which cannot be exploited by birds with shorter beaks. Short-beaked birds are perforce required to specialize on short-tubed flowers.

There is a well-established relationship between body weight and the calorific reward obtained per flower by humming-birds. Although humming-bird body weight varies tenfold between different species (from 2.7 g to 20 g), stabilizing selection for body weight seems to have occurred, based on the trade-off between energy requirements and aggressive defence of flower patches. Larger species are more likely to succeed in defence of a patch against smaller species of birds or insects, but will need more energy to do so. Plants adapted to large bird species may thus produce more flowers with greater rewards per flower in larger patches to make this defence worthwhile.

Such trends towards oligophily, which are expensive to the flower as well as the bird, may only succeed when increase in resource spent on flower quantity and quality is matched by a reproductive performance superior to that of less expensive species.

Oligotrophy by birds may result in greater quantities of pollen being carried more accurately, and over longer distances than is the case for more frugal flowers visited by smaller and less specialized species. Such expensive specialist strategies are also only likely to succeed when the diversity of pollinators is high. Feinsinger, Wolfe and Swarm (1982) and Feinsinger and Swarm (1982) show that birds on small, species-poor Caribbean islands are more generalist (polytropic), and bird-flowers are on the whole less specialized than for larger islands.

Specialization in bird-flowers may be driven by intraspecific competition between plants for visits. However, Pyke (1982b) studying the Australian warahtah (*Telopea speciosissima*) and Paton (1982) for three other species of Australian bird flowers (Fig. 5.5) suggest that the reproduction of these bird-flowers is limited by female resource rather than by pollination. Birds preferentially visit inflorescences with more flowers, and those in which the individual flowers have greater rewards. They also spend a longer time at these more productive inflorescences. However, there is no indication that such enhanced pollinator activity increases fruit-set; interestingly, Pyke (1982b) shows that in *Telopea* better fruit-set is solely a function of inflorescence size, and hence presumably of available resource. Artificially enhanced cross-pollination did not increase fruit-set in open-pollinated plants.

Some humming-birds may weigh as little as 2.5 g and are among the smallest of all vertebrates. The relationship of body surface area to mass is such that very small homoiothermic individuals experience difficulty in stabilizing body temperature. In contrast, invertebrates, which lack the lungs and efficient oxygen-carrying vascularization of terrestrial vertebrates, are limited by gaseous diffusion to small body sizes, and rarely grow as large as 2.5 g. Vertebrate flower visitors, whether birds, bats, small rodents or marsupials, are almost always heavier than invertebrate flower visitors. Being homoiothermic, they require additional energy to maintain high body temperatures. Thus, vertebrate flower visiting is almost restricted to tropical and subtropical areas where high ambient temperatures, long flowering seasons, and highly productive plants render this strategy feasible.

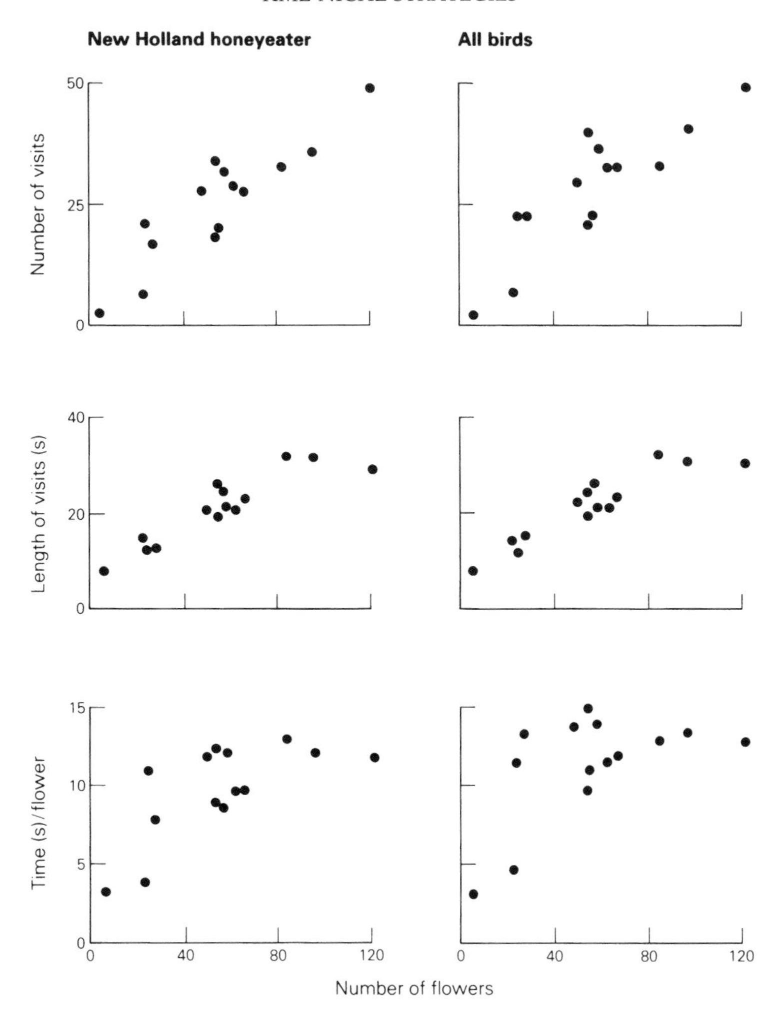

Fig. 5.5 Relationship between foraging behaviour of the New Holland honeyeater (left) and all visiting birds (right) and number of flowers open per plant of the gum *Eucalyptus cosmophylla* (Myrtaceae) in Cleland, Australia. (After Paton, 1982.)

TIME-NICHE STRATEGIES

By extending the feeding season of pollinators, it is of advantage to zoophilous flowers to fill as many time-niches in the community as ambient temperatures allow. Thus, in the non-seasonal tropics, different plants flower at different times, so that rewards for a class of pollinators are available at all times of year. Near-yearly time clocks (which may, however, vary from 9-month to 16-month cycles) for the flowering of each species ensure that this is so. Gentry (1974, 1976) has shown that of the 200 or more neotropical species of Bignoniaceae, most of which are lianes, not more than 20 usually coexist. In any one area there are usually about 50% red bird-visited species, and 50% yellowish butterfly- or moth-visited species. For each class, no two species have the same flowering time, there being about 10 time-niches in the year.

Even in tropical systems with a dry season, some species flower during the dry season, and may produce more copious and less concentrated nectar to provide water as well as fuel for visitors. Some may flower when they have shed their leaves (*Cochlospermum*, *Erythrina*) and quite often flowers are borne directly on bare trunks or twigs (cauliflory) as in *Couroupita*.

Specialist flower-visitors are more likely than generalists to need a succession of flower species in different time niches. Unlike the specialist humming-birds, which are largely restricted to neotropical regions, the honey-eaters of seasonal Australia are insect eaters as well as flower feeders, and eat insects for about half the year. In colder seasonal climates, where vertebrate pollinators do not occur, specialist flower feeders may only have very short emergence periods to cope with the relative lack of time-niches within the season. Many solitary bees have quite short emergence periods as adult flower feeders, as do syrphids, wasps, butterflies and moths. Even social bees only produce workers for part of the year, and queens are more generalist feeders, specializing on different types of flower from the workers and hibernating during the winter.

In tropical systems, there is a marked inverse correlation between the length of time that a species is in flower, and the number of flowers that are produced by an individual at any one time. Gentry (1974, 1976), working with neotropical Bignoniaceae, has described markedly different strategies with respect to the timing of flower production, and Bawa (1981) showed how these may influence patterns of speciation and diversity in different levels of the forest (Table 5.3). Species with a long flowering season, with only a few flowers produced each day have been termed 'steady-state' strategists, whereas those in which the flowers open all together have been called 'big-bang' strategists. Sometimes, several related 'big-bang' strategists will all flower together throughout the forest, as for some tropical figs (*Ficus*). This 'cornucopia' strategy encourages large communal visitors, such as parrots, to specialize on these flowers at this time (and later for fruit-feeders such as monkeys to disperse the seeds).

Because of the constraints imposed by geitonogamous pollination in multiple-flowered inflorescences, where pollen supply for outcrossing is not seriously limited, Harder and Barrett (1995) note that male sexual selection should select for 'steady-state' strategies where only one or a few flowers open on an individual at any one time. In such cases, most pollen

Table 5.3 The timing of flower production in tropical forest systems

Strategy	Flower production per plant per day	Flowering season	Chief pollinators	Species density	Distance of pollen travel	Position in forest	Species diversity
'Steady-state'	1 or few	Long	Bee, butterfly	High	Short, accurate	Low (floor)	Low
'Big-bang'	Most together ('cornucopia')	Short	Bird, bat	Low	Long, accurate (mostly selfed)	High (canopy)	High
'Multiple-bang'	Cornucopia simultaneously in several species (Müllerian mimicry)	Short	Bird, bat	Low	Variable, inaccurate (mostly selfed)	High (canopy)	High

which reaches stigmas is outcrossed and so wasteful 'pollen discounting' against which male function should select, is minimized.

Foraging strategies of pollinators will depend on the distribution of flowers in space and time. Linhart (1973) has shown that territorial humming-birds in Costa Rica (*Amazalia* spp.) favour feeding on species of *Heliconia* which grow aggregated together in clumps, with a 'big-bang' phenology, on the forest edge. Non-territorial humming-birds (*Glaucis hirsuta*, *Threnetes ruckeri*, *Phaethornis* spp.) range widely, and preferentially visit 'steady-state' *Heliconia*, with spaced individuals each producing one flower per day (Fig. 5.6). In these, pollen travel between

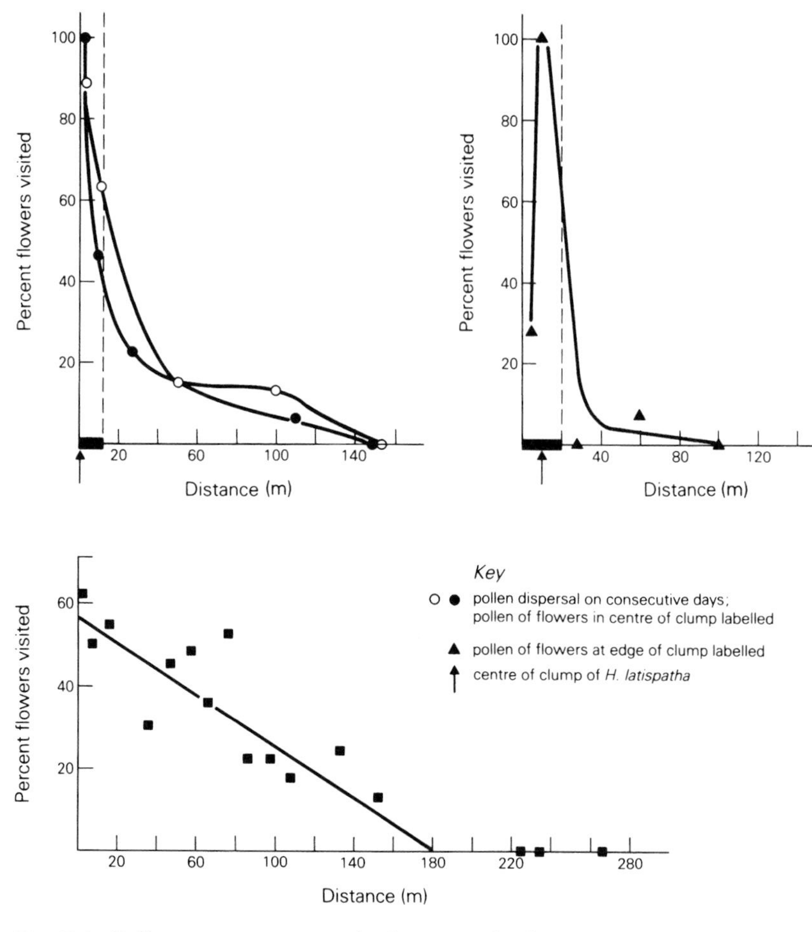

Fig. 5.6 Pollen movements in the hummingbird-pollinated *Heliconia latispatha* (top) (Musaceae) which has a clumped distribution, and *H. acuminata* (bottom) which has a scattered distribution, in Costa Rica. (After Linhart, 1973.)

flowers will at times be distant. However, as plant densities are low, plant neighbourhood sizes (p. 188) might be no higher than for big-bang species.

Other examples of distant travel between steady-state strategy plants are found in 'traplining' euglossine bees with specialized modes of scent collection from, for example, epiphytic orchids (Dressler, 1968; Williams and Dodson, 1972).

For tropical forest trees, pollinators can be expected to travel less far between steady-state plants with poorly rewarding patches, than between big-bang plants with highly rewarding patches. This pattern contrasts markedly with that of temperate plants such as *Delphinium*, in which pollen travel usually varies inversely with quality of reward (Waddington, 1981).

Thus, big-bang flowers will mediate distant pollen travel, and can reproduce successfully when growing at lower densities and higher diversities in the forest. In tropical forest, for instance in South-East Asia, such tree densities can be as low as one individual/100 ha, and tree diversities as high as 220 species/ha (Whitmore, 1984). Most of these trees have 'big-bang' strategies. They will suffer from very high levels of geitonogamous pollination, and in the case of 'cornucopia' strategies, much crossed pollen may be wasted on a non-receptive species. Many big-bang species are self-incompatible, thus minimizing the formation of selfed fruit-set, and so few fruits may be set. In a stable forest, where opportunities for seedling establishment will be few, a premium will be placed on the quality rather than the quantity of fruit-set.

POLLEN TRAVEL WITHIN AND BETWEEN PLANTS

Techniques

To understand microevolution, niche width, populational subdivision and the evolutionary potential of plant populations, it is necessary to investigate the genetic structure of those populations. This genetic structure is controlled by an interaction between the breeding system of the plants and natural selection. A major factor in the nature of the breeding system is pollen travel, for example:

1. What proportion of the pollen reaches a conspecific stigma (pollination efficiency)?
2. What proportion of the pollen reaching a conspecific stigma does so on another individual (degree of outcrossing)?
3. How far does pollen travel onto stigmas of other individuals (neighbourhood area), and how many other individuals can it encounter (neighbourhood size, or effective population number)?

Until about 1987, such studies required that pollen or pollen sources should be marked or identified in some way and Handel (1983) gave a useful review of techniques used in studying pollen travel. Many studies of pollen travel were based largely or entirely on observations of the foraging animal, which can be a very inaccurate and misleading technique.

Almost all the techniques which used marked pollen or sources involved some sort of bias. Frequently, the source, or the recipient of the pollen was assumed to be the nearest plant, or the first one visited, and carry-over was ignored, as was pollinator bias. Bertin (1988) summarizes a number of investigations where it has been shown that any such indirect measurements of pollen travel are poor estimators of gene travel by means of pollen.

From 1980, several investigators (Schaal, 1980; Moran and Brown, 1980; Mitton *et al.*, 1981; Shen, Rudin and Lindgren, 1981) pioneered the use of isozymes as genetic markers to monitor gene travel by pollen. Originally, there was a tendency to introduce plants carrying an allozyme not present in a population as artificial sources ('rare marker approach'). Seedlings were grown on and surveyed for the presence of the foreign allozyme.

Latterly, techniques involving paternity analysis have allowed more sophisticated measures to be made of spatial mating patterns. Multilocus procedures identify genotype combinations in seedlings not present in maternal genotypes or populations (paternity exclusion analysis) (Smith and Adams, 1983; Ellstrand, 1984). As this technique is exclusive, it is very useful for estimating outcrossing rates. However, the identification of male parents is rarely certain, and statistically based maximum likelihood procedures are needed (Meagher, 1986). These have been refined by Devlin and Ellstrand (1990) and Adams, Griffin and Moran (1992). Adams, Birkes and Erikson (1992) provide a review.

In recent years, random amplified polymorphic DNA (RAPD) techniques have also been used in gene flow studies (Nybom and Schaal, 1990), and these certainly add to the number of markers available to the investigator, while also risking an increase in the amount of linkage disequilibrium which confuses paternity analysis. In higher plants, restriction fragment length polymorphism (RFLP) techniques which are currently mostly restricted to cpDNA are irrelevant to paternity studies, as their inheritance is largely if not entirely maternal.

Pollen reaching a conspecific stigma (pollination efficiency)

Pollination efficiency can be considered either as a function of the proportion of pollen grains produced per flower, which reach a stigma of the same species of plant (approximation to male fitness, p. 25), or as a function of the number of conspecific grains which reach a stigma in relation to the number of ovules to be fertilized via that stigma (reproductive efficiency).

Male fitness

Male fitness is estimated as the reciprocal of the expenditure on male fitness per offspring fathered. For our present purposes it can be approximated as the product of the likelihood of a pollen grain (unit of male resource) being removed from an anther by a flower visitor, and the likelihood of that pollinator depositing that grain on a cross-compatible stigma.

Factors affecting male fitness are complex, and have been reviewed by Thomson and Thomson (1992) and Stanton *et al.* (1992). Such factors as the timing and pattern of anther dehiscence (Harder and Thomson, 1989), pollen grain number, the weather, flower visitation rates and patterns and pollen viability all affect male fitness.

For instance, when a flower of *Phlox pilosa* or *P. glaberrima* is visited by the sulphur butterfly *Colias eurytheme*, between 7% and 13% of the available pollen (15 000 grains/flower) is on average extracted on the proboscis of the butterfly, although some of this is lost when the proboscis recoils. Between 10% and 16% of the pollen remaining on the proboscis was estimated to be deposited on the next flower visited, and thus a single visit between flowers only transports about 0.6–2.0% of the available pollen (Levin and Berube, 1972).

With a narrow, coiled proboscis, butterflies are probably rather inefficient pollen vectors. Primrose (*Primula vulgaris*) is visited by a range of pollinators, of which syrphids and bombyliids are important. Piper and Charlesworth (1986) showed that on average, 31% of the pollen was removed by pollinators from the anthers of long-styled (pin) primrose flowers (see Chapter 7), but for short-styled (thrum) flowers 59% of the pollen was removed (Table 5.4). Interestingly, as 2.13 times as many of the smaller grains are produced on average by a pin anther compared with a thrum anther, there is no statistical difference in the average number of pollen grains removed from each type of flower in this heteromorphic species.

Using emasculated flowers, Piper and Charlesworth (1986) showed that both pin and thrum stigmas receive a similar proportion of legitimate (other-morph) pollen removed from anthers (3%), which is only 0.7–2.0% of all pollen produced. However, pin stigmas receive about ten times as much pollen as this (9% of all removed pollen), but most of it results from illegitimate and ineffective within-flower pollination.

Thus, in comparison with *Phlox*, pollinators remove a much greater proportion of primrose pollen from anthers, but deposit a much smaller

Table 5.4 Male fitness and reproductive efficiency in *Phlox pilosa, P. glaberrima* and *Primula vulgaris* (after Levin and Berube, 1972 and Piper and Charlesworth, 1986

Average attributes	*Phlox pilosa*	*Phlox glaberrima*	*Primula* pin	*Primula* thrum
Number of pollen grains per flower	15000	15000	36620	17166
Number of grains removed per pollinator visit	1942	1053	11229	10158
% grains removed/visit	13	7	31	59
Legitimate pollen loads/ stigma	320	100	267	305
% grains removed deposited on stigmas	16	10	2.3	3
% grains produced deposited on stigmas (male fitness)	2	0.6	0.7	2
Number of ovules/flower	3	3	85	85
Pollen/ovule ratio	5000	5000	430	202
Reproductive efficiency % (ovules per capsule/pollen load on stigma)	0.9	3	32	28

proportion of removed pollen onto stigmas, so that the overall male fitness does not differ between the two genera (Table 5.4).

Reproductive efficiency

Although only about 1% of *Phlox* pollen reaches a conspecific stigma, this still averages some 100–300 grains per stigma. However, the maximum number of seeds set in a *Phlox* capsule is only three (average about 2.8), so the number of grains available for full fertilization is vastly (more than 30 times) in excess of that needed.

In primrose, between 250 and 310 legitimate grains on average reach a stigma, but this is only about three times as many as are required for each ovule to be successfully fertilized. In this sense, primrose, although equally male fit, is much more reproductively efficient than is *Phlox*, in that more ovules are fertilized per pollen grain produced. In both genera most ovules are usually efficiently cross-fertilized, but in primrose the pollen/ovule ratio is between 200 and 420, whereas in *Phlox* it is about 5000.

Comparable figures for male fitness and reproductive efficiency could be displayed for a number of plant species, and this subject is well reviewed by Stanton *et al.* (1992). Wild radish (*Raphanus sativus*) has been particularly thoroughly examined in this regard.

Proportion of pollen reaching a conspecific stigma on another individual (degree of outcrossing)

When a pollinator visits a homogamous hermaphrodite flower, it is likely that it will deposit some of the pollen collected onto the stigma of the same flower (autogamy). Also, most pollinators generally feed from more than one flower on an individual plant during a visit. Therefore, a proportion of the pollen produced will be deposited onto stigmas of the same individual plant (geitonogamy). These are both self-pollinations. Whether these self-pollinations result in self-fertilization will depend on whether the plant is self-incompatible (Chapter 6). However, this chapter is concerned with patterns of pollen travel.

The extent to which pollen is outcrossed, that is deposited on the stigma of another individual plant, will depend on two factors:

1. the carry-over schedule, that is, the pattern of pollen deposition onto stigmas of the first to *i*th flowers visited by the pollinator after that from which pollen was gathered;
2. patterns of flower-visiting by pollinators within and between individual plants. This will depend on the rewards offered by a plant and its patch, on genet density, on the behaviour of the pollinator, and, crucially, on the number of rewarding flowers produced by an individual plant at any one time.

Carryover schedules

We can get a rough estimate of carryovers by a comparison of the distribution of pollinator flights (Fig. 5.7) with that of the travel of pollen in the same population (Fig. 5.8), for instance in the cowslip, *Primula veris*. In this species most bee flights between flowers were of less than 2 m, and whereas about half the pollen also travelled less than 2 m, a higher proportion of the sample travelled from 2 to 6 m minimum distance than did bees. Most of this further travel can be assigned to carryover of pollen not deposited on the first flower visited by bees.

The patterns of carryover, and its consequences, have been modelled by Crawford (1984b), partially based on Bateman (1947), Levin and Kerster (1971) and Primack and Silander (1975) and more recently by Harder and Barrett (1996). Crawford examines a situation in which the number of pollen grains on a pollinator is in dynamic equilibrium so that each time it visits a flower, a proportion of the pollen grains on it, p, is exchanged. It is assumed that pollen is deposited on a stigma of a flower before fresh pollen is collected from that flower (which is probably unrealistic), and that the plant is fully self-compatible. It is also assumed that n flowers are visited, in sequence, on a plant (genet). The proportion of all pollen collected from a plant (genet) geitonogamously selfed on that plant is described by:

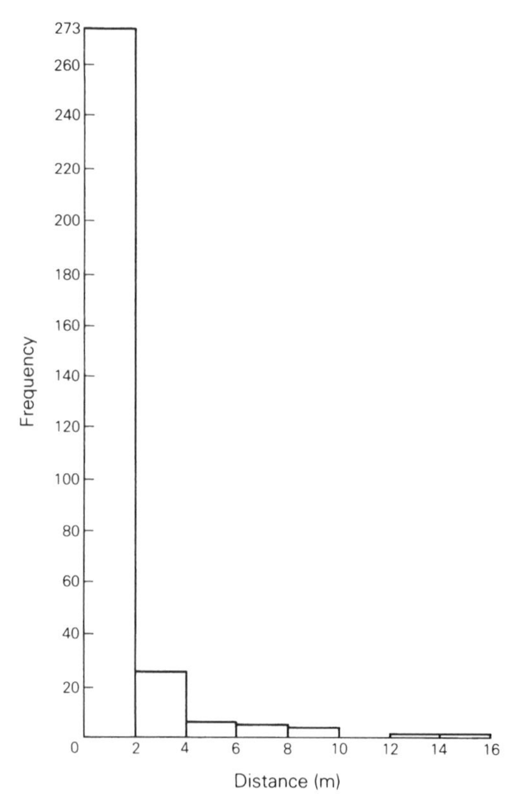

Fig. 5.7 Distances travelled by bees (*Bombus* spp.) between inflorescences of the cowslip, *Primula veris*, in a population in Northumberland, UK.

$$1 - 1/np\left[1-(1-p)^n\right]$$

The proportion of pollen collected from a genet deposited on the *i*th plant genet subsequently visited is described by:

$$1/np\left[1-(1-p)^n\right]^2(1-p)^{n(i-1)}$$

Thus, the amount of geitonogamous selfing, as opposed to crossing, that will occur is a function solely of the rate of pollen exchange that occurs at each flower visit and the number of flowers visited per genet by the pollinator.

Patterns of flower visiting

The number of flowers visited per plant individual (genet) will depend very considerably on the number of rewarding flowers presented to the

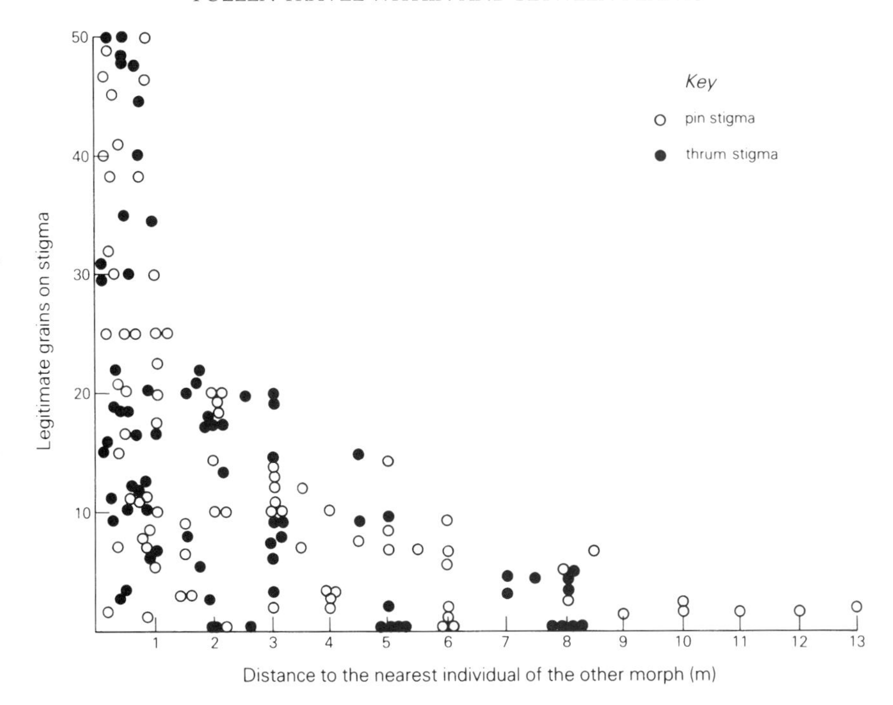

Fig. 5.8 Distances travelled by pollen between pin and thrum cowslip, *Primula veris*, as estimated by numbers of legitimate grains on stigmas and distance to nearest legitimate pollinator.

pollinator by a genet at any moment in time. For a standard reward per flower, ramets with many flowers, or genets with many flowering ramets, will receive much more geitonogamous self-pollination than do few-flowered genets. This will be reinforced by the tendency for the pollinator to fly less far between ramets, and to visit more flowers within a ramet, if the ramet has many rewarding flowers.

Ironically perhaps, plant genets that perform well are more likely to be self-pollinated. Crawford (1984b) demonstrates this clearly for the musk mallow, *Malva moschata*, for which there is a direct positive relationship between the amount of selfing and the number of flowers per genet. In this case, selfing was estimated in experimental populations using homozygous electrophoretic markers in the parents, and by scoring heterozygotes for acid phosphatase in the offspring.

Using acid phosphatase together with a flower-colour marker, Crawford confirmed this result in a wild population of the same species, showing a clear negative relationship between the amount of outcrossing and the number of flowers per genet.

However, such results are only relevant for self-fertile species. Well-visited, rewarding plants may not only self-pollinate more than do less rewarding plants, but they may also father the offspring of more mothers than do less rewarding plants. If they are fully self-incompatible, as in the sky-pilot *Polemonium viscosum* (Galen, 1992), male expenditure and display may in fact result in more male outcrossing. Galen's results fulfil the predictions of Lloyd (1984) that if expenditure on floral display and reward induces more visitors to visit a male parent, this above all other factors will potentially increase the mating success of that male parent.

Earlier work used pollen markers rather than genetic markers to assess pollen carryover. Gerwitz and Faulkner (1972) used source plants marked with ^{32}P as a radioactive marker for their work with *Brassica*, and showed that the proportion of radioactive pollen was decreased by about 30% for each subsequent non-radioactive flower visited. Similar patterns of results have been obtained for *Colias* butterflies visiting *Phlox* (Levin and Berube, 1972), *Bombus* and humming-bird species visiting *Delphinium* (Waddington, 1981; Waser and Price, 1981; 1983; Waser, 1988), artificial pollination using stuffed humming-birds (Pyke, 1981) and *Bombus* visiting *Erythronium americanum*, *Clintonia borealis* and *Diervilla lonicera* (Thomson and Plowright, 1980). More recently, pollen dispersal schedules have been thoroughly described for radish, *Raphanus sativus*, by Stanton *et al.* (1992) and for *Polemonium viscosum* by Galen (1992).

Typically, up to 50% of pollen is deposited on the first flower visited after pollen collection, and deposition on subsequent flowers decreases rapidly in a leptokurtic fashion; less than 1% of pollen collected usually survives on the pollinator after eight successive flower visits (Table 5.5). Carryover appears to be greater on butterflies, and less on humming-birds, than it is on bees, but information is at present very limited. For instance, Waser (1988) indicates that humming-birds may allow pollen to carryover more flowers than do bees.

Not only the pollinator, but also the type of flower will influence carryover patterns. Viscous nectar in deep floral tubes is harder for the pollinator to win than is dilute nectar in short tubes. (However, concentrated nectar is of course more rewarding, and it is better protected in long-tubed flowers.) Nevertheless, Harder (1986) shows that bees tend to switch patch more readily when more energy is expended in the gaining of nectar.

In all investigations in which pollen sources are experimentally localized, pollen carryover is likely to be underestimated (Handel, 1983). Nevertheless, given the information on carryover schedules, and on the number of receptive flowers on a plant that acts as a pollen donor, and the number of flowers on subsequent individuals that are visited by the pollinator, rates of outcrossing can be estimated. This information is given by Waddington (1981) for *Delphinium virescens*. Between two and

Table 5.5 Carryover of pollen on *Bombus* species visiting three North American flowers (after Thomson and Plowright, 1980). Values are numbers of grains

Flower	Sequence	Flower number visited after collection														
		1	2	3	4	5	6	7	8	9	10	11	12	13	14	15
Erythronium americanum	a	4	6	0	0	0	0	1	1	0	0	1	0	0	0	0
	b	14	11	2	5	5	0	3	3	2	0	5	0	0	3	0
	c	8	15	3	4	5	3									
	d	4	0	6	7	3	2	6	0	0						
	e	5	2	1	3	0	0	0	0	0	0	0	0	0	0	0
	f	4	7	3	3	5	1	7	0	0	0	0	0	0	1	0
	g	0	3	0	4	1	0	0	0							
Clintonia borealis	a	369	148	110	2											
	b	30	21	153	28	484	12	17	159							
	c	24	21	58	47	19	36	2	2	9	1	6	5			
	d	258	178	5	28											
	e	346	21	34	73											
	f	131	142	16	68	45	69	51	17	5	22					
	g	97	20	93	54	49	5	34	6							
Diervilla lonicera	a	80	1	1	5	18	2	0	2	0	1	0				
	b	48	10	32	12	21	23	11								
	c	24	63	8	8	19	0	4	0	10	1					
	d	32	5	4	8	1	0	3	1							

six flowers are usually visited per inflorescence, the number being highly dependent on the volume of nectar being donated by that inflorescence (p. 149). In this species, 44% of the pollen is deposited on the first flower visited after collection, and 83% is deposited on the first three flowers. Thus, for an inflorescence which is highly rewarding, on which a visitor will visit an average of five flowers, much geitonogamous selfing will occur. For a poorly rewarding inflorescence, on which less than three flowers may be visited on average, more cross-pollination is likely to occur (Fig. 5.9). Bees also fly further, more than twice as far on average, from poorly rewarding inflorescences. These results have been explained by Zimmerman (1983) by the marginal value theorem of foraging theory. Bee behaviour may well have evolved so that the bee will decide to leave an inflorescence to forage another patch when the expected reward to be gained from the next flower on the inflorescence falls below that of the average reward for the patch. (Of course, to state that bees should have evolved to this behaviour pattern is teleological.)

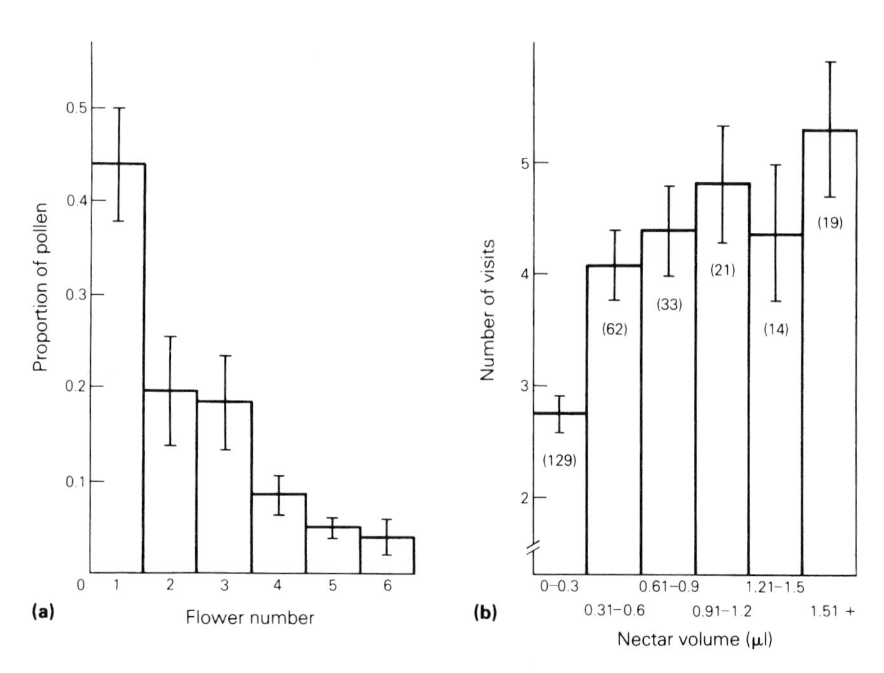

Fig. 5.9 Foraging strategies of *Bombus americanorum* on *Delphinium virescens* in Kansas, USA. (a) The mean proportion of pollen grains deposited on the stigmas of the 1st, 2nd . . . , 6th flower visited after the marked flower. (b) The relationship between nectar volume of the bottom-most flowers and the number of flowers visited per inflorescences Poorly rewarding inflorescences will be more outcrossed than well-rewarding inflorescences. (After Waddington, 1981.)

The goals of reproductive efficiency and the maximization of outcrossing are in direct opposition, so stabilizing selection may adapt a plant to produce a quantity of reward at an ESS threshold point between the optimal points for each goal (p. 151). However, by varying the quantity of reward with the age of the inflorescence, or even within a daily rhythm, both goals may be met with in different time-niches.

However, it is wrong to assume that obligate outcrossing should be the goal of every plant. Selfing also confers fitness attributes (p. 360). Holsinger (1992), using a 'mass-action' model, shows that mixed mating strategies (involving both outcrossing and selfing) may often be evolutionarily successful.

The work of Thomson and Plowright (1980) vividly illustrates the importance of flower number in controlling the degree of outcrossing (Table 5.5). Although the three species studied have pollen carryover schedules that are not dissimilar, it is very important to note that an individual of *Erythronium americanum* usually has a single flower open at any one time, whereas in *Clintonia borealis* a short raceme bears perhaps eight rewarding flowers at a time. However, *Diervilla lonicera* is a large shrub which may bear more than a thousand flowers simultaneously. Clearly, almost all pollen travel in the *Erythronium* is xenogamous (outcrossed), whereas nearly all between-flower transport of pollen in the *Diervilla* will result in geitonogamous selfs.

In this case too, there should be stabilizing selection for flower number and floral display. An individual should produce flowers showy and/or numerous enough to attract visitors, but those with very numerous flowers may attract many visitors, but will suffer the penalty of low levels of outcrossing. The offspring of a genotype which produces fewer flowers may be advantaged by its high levels of outcrossing, but the parent will be less reproductively efficient, on two counts. Perhaps for this reason, many large plants (e.g. trees, clonal grasses) have escaped this dilemma by being wind pollinated (p. 88).

In order to understand patterns of gene transport within and between plants in a population, it is not enough to describe pollen carryover. It is also necessary to know the number of flowers visited per plant, the number of receptive flowers carried by a plant, and the self-incompatibility or self-compatibility of the plant. It is rare indeed to encounter a piece of work where all this information is transmitted together.

Latterly, the pervasive use of isozyme markers in plant populational studies has transformed our knowledge of outcrossing rates in plants, and this information is discussed in more detail in Chapter 9. However, this work infers levels of outcrossing from proportions of plants in the population which are heterozygous. Such information may give us very misleading information as to levels of cross-pollination. Selfed pollen may fail to function, or the offspring of selfs may be less fit than the offspring of cross-pollinations, so that estimates of outcrossing may be much greater than actual levels of cross-pollination. Even where seedlings from known mothers are assayed for markers which demonstrate cross-fertilization (e.g. Barrett, Kohn and Crazan, 1992), variables in postpollination male fitness (most obviously in the self-incompatibility system) render estimates of cross-fertilization different from those of cross-pollination.

Distance of pollen travel

Information on pollen travel is vital to our understanding of the genetic structure of populations in terms of levels of heterozygosity, the maintenance and breakage of linkage groups, the establishment of novel mutants, and genetic impoverishment in small populations. Gene migration is also mediated by patterns of seed dispersal, and on this we are relatively ill-informed (p. 192).

Leptokurtic distribution

The outstanding feature of both anemophilous and zoophilous pollen travel is its leptokurtic distribution, a pattern that is common to most biotic movements (Handel, 1983). The leptokurtosis of zoophilous pollen travel is compounded by the clumped, non-random distribution of most plant populations and of most patterns of flower visitation.

Thus, the distribution of between-plant flights of bees and butterflies visiting three American *Senecio* species (see Fig. 5.2), the distribution of between-plant bee flights and pollen flow in *Primula veris* (see Figs. 5.7 and 5.8), and pollen dispersal by syrphids and moths in *Saxifraga hirculus* are examples quoted earlier in this chapter that clearly show this leptokurtic distribution. Other bee flowers, such as *Aconitum columbianum*, *Delphinium virescens*, *Trifolium repens* and *Lupinus texensis* show very similar patterns of pollinator travel between plants (Fig. 5.10). Patterns of gene travel in *Lupinus texensis*, *Cucumis sativus* (cucumber) and *Chamaecrista fasciculata* (Fenster, 1991a) are also similar, although genes travel further than does pollen because of carryover (Fig. 5.11).

In the case of *Asclepias exaltata* which is pollinated chiefly by strong-flying nymphalid butterflies, and which has a pollinium-mediated 'all-or-nothing' pollination mechanism, gene travel was shown by Broyles and Wyatt (1991) to greatly exceed estimates of pollinator travel, so that effective population sizes were very large.

Humming-bird-pollinated *Heliconia* species (Musaceae) are rare cases where the leptokurtic rule does not always hold for pollen and gene dispersal. Study of the distribution of marked pollen in Costa Rica by Linhart (1973) shows that territorial birds visit *Heliconia* species growing in clumped distributions (*H. latispatha* and *H. imbricata*) and in these the pollen travel is very markedly leptokurtic (Fig. 5.6). However, for ran-

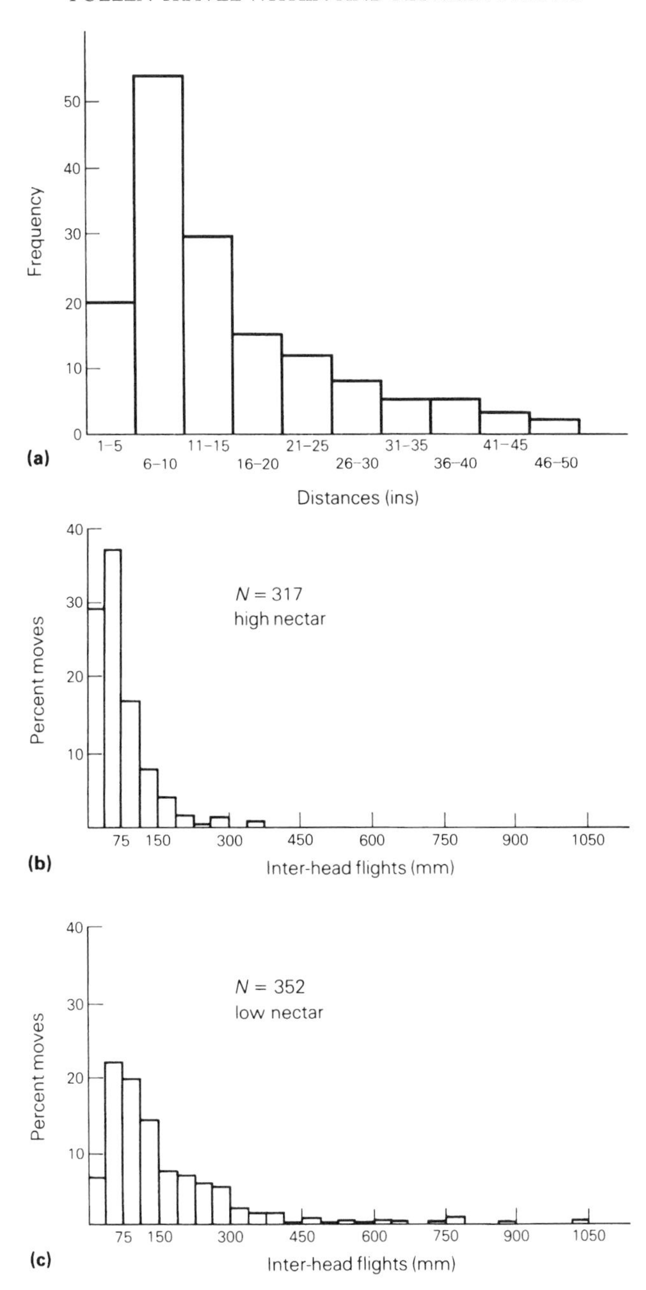

Fig. 5.10 Distribution of distances travelled between inflorescences by foraging bees in (a) *Aconitum columbianum* (after Pyke, 1978b) and *Trifolium repens* (after Heinrich, 1979), showing differences in pollinator travel between well-rewarding (b) and poorly rewarding (c) patches.

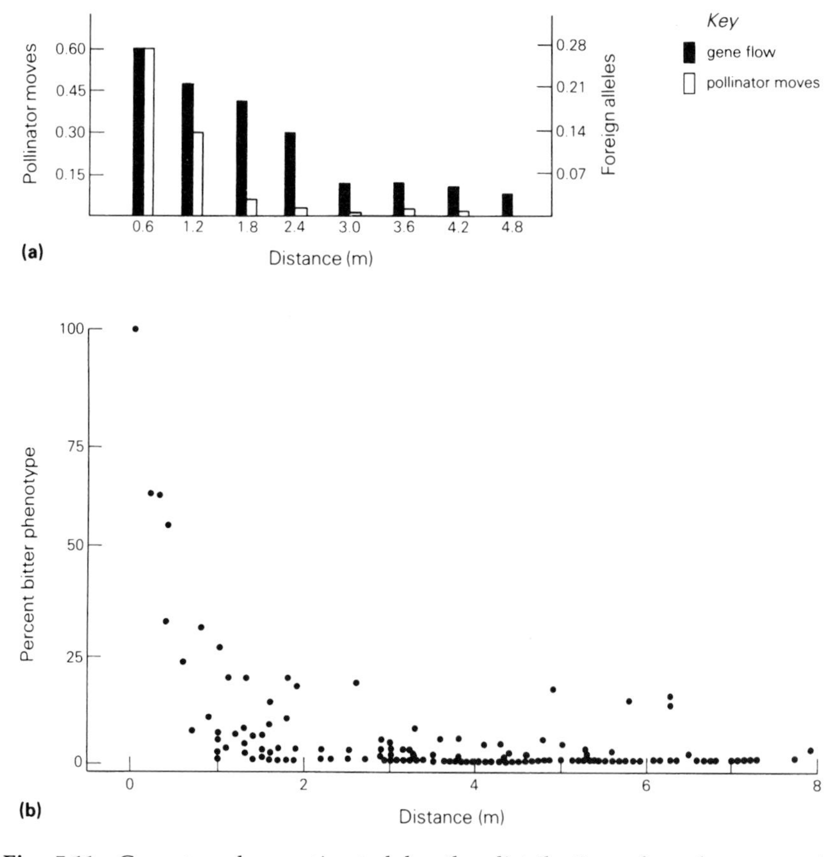

Fig. 5.11 Gene travel as estimated by the distribution of marker genes in seedlings of *Lupinus texensis* (a, after Schaal, 1980) and *Cucumis* (b, after Handel, 1983).

domly spaced *Heliconia* (*H. acuminata*) or those which in this case showed linear distributions (*H. tortuosa*), pollen dispersal showed a linear relationship with distance, or was even bimodal. These species were visited by non-territorial birds which showed a randomized far-ranging feeding pattern quite different from those that held territory.

The leptokurtosis typical of pollen and gene travel in most zoophilous flowers, certainly when visited by bees, butterflies and most birds, is found with respect to both within-plant and between-plant travel, and results from an interaction of several behavioural and physical features.

1. Most plants have a clumped distribution.
2. Within a plant population, some patches will be more rewarding than others at a given time, thus exaggerating this clumping.
3. Pollinators will concentrate on more rewarding patches.

4. When pollinators visit more rewarding patches, they will travel less far to the next flower or plant.
5. Pollen carryover itself tends to be leptokurtic in distribution (Table 5.5); thus most pollen is carried to the next flower visited.
6. For most flower visitors, length of flight between flowers is dependent on their behavioural strategy at that moment. Thus, for a foraging bee, between 90 and 99% of flights are short-distance foraging flights. The remainder are 'escape flights' over much greater distances, for the purposes of visiting new patches, escaping predators or returning to the hive or nest. For flower-visiting birds, foraging flights are much shorter than those involved in territoriality, aggression and courtship. Combinations of normally distributed flight patterns with different variances (i.e. foraging and escape flights) will automatically generate leptokurtic flight distributions.

Pollen transport by bees and plant density

Although the leptokurtic pattern of dispersal of pollinators, pollen and genes is almost universal, the distances involved do vary a great deal. This variation can be considerable even within a flower species. Thus, for the *Senecio* species studied by Schmitt (1980) mean bee flight distances vary from 0.32 m to 1.06 m, whereas mean butterfly flight distances vary from 2.30 m to 12.39 m. Clearly, the leptokurtosis of pollen dispersal is increased in this case by the different behaviour of the two classes of pollinator. Heinrich (1979a), working with clover, *Trifolium repens*, visited by *Bombus terricola*, shows that for plants protected from nectar foraging for two days, and thus with plenty of nectar, 99% of foraging flights were less than 2 m in distance. For unprotected plants, with less nectar, 99% of flights were less than 4 m (Fig. 5.10).

For bee-pollinated plants, it is usual to find that at least 80% of flights are less than 1 m in distance, and 99% are less than 5 m (Figs. 5.2, 5.7 and 5.10). Such is the nature of pollinator behaviour that occasional grains may travel much further. Halijah Ibrahim and I found that occasional grains of *Primula veris* may move at least 13 m, and there is evidence in this species of gene travel of at least 30 m (Ibrahim, 1979). Naturally, the mean distance of pollen dispersal is dependent not only on pollinator behaviour, but also on plant density. Travel will be much shorter in areas of high plant density, as originally emphasized by Levin and Kerster (1968, 1969a,b, 1974) and Levin (1979). We found pollen dispersal highly dependent on plant density in *Primula veris*, where in our experimental population, density of plants varied from 0.10 to 22.0 per m^2.

In sparsely populated areas, pollen travel will, by definition, be more distant. However, the number of plants which the pollen may reach may be no greater than in dense populations, and indeed may well be less. Also, sparse areas of dense populations will receive far fewer pollinator visits, as these areas will form patches of low reward, and this will minimize gene travel in these areas.

For many species, especially in tropical forest, plant density is always low (p. 165). Because dominant species are rare or absent, many pollinators in these warm climates fly long distances between energetically rewarding flowers, which, being of a 'steady-state' type, may each have only one or a few flowers open. This plant/pollinator strategy is termed 'trap-lining', and originates from work on Euglossine bees and orchids by Dressler (1968) and Williams and Dodson (1972). Later work by Janzen (1971), Frankie (1976), Frankie, Opler and Bawa (1976) and Heinrich (1975) extended these ideas to forest trees and to other classes of solitary bees.

Pollen transport by birds

Linhart (1973) shows that although pollen dispersal is leptokurtic for territorial birds visiting clumped *Heliconia*, the distances involved can be 50 or 100 times greater than for colonial bees (see Fig. 5.6). However, in one instance (*Amazalia saucerottei* visiting *H. latispatha*) only 1% of the pollen travelled more than approximately 20 m. It is not clear how much these differences are the result of patch sizes, or densities, and how much the result of bird behaviour.

Using both pollen travel (onto experimentally emasculated flowers of *Delphinium nelsonii*) and dye travel, Waser (1988) showed that pollen travel was both further and less leptokurtically skewed when mediated by humming-birds than by *Bombus* bees. However as these data are restricted to carryover, and no information is given as to how many flowers per genet are visited, this information cannot be translated into populational terms.

Pollen transport by wind

Wind-dispersed pollen also has a leptokurtic dispersal pattern, but the causes for this are quite different from that of animal-dispersed pollen (Gleaves, 1973; Whitehead, 1983).

1. The lateral component of pollen travel will be due to wind (Fig. 5.12), which is rarely steady, but blows in gusts usually of 1–10 m/s, whereas terminal velocities of falling pollen, depending on pollen size and density, are much lower, ranging from 2 to 6 cm/s; the distribution of speeds of wind gusts is itself leptokurtic, that is, the strongest gusts are the rarest. Thus, most dispersed grains will only be blown a short

Fig. 5.12 Pollen being distributed by wind from male cones of *Pinus nigra*. Pine pollen has two air sacs and is unusually well distributed in the air (×0.33).

distance, but a few will meet stronger gusts after dehiscence and will be blown much further.

2. Tree pollen, in particular, is released from various heights above ground and, as most trees have a somewhat conical canopy shape, the amount of pollen available for release will decrease with height above ground. The potential for lateral dispersal, that is of encountering a gust of wind, will increase with the height of the point of pollen release above ground. This is partly because it is windier higher up, and partly because the pollen is falling for a longer time. In order to disperse, pollen must escape the boundary layer of the tree or field, and this layer is more likely to be disturbed by strong gusts of wind.
3. Most wind-dispersed pollen is released in a conspecific canopy (trees) or stand (grasses, sedges) and it will thus tend to lodge on foliage, and stick or fall to the ground a short distance from the point of dispersal. Only a relatively few grains, or grains from relatively few

isolated plants, will be able to escape the shelter, boundary layer and obstruction conferred by the canopy and disperse any distance. Handel (1983) shows that the sedge, *Carex platyphylla*, receives mostly outcrossed pollen when occurring in small clones with less than 10 flowering ramets (culms), but if clones (genets) are larger than this, the amount of selfed (geitonogamous) pollination is considerable. However, this level of geitonogamy increases no further with increasing clone size.

The leptokurtosis of wind-pollination is also influenced by pollen collection onto stigmas. Whitehead (1983) shows this to be proportional to wind velocity and to the density and diameter of pollen, and is inversely proportional to the diameter of the stigma. Thus, pollen should be large and heavy, and stigmas should be wide (this is often achieved by feather-like ramifications of the stigma) for efficient collection. Analysing these properties, Tauber (1965) shows that at low wind velocities, beech (*Fagus sylvatica*) pollen is collected by stigmas five times more effectively than is birch (*Betula* spp.) pollen. However, at higher velocities, which are more commonly encountered by birch than for beech, birch collection rates become relatively more efficient.

As a result of these factors, most wind-dispersed pollen travels only relatively short distances. Nevertheless, it appears that pollen travels by wind much further than is usual for animal-dispersed pollen (Table 5.6). However, trees or grasses may have very large individuals, and the densities of genetically different individuals (genets, or clones) may be very low (Chapter 2), so that even far-travelling pollen may reach only a few other individuals.

The data in Table 5.6 must be interpreted with considerable care and caution, for different measuring techniques are used and different units of measure are employed. Some of these biases are overcome by expressing data as a percentage of the total detected, but these figures rely heavily on the distances at which measurements were actually made. Thus, cross-comparisons are dangerous, and the data can only safely be used to show the orders of distances that wind-dispersed pollen may travel, and its tail-off with distance. Generally, readings are minimum estimates only.

Very few estimates of gene travel have been published for wind-pollinated angiosperms, so it is difficult to determine how realistic are estimates resulting from measurements of pollen travel. Adams, Birkes and Erikson (1992) review information concerning gene travel within tree seed orchards, but this information is largely restricted to conifers and so falls outside the scope of this book. Also, such studies tend to concentrate on the diversity of paternities to a mother tree, and rarely follow the pollination schedules of known fathers.

Table 5.6 Percentage of detected wind-dispersed pollen encountered at various distances from source (adapted from Altman and Dittmer, 1964)

Species	Distance from source (m)																
	1	2	3	5	10	15	20	50	100	150	200	300	400	500	600	700	800	900
Grasses																		
Phleum pratense									17.8		11.5	7.8		3.2				
Dactylis glomerata											11.4		4.3		3.1		2.2	
Lolium perenne											17.0			8.7		5.6		
Pennisetum glaucum				90.8		8.1			0.7	0.4								
Zea mays			93.7		4.3		1.5	0.4										
Beta vulgaris												13.1		7.2			1.9	
Trees																		
Fraxinus excelsior					68.4		27.1	3.8		0.8								
Populus sp.							26.5			21.2			18.8					17.1
P. deltoides					47.3				25.5	18.9				8.2				0.1
Ulmus sp.										40.1		53.0						4.2
Conifers																		
Pinus cembroides				93.0				5.1	0.9	0.6								
P. echinata										17.0		15.0		10.0				
Cedrus atlantica						44.2		27.2	16.7	12.0	0.02							
C. libani				52.3				25.0	14.9	8.9								

It should be noted that in three cases where the gene travel of anemophilous flowering plants has been compared with its pollen travel, genes appear to travel much less far than does pollen. In rye-grass, *Lolium perenne*, over 99% of the genes travelled less than 20 m (Gleaves, 1973), whereas in beet, *Beta vulgaris*, no gene travel was observed beyond 10 m. Estimates of pollen travel by wind in these species are made in traps (Table 5.6), and may be overestimated in comparison with those of zoophilous travel which are more realistically made on stigmas.

The most thoroughly studied wind-pollinated flowering plant to date has been *Plantago lanceolata* (Bos, Harmens and Vrieling, 1986), which is rather untypical for an anemophilous species, being small and essentially non-clonal (and in fact it does receive some insect visitation). Here, pollen and seed travel are sufficiently limited for extensive populational subgrouping to result, and neighbourhoods (p. 188) commonly contain less than 20 individuals. In this case too, experiments suggest that pollen travels much further than do genes.

Two features stand out from an examination of Table 5.6. First, appreciable quantities of both grass and tree pollen can in some circumstances travel distances of 1 km or more, as sufferers from pollen allergies ('hay fever') living in plantless cities will testify. In all the cases quoted here, except for one, at least one in 200 grains sampled had travelled at least 100 m. This distance of dispersal is only known for bird pollination among zoophilous plants. The exception, *Zea mays* (maize), is a crop plant in which the nature of pollen, and its presentation and release, may have changed markedly since domestication. It has remarkably large, and somewhat sticky, pollen grains.

Second, the proportion of grains detected declines with distance very gradually, especially for the meadow grasses and the poplars examined. The leptokurtic curve is very shallow in contrast to that of zoophilous species, although in no case is there a linear relationship with distance. This very gradual reduction in pollen travel with distance will render the variances about the means of pollen travel very large, and will lead to very considerable estimates of neighbourhood area and, depending on plant density, neighbourhood size (p. 190).

There is a relationship between the area occupied by an individual clone (genet) and its pollination system. Most trees which are individually large, or which form suckers to make large clones, are wind pollinated. Equally, herbaceous perennials that ramify to make large clones (e.g. grasses, sedges and rushes) are also wind pollinated. Thus wind pollination, with its distant travel, may be interpreted as being advantageous in allowing more pollen to reach beyond the bounds of the clone (genet), thus enhancing levels of outcrossing. In contrast, by providing a bountiful 'patch', animal-pollinated trees will be disadvantageously self-pollinated (p. 165).

Where trees are zoophilous, they are usually small (*Sorbus*, *Crataegus*, *Malus*) and without vegetative reproduction, or they are habitually solitary, or non-dominant, as in the very diverse trees of tropical forest. Zoophilous trees are also usually self-incompatible or dioecious, so that the abundant geitonogamous pollinations do not result in selfed seed-set. In many cases the constraints of the pollination system of zoophilous trees has encouraged the evolution of agamospermy (Chapter 10).

In contrast, it is usual to find that wind-pollinated species are dominant or codominant in a locality, as this will aid the efficiency of their non-specific pollination.

In this way, an anemophilous trait should encourage the subsequent evolutionary development of large genets such as trees. However, it is also likely to be advantageous for large zoophilous genets to develop a wind pollination system. In this context, it is worth noting that early Angiosperms may have been zoophilous trees, the evolution of anemophily being secondary (Taktajhan, 1969).

Genetic migration between populations or subpopulations

This topic was termed gene flow by Slatkin (1985) and has been so-used by subsequent authors (e.g. Ellstrand and Elam, 1993). However, earlier theoretical workers such as R. A. Fisher and Sewall Wright referred to genetic migration between populations and I intend to retain this prior phrase. In my view any movement of genes by pollen or seed within as well as between populations is more properly termed gene flow.

It is in the nature of leptokurtic distributions that rare movements may occur over considerable distances, for instance between what might be sensibly termed different populations. For zoophilous plants, such distant movements may result from bee 'escape flights' (p. 179), and may often be mediated by pollinators with less focused foraging strategies such as lepidopterans (p. 157). As discussed in a later section (p. 192), seed dispersal may often involve rare movements over long distances, involving both gene migration and colonization events.

When such movements occur, they may have a disproportionate effect on the genetic structure of recipient populations, for the influence of such genetic migration tends to be irrespective of the rate of that migration. Populations, often of rare species, which are normally isolated may suffer from genetic impoverishment and low levels of heterozygosity (i.e. high levels of genetic fixation, estimated by F_{st}). If they receive an input from another genetically differentiated population, the genetic fitness of the recipient populations may change considerably, at least in the medium term (Chapter 2, p. 47). Ellstrand and Elam (1993) show that intraspecific migrations between populations of rare plants happen quite frequently in most species, irrespective of the breeding system or pollen vector, and that the genetic effects of such migrations stabilize over some ten subsequent generations.

Slatkin (1985) discussed models which estimate the rate of gene migration as numbers of migrating individuals, N_m, between 'island' populations, based on the fixation index F_{st} for polymorphic alleles in each.

Two conflicting principles govern the fitness parameters accompanying such migrations. As individuals from distinct populations are likely to be more unalike genetically than are those taken from within populations, the immediate 'hybrid' products of migrations between such populations should be more heterozygous, and thus probably more vigorous and cross-compatible (p. 384) than are most individuals in either population. Olesen and Warncke (1992) showed that more seed was set after distant pollen travel in marsh cinquefoil, *Potentilla palustris*, than after near-neighbour pollinations (Fig. 5.13). Working with the annual legume *Chamaecrista fasciculata*, Fenster (1991b) showed that mean progeny fitnesses increased with interparent distances within the (small) neighbourhoods (p. 192) found in this species. He also showed that the off-

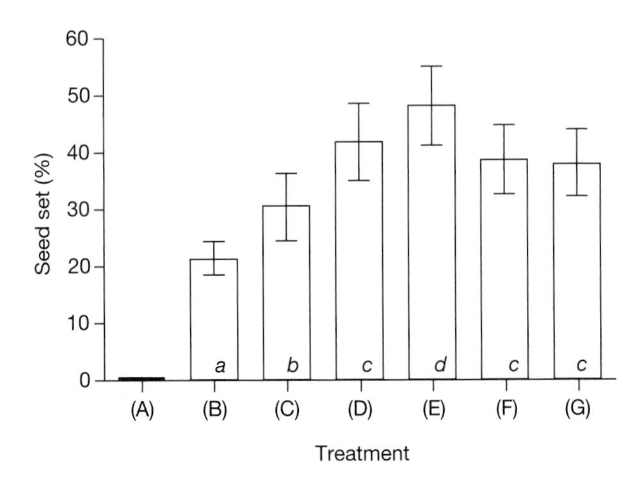

Fig. 5.13 Seed set of *Potentilla palustris* after varying treatments: (A) bagged and emasculated, (B) bagged, (C) bagged and hand-pollinated with self-pollen, (D) and (E) bagged, emasculated and hand-pollinated with pollen collected within a distance of 10 m and 60 m, respectively, (F) untreated control plants, and (G) naturally pollinated and hand-pollinated with pollen collected within a distance of 10 m. Treatments (A)–(G) include 34, 60, 27, 40, 47, 40 and 41 flowers, respectively. Standard error is indicated with a vertical bar. Bars with the same lower-case letters are not significantly different (Kruskal–Wallis one-way analysis of variance, $P < 0.05$). (After Olesen and Warncke, 1992.)

spring of crosses between parents from different neighbourhoods proved to be more fit than those made between interneighbourhood parents. Ledig (1986) cites a number of examples in trees where the offspring of crosses between different populations are more vigorous than those made within populations.

However, distinct populations may have undertaken genotypic diversification into different niches (the so-called 'Wahlund Effect' (Wahlund, 1928)), so that 'hybrid' products are maladapted to either. This principle probably underlies so-called 'outbreeding depression', where the products of distant matings are less fit than are the products of near-neighbour matings. Waser (1993) reviewed 25 such cases, and showed that a majority suffered from outbreeding depression, so that on average the products of distant matings tended to be less rather than more fit than the products of near-neighbour matings. In these cases, it is important to distinguish between offspring vigour, as might be measured, and prized, by an arboriculturist; and fitness, where the performance of offspring in the natural habitat should have been monitored.

In special circumstances, the mating system renders the likelihood of near-relative mating independent of distance. In the butterfly-pollinated milkweed *Asclepias exaltata*, Broyles and Wyatt (1991) show that the products of near-relative matings are less fit than are matings overall, but that there is no relationship between interparental distance and offspring fitness.

Ellstrand and Elam (1993) and Holsinger and Gottlieb (1991) also discuss how the sizes of donor and recipient populations crucially influence the probability of genetic migration between populations. Most of the information gleaned so far results from work on model populations of radish *Raphanus sativus*. Where 'source' and 'sink' populations are asymmetric in size, large populations tend to be both better donors, and better recipients of migrant pollen. When populations are relatively symmetrical in size, whether large or small, gene migration becomes very unusual. Presumably pollinators tend to stay within large rewarding patches, and if they leave a small patch, will be more likely to 'escape' to a rewarding large patch. Such patterns of distant gene migration crucially influence the survival of a rare species which is surrounded by populations of a more aggressive interfertile species (p. 199).

NEIGHBOURHOOD AND GENE FLOW

Neighbourhood is a rather difficult concept introduced by the great theoretical population geneticist Wright (1938, 1940, 1943, 1946, 1951). It concerns the area within a population in which panmixis, that is random gene exchange, can be said to occur. A number of factors, including the distribution of pollen travel and of seed travel, and the varying density of interbreeding individuals, may result in panmictic areas, neighbourhoods which are more limited than the whole population. Wright's aim was to describe a neighbourhood size N_e (number of individuals) that occurred in a neighbourhood area *A*, where N_e described the number of individuals responsible for a decay in genetic variance equal to that in a model population of size *N*. That model population *N* has a constant size, finite limits, non-overlapping generations, total random panmixis within those limits with all individuals interbreeding, equal numbers of both sexes (or hermaphroditism) and an equal chance for each individual to transmit genes to the next generation. If the model population shows decay of genetic variance in polymorphic genes of neutral effect, this will be a consequence of inbreeding effects in populations in which *N* is smaller than infinitely large (i.e. gene drift, Chapter 2).

Wright showed that *N* can be estimated as the number of individuals falling within a circle whose radius is equivalent to twice the standard deviation of the gene dispersal distance per generation in that population. As the area of a circle is πr^2, and $r = 2\sigma$ this circle is described by $4\pi\sigma^2$, where σ is the standard deviation of the gene dispersal distance for each sex of a unisexual species. This is approximated by $12.6\sigma^2$ and gives the neighbourhood area *A*. The neighbourhood size (number of individuals) *N* will be dependent on the density of individuals *d* (using the same area units) capable of interbreeding at a given time. Thus $N = 12.6\sigma^2 d$.

To obtain N_e, rather than *N*, one must first estimate *N*, and then N_e/N, allowing for unequal numbers of males and females, fluctuating population size and unequal parental contributions to the next generation, etc. Wright's equation was originally formulated for unisexual organisms, where σ is usually a function of the distances males and females may travel to mate (from birth), and it has been so successfully used, as by Kerster (1964) for the rusty lizard, *Sceloporus olivaceus*. However, for most plants, gene dispersal has two components, pollen travel and seed travel, and, furthermore, most plants are not unisexual (dioecious) but hermaphrodite. Levin and Kerster (1969a, 1971) have proposed that for totally outcrossed hermaphrodite plants, neighbourhood size is best described by $12.6\,(\sigma_p^2/2 + \sigma_s^2/2)d$ where σ_p^2 and σ_s^2 describe the variance of dispersal of the pollen and seed, respectively. This was used by Richards and Ibrahim (1978). In the case of a partially selfed species this can be modified to incorporate the amount of outcrossing *r* as follows:

$$12.6\left(\sigma_p^2 r/2 + \sigma_s^2/2\right)d$$

or:

$$6.3\left(\sigma_p^2 r + \sigma_s^2\right)d$$

Another modification has been used by Schmitt (1980) where two pollinators A and B with very different flight variances σ_A^2 and σ_B^2 share a proportion of the pollinating events a and $1 - a$. In this case:

$$N = 6.3\left[\left(a\sigma_A^2 + (1-a)\sigma_B^2\right)r + \sigma_S^2\right]d$$

(this equation is not as it appears in Schmitt's paper).

Kerster (1964) noted that Wright's original equations make the assumption that populations do not move in space. Thus, there must be no net movement of pollen or seeds in any given direction (as could happen with unidirectional winds, for instance on coastal sites; p. 154). He suggests that mean move distances will, by definition, equal zero. Thus if the variance of pollen moves, p_i, around a mean $\sigma_p^2 = \Sigma(p_i - p)^2/n_p$, and $p = 0$, then the variance can be calculated merely by the mean of the squares of pollen moves, $\Sigma p_i^2/n_p$. Naturally, the same holds for seed moves, $\Sigma s_i^2/n_s$, where n_p and n_s are sample sizes of measurements of pollen and seed moves respectively.

Levin (1978) noted that whereas the genetic contribution of pollen moves was haploid, that of seed moves was diploid, and thus twice that of pollen moves. Thus, his modified neighbourhood size equation becomes:

$$N = 6.3\left(\sum p_i^2 r/2n_p + \sum s_i^2/n_s\right)d$$

(In both Levin (1978) and Schmitt (1980), r, the proportion of outcrossing, is placed outside the equation brackets. However, it seems to me that r relates only to pollen travel and not to seed travel.) What is essentially this equation, but omitting r, is used by Schaal (1980); presumably the value of 3.6 in her equation is a misprint for 6.3.

Crawford (1984a) has suggested that σ^2, the total variance of plant movement between generations, is composed of moves of both the male and female components of reproduction. Male gamete dispersal is via pollen and contributes σ_p^2, whereas female gametes do not disperse in higher plants. The average gamete dispersal variance is therefore $^1/_2\sigma_p^2$, to which must be added the seed dispersal variance. Crawford suggests that $\sigma^2 = \sigma_p^2/2 + \sigma_s^2$, and thus $N = 12.6(\sigma_p^2 r/2 + \sigma_s^2)d$. If, as in Levin (1978), we assume that net moves are zero, this can be translated to:

$$N = 12.6\left(\sum p_i^2 r/2n_p + \sum s_i^2/n_s\right)d$$

or estimates that are twice those of Levin's. This seems to me the best way of interpreting this difficult concept.

The neighbourhood area, A, is merely N/d. This area is a circle (which assumes that the population is two-dimensional, not necessarily always accurately). A circle of this type should include between 63.2 and 86.5% of the parents of an individual at its centre for non-linear population. An interbreeding population can thus be considered as a very large number of overlapping circles of this type. Neighbourhood is valid in space, but not in time. Thus, although neighbourhood may be a useful way of considering the number of individuals that are able to interbreed panmictically in a given season, and thus give indications of potential inbreeding effects, it is less successful as an indicator of gene migration and recombination. Nearly all plant populations overlap heavily for time as well as space. Generation overlap, and neighbourhood overlap, will allow some genetic recombination to occur throughout individuals in a population capable of some genetic exchange, however non-panmictic they may be with respect to one another in a given generation.

Although I have used the concept of neighbourhood in an earlier paper on *Primula veris* (Richards and Ibrahim, 1978), I have become increasingly concerned about its general application and usefulness, for several practical reasons.

First, measurements of pollen or gene travel are always problematical, and in particular tend to vary very considerably from site to site, year to year, day to day, and hour to hour. In *P. veris*, estimates of σ^2_p vary by a factor of more than 15 at the same site (Richards and Ibrahim, 1978; Ibrahim, 1979).

Second, estimations of neighbourhood size are heavily dependent on the density of plants (d). In a few species, particularly those of dry places such as dunes and deserts, plants are regularly spaced due to the limitations of water or other resources, and in these density may be a highly repeatable function. However, in the very great majority of plant populations, density varies greatly from one location in a population to another. In our experimental populations of *P. veris*, density of plants and thus estimates of neighbourhood size varied by a factor of over 50, which made any estimate of neighbourhood size within a population almost meaningless, although neighbourhood area remains valid. Variation in density estimates does of course depend on quadrat size, which in this case was $1\,m^2$ (where clone size averaged about $10\,cm^2$, or one-hundredth of the quadrat size). Larger quadrats would mean less variation in estimates of density, but if quadrat size approaches or exceeds neighbourhood areas, as would be the case for *P. veris* at even $5\,m^2$, estimates of neighbourhood size would become highly misleading. In fact, because of considerations of varying density, every plant in each population will have a different estimate of neighbourhood size appertaining to itself, and overall population estimates of neighbourhood seem to be meaningless in most cases.

Similar results, and conclusions, to ours have been published by Cahalan and Gliddon (1985) for the related primrose, *Primula vulgaris*.

Apart from regularly spaced populations, the only instances in which neighbourhood otherwise might be a useful concept is when A equals the limits of the population *n*. This will often be the case in outbreeding plants with good pollen travel in which populations are spatially limited by habitat (biological islands). In these populations, neighbourhood calculations may be realistic, but may not be so informative as straight population counts. However, population counts take no account of gene dispersal.

Another reservation concerning the neighbourhood concept results from the assumption that the products of all matings are equally fit. As has been discussed on p. 187, this is frequently not the case, as with 'outbreeding depression', or its corollary. Fenster (1991b) in fact modifies his neighbourhood estimates to take account of the increased fitness of distant matings that he discovered. These increase the effective size of his neighbourhood estimates.

There are also examples in the literature where neighbourhoods have been indirectly inferred from patterns of genotype or allelic frequency subgrouping within and between populations (Golenberg, 1987). This procedure has a distinguished ancestry (Wright, Dobzhansky and Hovanitz, 1942), but in my view it is unsatisfactory as it can involve circularity, and it ignores the likelihood that localized subgroups override gene flow, reflecting adaptations to microenvironmental heterogeneity.

With these very considerable reservations in mind, it may nevertheless be instructive to report on a few neighbourhood estimates in plants that have been reported in the literature (Table 5.7). Some of these have been extracted from the review by Crawford (1984b), and include examples of both wind-pollinated trees and insect-pollinated herbs. In many cases, estimates of seed travel are scanty, or absent.

It will be seen that many estimates are so variable as to be virtually meaningless. In some cases, neighbourhood estimates (*N*) may be very small, indeed sometimes less than ten individuals in extent, especially where populations are sparsely distributed, or limited in extent. Very small plant populations are probably very common, as in *Carex* where a third of all populations may number less than 20 individuals (Chapter 2). In other instances, *N* may be very much smaller than *n*, as in *Primula veris*, where the study population exceeded 10 000 plants. Similarly, in *Lupinus texensis* and *Linanthus parryae*, populations often exceeded 10 000 and were thus very much larger than neighbourhood size calculations. It is noteworthy that all these species are herbaceous perennial entomophilous outbreeders, and at least some (*Primula veris*, *Phlox pilosa*, *Linanthus parryae*) are obligate outbreeders. Yet even in plants of this nature, effective population size, at least within a season, may be very small, and certainly small enough to lead to subgrouping effects, as has been shown by, for example, Zimmerman (1982), Bos, Harmens and Vrieling (1986) and Golenberg (1987).

Table 5.7 Estimations of neighbourhoods in plants

	A (m^2)	*N*	Reference
Trees			
Fraxinus pennsylvanica	42m[a]	16	Wright (1953)
Fraxinus americana	1766	4.4	Wright (1953)
Ulmus americana	1681m	253	Wright (1953)
Populus deltoides	1528m	230	Wright (1953)
Herbs			
Linanthus parryae	30	10–100[b]	Wright (1978)
Phlox pilosa	11–21	75–282	Levin and Kerster (1968)
Lithospermum carolinense	4–26	2–7	Kerster and Levin (1968)
Liatris aspersa	17–30	30–191	Levin and Kerster (1969a)
Liatris cylindracea	33	165	Schaal and Levin (1978)
Viola spp.	19–57	167–547	Beattie and Culver (1979)
Primula veris	20–30	5–200	Richards and Ibrahim (1978)
Lupinus texensis	3–6	42–95	Schall (1980)
Triticum dicoccoides	78[c]	d not given	Golenberg (1987)
Plantago lanceolata	0.4[c]	14–20	Bos, Harmens and Vrieling (1986)
Chamaecrista fasciculata	17	d not given	Fenster (1991a,b)
Asclepias exaltata	160	81	Broyles and Wyatt (1991)

[a]m = length of neighbourhood on linear model.
[b]some allowance made for effective density.
[c]estimates from genetic structure of population.

SEED DISPERSAL

As we have seen, estimations of neighbourhood require information about the variance of seed dispersal, and more recent formulae give seed dispersal values twice the weighting of pollen dispersal in neighbourhood calculations. However, relatively little attention has been given to the seed dispersal of plants in population and breeding system genetics, and there is often a tacit assumption (as in Kerster and Levin, 1968 or Schmitt, 1980) that seed dispersal is much less important than pollen dispersal. Yet neighbourhood estimates are almost doubled in Levin and Kerster (1969a), when seed dispersal variance is included in the calculations, and the estimates of neighbourhood based on gene travel in Schaal (1980) are more than twice those based on pollinator flight distances alone.

There are many estimates of seed travel in plants (reviewed in Harper, 1977), and these range from a few metres to many kilometres. However, this information is rarely sampled randomly, treated statistically, or used in estimates of gene flow and gene migration. In general, it seems likely that seed dispersal schedules will be even more highly skewed than are pollen dispersal patterns, with occasional seeds travelling very long distances, as in colonization events.

The importance of seed travel in determining gene travel and neighbourhood will depend largely on the nature of the seed and the fruit. In *Primula veris*, the variance of seed travel measured by direct means varied from 0.67 to 1.77 m^2. This species has rather large seeds with no specialized means of transport, and compares with estimates of the variance of pollen travel from 1.75 to 13.07 m^2. Rather similar values are obtained by Levin and Kerster (1968, 1969b) for *Phlox pilosa* and *Liatris aspera*, and for *Chamaecrista fasciculata* by Fenster (1991a). However, for the arctic–alpine *Saxifraga hirculus*, seed dispersal variance is only 0.0001 m^2, although pollen dispersal variance is 8 m^2. In such a case, the input of seed travel is negligible (and is in fact less than is vegetative spread) (Olesen and Warncke, 1990). Yet, one would imagine that the pappus-bearing cypselas of *Senecio* would disperse over long distances, and would add greatly to the neighbourhood estimates of Schmitt (1980). Little information is available for most animal-dispersed fruits. Those that pass through a gut, or are carried on fur and feathers and are dispersed by grooming, may habitually travel long distances, but non-randomly. Indeed, their final dispersal points may be very clumped indeed, as under a bird roost, at the entrance of a mouse burrow (*Primula vulgaris*; M. Wilson, personal communication), in an ant nest (*Viola* spp.; Beattie, 1978) or at a car park (burrs of *Acaena novae-zelandiae* on Lindisfarne, UK; Culwick, 1982). As far as I am aware, algebraic estimates of the effects of far-dispersed clumped seed on neighbourhoods have yet to be derived.

However, Beattie and Culver (1979) show that the variance of the dispersal of ant-dispersed *Viola* seed varies from one-sixth to two-thirds times that of the variance of pollinator flight distances. *Viola* seed is first dispersed ballistically, and then a high proportion of the seed is carried by ants, who are attracted by the gelatinous elaiosomes, to their nests, which provide favourable sites for germination and seedling establishment. Ant carry is, of course, random in direction with respect to the initial ballistic phase of dispersal, and is from one-third to one-sixth the mean dispersal variance of ballistically dispersed seed. The variance of seed dispersal varies from 8.2 to 10.2 m^2, much higher than that estimated for *Primula veris*.

The burred fruits of a New Zealand member of the Rosaceae, *Acaena novae-zelandiae*, which has become extensively naturalized on Lindisfarne dunes in the UK, can be transported in the fur and feathers of animals and on the clothes of humans. They mostly disperse a very short distance through the agencies of the weather and small rodents. Culwick (1982) found that the variance of the dispersal of marked single fruits varied from 15 to 30 cm^2, or only about 0.02 m^2. However, when the distance of 50 isolated burrs from the nearest fruiting plant was measured, the variance of the minimum dispersal distance of these proved to be 20 m^2. Some burrs travelled at least 20 m, and the distribution of isolated seedlings suggests that single dispersal events of hundreds of metres may occasionally happen.

The leptokurtosis on the distribution of animal-dispersed fruits may be very extreme indeed. Thus petrels and albatrosses (Procellariidae) have almost certainly carried *Acaena* burrs, found in their downy feathers, thousands of kilometres across southern oceans, and the genus is distributed on all the isolated sub-antarctic islands. In 1981, a Newcastle University botany class discovered a single seedling of *A. novae-zelandiae* beside the path of an upland (300 m) site, 50 km from the nearest site on the coast at Lindisfarne. It immediately occurred to us that during the previous year the equivalent class had visited Lindisfarne and the upland site on successive days, and had been much troubled by the itchy burrs that had adhered to socks etc. from the previous day. This new upland colony has since expanded (1983) to 20 plants distributed over 200 m of track. In recent years, several new coastal populations have also been established, doubtless due to the increased mobility of holiday-makers. Recent dispersal events in the county thus vary from 5 to 50 km, and variance estimates based on these accidental events are exceedingly large.

Abundant evidence is available for the dispersal of seeds and fruits of many kinds (Salisbury, 1942; Altman and Dittmer, 1964) and there is no doubt that seed dispersal can at times be very distant, even intercontinental, with huge variance estimates. It is likely that accurate measurements made on seed dispersal do not take into account such rare, distant dispersal events.

ASSORTATIVE MATING

Up to this point, there has been an assumption that pollen is potentially able to fertilize any plant in the population equally, and if panmixis is restricted and populational subdivision occurs, these are functions of spatial separation alone.

However, it has been realized for many years that pollinators may discriminate between different flowers in the same population, and that features that lead to this discrimination are often heritable, for example flower colour (Kay, 1978, 1982; Levin and Watkins, 1984), shape (Levin and Kerster, 1971, 1974) or height above ground (Levin and Kerster, 1973; Eisikowitch, 1978; Levin and Watkins, 1984). It has often been suggested that intrapopulation variability of this kind, which leads to assortative mating by pollinators, may result in gene barriers that allow sympatric speciation to occur. Many books on the processes of plant speciation have been written, and in some of these the role played by pollinators is given due consideration, especially when written by Grant (1963, 1981), based on his work on *Aquilegia* (Grant, 1952) and particularly *Gilia* (e.g. Grant and Grant, 1965).

Flower-colour polymorphisms

Darwin (1877) pioneered work on intraspecific discrimination by pollinators, but he restricted himself to heterostylous and diclinous species, and to some domesticated flower mutants. Detailed field observations on pollinator discrimination especially between flower-colour polymorphisms, are relatively recent, and have been reviewed by Kay (1978, 1982), and Mogford (1978). Although flower-colour polymorphisms were intensively studied as early as 1942 (Epling and Dobzhansky, 1942) in *Linanthus parryae*, their pollination seems to have been ignored until Lloyd (1969) worked with *Leavenworthia* and Levin (1969) studied *Phlox*.

Kay (1982) provides detailed evidence of assortative within-population pollination for four species with flower-colour polymorphisms (Mogford, 1978 does so for a fifth), and for two species with dioecy, as well as providing a review of the subject.

The ecological and geographical distribution of gene frequencies for flower-colour polymorphisms may show contrasting patterns. In the purple-flowered marsh thistle, *Cirsium palustre*, white-flowered plants are widespread, but become more common with altitude. This distribution is echoed by some other species in the Alps, and other mountainous regions (Mogford, 1978).

In the UK, the mountain pansy, *Viola lutea*, in southern and low-level populations is purple, whereas in upland and northern populations it is yellow. Polymorphic populations, including many plants with different coloured petals, occupy broad intermediate zones. In contrast, white flowers (although with purplish veins) of the predominantly yellow flowered wild radish, *Raphanus raphanistrum*, are restricted in the British Isles to the south and east (Kay, 1978).

Polymorphic populations may be very local. Among very many examples, I can instance the white/pink population (60% white) of *Geranium robertianum* in a small wood at Capheaton, Northumberland, or the white/pink population (30% white) of the storksbill, *Erodium cicutarium*, at Seaton Sluice, Northumberland, UK. It is likely in these isolated instances that white-flowered genes have become established in populations by chance events (genetic drift).

In a few cases, all populations of a plant tend to show flower-colour polymorphism. I believe that all British populations of the fritillary, *Fritillaria meleagris*, are polymorphic for purple and white (Fig. 5.14). This bulbous vernal genus displays polymorphism in a high proportion of its

Fig. 5.14 Purple and white-flowered plants of the fritillary, *Fritillaria meleagris* (Liliaceae) in southern England (×0.5).

100 or so species and their populations. Sometimes these polymorphisms are for a wide range of colours between pale green and almost black.

It is curious and remarkable that a high proportion of the Mediterranean vernal bulbous and cormous genera, such as *Anemone*, *Ranunculus asiaticus*, *Iris*, *Crocus*, *Scilla* and *Orchis* demonstrate flower-colour polymorphisms in a high proportion of populations. This seems never to have been explained satisfactorily. Possibly the range of potential flower visitors is thus maximized for these very early flowering species. It would be interesting to compare the numbers of insect species visiting monomorphic and polymorphic populations. (Compare, however, p. 107.)

Pollination

For devilsbit scabious (*Succisa pratensis*), marsh thistle (*Cirsium palustre*) (both purple/white), crown daisy (*Chrysanthemum coronarium*) and wild radish (*Raphanus raphanistrum*) (both yellow and white), Kay (1978, 1982) convincingly shows that pollinators discriminate between colour morphs in wild populations. In the last two species in particular, pollinator visits are of crucial importance, for both are annuals, and *Raphanus* at least is self-incompatible.

By marking individual pollinators, Kay has been able to study discrimination not only by a species of pollinator, but by an individual. In so doing, he has revealed, somewhat unexpectedly, that different individuals within a pollinating species at a single site frequently show differential discrimination of flower colours. Thus, individuals of hive bees (*Apis*) visiting *Succisa* may specialize on purple flowers or on white flowers, or do not discriminate between them. Nevertheless, from Kay's results, some generalizations about pollinator preferences can be made. For *Cirsium* and *Succisa*, butterflies of several species nearly always visit purple rather than white flowers. For *Chrysanthemum* and *Raphanus*, hive bees nearly always visit yellow (insect purple) rather than white (insect white) flowers. However, generalizations concerning one plant break down when another is investigated. Butterflies (chiefly *Pieris* spp.) very markedly prefer yellow flowers of *Raphanus*, whereas some individuals also prefer yellow flowers of *Chrysanthemum*, others prefer white flowers, and yet others do not discriminate between the colours. Kay suggests that *Pieris* species may use the yellow flowers of some Brassicaceae (such as *Raphanus*) as a recognition signal for choice of plant for egg laying, and thus innately prefer yellow-flowered plants for nectar feeding as well.

Kay (1978) also notes differential discrimination on *Raphanus* between the long-tongued *Bombus pascuorum* (white flowers) and the short-tongued *B. terrestris* (yellow flowers). This he suggests may relate to the more abundant, and short-tubed, charlock (*Sinapis arvensis*), which is always yellow flowered, but superficially closely resembles *Raphanus*. It is possible that many individuals of *B. terrestris* major on *Sinapis*, and cannot differentiate between this species and yellow *Raphanus*. In contrast, the long-tongued *B. pascuorum* may find the short-tubed *Sinapis* relatively unrewarding, as many flowers will have been visited by other species. It may therefore avoid yellow cruciferous flowers, but recognize the white flowers of long-tubed *Raphanus* as a more rewarding source.

Pollinator discrimination between flower-colour morphs leading to assortative mating (which has been tested in progeny) is a common and widespread phenomenon in polymorphic populations, although it is probably never complete enough to lead to a total breeding barrier.

Other characters leading to population subdivision

Levin (1969) gives an example of discrimination by pollinating butterflies between cultivars of *Phlox drummondii* with different corolla shapes. He additionally observed discrimination between different flower-colour morphs in this species. Levin (1972) also shows that interspecific differences in corolla outline between two sympatric species of *Phlox*, *P. bifida* and *P. divaricata*, are probably largely responsible for the very considerable discrimination between these species shown by their lepidopteran pollinators. Only 1–2% of outcross pollinations are between species in mixed populations. Very similar values (1.6% pollinations interspecific) were obtained by Levin and Berube (1972) for artificial mixed populations for another pair of *Phlox* species, *P. pilosa* and *P. glaberrima* pollinated by the butterfly *Colias eurytheme* (a 'sulphur', or 'clouded yellow'). These species have flowers with very different tube lengths and anther and stigma positions.

It seems that lepidopteran pollinators may respond more strongly to flower shapes with complex outlines. Herrera (1993) compared levels of fruit-set for morphs with different corolla outlines among populations of the long-spurred shrubby local endemic *Viola cazorlensis* (Spain), showing that narrow-petalled flowers were preferentially visited by the sphingid *Macroglossum stellatarum*.

Intraspecific discrimination according to the height of the flowers above ground occurs as a result of constancy in foraging height (p. 155). This phenomenon has been demonstrated by Levin and Kerster (1973) for dwarf and tall forms of purple loosestrife, *Lythrum salicaria*, and by Eisikowitch (1978) for artificially mixed populations of the tall inland love-in-the-mist *Nigella arvensis* ssp. *tuberculata*, and its dwarf maritime ecotype ssp. *divaricata*.

Population subdivision frequently occurs due to differences in flowering time, and such differences are probably among the commonest stimuli to sympatric speciation. Such familiar interfertile species pairs in the British flora as the red and white campions (*Silene dioica* and *Silene latifolia* (= *alba*)), the primrose and cowslip (*Primula vulgaris* and *P. veris*) and herb bennet and water avens (*Geum urbanum* and *G. rivale*) are separated by flowering time as well as at least partially by habitat. Hybridization is commonest in the north of Britain where overlap in flowering time is the greatest (Stace, 1975). Mutants with abnormal flowering seasons quite frequently occur in wild populations (the winter flowering form of hawthorn, *Crataegus monogyna*, known as 'Glastonbury Thorn' is a familiar example). Such mutations often respond to 'Wallace Effects' where hybridization between two ecodemes is disadvantageous, so that mutants which separate the ecodemes by season are selectively favoured. This has been found to occur in Welsh mine populations of the grasses *Agrostis capillaris* and *Anthoxanthum odoratum* (McNeilly and Antonovics, 1968), where metal-tolerant races flower some six to eight days earlier than non-tolerant plants growing in adjacent non-toxic sites. This difference is sufficient to provide some reproductive isolation for the tolerant ecodeme, and helps to preserve the genetic integrity of the tolerant and non-tolerant races. Zinc-tolerant races of the bladder campion, *Silene vulgaris*, similarly flower several weeks earlier than non-tolerant races in Germany (Broker, 1963). Unlike the grasses, this species is insect pollinated.

Behavioural (ethological) discrimination by pollinators on flower variation within populations can doubtless encourage speciation (Waser and Price, 1983). In the bee-pollinated *Pedicularis* (Macior, 1982) and the pseudocopulatory *Ophrys* (Stebbins and Ferlan, 1956), it is probable that pollinator discrimination has played a major role in speciation. Flower-colour variation may also have led to speciation in genera such as *Aquilegia* (Grant, 1952) and *Mimulus*, where variation between white, blue and red, and yellow and red, respectively, can differentially attract such different classes of pollinators as moths, butterflies and birds, or bees and birds.

Ellstrand and Elam (1993) discuss how such breeding barriers can break down for small threatened populations of rare species which may then suffer from hybridization with more widespread and abundant (and in some cases introduced) relatives. In these cases, plants may be sufficiently rare as to no longer form a major food source for pollinators, so that any visitation which results in cross-pollination may be made by a pollinator which is chiefly visiting the cross-fertile relative. They cite 22 such cases in California. In the British Isles, *Viola persicifolia, Ranunculus reptans* and *Prunella laciniata* are among rare species whose specific integrity has been upset by hybridization (with *V. canina, R. flammula* and *P. vulgaris*, respectively).

CHAPTER 6

Multi-allelic self-incompatibility

INTRODUCTION

Self-incompatibility (s-i) is best defined as the inability of a fertile hermaphrodite seed plant to produce zygotes after self-pollination (De Nettancourt, 1977). It is a mechanism that usually ensures obligate outbreeding, but it may carry the penalty of reproductive inefficiency. Over 316 species, the average fruit-set per flower in self-compatible species is 72.5%, but in self-incompatible species it is only 22.1% (Sutherland and Delph, 1984) (Table 2.4).

Self-incompatibility results from the failure of selfed pollen grains to adhere to, or germinate on, the stigma, or the failure of selfed pollen tubes to penetrate the stigma, or grow down the style. With a few possible exceptions (e.g. *Borago officinalis*, Crowe, 1971), all self-incompatibility operates before fertilization. This definition thus rules out failure of sexual fertilization due to sterility, embryonic lethality or breeding barriers (p. 239).

We must suppose that cultivators have realized for thousands of years that many plants are self-incompatible, that is self-sterile but cross-fertile. However, early botanists were principally concerned with the systematics and medicinal properties of plants, and little attention was paid to their reproduction. Because most flowers were seen to be hermaphrodite, it was assumed that they self-fertilized (Kolreuter, 1763; Sprengel, 1793). It was Kolreuter, however, who has been credited with the first published observation of self-incompatibility, in *Verbascum phoeniceum*. Further casual observations of self-incompatibility were made during the 19th century, but there was a tendency to regard these as the result of inbreeding, and their true significance was missed. Darwin (1876) studied self-incompatibility in five species, but he ignored the existence of cross-incompatibility, and the importance of this phenomenon in understanding the inheritance of self-incompatibility. As he was also in ignorance of the principles of genetics, this is perhaps understandable. He did, however, note that self-incompatibility favours outbreeding, and that fertilization is prevented when the sexual elements are identical. There is a good review of early work on self-incompatibility in Arasu (1968).

The rediscovery of Mendel's paper on inheritance at the turn of the century stimulated further work on incompatibility, and early attempts were made to fit patterns of self-incompatibility into the new genetics (De Vries, 1907; Correns, 1913). In 1917, Stout coined the world self-incompatibility (self-sterility, which now has a wider meaning, had been used until then). It was not until the 1920s that a group of papers, which were produced independently, first showed how multi-allelic gametophytic self-incompatibility was inherited (Prell, 1921; East and Mangelsdorf, 1925; Lehmann, 1926; Filzer, 1926).

It is curious that, even today, little attention is paid to self-incompatibility except by plant breeders and research workers concerned with breeding systems. With the exception of Clapham, Tutin and Warburg (1962), no regional floras give information on the breeding system in general or self-incompatibility in particular. There have been few concerted attempts to collect information on self-incompatibility in plants in a systematic way, as has been done for chromosome numbers, pollen morphology, anatomy, cuticle structure, secondary chemical products, ecology and chorology. The exceptions are East (1940) and Brewbaker (1957). Textbooks on genetics rarely give self-incompatibility in plants more than a passing mention (Williams, 1964 is a distinguished exception). Usually, any treatment in depth refers to the much more uncommon diallelic, heteromorphic conditions such as that of *Primula* (Chapter 7). Even modern research papers on pollination biology and gene flow usually manage to omit any information about self-incompatibility. Confusions between self-pollination and self-fertilization are rife in the literature.

At the simplest level, it is extremely easy to establish whether a plant is self-incompatible or not. If individual plants are isolated, they will set little if any seed if they are self-incompatible (s-i). If they are self-compatible (s-c) they will usually set seed, especially when they have been artificially self-pollinated (between flowers of the same plant). If s-i is suspected, it is readily confirmed by making a cross with another individual genet (i.e. a plant that has arisen from another seedling). This should usually result in seed-set.

If no seed is set after cross-pollination, disease or predation of seeds (examine when young), or sterility (examine pollen stainability) should be suspected.

Self-compatibility or self-incompatibility are often considered to be synonymous with the breeding system (as in 'breeding system genetics'). However, they form only one function in a suite of characteristics which control the breeding system, all of which are important, and all of which must be considered together:

1. Within-flower position of anthers and stigma (herkogamy); is within-flower pollination (autogamy) possible without an animal visit?

2. Within-flower timing of anther dehiscence and stigmatic receptivity (dichogamy); is within-flower pollination possible with respect to timing?
3. Amount of pollen carry to stigmas between flowers of the same plant (geitonogamy) and different plants (xenogamy) (together, allogamy); degree of pollen carryover (Chapter 5);
4. Number of different plants (genets) normally reached by cross-pollination from a single source (neighbourhood size, Chapter 5).
5. Behaviour of the pollen on stigmas of the same genet (s-c or s-i).
6. Proportion of the ovules behaving sexually (possibility of agamospermy, Chapter 10).

It is frequently found that an s-i species, which is an obligate outbreeder, nevertheless undergoes substantial self-pollination. Conversely, a s-c species may rarely, if ever, be self-pollinated and is thus effectively an outbreeder (this is true of many orchids, Chapter 4). As a result of herkogamy or dichogamy, many s-c species undergo no within-flower, or even within-ramet, self-pollination.

THE BASIS OF MULTI-ALLELIC GAMETOPHYTIC SELF-INCOMPATIBILITY (gsi)

Early workers in the field soon established the principle that recognition of self led to incompatibility, and that recognition factors were heritable (East, 1915a,b, 1917a,b, 1918, 1919a,b,c; Correns, 1913, 1916a; Stout, 1916, 1917, 1918a,b; East and Park 1917; Baur, 1919). Work on *Nicotiana* by Anderson (1924) and especially by East and Mangelsdorf (1925) showed in some detail how one form of self-incompatibility (which we now know as one-locus gametophytic) was inherited. There is a useful early review by Sirks (1927), and for fruit trees by Crane and Lawrence (1929).

The following principles were established on the basis of this work.

1. The same alleles controlled recognition factors operating in the pollen and the stigma.
2. These factors operated independently in the diploid stigma, and in the haploid pollen grains; thus pollen from the same diploid father could show two types of reaction on a single stigma (semi-compatible) (Fig. 6.1).
3. When the pollen grain and the stigma carried the same factor, incompatibility ensued; thus successful crosses arose between gametes carrying different factors, and plants were by definition heterozygous for incompatibility factors.

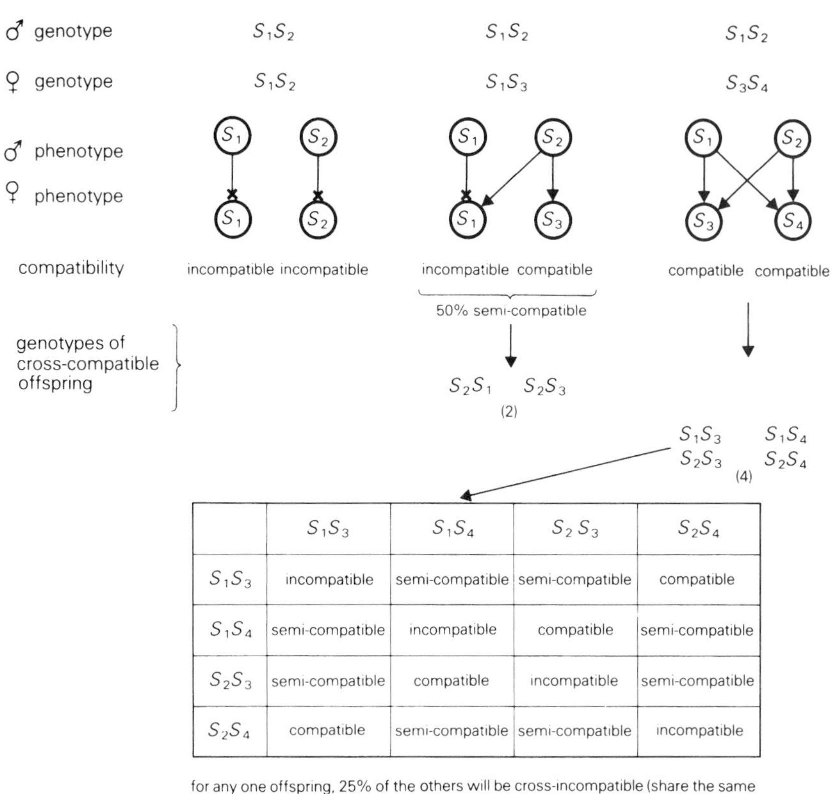

	S_1S_3	S_1S_4	S_2S_3	S_2S_4
S_1S_3	incompatible	semi-compatible	semi-compatible	compatible
S_1S_4	semi-compatible	incompatible	compatible	semi-compatible
S_2S_3	semi-compatible	compatible	incompatible	semi-compatible
S_2S_4	compatible	semi-compatible	semi-compatible	incompatible

Fig. 6.1 Behaviour and inheritance of *S* alleles in one-locus gametophytic self-incompatibility.

4. Recognition factors were inherited at many alleles at the same locus; this was known as the *S* locus, and alleles were characterized by S_1 S_2 $S_3 \ldots S_n$.

Work over the following 60 years has revealed many exceptions to these early principles. We are also much closer to understanding the physiological basis of self-incompatibility, which has proved surprisingly difficult to elucidate.

The experimental basis of early work on gametophytic systems usually took the following pattern:

1. cultivation of as many genets as possible of a population or species that showed self-incompatibility;
2. creation of as many pair-wise crosses within this material as possible;

3. examination of pollen-tube growth and/or seed-set of these pair-wise crosses;
4. classification of parents into groups, within each of which plants are cross-incompatible. These are assumed to share the same *S* genotype; if pollen growth is examined, semi-compatible crosses can also be identified (Fig. 6.1);
5. determination of the number of alleles residing at the *S* locus by the formula: number of distinct *S* genotypes = $n(n - 1)/2$ where n is the number of alleles in the system (Lewis, 1955). This equation must be solved quadratically. It is also possible to show that the proportion of all crosses that are incompatible equals $2/n$;
6. growing to maturity families of siblings from one or more crosses, and determining the number of cross-compatible classes that occur between siblings of a single cross.
   (a) If the parents have totally different genotypes, four cross-compatible classes should arise among the offspring; for each offspring, a quarter of the others will be incompatible, half semi-compatible and a quarter fully compatible (Fig. 6.1):

   (b) If the parents share an *S* allele (semi-compatible), two cross-compatible classes should arise among the offspring:

   If more cross-compatible classes arise among the offspring family, it can be assumed that more than one incompatibility locus is involved (Fig. 6.1). If fewer cross-compatible classes arise among the offspring family, it is likely that the system is under sporophytic control.

Frequency dependence

One of the features of the *S*-allele system is the large number of alleles that can coexist, apparently at one locus, within a population. In *Oenothera organensis*, Emerson (1939) reported the occurrence of 45 alleles in a relict population of approximately 500 plants. Extrapolations from the proportion of *S* alleles discovered to be novel when new plants were added to the system allowed Lewis (1955) to suggest that up to 400 alleles may coexist within the world-wide distribution of the clovers *Trifolium pratense* and *T. repens*.

If this estimate is correct, further work has suggested that it is unusually high. O'Donnell and Lawrence (1984) have thoroughly examined *S* allele frequency in a number of populations of the poppy *Papaver rhoeas*. In three large but well-separated populations in the English Midlands they discovered 32, 35 and 38 *S* alleles respectively, estimating that each probably contained approximately 40 *S* alleles. The correspondence between the *S* alleles in each population was very high, and they considered that each population was virtually identical in its *S* allele structure, suggesting that all poppy populations in this area were effectively members of one 'superpopulation'. Examination of Spanish poppy populations showed that on average they possessed a 53% resemblance to British poppy populations for *S* alleles, but surprisingly few new alleles were added to the system, so that Lawrence was able to suggest that within the whole species less than 66 *S* alleles probably occur.

Clearly, frequency dependent selection should act on *S* alleles. Alleles rare in pollen will be favoured as they will rarely encounter the same alleles in stigmas. Conversely, common alleles will be selected against as they will frequently alight on cross-incompatible stigmas. Lewis (1949a) concluded that whereas selection pressures will act on most genetic loci to reduce the number of coexisting alleles (most of which will be disadvantageous), the opposite will be true for incompatibility loci. If it is assumed that all alleles are equally viable, frequency-dependent selection will encourage the coexistence of large numbers of alleles, each at the frequency of $1/n$. However, the advantage borne by a novel allele will become decreasingly small as the number of alleles in a population increases. Consequently, it is unlikely that selection will favour the maintenance of *S* allele numbers greatly in excess of 50 within a population.

O'Donnell and Lawrence (1984) show for *Papaver rhoeas* that not all alleles do in fact occur within a population at an equal frequency. They suggest that such anomalies may be due to differential fitnesses between *S* alleles caused by 'hitchhiking' (p. 23), where certain alleles are closely linked to advantageous or disadvantageous alleles of unrelated genes. In such cases, where *S* allele numbers in populations are small, and rare *S* alleles are thus strongly advantageous, the linkage disequilibrium of such rare *S* alleles with unrelated disadvantageous genes could cause the latter to persist in populations against selection pressures. This is a potential penalty of small population size (p. 49) which is an artefact of the mating system.

GAMETOPHYTIC AND SPOROPHYTIC SYSTEMS

Even before the basis of gametophytic self-incompatibility (gsi) was first established, Darwin (1862, 1877) and Hildebrand (1863) had worked with self-incompatible heteromorphic *Primula*. They were able to raise some offspring from selfs and showed that pin selfs gave pin offspring, but that thrum selfs gave both pin and thrum offspring. The significance of these simple results was noted much later, after the rediscovery of Mendel's genetics. Thrum selfs can generate both morphs and must be heterozygous for the genes controlling the mating system. Thus they must produce pollen grains of two genotypes (*S* and *s*). Yet, all thrum pollen grains behave in the same way, as thrums (*S* phenotype). We can conclude in this case that the control of pollen grain behaviour is invested not in the pollen grain itself, as is the case for gametophytically controlled systems, but rather in the male parent. In other words, behaviour is controlled by the diploid sporophyte (sporophytic control).

Work by Correns (1912, 1913) on *Cardamine pratensis* showed that a similar inheritance of a di-allelic sporophytic incompatibility system could occur in the absence of distyly or heteromorphy. However, it was thought for many years that sporophytic systems were di-allelic and gametophytic systems multi-allelic. It was not until 1950 that Gerstel, and Hughes and Babcock demonstrated multi-allelic sporophytic incompatibility (ssi) in *Parthenium argentatum* and *Crepis foetidus*, respectively. Similar systems were later shown to operate in *Cosmos bipinnatus* (Crowe, 1954), *Iberis amara, Raphanus sativus* and *Brassica campestris* (Bateman, 1954) and by later workers in *Brassica oleracea* (cabbage) and *Raphanus raphanistrum* (wild radish). These species all belong to the families Compositae (Asteraceae) and Cruciferae (Brassicaceae), respectively. It seems to be accepted today that multi-allelic sporophytic systems are largely confined to these two large families of plants, which contain many species of agricultural significance, and that all s-i plants in these families have sporophytic systems. However, examples of sporophytic self-incompatibility (ssi) are also known from a few other families, for instance Convolvulaceae and Capparidaceae.

Evolution of gsi and ssi systems

It is remarkable that the families Cruciferae and Compositae have no obvious phylogenetic relationship. It has been suggested that sporophytic systems are secondary, having been derived from self-compatible plants, or (which I consider less likely) from gametophytic systems directly. If this is the case, similar systems appear to have arisen independently early in the developmental history of these two plant families.

In contrast, di-allelic sporophytic systems occur in some 13 other plant families, usually in conjunction with a floral dimorphy. These systems are also considered to be secondary in origin, having arisen from self-compatible plants (Chapter 7). It is not possible for a di-allelic system to function under a gametophytic control, the minimum number of alleles permissible at a gametophytic locus being three, as shown below for a hypothetical di-allelic gametophytic system:

1. *SS* (male) × *ss*: *S* pollen can function giving *Ss* offspring only;
2. *ss* (male) × *Ss*: no pollen can function.

Gametophytic multi-allelic systems (gsi) are far more widespread than sporophytic systems (ssi). Our knowledge of their distribution is still very incomplete, but Darlington and Mather (1949) estimated that half the species of Angiosperms were s-i, and Brewbaker (1957) recorded s-i in at least 71 families, and in 250/600 genera. In a much smaller sample of 25 species of tropical forest tree, Bawa (1979) showed that 88% were s-i. Certainly, a high proportion of Angiosperm species shows gsi, and this may be higher than 50% overall.

What is certain is that gsi is found in most orders of flowering plants, and is found in what are traditionally considered to be the most primitive orders in the dicotyledons and the monocotyledons, the Magnoliales, Winterales, Hamamelidales and Nympheales. This provides reasonably strong circumstantial evidence that gsi is primitive to the Angiosperms, particularly as comparable systems have been discovered in the Pteridophytes and Gymnosperms (Bateman, 1952). Such a view has not been seriously challenged since it was first proposed by Whitehouse (1950), who, however, differed from Bateman about the point of the first evolution of s-i. Whitehouse considered that s-i arose very early in the history of the Angiosperms, but not before, and was associated with the development of the hermaphrodite flower and the closed carpel. He believed that it played a crucial role in the early expansion and subsequent success of the Angiosperms, and doubtless in this at least he was correct.

Generally speaking, it has been considered that all other mechanisms of s-i are secondary, having arisen from one-locus gametophytic systems as follows:

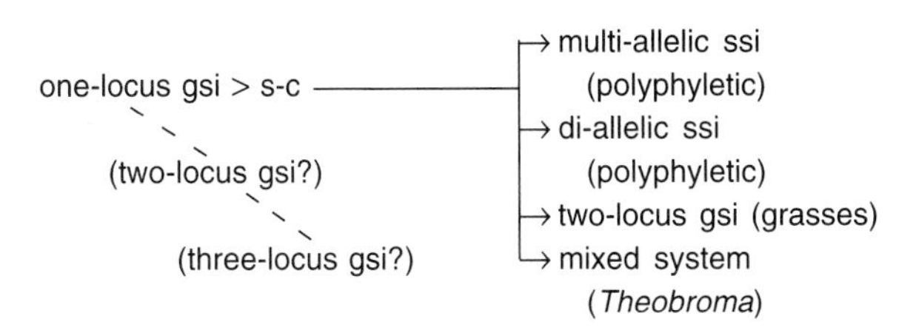

Identification of s-i systems

In order to identify the system operating in a self-incompatible plant, attention should be paid to three features:

1. The number of cross-compatible types arising in a parental population;
2. The occurrence of semi-compatibility; i.e. does all pollen from a single parent always behave the same way on a single female parent?

3. The number of cross-compatible classes that arise in the family resulting from a single cross.

Systems can be identified as follows:

1. Number of cross-compatible groupings among parents:
   (a) one – di-allelic sporophytic;
   (b) three or more – multi-allelic.
2. Occurrence of self-incompatibility;
   (a) all crosses fully compatible or fully incompatible – sporophytic;
   (b) some crosses semi-compatible (examine pollen) growth – one-locus gametophytic (Fig. 6.1);
   (c) some (very few) crosses quarter-compatible – two-locus gametophytic (Fig. 6.2).
3. Number of cross-compatible groupings in a single family from a fully compatible cross:
   (a) four – one-locus gametophytic;
   (b) more than four – two-locus gametophytic or three-locus gametophytic (see Fig. 6.2);
   (c) fewer than four – sporophytic.

The reason why sporophytic systems usually generate fewer cross-compatible classes among offspring is discussed later (p. 224). It is due to dominance interactions between *S* alleles.

There are features of pollen, the stigma, and pollen germination and growth in vitro, and in incompatible crosses that separate gametophytic and sporophytic systems (Table 6.1). Although it would involve a circularity to state that these can identify the system, their correspondence with various systems is very high, and these features provide a valuable confirmation as to which system is operating. They also provide valuable clues as to how the various systems function (Brewbaker, 1959; Heslop-Harrison, 1975b).

The close correspondence within gametophytic or sporophytic systems for these various features allows one to suggest the following points.

1. Gametophytic and sporophytic systems differ fundamentally in their operation.
2. These features might be functional with respect to the operation of each system and coadapted.
3. Gametophytic and sporophytic systems have each arisen on only one occasion (monophyletically). Although this is probably true for one-locus gametophytic systems, the sporadic, scattered occurrence of di-allelic and multi-allelic, heteromorphic and homomorphic sporophytic systems between unrelated taxa strongly suggests that these are in fact polyphyletic. If this is the case, we must assume that those features of

♂ Parent genotypes	Pollen phenotype combinations		♀ Parent genotype: $S_1S_2Z_1Z_2$				Compatibility		Offspring genotypes
			S_1Z_1	S_1Z_2	S_2Z_1	S_2Z_2			
$S_1S_2Z_1Z_2$	S_1Z_1	self	incompatible				incompatible	fully self-incompatible self	
	S_1Z_2			incompatible			incompatible		
	S_2Z_1				incompatible		incompatible		
	S_2Z_2					incompatible	incompatible		
$S_1S_2Z_1Z_3$	S_1Z_1	cross	incompatible				incompatible	50% semi-compatible cross	$S_1S_1Z_1Z_3$ $S_1S_2Z_3Z_2$ $S_1S_1Z_2Z_3$ $S_2S_2Z_1Z_3$ $S_1S_2Z_1Z_3$ $S_2S_2Z_2Z_3$ (6 cross-compatible classes)
	S_1Z_3		✓	✓	✓	✓	compatible		
	S_2Z_1				incompatible		incompatible		
	S_2Z_3		✓	✓	✓	✓	compatible		
$S_1S_3Z_1Z_3$	S_1Z_1	cross	incompatible				incompatible	75% semi-compatible cross	$S_1S_1Z_1Z_3$ $S_1S_3Z_1Z_3$ $S_1S_1Z_2Z_3$ $S_1S_3Z_2Z_3$ $S_1S_2Z_1Z_3$ $S_2S_3Z_1Z_1$ $S_1S_2Z_2Z_3$ $S_2S_3Z_1Z_2$ $S_2S_3Z_1Z_3$ $S_2S_3Z_2Z_3$ (10 cross-compatible classes)
	S_1Z_3		✓	✓	✓	✓	compatible		
	S_3Z_1		✓	✓	✓	✓	compatible		
	S_3Z_3		✓	✓	✓	✓	compatible		
$S_3S_4Z_3Z_4$	S_3Z_3	cross	✓	✓	✓	✓	compatible	fully compatible cross	$S_1S_3Z_1Z_3$ $S_1S_4Z_1Z_3$ $S_1S_3Z_2Z_3$ $S_1S_4Z_2Z_3$ $S_2S_3Z_1Z_3$ $S_2S_4Z_1Z_3$ $S_2S_3Z_2Z_3$ $S_2S_4Z_2Z_3$ $S_1S_3Z_1Z_4$ $S_1S_4Z_1Z_4$ $S_1S_3Z_2Z_4$ $S;S_4Z_2Z_4$ $S_2S_3Z_1Z_4$ $S_2S_4Z_1Z_4$ $S_2S_3Z_2Z_4$ $S_2S_4Z_2Z_4$ (16 cross-compatible classes)
	S_3Z_4		✓	✓	✓	✓	compatible		
	S_4Z_3		✓	✓	✓	✓	compatible		
	S_4Z_4		✓	✓	✓	✓	compatible		

Fig. 6.2 Behaviour and inheritance of two-locus gametophytic self-incompatibility in grasses. Number of haploid genotypes considered (left-hand column) are $4S \times 4Z = 16$.

Table 6.1 Morphological and physical features that differentiate multi-allelic sporophytic and gametophytic systems

Features	gsi	ssi[a] plus grasses
Pollen grain:	Binucleate	Trinucleate
respiration	Low	High
viability	Long	Short
growth *in vitro*	Easy	Difficult
Stigma papillae	'wet' with gappy cuticle	'dry' with entire cuticle
Site of incompatible tube inhibition	Style	Stigma surface
Site of callose deposition in incompatible pollen	Intine	Exine

[a] Excluding di-allelically controlled ssi systems.

a sporophytic system common to all taxa possessing such a system are necessary for it to operate correctly.

4. As binucleate gsi pollen only expresses incompatibility recognition after the second pollen-grain mitosis occurs in the pollen tube, it has been suggested that the timing of gene expression for gametophytic incompatibility in the pollen must occur late (Brewbaker, 1959). In sporophytic systems, it is argued, the second pollen-grain mitosis can occur much earlier, for the phenotype of the pollen grain is expressed by the anther, not the pollen. This view has been challenged by Lewis (1949b) and Pandey (1959, 1970), whose mutagenic work suggests that expression of gametophytic pollen identity occurs before either pollen-grain mitosis.
5. The Gramineae (= Poaceae, grasses) show all the characteristics of a sporophytic system, with trinucleate grain-type behaviour. However, they display a two-locus gametophytic control of incompatibility (p. 216). This suggests that self-incompatibility in the grasses (a very derived type of plant family) has arisen secondarily, in the same way as the sporophytic systems have, not being part of the evolutionary continuum of other gametophytic systems. It is not a precondition of multilocus gametophytic systems to have trinucleate-type behaviour, for some members of the Ranunculaceae have three-locus gametophytic systems with binucleate grain-type behaviour (Lundqvist *et al.*, 1973). Perhaps trinucleate grain-type behaviour is characteristic of all secondary systems, and only the evolutionary developments of the ancestral gametophytic system have binucleate grains.

GENETIC CONTROL OF GAMETOPHYTIC (gsi) SYSTEMS

The most frequent form of s-i in the Angiosperms, perhaps occurring in half the plant families, is one-locus multi-allelic gametophytic control. This system, first fully elucidated by East and Mangelsdorf (1925), for *Nicotiana sanderae*, has been considered primitive to the Angiosperms by Whitehouse (1950), and it certainly occurs in many primitive taxa. Essentially, between three and over 400 alleles appear to occur at a single locus. Diploids are heterozygous for two of these alleles, both of which are expressed in the stigma and style. Pollen grains show independent expression, segregating at 1:1, and thus crosses can be incompatible, semi-compatible (with half the grains growing normally) or fully compatible, depending on the genotype of each parent (see Fig. 6.1). Because matings can only occur between pollen grains and stigmas which contain different *S* alleles, offspring must be heterozygous.

There has been a good deal of discussion as to the nature of the *S* gene locus, and the generation of new *S* gene mutants. Two basic models of the structure of the *S* gene have been proposed. In the first, originally put forward by Lewis (1942b, 1949a), and later championed by Lundqvist (1960, 1964), it is assumed that the gene has two equal parts, both of which control specificity, one in the pollen and the other in the style.

However, a number of workers have shown that pollen-part s-c mutants tend to be associated with centric chromosome fragments, or with cytologically evident duplications in members of the Solanaceae (summarized in de Nettancourt, 1977). It has been suggested (Pandey, 1967) that such fragments or duplications contain *S* gene material of a type different from that inherited on the basic genome. This fragmental *S* gene material may be dominant to the genomic *S* allele, and confers pollen-part self-compatibility. In other cases (de Nettancourt, 1975), duplications of nucleolar organizer chromosome segments may reinstate ribosomal RNA formation in the potentially incompatible pollen tube. Although the association of such phenomena with s-c mutants is common in the Solanaceae, they have not been associated with stylar-part mutants, and they have not been recorded in other plant families. As yet their general significance is unclear.

The tripartite model of S *gene structure*

After work which involved the experimental irradiation of pollen at different ages (Lewis and Crowe, 1953, 1954), Lewis produced a new model (1960) which anticipated the classical regulatory models of gene control by Jacob and Monod (1961). Lewis envisaged a single specificity component of the gene, closely linked to two activator components, one for the pollen and one for the style. Pandey (1956, 1957), van Gastel (1972), de Nettancourt (1972), and van Gastel and de Nettancourt (1974, 1975) have identified mutants that confer self-incompatibility only to the male function (pollen-part mutants) or only to the female function (stylar-part mutants) or to both (specificity mutants) (Table 6.2).

Although the tripartite model of *S* gene structure and action has a logical elegance, and has attracted some experimental support, Lewis (1960) was unable to obtain recombinants between the three hypothetical segments of the gene. This may merely suggest that they are very closely linked.

S gene mutants are usually detected by noting occasional self-compatible pollen grains or seed-set in an otherwise fully s-i self. Care must be taken that such occasional self-compatibility does not have other causes, such as pseudocompatibility, unreduced pollen grains (which are diploid and may show dominance effects), the effects of timing, high temperatures etc., which are dealt with later in this chapter. Estimates of frequencies of spontaneous *S* gene mutation are given in de Nettancourt (1977) and vary from 0.2 to 4.3 per million grains. Such estimates do not distinguish between the three possible types of s-c pollen mutants unless the rate of occurrence of s-c is compared amongst within-ramet and between-ramet selfs (see Table 6.2).

In particular, it is of interest to enquire whether specificity mutants change the specificity of the *S* allele (S_1 to S_3), or lose it (S_1 to S_0). Both of these types of change would lead to self-compatibility in the first generation. However, the progeny of mutant selfs, selfed, will be s-c if specificity is lost (S_0). Conversely, if specificity is changed (S_1S_2 to S_1S_3) so that, for instance, S_3 mutant pollen can self-germinate on a S_1S_2 stigma, the seedlings from such selfs will be S_1S_3 or S_2S_3 and these offspring will revert to s-i. This has been called revertible mutation. As de Nettancourt (1977) shows revertible mutation can only be satisfactorily explained by the tripartite model of *S* gene structure, where changes rather than losses in specificity have occurred.

As yet, it is still unclear whether the tripartite model of *S* gene structure is robust enough to account for the following findings with respect to gsi.

1. Large numbers of different alleles can apparently coexist at an *S* gene locus. At the low frequencies at which each occurs, stochastic effects are likely to result in an attrition of alleles (Wright, 1939). Consequently, it is necessary to invoke high levels of *S* allele mutation if the maintenance of *S* allele diversity within populations is to be satisfactorily explained (Fisher, 1961).
2. When mutagens are used to enhance rates of the occurrence of s-c mutants, these mutants, often associated with gross chromosomal changes, seem only to involve a loss of specificity (S_1 to S_0) (Nasrallah

Table 6.2 Detection of pollen-part, stylar-part and specificity mutants

	Type of mutant		Specificity	
	Pollen-part	Stylar-part	*S* allele changed	*S* allele lost
Self	Self-compatible	Self-compatible	Self-compatible	Self-compatible
Cross with another ramet of same genet:				
male parent	Cross-compatible	Cross-incompatible	Cross-compatible	Cross-compatible
female parent	Cross-incompatible	Cross-compatible	Cross-compatible	Cross-compatible
Self on seedling from mutant self	Self-compatible	Self-compatible	Self-incompatible	Self-compatible

et al., 1969; de Nettancourt, 1977). Both pollen-part and stylar-part mutants of this kind have been reported.

3. In the absence of mutagens, the spontaneous mutation of new *S* specificity alleles (e.g. S_1 to S_3) has indeed been reported in *Trifolium* (Denward, 1963; Anderson *et al.*, 1974), *Nicotiana* (Pandey, 1970), and *Lycopersicon* (de Nettancourt *et al.*, 1971; Hogenboom, 1972). These usually occur when experimental populations have been subjected for some generations to forced inbreeding (by heat treatment, hormone treatment or bud pollination). For *Lycopersicon*, de Nettancourt *et al.* (1971) showed that a single new *S* allele (S_3) arose repeatedly in the styles of a clone homozygous for S_2 (because of forced inbreeding), and the S_3 mutant could occasionally revert to S_2 again.

If the tripartite *S* gene structure of Lewis (1960) is to be accepted, it is simplest to assume that the pollen-part and stylar-part activator components each only have one functional state. If mutations occur to these components, they will be disabled, so that they can no longer express the specificity component. The result is a generalized self-compatibility (S_0). It is very unlikely that further mutations to these disabled activator components could ever precisely restore their function, so that backmutations of the gsi system to renewed activity are unlikely ever to occur.

However, new *S* alleles can apparently arise by mutation to the specificity component of the gene, and these have two immediate advantages: firstly, they will allow selfing to occur in the generation in which they arise so that they stand a good chance of immediate establishment. Also, being rare, they will bear a frequency-dependent advantage initially.

For such novel specificity mutants, several models of origin can be proposed, none of which is entirely satisfactory.

1. MODEL 1. The *S* locus consists of a single cistron. *S* alleles differ by minor changes in the nucleotide sequence within a relatively short section of DNA giving gene products that only differ from each other by a few amino acids. The main problems with this delightfully simple model are that the mutation rate would be expected to increase with exposure to mutagens (which seems not to be the case for specificity mutants). Also, nonsense and frameshift type mutants leading to a loss of function for the specificity component of the gene would also be expected to occur. However, such s-c S_0 mutants seem only to occur as pollen-part or stylar-part mutants.
2. MODEL 2. The *S* locus consists of many cistrons between which meiotic and somatic (male side) or somatic (female side) recombinations occur. *S* alleles would thus depend on complementation between cistrons, which might be very few. The number of allelic possibilities would be 2^n where n is the number of cistrons within the locus. Thus only eight cistrons could generate 256 *S* alleles (Fisher, 1961). Such a

recombinational model is very attractive, but new specificity allele mutants could only originate in plants heterozygous at the *S* locus. Therefore, this model fails to explain why inbreeding particularly generates new *S* alleles, and how *S* homozygotes, specifically, are able to do so (p. 214).

3. MODEL 3. Mulcahy and Mulcahy (1983) have produced an explanation of *S* gene control, based on (a) cistronic complementation between loci which they claim need not be linked, and (b) heterotic effects between the pollen and the style. They suggest that gametophytic incompatibility is based entirely on heterosis. The greater the genetic difference between the cistronic components of the *S* gene in the pollen and the style, the more likely the pollen tube is to reach the ovule. This is an ingenious and attractive model which deserves further exploration, for it explains a number of worrying features of self-incompatibility. As stated, however, it seems implausible because of several features:
   (a) gametophytic incompatible pollen usually germinates and grows just as well as compatible pollen in initial phases;
   (b) the nature of the failure of incompatible pollen, as observed microscopically, strongly suggests an oppositional rather than a complementary mechanism (p. 223);
   (c) if unlinked, the components of the *S* gene would segregate after meiosis, so that, for instance, semi-compatibility could not occur, and one could not detect the same *S* allele among siblings;
   (d) for *S* loci with many alleles, very many cistrons would be necessary, and, as nearly all would be homozygous between different alleles, heterotic effects would be minimized. Lawrence *et al.* (1985) have published a fully-argued rebuttal of the Mulcahys' hypothesis.
4. MODEL 4. In recent years, transposable genetic elements (transposons) have proved to be very widespread in a variety of organisms (perhaps even all) (Calos and Miller, 1980). Transposons are fragments of DNA which are nomadic, having the capability to integrate themselves into, and excise themselves from, the chromosomes. Their activity may be generalized, generating an increased rate of chromosomal structural reorganization (Woodruff and Thompson, 1980), or highly specific, as shown by the pioneering work of McClintock on maize (discussed in Fincham and Sastry, 1974). They may have at least three effects:
   (a) inhibition in the expression of a gene with which they become associated;
   (b) transport of such a gene to another site, giving rise to position effects;
   (c) high levels of chromosome breakage, chromosomal reorganization and somatic recombination.

Transposons are now thought to be responsible for many previously mysterious phenomena, not least the very high mutability and very large number of phenotypic expressions of antibodies in response to various antigenic stimuli in man and other animals.

Thus, it is possible to envisage a model of the specificity component of the *S* locus in which one or more transposons travel through the locus, giving a different phenotypic response (*S* allele) in each of its different positions. This model is very attractive, and explains many features of the gametophytic incompatibility system. It does not in itself explain why new alleles should only apparently arise after inbreeding. However, mutation rates in general, and transposon-generated changes in the DNA in particular, are themselves under genetic control, and can be influenced, for instance, by hybridity. Extreme heterozygosis can destabilize the genome by releasing transposon activity. It is possible that elements controlling transposon activity at the *S* locus are recessive, only being expressed when homozygous. If they are closely linked to, or part of, the *S* locus themselves, they may only be expressed after a cycle of inbreeding, when *S* alleles become few or homozygous. If this mechanism does exist, and there is no evidence for it, it would be of great selective advantage to outbreeders in which populations become so small that *S* alleles become critically few, as has been discussed by Wright (1939). It would encourage the rapid creation of new *S* alleles during such population bottlenecks, which occur very frequently in some species. It would also explain the occurrence of very many *S* alleles in very small inbred populations, which appears mathematically very unlikely, as in *Oenothera organensis* (Emerson, 1939; Wright, 1939).

Two-locus gametophytic incompatibility systems

Two-locus (sometimes called 'bifactorial') incompatibility systems under gametophytic control were first described in the grasses (Lundqvist, 1956, 1961, 1968), and have otherwise only been discovered in a few aberrant members of the Solanaceae (Pandey, 1957, 1962). The majority of grasses are self-incompatible (although this is untrue of all cereals except rye), and most that have been investigated have this mechanism (*Lolium* may have a three-locus system, p. 218).

A two-locus system is most readily detected by the number of cross-compatible classes arising among full siblings. As a fully compatible cross will usually involve two heterozygous individuals containing between them four alleles at any locus (if they are diploid), the number of cross-compatible classes among siblings is calculated by 4^n where n is the number of loci. Thus in the two-locus grasses, up to 16 sibling classes can occur (Fig. 6.2).

Two-locus systems in the grasses are most probably secondary (p. 210), having arisen from self-compatible derivatives of one-locus gametophytic systems. In particular, they share with the secondary sporophytic systems a trinucleate pollen grain, and the various morphological and physiological peculiarities of sporophytic systems with trinucleate grains (see Table 6.1).

The gametophytic nature of two-locus systems in grasses is most simply observed when full siblings are crossed. In such crosses the pollen grains do not all behave in the same way, as they would for a sporophytically controlled system. Rather, semicompatibility renders either a quarter or a half of the pollen grains involved in each cross incompatible (for one-locus gametophytic systems semicompatibility always renders half the grains incompatible) (Fig. 6.2).

The two incompatibility loci in grasses are termed *S* and *Z*. They are unlinked, segregating independently from each other after meiosis, and each is polyallelic. There is co-operation between alleles at different loci in the pollen, but independent reactions of *S* and *Z* alleles in the style. Each combination of alleles establishes a specificity in the haploid pollen, and rejection occurs when this specificity is matched by one of the four possible combinations of *S* and *Z* in the diploid style (see Fig. 6.2). Thus, the effects of the two loci are multiplicative, the number of incompatibility genotypes being the product of the number of alleles at each locus. Lundqvist (1964) calculates that for *Festuca pratensis*, with 6 alleles at the *S* locus and 14 alleles at the *Z* locus, 84 haploid genotypes occur. It follows that this system is far more efficient than a one-locus system, where for 20 alleles (6 + 14) only 20 haploid genotypes are possible. It is important for the efficiency of the system that *S* and *Z* are unlinked. If linked, they would behave essentially as a single locus with additive rather than multiplicative effects, although some recombination between the loci might occur. Loosely linked multilocus systems such as this have not been recorded.

The efficiency of the two-locus system means that it should be more resistant to decay than a one-locus system. For instance, the viable minimum number of alleles which are necessary to occur at each locus within a population is much lower, probably as low as three at each (nine pollen genotypes). The *S* and *Z* loci act in both a complementary and an independent manner. Thus, if one locus produced an s-c mutant, the other would continue to convey incompatibility, and the s-c mutant would not be advantaged selectively, as would be the case where selfing is favoured in a one-locus system. However, this intuitive approach is not supported by the models of Mayo and Hayman (1968), who consider that times to extinction of two-locus alleles when inbred in very small populations are similar to those for one-locus systems. Thus, in very small populations, two-locus systems can probably break down, leading once again to the evolution of self-compatibility.

Lundqvist (1962) has suggested that interaction between the *S* and *Z* loci may occur, giving dominance effects between an *S* allele and a *Z* allele in a pollen grain, as would be the case if the pollen grain were diploid (p. 219). There is as yet no clear proof of this, but if the *S* and *Z* loci originally arose by duplication of a single locus, and thus act in the same way, such interactions might be expected. The apparent similarity of mode of action by *S* and *Z* loci, and their complementary activity with respect to each other, support the view that they are indeed unlinked duplicates.

De Nettancourt (1977) favours the view that the *S* and *Z* loci have a similar structure to one-locus gametophytic systems, that is that they are probably tripartite in structure. Because of the complementary action of the loci, it is very difficult to undertake mutation studies in two-locus systems, for mutants will usually be unexpressed phenotypically. There is in fact no reason why the structure of the two-locus system should be the same as that of the one-locus system, if they have different origins (p. 207). Lundqvist (1964) followed Lewis (1954) rather than Lewis (1960) in assuming a bipartite structure for two-locus systems, but in fact there seems to be little evidence in either direction.

Three- and four-locus gametophytic incompatibility systems

Gametophytic systems with more than two multi-allelic loci were first claimed by Lundqvist *et al.* (1973) for *Ranunculus acris* and *Beta vulgaris*, and later by Spoor (1976) and McCraw and Spoor (1983a,b) for *Lolium* species, and by Murray (1979) for *Briza spicata*. As we have seen already, multilocus systems generate more cross-compatible classes among the offspring of a single cross. Using the formula 4^n for the number of such classes, where n is the number of loci, it will be seen that three-locus systems should generate 64 cross-compatible offspring classes from a single cross, and four-locus systems can generate 256 such classes. A conventional two-locus system can only generate 16 classes. Naturally, it is necessary to grow up large numbers of offspring, and make many crosses before such effects are detected. Thus, for *Ranunculus acris*, families with 20, 19 and 18 cross-compatible classes were detected (only slightly more than 16 and many fewer than 64).

If multilocus systems do occur, we would also expect semicompatibility among crosses between full siblings as percentages of successful pollen-tube growth to depart from 75% or 50%. In three-locus systems they should include crosses showing 62.5%, 75% or 87.5% semicompatibility. One-way incompatibility (non-reciprocity) is also likely to be a feature of multilocus crosses between full siblings. One-way incompatibility, in which a reciprocal cross between two individuals may be incompatible in one direction but not the other, should be caused by incompatibilities arising in the absence of a total match of alleles between the pollen and the style (McCraw and Spoor, 1983a). This would probably occur through dominance effects between alleles at different loci within the pollen, resulting in a limitation of male, as against female phenotypes.

It must be emphasized that the existence of multilocus (more than two locus) systems is based on rather slender evidence, and its occurrence in grasses such as *Lolium* has been severely criticized (Lawrence *et al.*, 1983). Other interpretations could explain the existence of more than 16 cross-compatible offspring categories arising from a single cross involving a two-locus system, for instance if dominance interactions occurred between alleles in the style.

Pollen in *Lolium* is trinucleate, and the system is otherwise similar to those in other grasses. However, in *Ranunculus* pollen grains are binucleate and their mode of action is apparently of the binucleate gsi type (Table 6.1). Thus, whether or not *Ranunculus* has a three-locus system (and various features of crosses between full siblings of *R. acris* suggest that this may indeed be the case), it certainly has a two- or three-locus system which seems to have evolved from primitive one-locus gsi systems directly. Lundqvist (1994) has suggested that *Ranunculus repens* (an allotetraploid) has no fewer than four incompatibility loci working disomically in a complementary fashion, but that these originated from two diploid parents the incompatibility systems of each of which were controlled by different two-locus systems.

The rarity of such apparently multilocus systems renders it highly unlikely that such systems were aboriginal to the Angiosperms, as Lundqvist *et al.* (1973) have suggested. It is more likely that, unlike two-locus systems in the grasses which seem to be secondary, multilocus systems have arisen from the original gsi one-locus systems by duplication of the *S* locus, thus increasing their efficiency (see also p. 227).

Polyploidy in gametophytic systems

Up to this point, only diploids, or at least plants disomic for the incompatibility locus, have been considered. For these, the pollen grain is effectively haploid, and thus interactions between alleles of one-locus gametophytic systems are not possible within a pollen grain.

For polyploids in which the pollen grain is disomic, or even polysomic for the *S* locus, a different condition operates. The different *S* alleles within the pollen grain interact, showing either dominance with respect to each other or independent expression. Thus, the pollen grain S_1S_2 has the possibility of three different phenotypes, S_1, S_2 and S_1S_2. It follows that partial homozygotes can arise, as from the cross $S_1S_2S_3S_4$ (male) × $S_1S_3S_4S_5$ where S_2 is dominant to S_1 in the pollen. Thus, a pollen grain S_1S_2 would have the phenotype S_2, and would be compatible with $S_1S_3S_4S_5$. It could therefore cross with a female gamete S_1S_3, giving the offspring $S_1S_1S_2S_3$, partially homozygous for S_1. In such a system, recessive alleles which are less likely to recognize self in the style, are favoured and will tend to be both more common and more often partially homozygous, with a concomitant loss of dominant alleles. If many alleles are lost, fertility of the population will be impaired and s-c will be favoured.

In practice, new polyploids, such as those artificially induced by the use of colchicine, of species with one-locus gametophytic systems seem usually to lose incompatibility on the male side (Crane and Lawrence, 1929; Crane and Lewis, 1942; Stout and Chandler, 1942). It is typical of these to be s-c as male parents, but s-i as female parents, as exemplified by differences in reciprocal crosses between tetraploids and their diploid parents. This raises an interesting point with respect to occasional self-compatibility in diploids, which may at times be caused by non-reductional (diploid) pollen grains. The offspring of these selfs will of course be sterile autotriploids.

Autotetraploids will be partially homozygous (e.g. $S_1S_3 > S_1S_1S_3S_3$) and thus may give rise to either homozygous (S_1S_1 or S_3S_3) or heterozygous (S_1S_3) pollen. Lewis (1947, 1949a) showed that only heterozygous grains showed such novel male-side self-compatibility, there being semicompatibility (50%) in the newly formed autotetraploid when selfed. There seems to be no rational explanation for this finding at present, which suggests that we are still some way from a complete understanding of the mechanism of gametophytic incompatibility. A convincing model would also have to explain why autotriploids fail to show self-compatibility (van Gastel, 1974), and why autopolyploids in the monocotyledons do not apparently show self-compatibility (Annerstedt and Lundqvist, 1967).

Autopolyploidy may lead to partial self-compatibility immediately, and increasing homozygosity for *S* alleles and loss of *S* alleles may eventually encourage a total loss of self-incompatibility in autoploids. However, autopolyploidy is a rare condition in natural populations. Most polyploids, perhaps half of all Angiosperm species, are alloploids that have obtained genomes, for each of which they are usually diploid, from two to several different species. Whether the polyploids are disomic or polysomic for the incompatibility locus/loci will, of course, depend on how many of the parents carry functional self-incompatibility loci, and whether these loci are genetically and positionally homologous between parents (Fig. 6.3). Only those incompatibility loci that are functionally disomic in a complex polyploid will be likely to survive. However, gametophytic multi-allelic incompatibility is frequently found in allopolyploids, and we must assume that in these the *S* locus is in fact disomic.

Nevertheless, self-incompatibility is much less common in polyploids than in diploids (e.g. Stebbins, 1950), perhaps because newly arisen, reproductively isolated polyploids are more likely to establish if self-

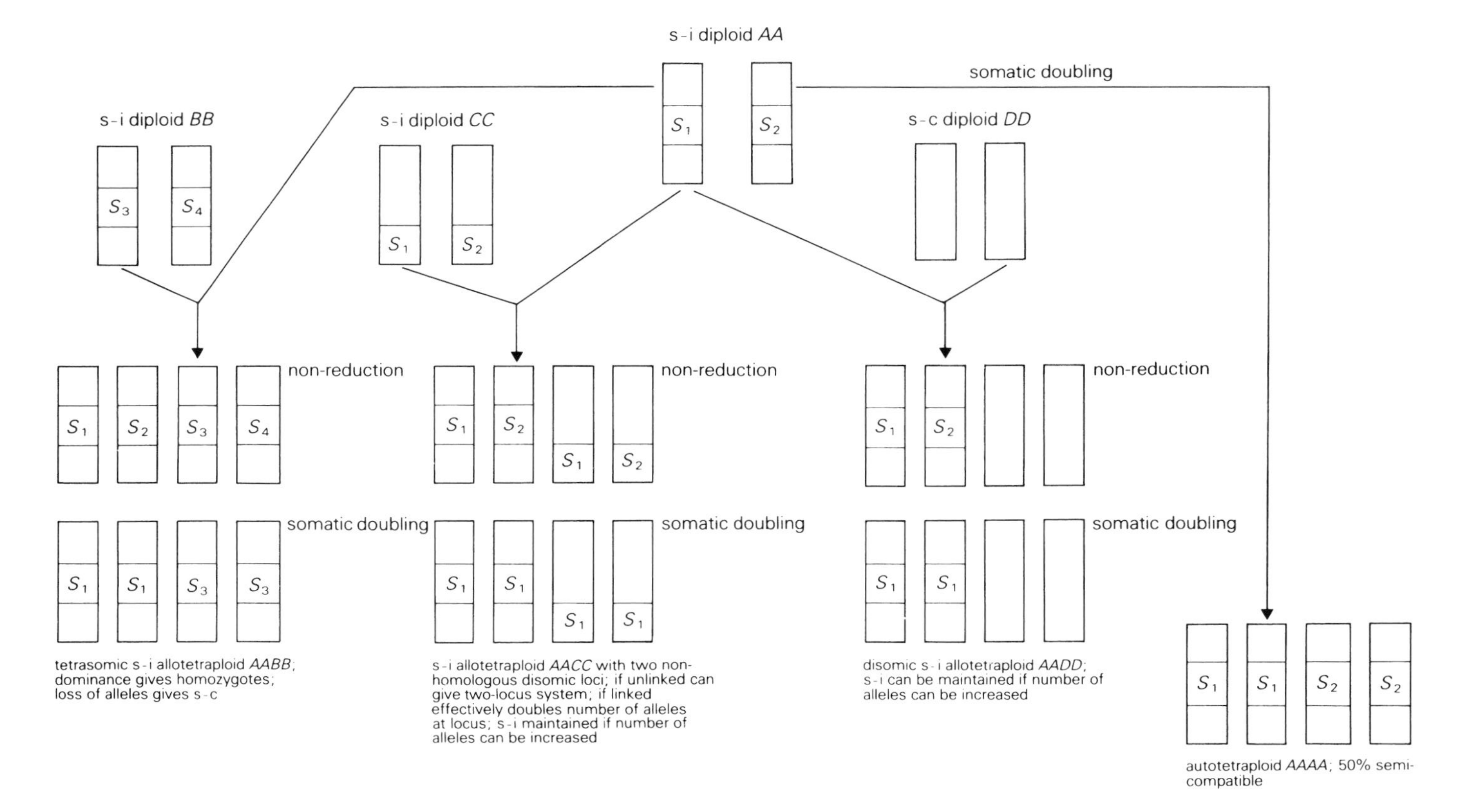

Fig. 6.3 Maintenance and loss of gametophytic s-i in polyploids.

compatible. Polyploidy may be favoured in selfers as polysomic loci store genetic variability more efficiently than do disomic loci (p. 382).

A further problem facing the new s-i allopolyploid is that only two alleles will have entered the new allopolyploid gene pool (see Fig. 6.3). These are too few for an s-i system to function correctly (p. 206). This problem will only be overcome if more than one alloploid, each with different *S* alleles, occur together, or if a new alloploid crosses with unreduced gametes from parental diploids exhibiting other *S* alleles in the surrounding population.

That such processes do happen is substantiated by the frequent occurrence of self-incompatible polyploids, including such well-known species as *Trifolium repens* and *Lotus corniculatus*.

Breakdown of gametophytic incompatibility

So far, it has been assumed that gametophytic self-incompatibility in diploids is absolute, only being lost through self-compatible pollen-part mutations, or as a result of diploid pollen grains.

However, there are other circumstances through which an otherwise functional gametophytic s-i system can fail but continue to be inherited in the offspring. These have been called pseudocompatibility (Pandey, 1959) and are well reviewed in de Nettancourt (1977), from which this account is taken, and Pandey (1979). They include the following conditions.

1. Bud pollination. In *Nicotiana* (Pandey, 1959) and *Petunia* (Shivanna and Rangaswamy, 1969) substantial quantities of selfed seed can be obtained by pollinations with mature pollen on to immature stigmas. It is concluded that the *S* gene is not yet expressed in the young style, so allowing the growth of incompatible pollen tubes.
2. Delayed pollination. Ascher and Peloquin (1966), working with *Lilium*, have shown that fresh selfed pollen can also achieve the breakdown of self-incompatibility when placed on overmature stigmas, although similar effects cannot be obtained in some other gsi genera. Decay of *S*-gene products in the ageing style may permit some incompatible pollen tube growth.
3. End-of-season effects. It has been noted for several members of the Solanaceae that abnormally late flowers, often on a moribund plant, can be self-compatible.
4. Stylar irradiation. Linskens, Schrauwen and van der Donk (1960) for *Petunia*, and Hooper and Peloquin (1968), for *Lilium* have shown that X-radiation of styles immediately after self-pollination prevents self-incompatibility. The optimum dose for *Petunia* was 2000 rads, but above 6000 rads for *Lilium*. Radiation before pollination, or 24 h after

pollination did not give this effect. It is suggested that temporary gene inactivation is involved, although there is no indication as to exactly how this can operate.

5. Mentor effects. These describe the induction of self-compatibility in incompatible pollen when it is mixed with foreign pollen. The foreign pollen does not achieve fertilization as it has been previously killed (usually by doses of radiation, Knox, Willing and Ashford, 1972), or is distantly related and thus interspecifically incompatible, although sharing the same self-incompatibility system (as in *Lotus*, e.g. Miri and Bubar, 1966). Although it is not unexpected that such effects may occur in sporophytic s-i, in which the incompatibility factor is known to be carried on the outside of the pollen grain (p. 237), it is very surprising that it should operate in gametophytic systems such as *Lotus*. It must be presumed that a diffusible substance from the non-functional compatible pollen overcomes the incompatibility inherent in the selfed pollen (see below).
6. High temperatures. Many workers have discovered that high temperatures, usually in excess of 30°C and as high as 60°C, remove self-incompatibility. This appears to be a widespread and general phenomenon when high temperatures are applied to the style immediately after self-pollination, and occurs in both gametophytic and sporophytic systems. There is some suggestion that the temperature effect is specific to certain isozymes in the style (Pandey, 1973) and that variation occurs between plants with respect to heat sensitivity.
7. Biological inhibitors. Inhibitors of RNA synthesis such as actinomycin D and 6-methylpurine, and enzyme inhibitors such as puromycin and *p*-chloromercuribenzoate, have all been shown to limit self-incompatibility to various extents.
8. At least in *Petunia*, fertilization *in vitro* can break self-incompatibility when pollen grains are allowed to germinate in juxtaposition with cultured ovules. Other mutilative and manipulative experiments have shown equally clearly that self-incompatibility is mediated by various fractions of the stigma and style.

All of these methods of overcoming self-incompatibility show that self-incompatibility is an active phenomenon, which must be bypassed or inhibited in some way if it is to be broken. This argues in favour of an oppositional system of incompatibility control, and against the complementary system suggested by Bateman (1952) and Mulcahy and Mulcahy (1983). Complementary systems would promote compatibility. In their absence, pollen-tube growth would be inadequate, the resulting incompatibility occurring passively.

However, if inhibitors of floral abscission such as auxin (3-indole-acetic acid) or its analogue, α-naphthalene acetic acid, are applied to the calyx or the base of the flower in solution at regular intervals after an incompatible pollination, the flower remains in good condition for a longer time. Incompatible pollinations can then often result in seed-set. Applications of auxin to the style or stigma do not promote self-fertilization, and indeed inhibit compatible fertilization, so there is no direct effect of auxin on incompatibility. These results, which have been obtained for many plants with gametophytic systems (de Nettancourt, 1977) may merely allow *S*-gene products to decay in the ageing style, permitting growth of late-germinating incompatible pollen. Thus, although the effects of auxin appear to indicate the operation of a passive system of incompatibility, it may merely indicate the breakdown of an active, oppositional system with time.

GENETIC CONTROL OF MULTI-ALLELIC SPOROPHYTIC (ssi) SYSTEMS

Sporophytic systems differ fundamentally from gametophytic systems in that the control of the behaviour of the pollen grain comes from the sporophytic anther that gave rise to it, not from the grain itself. This has the following genetic consequences.

1. All the pollen grains from a single male parent will show the same behaviour on a given female parent (i.e. semicompatibility does not occur).
2. Because the control of the behaviour of the pollen grain comes from the diploid anther, dominance will usually be expressed, i.e. the compatibility phenotype of the pollen grain will express only one of the alleles in the anther. Thus, the individual haploid pollen grain may carry an allele different from its incompatibility phenotype. For instance, if S_1 is dominant to S_2 in the male parent S_1S_2, a grain carrying S_2 will have the phenotype S_1. (If there is incomplete dominance, both paternal alleles may be expressed (i.e. independent interaction).)
3. One of the consequences of this is that homozygotes will occur for the *S* locus. Thus if $S_1 > S_2 > S_3$, the cross S_1S_2 (male) × S_2S_3 will be compatible as the pollen will show the phenotype S_1. But S_2 carrying grains can fertilize S_2 carrying egg cells to give S_2S_2 homozygous offspring.
4. Another consequence is that the number of cross-compatible classes among the offspring of a single cross will be less than 4^n, where n is the number of loci (i.e. four classes for one-locus diploid gametophytic systems). Thus, if the following crosses occur, the number of cross-compatible classes among the offspring will be as shown ($S_1 > S_2 > S_3 > S_4$):

Male		Female	Offspring genotypes	Offspring male phenotypes	Number of cross-compatible classes in offspring
S_1S_2	×	S_3S_4	S_1S_3	S_1	
			S_1S_4	S_1	2
			S_2S_3	S_2	
			S_2S_4	S_2	
S_1S_2	×	S_2S_3	S_1S_2	S_1	
			S_1S_3	S_1	2
			S_2S_2	S_2	
			S_2S_3	S_2	
S_1S_2	×	S_2S_2	S_1S_2	S_1	
			S_1S_2	S_1	1
			S_2S_2	S_2	
			S_2S_2	S_2	
S_1S_1	×	S_2S_2	S_1S_2	S_1	
			S_1S_2	S_1	0
			S_1S_2	S_1	
			S_1S_2	S_1	

This model describes the operation of multi-allelic sporophytic systems which assume a simple dominance hierarchy between alleles in the anther, but independent expression of alleles in the stigma. Such a simplistic system is probably never true for any sporophytic system, although it is approached most closely in *Brassica* (Richards and Thurling, 1973; Ockendon, 1974).

More frequently, a complex series of interactions between S alleles occurs in both the anther and the stigma, the details of which commonly differ between the anther and stigma even in the same population. Four types of interactions between alleles are recognized, which are listed in order of importance; all can occur within the same species:

1. dominance $S_1 > S_2 > S_3 > S_4$; genotype S_1S_2; phenotype S_1;
2. independent $S_1 = S_2$; genotype S_1S_2; phenotype S_1S_2;
3. interaction $S_1 \ldots S_2$; genotype S_1S_2; phenotype S_3;
4. mutual weakening S_1———S_2; genotype S_1S_2; phenotype $S_{(1)}S_{(2)}$ (more or less self-compatible);

It is most common to find that dominance is the most important system in the anther, and independent action as the commonest mechanism in the stigma (as in the examples given above), but considerable variation occurs. To take two simple examples which are frequently quoted, for instance by de Nettancourt (1977), from *Iberis amara* (Cruciferae) (Bateman, 1954) and *Cosmos bipinnatus* (Compositae) (Crowe, 1954), respectively (alleles in circles):

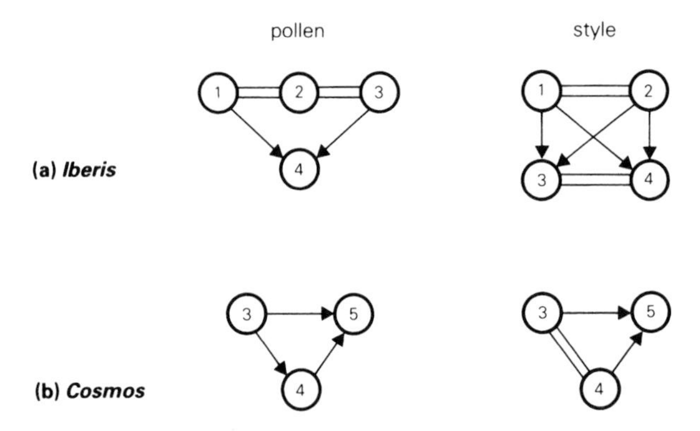

To illustrate how these systems work, let us take by way of illustration one reciprocal cross, and one reciprocal self from each example:

	Male		Female		
	Genotype	Phenotype	Genotype	Phenotype	Compatibility
Iberis					
cross	S_2S_3	S_2S_3	S_3S_4	S_3S_4	incompatible
	S_3S_4	S_3	S_2S_3	S_2	compatible
self	S_3S_4	S_3	S_3S_4	S_3S_4	incompatible
Cosmos					
cross	S_3S_4	S_3	S_4S_5	S_4	compatible
	S_4S_5	S_4	S_3S_4	S_3S_4	incompatible
self	S_3S_4	S_3	S_3S_4	S_3S_4	incompatible

From this we can make the further following deductions.

1. Self-pollinations will always be self-incompatible unless the dominance order between alleles differs in direction between the pollen and the style (a rare occurrence, if indeed it ever does occur).
2. Reciprocal differences in compatibility will commonly occur between the same two parents (such reciprocal differences never occur in diploid gametophytic systems, but can occur in polysomic gametophytic systems).
3. Cross-incompatibility will frequently occur between two individuals with different genotypes (cross-incompatibility will only occur in diploid gametophytic systems when individuals share the same *S* genotype).
4. Although semicompatibility cannot occur, reciprocal differences and cross-incompatibility between different genotypes will only occur when the two parents share one *S* allele.

5. Dominant alleles will be at a selective disadvantage in comparison to alleles of independent action, which will be at a disadvantage with respect to recessive and mutual weakening alleles. This is a complicated relationship, because of chains of dominance relationships between many alleles, which may differ between the pollen grain and the style.

S *and* G *systems*

On occasion, workers using *Raphanus sativus* and *Brassica campestris* have noted anomalous self-fertilities which are not easily explained, despite the complexity of the ssi mode of action. These have led Lewis, Verma and Zuberi (1988) and Zuberi and Lewis (1988) to propose that these Cruciferae do in fact have a sporophytic–gametophytic mode of action. They suggest that linked to the ssi *S*-locus resides a locus *G* which acts gametophytically in complementation to the *S*-locus. In their model, crosses which are predicted to be cross-incompatible prove to be cross-compatible because parents differ for *G*-alleles. However, this complementation only seems to operate for certain *S*-alleles.

These authors suggest that *G* is a relict from the aboriginal gsi system, which has persisted through into secondary ssi systems. Only two *G*-alleles have been identified, confirming the degraded nature of *G*-incompatibility. In fact, we now know that the *SLG* locus of *Brassica* (p. 239) is expressed gametophytically in the pollen grain as well as sporophytically in the anther tapetum. It is possible that the *S* and *G* loci identified from controlled crosses are homologous to the *SLG* and *SRK* loci identified by molecular means, although this is by no means certain.

A very curious sporophytic–gametophytic s-i is also known for cocoa (*Theobroma*), where a sporophytically controlled mechanism is mediated by gametic fusion (Knight and Rogers, 1955; Cope, 1962). This is more thoroughly described in the section on late-acting s-i (p. 241).

Distribution of S *allele frequency in ssi systems*

Alleles that show dominance to the greatest number of alleles on the male and female side will have the greatest disadvantage. This is because they are most frequently expressed, phenotypically, and are thus most likely to show a cross-incompatible reaction through self-recognition.

Ockendon (1974, 1977, 1980) in an analysis of the distributions of *S* alleles among commercial strains of Brussels sprouts and broccoli (*Brassica oleracea*) showed that *S* alleles are few in number within varieties, very uneven in distribution between varieties, and that the many scarce alleles were dominant and the few common alleles were recessive to them. These factors all militated against high seed-set and hampered the successful breeding of certain new strains. From a breeding standpoint, a great premium was placed on the introduction of one or more rare dominants into a strain.

This uneven distribution, so typical of a sporophytic system, contrasts vividly with the even frequency of *S* alleles in most diploid gametophytic systems, where every allele is expected to have the same frequency. However, in gsi systems where few clones of a crop plant are grown, artificial selection of crop genotypes can also seriously diminish *S* allele diversity, as in hazel (*Corylus avellana*) where only 11 *S* alleles were identified among 55 cultivars (Thompson, 1979). Cherries and apples provide other familiar examples where cross-compatible mates must be carefully chosen in orchard plantings.

For many features, a diploid gsi system tends to be more efficient than an ssi system; that is, although self-incompatibility is general to both systems, there will be much less cross-incompatibility in a gametophytic system – more individuals in the population will be available for crossing. However, by restricting sibling mating, a sporophytic system will reduce mating between close relatives and thus may lead to more outbreeding than a gametophytic system.

It is difficult to estimate the number of *S* alleles occurring within a population, or a species, in a sporophytic system. Although a sporophytic system resembles a gametophytic system in that there is nearly always total self-incompatibility, one cannot assume, as in the latter, that incompatible crosses are between the same genotype. They will often be between genotypes that share one allele. At the same time, crosses between genotypes sharing an allele will often be compatible in one direction, and if the common allele is recessive to both the others, such a cross will be compatible in both directions. Other factors that must be allowed for are interaction between alleles (giving more compatibility) and homozygotes (which will also tend to give more cross-compatibility, by reducing the frequency of allele sharing).

The only genotypes that can be relied on to be always cross-compatible in a sporophytic system are those that share no *S* alleles. These are groups of plants that are always fully cross-compatible with all others in the system for both reciprocal crosses. These will equal $n/2$, where n is the number of alleles. Thus, for a system containing eight *S* alleles, the following genotypes will always be fully cross-compatible: S_1S_2, S_3S_4, S_5S_6, S_7S_8, (four out of 28 possible different genotypes), the number of possible genotypes being calculated by $(n^2 - n)/2$ (that is where $n = 8$, $(64 - 8)/2 = 56/2 = 28$). The remaining 24 genotypes should show some cross-incompatibility with some of these four, and with each other due to dominance or independent action.

A very thorough series of reciprocal crosses between all genotypes present is necessary if the number of fully cross-compatible genotypes

Table 6.3 The number of *S* alleles identified in various plants with sporophytic incompatibility

Species	Number of plants analysed	Number of *S* alleles estimated	Reference
Iberis amara	47	22	Bateman (1954)
Brassica oleracea	488	19	Ockendon (1974)
(Brussels sprouts)	(16 varieties)		
All varieties		60	Ockendon (1974)
Raphanus	45	9	Sampson (1967)
raphanistrum	(5 populations)		
	(total)	34	
Sinapis arvensis	10	14 (24)	Ford and Kay (1985)

present is to approximate to $n/2$. In fact, due to interactive effects, and homozygotes, this equation will always overestimate the value of n (the number of *S* alleles), as will dominance effects unless these are very fully understood. In practice, more sophisticated techniques for the estimation of the number of alleles are used, which vary according to the system, and the level of understanding of interactive effects between alleles within each system. Nevertheless, the number of *S* alleles identified in the few cases where this has been attempted is quite high (Table 6.3).

The $n/2$ relationship breaks down when less than four (two or three) alleles are involved, notably in two-allele systems, where simple dominance renders the number of cross-compatible classes equal to the number of alleles (Riley, 1936, and Chapter 7).

Breakdown of multiallelic ssi incompatibility

A certain amount has been published on the physiological breakdown of sporophytic systems, particularly in *Brassica*. Early work showed that self-incompatibility can be overcome in this genus by both bud pollination (Attia, 1950) and delayed pollination (Kakizaki, 1930), as is also found in some gametophytic systems (p. 222). Self-fertilizations using these techniques are routinely carried out in plant breeding stations to obtain inbred lines from which F_1 hybrids can be made. Self-incompatibility can also be broken by high temperatures (Richards and Thurling, 1973), perhaps because enzymes that mediate the action of the *S* gene in *Brassica* have lower optimal temperatures than those that control pollen-tube growth and fertilization. In *Primula*, with a diallelic sporophytic mating system, Lewis (1942b) has described the breakdown of self-incompatibility at high temperatures, and I have found (Richards and Ibrahim, 1982; A. J. Richards, unpublished observations) that thrum *Primula* 'Polyanthus' shows high levels of self-fertilization when maintained at 35°C for 24 h after pollination (p. 275). Once again, deactivation of one or more enzymes at high temperatures can be plausibly implicated. De Nettancourt (1977, p. 109) makes the telling point that processes involved in the incompatibility system seem as a rule to be more sensitive to abnormal external stimuli than those that control pollen germination and fertilization in general, so that it is possible for incompatibility to be successfully overcome through environmental manipulation.

Other external, but artificial, stimuli implicated in the breaking of *Brassica* self-incompatibility include high concentrations of gaseous CO_2 (at 3–5%, Nakanishi and Hinata, 1973), electrical stimuli (100 V, Roggen, van Dijk and Dorsman, 1972), and stigmatic abrasion using a wire brush (Roggen and van Dijk, 1972). In the last cases, damage to, or chemical transformation of, the cuticle of the stigmatic papillae may promote the penetration of selfed pollen tubes. As yet we know little about the mutational breakdown of ssi. In the case of rarely encountered self-compatible mutants in *Brassica* (Nasrallah and Nasrallah, 1993), it seems that there is a failure of a gene controlling RNA transcription of the *SLG* locus glycoprotein (but not *SRK*) onto the stigma papilla (p. 239). This gene, which is unlinked to the *S* locus, has been termed *SCF1*.

THE FUNCTIONING OF SELF-INCOMPATIBILITY

A great deal of work on the physiology and ultrastructure of self-incompatibility has been published since about 1965. Despite this, a final comprehension of the mechanisms by which self-incompatibility functions has proved peculiarly difficult to achieve, and we are still some way from this goal. There seem to the two reasons for this. First, the mechanisms appear to be complex, with a number of different functions interacting. Secondly, it seems that different s-i systems operate in quite different ways, reinforcing the supposition that these systems arose independently of one another.

We need to know how plant cells recognize each other. Animal cells, which are naked, surrounded only by a membrane, tend to accept cells of the same genotype but to reject or attack foreign cells of the same or different species. Thus, for a successful animal mating, the innate antagonism of the internal female environment against the male sperm must be suppressed. Mechanisms of antagonism, and its suppression, which are mediated through surface-carried immunoglobulins, are quite well understood, and the science of immunology has made great advances in manipulating these mechanisms, leading to the successful control of disease, organ transplants and reproductive incompatibility.

In contrast, plant cells are usually enclosed in cellulose boxes. Membranes in different cells are most likely to transmit and receive chemical messages via water-soluble substances, although stigmatic and stylar cell contents may connect by plasmodesmata, most notably in the so-called 'key junctions' in the style of *Petunia*. However, the pollen tube is only separated from the stylar intercellular environment by a pectin sheath, which is permeable to, among other things, RNA and ribosomes, and in this it more closely resembles animal cells.

Plant cells can recognize foreign cells, for example when they are attacked by a fungal pathogen. They can also distinguish between different types of foreign cells, so that an orchid root will accept invasion by an endomycorrhizal fungus but will reject other soil-borne fungi. A legume root nodule will accept invasion by the nitrogen-fixing *Rhizobium*, but may reject other bacteria.

In fact, a fungal spore which produces a hypha on germination presents a challenge to stigma papillae of a very similar kind to that provided by an angiosperm spore (pollen grain) which produces a pollen tube on germination. The evolution of the stigma, designed to welcome germinating spores, was a necessary concomitant to the enclosure of the sexual process in flowering plants. However, many spores arriving at the stigma are potentially hostile. In order to function successfully, early stigmas had to evolve recognition systems whereby only conspecific spores were succoured. Interestingly, Dickinson, Crabbe and Gaude (1993) and Dickinson (1994) suggest that *S*-specific RNases recently discovered in the style are related to other stylar RNases involved in disease control.

Another recent discovery is that *S*-specific RNases change the behaviour of unilateral interspecific incompatibility in *Nicotiana* (Murfett *et al.*, 1996). It has been known for many years (Lewis and Crowe, 1958) that when related self-compatible and self-incompatible species are crossed, hybrids rarely result when the self-compatible species forms the male parent, even when the reciprocal cross is fully fertile. This phenomenon has been termed unilateral incompatibility (UI). UI suggests that pollen from the self-incompatible parent provides a recognition factor which allows pollen tube growth to take place in both intraspecific and interspecific crosses and that this factor has been lost in the self-compatible relative. Interestingly, UI holds for both gsi and ssi species. The work of Murfett *et al.* (1996) now provides evidence that interspecific incompatibility (UI) has at least part of the same molecular basis as intraspecific incompatibility (gsi, and by implication, ssi). Perhaps, for ssi species, this concerns the G system (p. 227). Thus, arguably, *S*-specific RNases employed in intraspecific incompatibility have evolved from those involved in disease recognition, and those mediating interspecific incompatibility, with relatively little evolutionary modification.

Nevertheless, the self-incompatibility mechanism is without parallel, in the sense that it rejects cells of the same genotype (at a specific locus) in an oppositional way (p. 223), but accepts cells of different genotypes.

What is very remarkable about self-incompatibility is that plant cells are continuously in contact with cells of the same genotype without rejecting them. It is only the pollen grain or the pollen tube that is rejected by the stigma or style. Thus, to understand self-incompatibility, we must search for gene products that are unique to the pollen grain, pollen tube and gynoecium.

An understanding of the mechanisms of self-incompatibility should produce great practical benefits to plant breeding. Successful plant breeding often depends on the development of 'pure breeding lines' by selfing, which are then crossed to produce F_1 hybrid seed. Both procedures depend on an ability to manipulate the breeding system. For s-i crop plants, the production of pure breeding lines has proved a major obstacle. It would be very convenient if a chemical treatment that leads to instant, reversible self-compatibility could be developed.

One-locus gametophytic systems

The basic features of the one-locus gametophytic incompatibility system can be briefly listed as follows.

1. Control of the behaviour of the individual pollen grain depends on the genotype of that haploid grain.
2. Pollen grains become hydrated externally from the 'wet' stigmatic exudate.
3. In most cases, pollen tubes germinate, grow and penetrate the stigma equally well in compatible and incompatible pollinations; tubes grow over the surface of the papilla and penetrate the stigma at the base of the papillae by dissolving the middle lamella of the cell wall.
4. All pollen tubes then grow between cells into the style (except in *Oenothera* in which incompatible pollen tube inhibition takes place in the stigma). For incompatible pollen tubes as they grow through the style, at least in the Solanaceae, rough endoplasmic reticulum (ER) then appears in concentric circles in the tip of the tube; the walls of the tube become thinner, and the callosic inner wall disappears; at the same time, numerous particles about 0.2 μm in diameter appear in the cytoplasm at the tube tip, the tip swells markedly, and then bursts, releasing the particles; the tube then ceases to grow (de Nettancourt *et al.*, 1974; de Nettancourt, 1977).

   This behaviour closely resembles that which occurs in a compatible tube at the time of fertilization within the embryo-sac (p. 60), when the apex of the tube bursts to release the sperm cells. Thus, de

Nettancourt (1977) very plausibly suggests that the gametophytic self-incompatibility reaction 'may perhaps be equated to an anticipation of a release-phenomenon scheduled to take place, upon a signal from a synergid, at the time the pollen tube has reached the ovule'. Unfortunately it is not yet clear how widespread tube apex rupture-mediated s-i is outside the Solanaceae. It does not apparently occur in *Lilium longiflorum*.

It is worth noting that the incompatibility reaction does not occur until after the second pollen mitosis has taken place in the tube (p. 210). All the features exhibited lead to the suggestion, at the simplest level, that an incompatibility factor(s) is synthesized by the pollen tube as it grows down the style, and that this recognizes another factor present in the stylar tissue, which stops incompatible pollen tube growth, oppositionally.

Most evidence suggests that the stylar factor is already present in the unpollinated style, but that the pollen factor is transcripted and translated via RNA during pollen tube growth (Pandey, 1967; Linder and Linskens, 1972; van der Donk, 1974; de Nettancourt, 1977). It is possible to simulate the effects of incompatibility *in vitro*, using crude stylar extracts, at least in *Petunia* (Sharma and Shivanna, 1982). In *Lilium*, similar results can be obtained using secretions from the stylar canal (Dickinson *et al.*, 1982).

The nature of this system is such that incompatibility factors will not involve those enzymes mediating pollen germination, pollen-tube penetration, or pollen-tube growth (such as cutinase-type esterases or pectinases). Rather, we must target mechanisms that directly inhibit growth in the style. We must also look for mechanisms which account for the known genetic features of gametophytic incompatibility; namely

1. the recognition, and disablement, of a *S*-specific factor in the pollen tube by a style which also expresses the same factor;
2. a tripartite locus with a specificity segment, and very closely linked activity segments for the pollen and style;
3. the ability of the specificity segment to code for one of very many allelic forms;
4. an absence of interactions between compatible and incompatible pollen tubes in the same style (i.e. 'non-diffusibility' leading to semicompatibility, compare p. 202);
5. loss of incompatibility in a heterozygous diploid pollen grain.

It is a striking feature of gsi, at least in the Solanaceae, that it seems to function simply, using a single reaction.

The molecular basis of gsi

Recently, remarkable progress has been made towards an understanding of the molecular basis of one-locus gsi, at least on the stylar side, in the Solanaceae. Anderson *et al.* (1986) were the first to report the cloning of

S-linked genes, in *Nicotiana*. The gene products proved to be glycoproteins with ribonuclease (RNase) activity. Later work showed that these RNases were expressed at disappointingly low levels in *Nicotiana*, but were much more abundant in the styles of *Petunia*.

Lee, Huang and Kao (1994) have been able to clone the gene coding for *S*-RNase from *Petunia* styles, and then to synthesize antisense *S*-RNase which has been transformed back into the same *S*-genotype. These plants became self-compatible. In further experiments, S_1S_2 genotypes were transformed with *S*-RNase from another genotype (S_3-RNase). This caused them to reject pollen from individuals containing S_3 with which they had previously been cross-compatible.

At the same time, Murfett *et al.* (1994) reported experiments where stylar-part promoters from tomato were used to amplify *S*-RNase expression in *Nicotiana* styles, allowing *S*-specific RNase to be transformed into other individuals. Once again, recognition and pollen rejection patterns in transformed plants supported the hypothesis that *S*-gene function on the female side in the Solanaceae is fully mediated by *S*-specific RNase.

Dickinson (1994) also reports the findings of Adrienne Clarke and her co-workers, that stylar *S*-RNase from a s-c mutant of tomato (*Lycopersicon peruvianum*) was enzymically inactive, the mutation having changed a histidine residue at the RNase active site on the polypeptide. It seems likely that this study involved a stylar-part disabling mutant (p. 212).

Dickinson notes that stylar RNase is now strongly implicated in the functioning of gsi in the Rosaceae and Scrophulariaceae as well as in the Solanaceae, but that in *Papaver* a smaller *S*-specific polypeptide without RNase function has been discovered.

A possible model for gsi *S*-allele function

As yet we lack firm evidence as to how *S*-gene recognition functions on the male side, or exactly how stylar *S*-specific RNases disable incompatible pollen tubes.

Ribonucleases generally have the function of mediating the turnover of RNA. Pollen tubes grow rapidly, requiring fast gene translation, and pollen tube apices are richly supplied with ribosomes and RNA which pass freely across the sheath between the pollen tube and interstylar tissue (p. 58). Ribosomal RNA in the pollen tube apex has been shown to be degraded by *S*-RNase of stylar origin (McClure *et al.*, 1990). Thus, it is reasonable to assume that RNA turnover in the pollen tube might be controlled entirely by stylar-part *S*-specific RNases of female origin. In this hypothesis, I am suggesting that the pollen-part *S* gene operator blocks male *S*-gene translation, so that the only *S*-gene product on the male side is mRNA.

Suppose that a stylar *S*-specific RNase is only enabled to recognize pollen tube RNA coded for by exactly the same *S*-specific allele. Suppose also that pollen-part *S*-specific RNA also codes for another vital gene product, such as ribosomal RNA, or glutamate translation.

In this hypothesis, incompatible pollen tubes would progressively fail to produce a vital metabolite, leading to metabolic imbalance, osmotic disruption and pollen tube rupture.

This hypothesis requires that stylar RNase fails to recognize *S*-specific stylar-part mRNA, so that it continues to be produced. Perhaps the stylar-part operator so protects it, whereas the pollen-part operator does not.

This hypothesis has the advantage of accommodating the molecular, genetic and ultrastructural observations associated with gsi. It also tolerates the suggestion that gsi might have evolved from a pathogen-recognition system (Dickinson, 1994). Presumably, anti-pathogen stylar RNases work by recognizing and destroying foreign mRNA.

As yet, this hypothesis has no experimental support, in part because it is very difficult to isolate pollen tube-specific RNA which has been produced within the stylar environment. It is also not clear how *S*-specificity shared between stylar RNase and pollen tube mRNA could lead to recognition. However, it seems inevitable that some kind of nucleic acid-enzymic self-recognition must be involved in the functioning of gsi.

Two-locus gametophytic systems in grasses

It has already been noted that grasses (Gramineae) differ from other gsi systems, not only in having two-locus control, but also in showing cytological and physiological features of the pollen grain which are much more similar to those in sporophytic systems (i.e. they are trinucleate, with trinucleate type behaviour; Table 6.1). It seems likely that gsi in the grasses has evolved secondarily from s-c ancestors as have ssi systems. Consequently, one might expect grass gsi to operate in a different way from one-locus gsi.

The physiology and ultrastructure of grass gsi has been reviewed by Heslop-Harrison and Heslop-Harrison (1982) and Heslop-Harrison (1982). There are two major morphological differences between the function of grass gsi and one-locus gsi.

1. Although s-i pollen germinates well, and the pollen tubes start to grow normally, tube growth usually ceases as soon as the tube touches the stigma papillae; occasionally, penetration of the stigma cuticle occurs, but then tube growth ceases soon afterwards, in the intercellular spaces of the stigma. Incompatible tubes never enter the style as they do in one-locus gsi systems.
2. The s-i tube does not swell and burst at the apex; instead contact of the incompatible tube with the papilla is rapidly followed by the appearance of nodules, probably formed from microfibrillar pectins in the

wall at the extreme apex of the tube. From this initial response, there follows a rapid accumulation of pectin at the apex of the tube until the whole apex of the tube is occluded. As in one-locus gsi, the subsequent cessation of growth is accompanied by a secondary concentration of inner-sheath callose around the tube tip, which forms a useful marker when fluorochrome dyes and ultraviolet microscopy are employed. As the arrested tube can be shown to continue to respire normally, it appears that cessation of growth is implemented solely by the pectic occlusion of the tube tip.

These facts have led the Heslop-Harrisons to propose the following model of gene action for s-i grasses.

1. The incompatibility factors on the female side are glycoproteins with binding properties, which are located on the papilla surface.
2. These glycoproteins bind specifically to sugar moieties present in the long-chain carbohydrates in the wall of the apex of the pollen tube; such binding is highly specific to incompatible pollen tubes.
3. Such tight binding of the tube tip mechanically prevents the growth and stretching of the tube tip, thus interfering with the extension of polysaccharide microfibrils. As a result, there is a build-up of wall precursor particles, including microfibrillar pectins. These cannot be dispersed into the wall, which has stopped growing. The apex, the wall of which is already trapped and stuck fast by the papilla, is thus internally 'gummed up' with pectins. Thus the apex, being disorganized, stops growing. As Heslop-Harrison (1983) states 'the key event is the disturbance of pectin insertion in the apical cap of the tube, and this leads to a new view of the nature of the response'.

The basis of this model is the lectin-like properties of the glycoproteins which specifically recognize sugar moieties associated with the pectin of the pollen-tube wall. It is interesting that Heslop-Harrison should consider that sufficient diversity can occur in such sugar arrays to account for the action of multiple *S* alleles. Neither does this model account for the requirement that the same *S* allele should code for both a glycoprotein and a non-protein related sugar, although the sugar moiety could, I suppose, be the same in both. However, the position of the reaction, on the stigma surface, does allow for a very precise ultrastructural analysis of the phenomenon.

Multilocus sporophytic systems

In *Brassica*, inhibition of incompatible pollen also occurs at, or very near to, the stigma surface (de Nettancourt, 1977; Heslop-Harrison, 1983). Thus, *S*-specific recognition on the female side is borne superficially on the stigma, as in grasses, rather than in the style as in one-locus gsi systems. *S*-specific glycoproteins, products of the *SLG* locus, which have lectin-like properties, have been detected in the stigma and may correspond to substances with 'antigenic' properties identified earlier by Lewis (1952) and Linskens (1960).

The outstanding feature of sporophytic systems is that the *S* phenotype of the pollen is controlled by the diploid anther. All pollen grains from the same plant exhibit the same phenotype. The phenotype of the grain is independent of its genotype, and is imposed upon it from the anther. Such a mechanism can only operate by the anther physically endowing the pollen grain with a substance externally, after pollen grain formation.

Exine-held tapetal glycoproteins

Cytochemical, ultrastructural and electrophoretic observations all show that complex substances manufactured by the tapetum of the anther are later found in association with the exine of the pollen grain wall.

(In contrast, intine-bound proteins only show activity after pollen grain hydration, and are presumed to be of gametophytic origin. Although they very probably play a role in governing the hydration and tube growth of the pollen, they may not be implicated in *S*-recognition (Knox *et al.*, 1976; Heslop-Harrison, Knox and Heslop-Harrison, 1974; de Nettancourt, 1977; Heslop-Harrison, 1978). However, they may be involved in G-type recognition (p. 227).)

Proteins derived from the anther tapetum are diverse. Seven proteins have been identified in the exine of *Cosmos bipinnatus* (Howlett *et al.*, 1975) and in *Brassica oleracea* (Knox *et al.*, 1976). In the Compositae, tapetal proteins are inserted into micropores in the tectum (between the extine and intine), whereas in the Cruciferae they are deposited into cavities in the exine. At least some of these tapetal products are glycoproteins which diffuse rapidly from the exine when the pollen becomes hydrated. Even in the absence of incompatible pollen, tapetal proteins from self anthers can elicit responses in the stigma typical of an incompatibility reaction, such as the production of callose (Heslop-Harrison, Knox and Heslop-Harrison, 1974).

The diffusability of exine-held tapetal glycoproteins (*SLG* products) and their importance in the incompatibility reaction is also well demonstrated by so-called 'mentor effects'. If dead compatible pollen is mixed with living incompatible pollen, the incompatibility of the latter can be overcome (reviewed in de Nettancourt, 1977). In some cases, incompatibility-related pollen diffusates can be allelopathic to the pollen of other species, as for instance where the pollen of the Compositae species *Parthenium hysterophorus* and *Hieracium floribundum* alight on the stigmas of other species (Thomson, Andrews and Plowright, 1981).

Exine-held tapetal glycoproteins are sealed into place by sticky coatings of lipoprotein, also derived from the tapetum, termed 'tryphine' or, by German authors, especially when coloured by carotenes, 'pollenkitt'. This tryphine, which also surrounds the germinating pollen tube, plays a vital role in the attachment of the pollen grain to the 'dry' stigma. The attachment of incompatible pollen to stigma papillae may not always be as effective as that of compatible pollen, so some binding recognition by the tryphine may well occur.

Pollen hydration and germination

Once the dry pollen is attached to the pellicle of the stigma papilla in *Brassica*, it becomes hydrated. During the course of this initial hydration, most incompatible pollen grains are inhibited from germinating (Zuberi and Dickinson, 1985a). However, at high humidity, this inhibition is often overcome so that incompatible pollen germination and stigma penetration occurs. In these cases, incompatible tubes often elicit a callose response in the stigma which blocks their further progress. As shown by Ferrari, Lee and Wallace (1981), *Brassica* pollen can germinate almost anywhere at high humidity, including petals and leaves, as well as in water.

When *Brassica* stigmas are washed with water before or immediately after pollination, all pollen usually fails to hydrate or germinate (Zuberi and Dickinson, 1985b), although this failure was also overcome by high humidity. However, germinating pollen on washed stigmas at high humidity usually failed to enter the stigma normally.

If incompatible pollen does enter the stigma, Hodgkin and Lyon (1984) suggest that its presence stimulates the production of a low-molecular-weight phytoalexin-like inhibitor. The synthesis of this inhibitor can itself be overcome by the use of cycloheximide, suggesting that it has a proteinaceous basis. Typically, the progress of incompatible tubes in the stigma is blocked by callose (Heslop-Harrison, Knox and Heslop-Harrison, 1974), the production of which may be stimulated by the inhibitor.

The physiology of ssi in *Brassica*

Clearly, the functioning of ssi has a complex basis whereby several independently mediated systems show mutual reinforcement. In this, the ssi system differs from the gsi system which functions by a single reaction. For an oppositional incompatible ssi reaction to occur, it seems likely that:

1. incompatible pollen will be less likely to stick to the stigma;
2. genetically identical *SLG* locus products (glycoproteins) and/or *SRK* locus products (kinases) expressed both at the stigma papilla surface,

and at the surface of the pollen grain, recognize one another in a self, and this somehow mediates a 'signal transduction chain that leads ultimately to pollen rejection' (Nasrallah and Nasrallah, 1993);
3. the germination of incompatible hydrated pollen is inhibited, perhaps because hydration is normally incomplete for incompatible grains (hydration being mediated by water-soluble substances on the stigmatic surface).

The molecular basis for ssi in *Brassica*

In recent years, great strides have been made in the elucidation of the molecular basis of ssi incompatibility recognition in *Brassica*, and we know far more about the molecular genetics of this phenonemon, compared with its physiology. The subject is well reviewed by Nasrallah and Nasrallah (1993) to which the reader is referred for further reading.

The so-called *S*-locus in *Brassica* is in fact composed of two loci, one of which produces a glycoprotein (*SLG*) and the other a receptor kinase (*SRK*). These loci are linked, being separated by some 200 kb of DNA. The SLG glycoprotein has 431 amino acids, including a small signal peptide. These products are expressed both at the stigma papilla surface and on the surface of the pollen grain (originating from the anther tapetum). Expression has strict timing control, which explains why bud selfs are successful.

The two genes have a similar molecular structure, typically having some 90% of sequences in common within their *S*-domains, which has led to suggestions that they may be of duplicative origin.

It is not yet certain that both the products of *SLG* and *SRK* need to be present for an incompatible recognition to be elicited. Three *S*-gene 'haplotypes' (i.e. *SLG*/*SRK* genotypes) have so far been sequenced, and the molecular divergence between these has led to suggestions that the system may be in excess of 21 million years (ma) old (note, however, that it is likely that the gsi-system may date from approximately 100 ma).

The *SRK* kinase and other SLR (S-locus related) gene products have intriguing similarities to kinases involved in the mediation of disease resistance in plants. As for the earlier evolution of gsi systems, this has led to the suggestion that ssi has evolved from mechanisms which repelled fungal invaders at the stigma.

LATE-ACTING SELF-INCOMPATIBILITY

Late-acting self-incompatibility has been defined by Seavey and Bawa (1986) as those conditions where the functioning of s-i is located within the ovary, either prior to fertilization, or as the result of the abortion of the selfed ovule or fruit. These authors list 21 examples (from 16 families!) where s-i is known to be of this kind. Until recent years, this topic has received very little attention, perhaps because most examples seem to be restricted to tropical or subtropical floras.

It is clear that late-acting s-i covers a 'mixed bag' of phenonema, which are of secondary and polyphyletic origins, and none of which seem to have given rise to any major evolutionary lines.

These mechanisms can be classified as follows.

- Prefertilization
  1. Early flower abortion after self-pollination thought to be mediated by interactions between pollinium auxin and stigmatic eleutherocytes (*Dendrobium*).
  2. Self pollen tubes fail before fertilization, after exhibiting distorted growth patterns on reaching the ovary (*Melaleuca*, Barlow and Forrester, 1984).
  3. Self pollen tubes grow normally until they reach the ovule, at which point they cease to grow (*Lotus corniculatus*, Dobrofsky and Grant, 1980).
  4. Self pollen tubes are arrested at the nucellus layer within the ovule, immediately prior to fertilization (*Acacia retinoides*, Kenrick and Knox, 1985).
  5. After s-i pollinations, male gametes are released, but fail to fuse with egg cells and PEN (*Theobroma*, Cope, 1962).

- Post-fertilization
  6. Ovules abort at an early stage after fertilization, perhaps as a result of an interaction between the pollen tube and the ovule integument (*Gasteria*, Sears, 1937). In a similar reaction, young zygotes and endosperms fail to develop further, and ovule abortion occurs (*Rhododendron*, Williams *et al.*, 1984).
  7. After self-pollination, endosperms degenerate after the embryo reaches the globular phase, followed by ovular abortion (*Lilium candidum*, Brock, 1954; *Felicia*, Jordaan and Kruger, pers. comm.).
  8. After self-pollination, endosperms develop normally, but embryo development fails and the ovule aborts (*Asclepias*, Sparrow and Pearson, 1948). Seavey and Bawa (1986) mention several other examples.
  9. Small embryos and endosperms form normally, but at this point the flower is abscised (two species each of *Chorisia*, Bombacaceae and *Tabebuia*, Bignoniaceae, Gibbs and Bianchi, 1993).
  10. Fruit abortion takes place if too few ovules develop from cross-compatible pollinations (*Theobroma*; *Campsis radicans*, Bertin, Barnes and Guttman, 1989).

In some partially s-i members of the Leguminosae (e.g. alfalfa, *Medicago sativa*), s-i tubes grow more slowly than do those from crosses, and many fail to reach the ovule (reported in Seavey and Bawa, 1986). Ovules that do result from self-fertilizations are then more likely to abort. In this case, the control of the incompatibility may be complementary rather than oppositional.

Only in the case of *Theobroma* is the genetical control of the incompatibility known. In this case, the genetics are highly complex (Simmonds, 1976). *S* alleles occur at a locus in a multiallelic series, and show dominance reactions one to another in a sporophytic fashion. Consequently, reciprocal differences in cross-fertility between parents, and plants homozygous at the *S*-locus can occur. However, control of fusion is under gametophytic control and it seems likely that the system is essentially gametophytic on the male side and sporophytic in its nature on the female side.

In discussing possible modes of action for late-acting s-i (LSI), Seavey and Bawa (1986) emphasize the need to distinguish between post-fertilization LSI and ovular abortion resulting from inbreeding depression effects on selfed embryos. They point out that the latter will rarely fail in a consistent way; a staggered abortion of only some selfed ovules is more likely. For instance C.F. Page (personal communication) points out that in many conifers, selfs result in reduced levels of seed set compared with crosses, and that seed which is set from selfs tends to germinate poorly and to give rise to weak seedlings. Such complex consequences of selfing can be assigned to inbreeding depression.

Care should also be taken to ensure that abortion is restricted only to selfed ovules. In many plants, adjustments to maternal load by ovular or ovarian abortions are commonplace (p. 36), but are not specific to selfed ovules.

Clearly, the very diverse modes of operation of LSI imply that each may employ different recognition systems. Whereas some prefertilization mechanisms (examples 2, 3, 4, and possibly 5) may merely involve late-acting gametophytic RNase-type systems, postfertilization mechanisms must work in another way. It is possible that gametophyte–sporophyte self-recognition can occur between the pollen tube and the ovule (example 6). Equally, sporophyte–sporophyte self-recognition may cause failures where embryos (example 8) or endosperms (example 7) react with their surrounding nucellus. In yet other cases, plant growth substances may send messages of dosage (example 10) or recognition (example 9) from the ovary to the pedicel, or from the pollen to the pedicel, respectively.

Late-acting self-incompatibility is a subject still in its infancy and we may expect that further progress will be made in its understanding over the following decades.

CHAPTER 7

Heteromorphy

INTRODUCTION

Heteromorphy occurs where two or three distinct hermaphrodite floral morphs coexist in populations at roughly equal frequencies. Typically, these morphs are cross-compatible, but are within-morph incompatible. Thus, if there are two floral morphs present, there are also two corresponding mating types. In the majority of cases, the morphs have a reciprocal herkogamy with stigmas and anthers at corresponding positions within the corolla tube (heterostyly) (Fig. 7.1). Distylous systems have two morphs, one of which is long styled and often has anthers sunken in the corolla tube ('pin'), whereas the other is short styled with anthers positioned at the mouth of the corolla tube ('thrum'). In a few genera, three floral morphs occur each with a different style length (tristyly), and in these cases there are three mating types.

In dimorphic systems, genetically, one of the morphs, usually the thrum, is heterozygous for the mating system chromosome *Ss*, whereas the other is homozygous *ss*. (However, the pin is heterozygous in *Hypericum aegypticum*, Ornduff, 1979). Although individual thrum pollen grains carry only an *S* or an *s* chromosome, nevertheless all behave in a thrum manner (phenotype S), so that it is clear that the control of the mating system is sporophytic.

The *S* and *s* chromosomes usually behave in an allelic way, so that the control of the mating system is effectively diallelic. However, pins and thrums differ from one another by a number of characters, and at least some of these can be shown to be recombinable, so that it is clear that *S* and *s* in fact represent a coadapted linkage group, or 'supergene'.

Because about half the individuals in a dimorphic species are unavailable for mating, heteromorphy should be a rather inefficient outcrossing mechanism in comparison with multiallelic s-i. However, Darwin (1877) proposed that the heterostyly should encourage disassortative (legitimate) pollen flow between the morphs, thus increasing the efficiency of pollen usage, and modern evidence (p. 250) suggests that he was right to so suggest.

Although heteromorphy is expressed in a variety of morphological features which vary between genera and families (Table 7.1), there are

Table 7.1 Characters that can vary between heteromorphs (after Ganders, 1979)

Character	Comment	Examples
Style length	Not always exactly reciprocal to stamen length, difference longer in pin *Pentas*, in thrum *Cordia*	*Primula* and most others including *Linum*
Stylar conducting tissue	Different cell shape and area	*Primula*
Style pubescence	In pins	*Oxalis*
Style colour		*Eichhornia*
Stigma size	Bigger in thrum	*Jepsonia*
Stigma shape	Flatter in thrums	*Primula*
Papilla shape	Almost always longer in pins	*Primula* and most others (39/53, Dulberger, 1975)
Stamen length (anther position)	See style length	*Primula* and most others, but not *Linum*
Anther size	Thrum bigger	*Lithospermum*, *Pulmonaria*, *Hottonia*
Pollen grain size and number	Pin smaller, more numerous	*Primula* and most others (50/55, Dulberger, 1975)
Pollen sculpturing		Plumbaginaceae, *Linum*
Pollen shape	Thrum ovoid, pin dumb-bell shaped	*Lithospermum*
Pollen colour	Thrum green, pin yellow	*Lythrum californicum*
Starch in pollen	Thrum with starch	*Lythrum*, *Jepsonia*
Corolla diameter	Thrum larger	*Fagopyrum*
Corolla pubescence inside tube	Pin pubescent	*Lithospermum obovatum*

nevertheless a series of features by which pins and thrums differ which are commonly shared between unrelated genera. This suggests that as the heteromorphic syndrome evolved in unrelated genera, common selection pressures tended to select for analogous characteristics.

Character	Pin	Thrum
Style length	long	short
Stigma papilla length	long	short
Stylar cells	long	short
Anther position in tube	low	high
Pollen size	small	large
Pollen number	many	few
Genetic control	homozygous	heterozygous

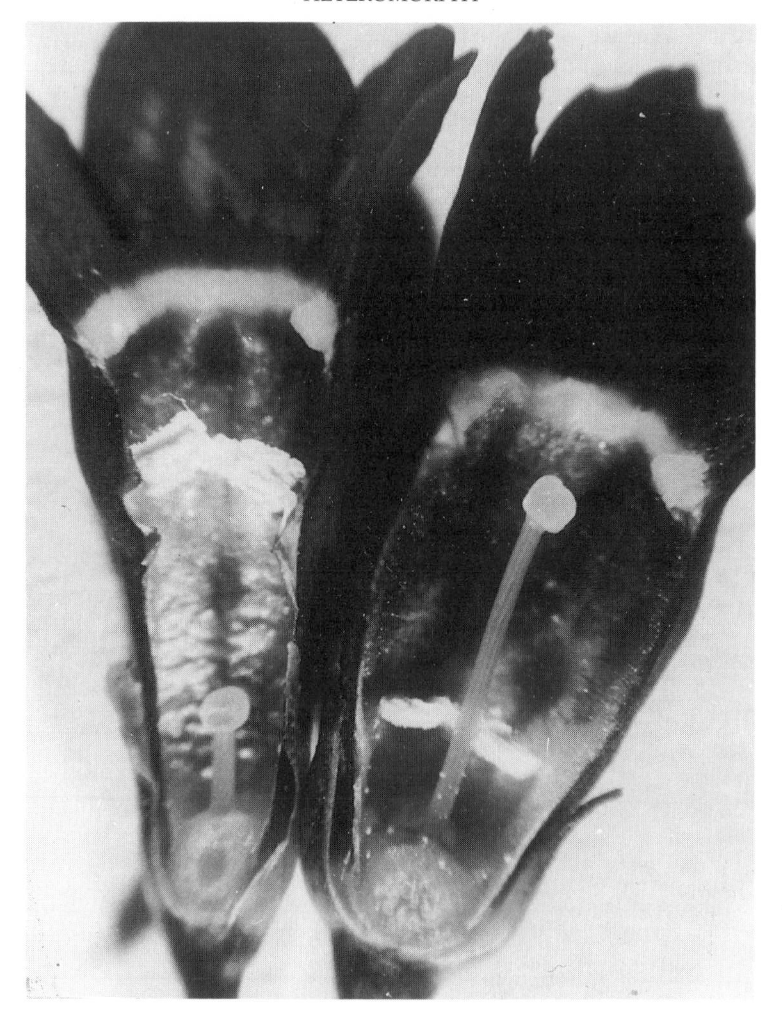

Fig. 7.1 Heterostyly in *Primula wilsonii*. Short-styled (thrum) flowers with high anthers (left); long-styled (pin) flowers with low anthers (right).

DISTRIBUTION OF HETEROMORPHY

Ganders (1979) shows that heteromorphy occurs in 25 families of flowering plants, spread among 18 orders. He lists 155 genera in which heteromorphy has been reported, of which no less than 91 belong to the Rubiaceae (Bawa and Beach, 1983). Other possible cases are listed in Barrett (1993) and Barrett and Cruzan (1994). Heteromorphy has a scattered distribution in the Angiosperms. Some major families have only

(a)

(b)

Plate 1 Parallel evolution of flowers adapted to bird-visiting belonging to three unrelated plant families, native to three continents. (a) *Passiflora incarnata* from central America – hummingbird flower. (b) *Aloe cameroni* from central Africal – sunbird flower. (c) (overleaf) The bird-pollinated inflorescence of *Strelitzia*. The coloured organs are bracts (orange) and anthers (blue).

(c)

(a)

Plate 2 (a) Section of the flower of *Incarvillea zhongdianensis* from Yunnan, China. In this spectacular bee-pollinated flower, the anthers mature in sequential pairs during the long flowering period. In the large chasmogamous flower they lie in a position which protects them from the monsoon rain, only being 'rocked' into a position where they can transmit pollen by the passage of the bee. (b) (overleaf) The 'brush' flower of *Calliandra ingens* (Mimosoideae) being visited by the 'painted jezebel' butterfly.

(b)

single heteromorphic representatives (*Jepsonia* in the Saxifragaceae, *Sebaea* in the Gentianaceae). Other families which contribute many heteromorphic genera and species include the Boraginaceae, Plumbaginaceae and Primulaceae. *Primula* has 426 species (Richards, 1993), 91% of which are heterostylous (Wedderburn and Richards, 1992).

Nearly all distylous species also have a diallelic, two mating-type incompatibility. Ganders expressly omits the few exceptions, species that are heteromorphic, but in which a diallelic incompatibility is lacking. He mentions only *Narcissus tazzeta* and the variably heterostylous *Amsinckia*. In the latter genus, there may be some slight incompatibility through competition between selfed and crossed pollen; the control of heterostyly is complex and probably polygenic, with pin-type and thrum-type plants occurring in different ratios and with different levels of heterostyly (Ganders, 1975).

Barrett, Lloyd and Arroyo (1996) show that five Spanish *Narcissus* species are distylous, and that one of these, *N. triandrus* has tristylous populations as well. All of these are to a greater or lesser extent self-incompatible (Barrett and Cruzan, 1994), but act as if possessing multi-allelic self-incompatibility, for most inter-morph crosses are compatible. Possibly, some have a late-acting self-incompatibility.

In recent years, many more examples of self-compatible heterostyles have become available, including members of the genera *Cryptantha*, *Decodon*, *Melochia*, *Nivenia*, *Oplonia* and *Quinchamalium*. References to these can be found in Barrett and Cruzan (1994).

Equally, the corollary, where a diallelic two mating-type breeding system occurs in the absence of an accompanying heteromorphy, is also rare, being recorded very dubiously at an early date for *Cardamine* and *Capsella* (Correns, 1916b).

Heteromorphy is found in all habitats from desert to aquatic, and from the tropics to the arctic. Generally, it is closely associated with perennial, often herbaceous plants with long fused corolla tubes (*Fagopyrum esculentum*, an annual, is an exception, and is the only notable crop-plant with heterostylous flowers). Consequently, heterostyly, is not found in plant families with a 'primitive' flower structure, and is very rare in the monocotyledons (*Nivenia*, Iridaceae, five *Narcissus* and various aquatic tristylous members of the Pontederiaceae are the only examples). Barrett (1993) places the occurrence of heterostyly in the context of the relationships between Angiosperm families.

Heterostyly is likely to be favoured in species with relatively specialized, animal-mediated cross-pollination with accurate pollen presentation and reception. Thus, heterostylous species nearly all have fused petals which form a bell, trumpet or tube. For pollination to be effective, especially on thrum stigmas, pollinators with narrow mouthparts which can penetrate a tube to feed on basally presented nectar will be encouraged.

Thus, most pollinators of heteromorphic flowers are social bees, moths, butterflies or birds which can successfully visit tube flowers; the 4 cm tubes of *Pentas* are quoted by Ganders (1979), but ironically those *Primula* species with the longest corolla tubes (*P. sherriffae* and *P. verticillata*) tend to be homostylous (Al Wadi and Richards, 1993; Tremayne and Richards, 1993).

Typically, the diameter of the limb of the flower is relatively small; zygomorphy is rare in heteromorphic flowers, as are large numbers of stamens. It is notable that in heteromorphic plants without heterostyly, such as *Armeria*, the corolla is bowl-shaped, and the petals are almost unfused, not forming a tube.

FORMS OF HETEROMORPHY

By far the commonest heteromorphic syndrome involves a reciprocal distyly of the primula type with 'pin' and 'thrum' flowers, and papilla length and pollen size dimorphies in association. This is typical of such diverse genera, each from a different plant family, as *Primula, Pulmonaria, Limonium, Menyanthes, Jepsonia, Fagopyrum, Forsythia, Sebaea* and *Nivenia*. However, a large number of other characters can also vary dimorphically between the mating types in certain genera (Table 7.1). Presumably, genes controlling these characters are also linked to the mating system S/s chromosome. In some cases, these additional dimorphic characters may be adaptive with respect to the functioning of the mating system itself (Dulberger, 1975, 1993), or may assist disassortative mating between the mating types, but in some cases they may merely represent neutral mutants which have 'hitchhiked' in close linkage with breeding system genes (p. 16).

Where closely related genera are heteromorphic, they tend to show corresponding heteromorphic features. In two of the other distylous genera in the Primulaceae, *Dionysia* and *Hottonia*, features of the morphology and the functioning of the mating system are exactly as in certain primulas, and Richards (1993) argues that the forerunners of these genera probably evolved from early heterostylous lines of *Primula*. Equally, all heteromorphic Plumbaginaceae, whether distylous as in *Limonium, Plumbago* and *Acantholimon*, or homostylous as in *Armeria*, all possess a distinctive dimorphy of the pollen exine (Baker, 1966) which seems to be functional with respect to the operation of incompatibility recognition (p. 274). Thus, it seems likely that heterostyly may have only evolved once in each of these examples, and it seems likely that the various heteromorphic genera within each of the Oxalidaceae, Pontederiaceae, Menyanthaceae and Oleaceae also arose from a single heterostylous ancestor. In other cases, heterostyly in related genera may be polyphyletic. The diallelic heterostyly in *Pulmonaria* has a number of individual features (p. 272), suggesting that it has evolved independently from that in *Anchusa*, *Alkanna* and *Cordia*.

In the commonest departure from the 'primula syndrome' of heterostyly, there is no heteromorphy for anther position, so that reciprocal herkogamy does not occur. This condition is best known in *Linum*, but also occurs in *Villarsia*, *Quinchamalium*, *Anchusa*, *Epacris* and *Chlorogalum* (Barrett, 1993). It is also found in one species of *Primula*, *P. boveana*. Al Wadi and Richards (1993) argue that in this primitive self-compatible species, anther position monomorphy may represent an intermediate condition in the evolution of full distyly (p. 265).

Pollen size and number dimorphy is common to many heteromorphic species (Dulberger, 1993), but it is not found in distylous *Linum*, and some distylous Plumbaginaceae and Boraginaceae. It is also absent in *Vitaliana primuliflora* (Primulaceae) (Wendelbo, 1961b), a distylous species which seems to have evolved distyly from homomorphic *Androsace*, independently from *Primula* (Richards, 1993). Dulberger quotes several species (e.g. *Lythrum californicum*) in which some populations are dimorphic for pollen size, but others are not.

Tristyly

The correspondence between features in systems which must have evolved independently in unrelated groups suggests that the coevolution of certain heteromorphic features is favoured if the heteromorphy is to be successful. This argument is reinforced by an examination of tristyly. A full tristyly is found in only three families, all unrelated, two of which are dicotyledons (Oxalidaceae, Lythraceae) and one of which is monocotyledonous (Pontederiaceae). (In *Narcissus*, monomorphic, distylous and tristylous species are found (Barrett, 1993) but in each case the morphs are not linked to a mating system.)

Tristyly with three mating types has certainly arisen on three occasions (Ganders, 1979). Yet, in all examples, there is a strong correspondence in basic features (Fig. 7.2). The three style lengths are each accompanied by the same patterns of stamen position in all three families. There is a correspondence in pollen size (long stamens always have the largest pollen), in papilla lengths (longest styles have stigmas with the longest papillae) and in cross-compatibility relationships (only pollen from stamens of the same position as the stigma is compatible on it).

Two loci (supergenes) control the tristylous systems (Fig. 7.2). These are termed *S* and *M*. Short-styled plants are *Ss*. Plants that are homozygous *ss* are epistatic to *M* with *ssM-* being mid-styled forms and *ssmm* being long-styled morphs. In *Lythrum*, and some *Oxalis*, *S* and *M* are unlinked. The frequency of the long- and mid-styled morphs will thus depend on the

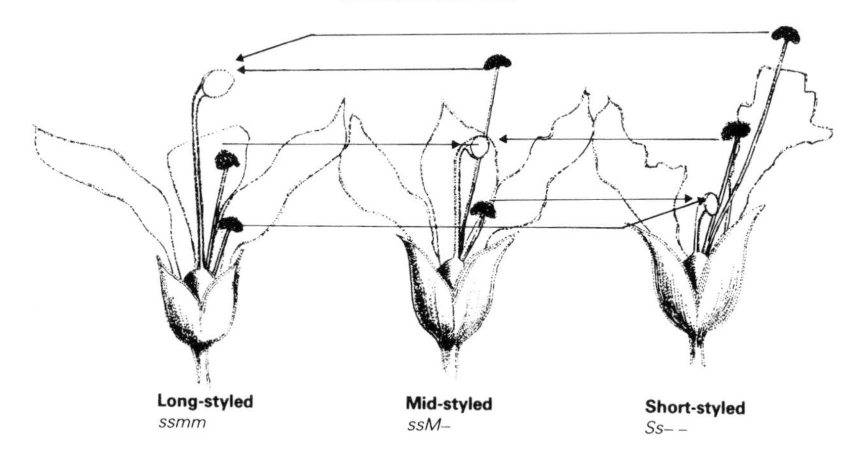

Fig. 7.2 Tristyly in *Lythrum salicaria*. Stigmas are pale and anthers dark. Arrows indicate compatible pollinations between morphs.

product of the frequencies of *ss* homozygotes and the allele *M*. In *Oxalis rosea*, *S* and *M* are rather closely linked. In this case, there is in effect an extension of the supergene, there being two extended linkage groups *sM* and *sm* which are relevant to the phenotypes in the system (frequencies of *SM* and *Sm* are irrelevant). The mating system should ensure that *sM* and *sm* chromosomes are equally frequent, and that *S* has a frequency of 0.33.

The tristylous system is also remarkable in that heteromorphic characters such as pollen size, papilla size, style length and stamen length are each apparently controlled by two linkage groups *S* and *M*, and thus presumably loci which control each of these features are repeated in both linkage groups. This leads to the suggestion that the *M* linkage group is a duplicate of the *S* group, and that tristylous plants have thus evolved from heterostylous (distylous) systems (see Charlesworth, 1979 for another opinion). This is very probably the case in both *Oxalis* and *Lythrum*, both of which also have distylous species. However, the Pontederiaceae have no distylous counterparts, and tristyly in this family appears on the face of it to have arisen *de novo*. The evolution of tristyly is comprehensively reviewed by Weller (1993).

The correspondence in general features between the different tristylous genera, and between tristylous and distylous species in different genera is remarkably strong. In each case, the long-styled morph is usually recessive, and has small pollen and long stigma papillae in both tristylous and distylous systems. Also, compatible crosses are only made between anthers and stigmas of the same level in both systems.

FUNCTION OF HETEROMORPHIC CHARACTERS

Reciprocal herkogamy

Darwin (1877) first suggested that reciprocally placed stigmas and anthers within the corolla tubes of distylous and tristylous flowers was a pollen-saving device, as it promoted disassortative pollination between the mating types. Flower visitors to a distylous species would be more likely to receive thrum pollen on their abdomens, and pin pollen on their heads, so that when they visited the other flower morph, the majority of pollen placed on stigmas would be legitimate. Darwin himself tested this hypothesis by looking at pollen grains on different parts of bees. Similar studies have been made for hivebees visiting buckwheat (Table 7.2) by Rosov and Screbtsova (1958).

In species with a pollen size dimorphy, it is easy to study legitimate and illegitimate pollen loads on stigmas. However, care must be taken that legitimate and illegitimate pollen adhere to stigmas equally readily. This is not the case in *Armeria* (Baker, 1995), nor probably in other Plumbaginaceae, so that pollen flow studies in this family are invalid.

Early reports suggested that disassortative pollen transport did not occur, but that pin stigmas in particular suffered large levels of illegitimate pollination (reviewed in Ganders, 1979). Ganders himself (1974) was the first to realize that distyly should have become adapted only to pollen incoming onto the flower. Studies with emasculated flowers of *Jepsonia heterandra* (Table 7.3) showed that thrum flowers mostly received legitimate (pin) pollen. Pin flowers received an excess of illegitimate (pin) pollen, but less than was expected if pollination were random. By comparison with unemasculated controls, Ganders showed that pin flowers in this species receive a good deal of illegitimate within-flower pollination. Similar studies were later made on tristylous *Pontederia cordata* (Price and Barrett, 1984; Barrett and Glover, 1985) and with primroses, *Primula vulgaris* (Piper and Charlesworth, 1986), and in each case strong dis-

Table 7.2 Position of pin and thrum pollen grains on honeybees (*Apis*)

Position of pollen on *Apis*	Mean numbers of pollen grains	
	Pin	Thrum
Thorax	402 500	252 500
Abdomen	180 000	762 500

Table 7.3 Pollination efficiency of heterostyly in *Jepsonia heterandra* (after Ganders, 1974)

	Stigmatic pollen load				
	Legitimate		Illegitimate between-flower (emasculated)		Illegitimate within-flower (non-emasculated minus emasculated)
Stigma	Observed	Expected	Observed	Expected	
pin	19.9	15.2	30.7	35.4	49.4
thrum	75.0	63.1	15.3	27.2	9.7

Table 7.4 Average numbers of legitimate and illegitimate pollen grains found on the stigmas of open-pollinated primrose (*Primula vulgaris*) in previously emasculated and intact flowers (after Piper and Charlesworth, 1986)

		Type of pollen (Mean number of pollen grains per stigma)		
Type of stigma	Treatment	Pin	Thrum	Total
Pin	Intact	2377	259	2636
	Emasculated	183	267	450
Thrum	Intact	256	583	839
	Emasculated	305	145	450

assortative pollination between the morphs was found after emasculation (Table 7.4).

In *Pulmonaria affinis*, Richards and Mitchell (1990) showed that strong disassortative pollination occurs onto thrum stigmas even in the absence of emasculation. Class studies have since shown that in some years, pin stigmas are also disassortatively pollinated. This species has a rather open bell-shaped flower in which pin and thrum pollen appears to be equally available to pollinators.

Present evidence supports Darwin's hypothesis as to the adaptive function of the reciprocal herkogamy. However, the situation is certainly more complicated than Darwin envisaged, as he did not consider the relevance of pollen number dimorphies, nor did he grasp the full implication of species such as *Linum grandiflorum* where the anther position is monomorphic. Distyly clearly has a function even when anther positions do not vary between morphs; as will be argued later, distyly in these circumstances is best interpreted as a mechanism which prevents within-flower pollination, rather than as a means to mediate disassortative pollen flow between flowers of different mating types.

Pollen size/number heteromorphy

Intuitively, it might be thought that thrum pollen is usually larger than is pin pollen in order to provide extra resource which allows the thrum pollen tube to grow down the long pin style (Lewis, 1942b). This disagrees with what we know of the means by which pollen tubes grow (p. 59), and Ganders (1979) shows no relationship between pollen size ratios and style length ratios between pins and thrums in 24 species. In *Primula*, pollen varies in volume within a species by a factor of approximately three, but pollen varies in volume between species by a factor of at least 500, and some of the smallest pollen grows down the longest styles (Richards, 1993).

Ganders (1979) was the first to show an inverse correlation between pollen size and pollen production for a number of dimorphic species. Dulberger (1993) provides data for 19 species in 12 genera which shows a very good correlation (Fig. 7.3). Generally, anther size does not differ between pins and thrums; the few exceptions listed by Dulberger (1993) do not appear in this data set. It may be supposed that a reduction in the size of pin pollen allows pins to produce more pollen without any increased demand for male resource (Ganders, 1979; Piper and Charlesworth, 1986).

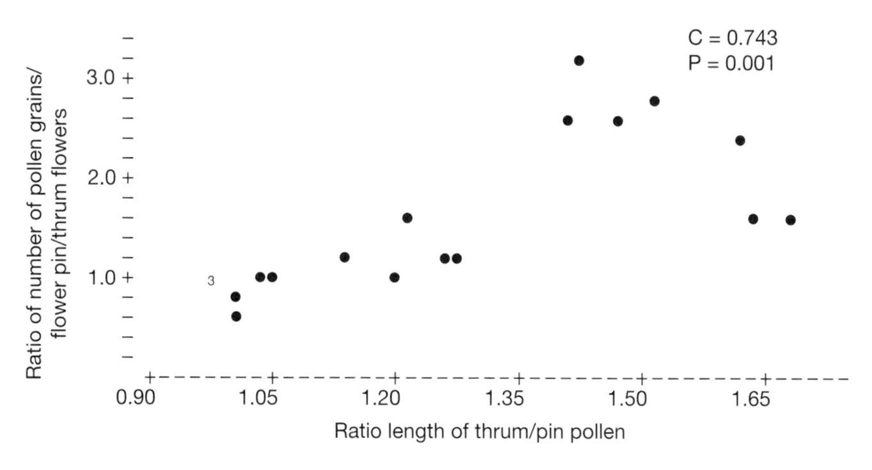

Fig. 7.3 Relationship between relative lengths of pollen grain and relative numbers of pollen grains produced for pin and thrum flowers for 19 species. (After Dulberger, 1993.)

Table 7.5 Average numbers of pollen grains produced by, removed from, and legitimately pollinated onto the stigmas of emasculated pin and thrum flowers of primrose *Primula vulgaris* (after Piper and Charlesworth, 1986)

	Pin flower	Thrum flower
Pollen produced by anthers	36620	17166
Pollen left within anthers	25391	7008
Pollen removed from anthers	11229	10158
Legitimate pollen on stigmas	267	305

John Piper studied the removal of pollen from anthers in primrose, *Primula vulgaris* (Table 7.5). He showed that whereas pin anthers with small pollen produce more than twice as many grains as do thrum anthers, there is no significant difference in the number of grains that insects remove from the 'hidden' pin anthers, in comparison with the 'available' thrum anthers. Piper also found no difference in numbers of legitimate pollen grains on the stigmas of emasculated pin and thrum flowers of primrose in the wild. Together, this information suggests that the small size and large number of pin grains have evolved in response to selection pressures to equalize male fitnesses between pins and thrums.

Not all heterostylous species differ in pollen size and pollen number between morphs, and in a few cases in *Amsinckia* and *Linum*, thrums actually produce more pollen than do pins (Dulberger, 1993). In general, such plants do not have pin pollen 'hidden' within a narrow floral tube, so that pin pollen may be at least as available to pollinators as is thrum pollen in these cases. Selection for variation in pollen grain size and number between pins and thrums should reflect differences in the availability of pollen to pollinators between the morphs.

Stigma papilla length heteromorphy

In general, stigmas of pin flowers have longer papillae than do stigmas of thrum flowers (Fig. 7.4). Also, in tristylous species, the length of the papilla tends to vary with the length of the style (Ganders, 1979). Lewis (1949a) was the first to conjecture that papilla lengths are merely developmental correlates of style lengths, pin and thrum gynoecia having the same number of cells which become more elongate in pin flowers.

This supposition is supported by studies on the length of stylar cells (e.g. Al Wadi and Richards, 1993) (Fig. 7.5) which also tend to be proportionately longer in pin styles than in thrum styles. These authors tested this hypothesis in homomorphic, self-compatible *Primula verticillata*, for which a good deal of continuous variation for style length occurs. In this species they also found that papilla length showed a correlation with both

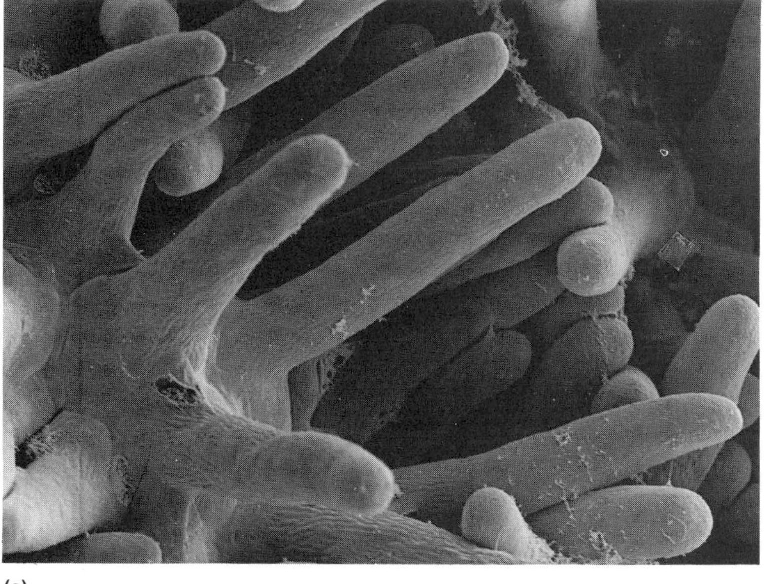

(a)

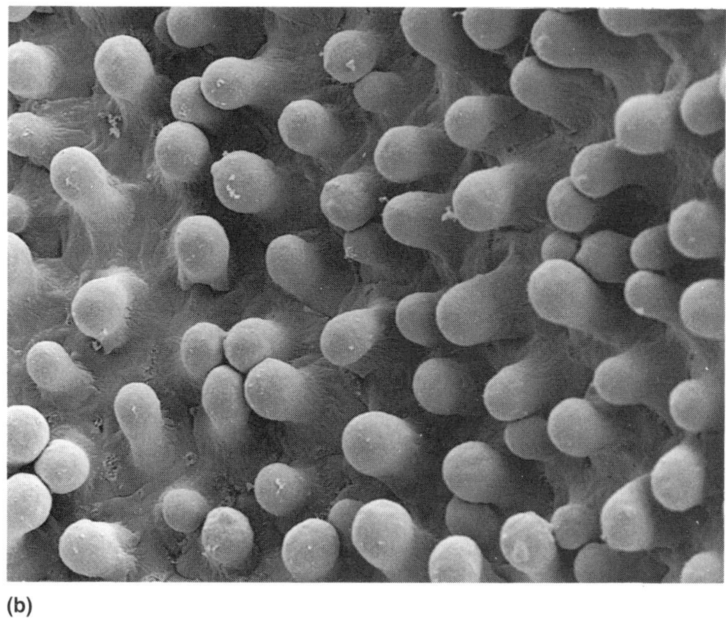

(b)

Fig. 7.4 (a) The long stigma papillae of a pin flower and (b) the short stigma papillae of a thrum flower in *Primula edelbergii*.

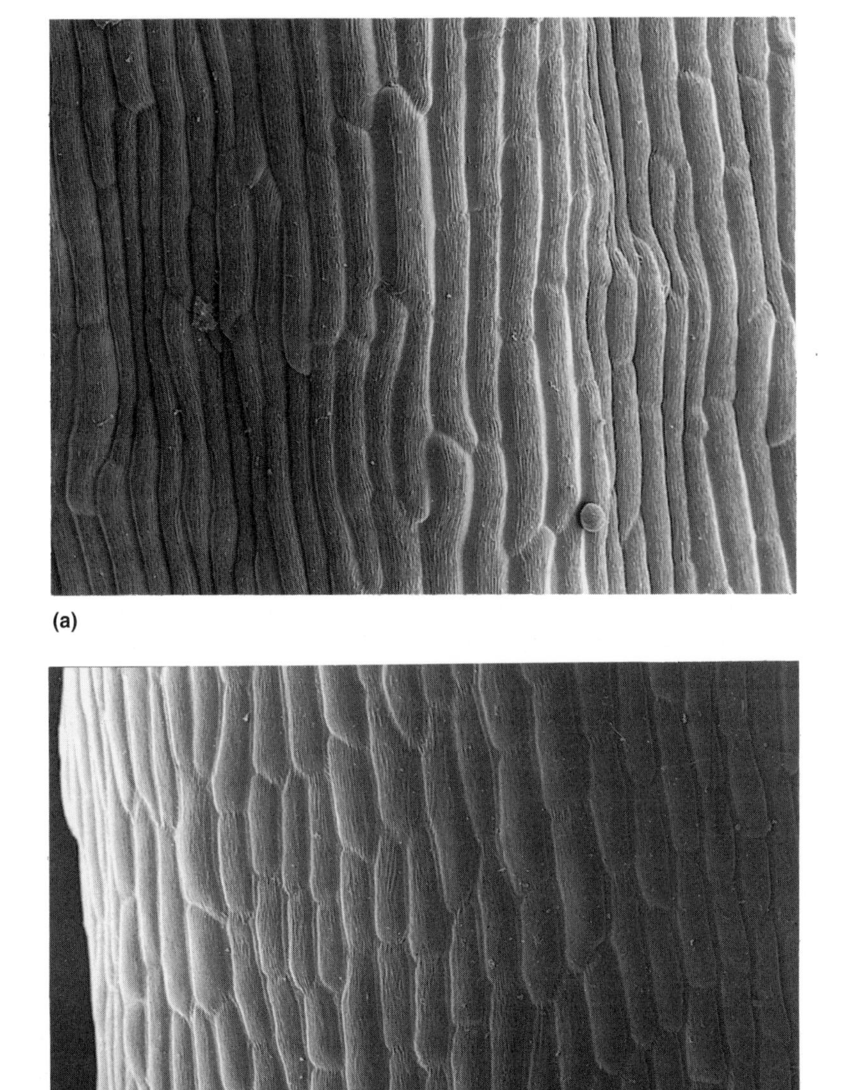

Fig. 7.5 (a) The long style cells of a pin flower and (b) the short style cells of a thrum flower in *Primula edelbergii*.

style length and stylar cell length, showing that papilla length variation with style length is not a prerequisite of the heterostylous condition. Further support is found in the work of Mather (1950), who found that modifier genes which shorten the length of pin styles in *Primula sinensis* also shorten stigma papilla lengths (see p. 263).

This explanation for papilla length dimorphy cannot be of universal significance, as is shown by dimorphic but homostylous species such as the thrift, *Armeria maritima*. In this species, the two mating types ('papillate' and 'cob') differ in their stigma papilla lengths, as their names suggest (Fig. 7.6), as well as in pollen exine sculpturing. Interestingly, the papillate morph is homozygous for the mating system chromosome (*ss*) as is usually the case for long-papillate pins in heterostylous species (Baker, 1966). Mattson (1983) shows that papilla length is strongly implicated in the functioning of the incompatibility reaction in *Armeria* (p. 274). *Armeria* is related to distylous *Limonium* species, but in that genus pin-flowered plants break the general rule by having the shorter stigma papillae, and by being heterozygous. This strongly suggests that distylous *Limonium* evolved from homostylous *Armeria*-progenitors in which the functioning of the incompatibility was already dependent on papilla length. In this case, style length may have evolved secondarily to, and independently of, stigma papilla length (Baker, 1966).

In *Primula vulgaris*, Shivanna, Heslop-Harrison and Heslop-Harrison (1983) show that thrum pollen germinates more readily at high humidity, which might be provided by the microenvironment at the surface of the long-papillate pin (legitimate) stigma (p. 275). Thus, there is a suggestion that the long stigma papillae typical of the pin might at times be implicated in the incompatibility reaction, perhaps having evolved secondarily so to do.

(a)

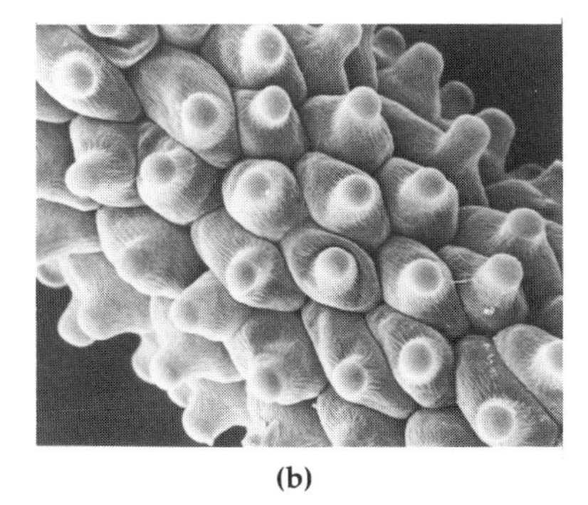

(b)

Fig. 7.6 Scanning electron micrograph (×500) of the stigmatic papillae of (a) cob and (b) papillate *Armeria maritima*.

STRUCTURE OF THE HETEROMORPHY LINKAGE GROUP; RECOMBINATIONAL HOMOSTYLES IN *PRIMULA*

Recombination of the linked loci controlling different heteromorphic features within the supergene has been studied in *Primula* (Ernst, 1933, 1936, 1955, 1957; Mather and De Winton, 1941; Mather, 1950; Dowrick, 1956), and to a lesser extent in *Limonium* and *Armeria* (Baker, 1966; Vekemans *et al.*, 1990), *Turnera* (Shore and Barrett, 1985) and *Pemphis* (Lewis, 1975). Most of our information comes from the massive, life-long study by Ernst, which has been comprehensively reviewed by Lewis and Jones (1993).

In *Primula*, three recombinable loci have so far been identified within the supergene (but see p. 264):

1. *G/g*, controlling style length, stylar conducting tissue, stigmatic papilla type, female incompatibility;
2. *P/p*, controlling pollen size, male incompatibility;
3. *A/a*, controlling anther height.

As the thrum is heterozygous *Ss*, thrum characteristics are dominant, and the thrum has the genotype *GPA/gpa*, the locus order being clearly shown by the absence of putative double recombinants, which would be self-fertile heterostyles (Lewis and Jones, 1993) (but see p. 263). Thus, the pin will be *gpa/gpa*. Recombination within the supergene can only occur in the heterozygous thrums, and seems to occur equally readily with thrums as male or female parents to the cross.

The commonest recombinants to be discovered are *Gpa* and *gPA*, secondary, self-fertile homostyles. *Gpa* recombinants have thrum female characteristics, and pin male characteristics (see Fig. 7.7), so that they automatically self-pollinate, and are self-fertile. Both the stigma and the anthers occur at the same level down in the floral tube, and such plants are known as short homostyles. Because they have thrum *S* phenotype female incompatibility, and pin *s* phenotype male incompatibility, they are self-fertile, and automatically self-pollinate, so setting high levels of selfed seed.

In contrast, *gPA* recombinants have pin female characteristics, and thrum male characteristics. The stigma and the anthers are produced at the mouth of the floral tube, and such plants are known as long homostyles (Fig. 7.8). They have pin *s* phenotype female incompatibility, and thrum *S* phenotype male incompatibility, so that once again they are self-fertile, although as the stigma and anthers are placed at the mouth of the corolla tube they are also cross-fertilized to some extent (about 10%, p. 290).

Being reciprocal recombinants, self-fertile short homostyles and long homostyles should arise at the same rate. In fact, Ernst obtained a significant excess of long homostyles, suggesting that short homostyles may be

Fig. 7.7 Flowers of a short-homostyle primrose, *Primula vulgaris* (variant with red flowers). This individual has stigmas in the short 'thrum' position, and included anthers in the 'pin' position as the result of genetic recombination between the loci controlling these features. Unlike pins and thrums, this individual was highly self-fertile.

less viable. Many of Ernst's crosses were made using recombinant individuals and resulted in the production of some back-recombinants. However, nearly half the new types he obtained could not be explained by recombinational models, and these he explained as mutations. Lewis and Jones argue that these non-recombinational off-types occur at too high a level to be explained by mutation. For instance nearly 3% of the progeny of his self-incompatible short homostyles *GPa/gpa* were normal thrums *GPA/gpa*. Lewis and Jones (1993) suggest that such off-types were caused by interactions between homoeologous loci in this polyploid plant. In my view it is more likely that modifier genes caused Ernst some problems in the correct scoring of his material (p. 260).

Self-incompatible long homostyles *gpA/gpa* (with small pollen) and self-incompatible short homostyles *GPa/gpa* (with large pollen) were also reported by Ernst. These occur more rarely than do self-compatible homostyles. Lewis and Jones (1993) estimate map distances of 0.19% between *P/p* and *A/a*, and of 0.37% between *P/p* and *G/g*. Such plants have no selfing advantage over heterostyles, and they have lost the advan-

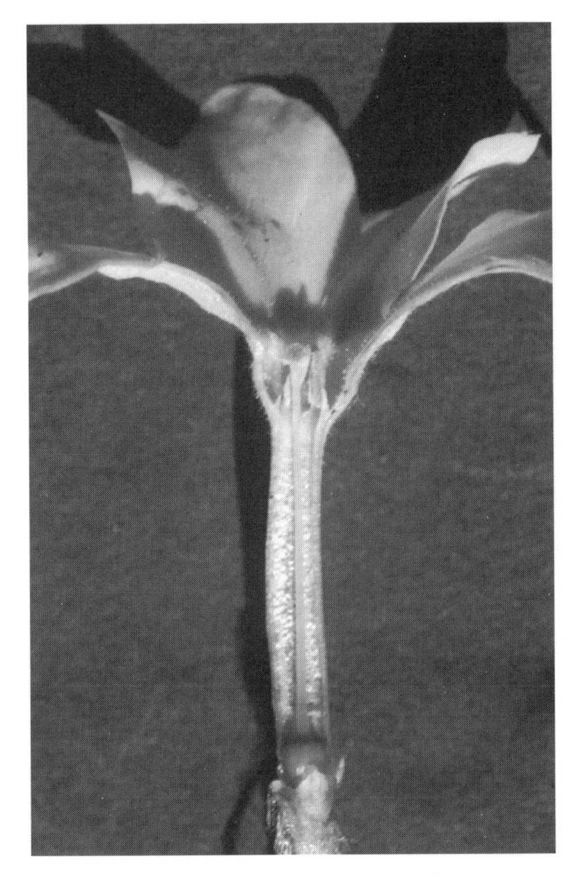

Fig. 7.8 Flowers of a long-homostyle primrose, *Primula vulgaris*, from the wild population at Sparkford, Somerset. Stigmas are in the long 'pin' position and anthers in the high 'thrum' position as the result of genetic recombination between loci controlling these features. Such individuals are highly self-fertile. (Photo by M. Wilson.)

tage of disassortative pollination, so they may not persist in wild populations.

The recombinational nature of secondary homostyles

Ernst (1936) originally considered that these four types of homostyle represent mutations from the distylous condition. Dowrick (1956), in a survey of homostyly in *Primula*, suggested that they might be better explained as recombinants of loci within the mating system supergene, causing Ernst (1957, 1958) to re-examine all his data. Ernst (1950) had in fact already provided results which confirmed this hypothesis. When he crossed homostylous *Primula japonica* with the heterostylous *P. pulverulenta* and *P. burmanica*, incompatibility reactions showed that *P. japonica* acted as if it were a pin parent when it was female, but as if it were a thrum parent when it was a male to these crosses. These results strongly suggest that the homostyle *P. japonica* has evolved from heterostyle ancestors by recombination, retaining all the morphological and incompatibility features of the original heterostyles.

Later, similar experiments (Wedderburn and Richards, 1992) were undertaken among distylous and homostylous members of *Primula* section Aleuritia. This work (Table 7.6) also shows convincingly that the homostyle species involved were indeed secondary homostyles which had evolved from heterostylous species by recombination. In this group, pin × pin selfs are slightly self-fertile, but thrum × thrum selfs usually set no seed whatever. Although crosses between two heterostyle species show that some interspecific sterility barriers occur, at least unilaterally, crosses between the heterostyle and the homostyle species act exactly as one would predict if the homostyle were acting as a thrum male parent and as a pin female parent.

Table 7.6 Average proportions of ovules set as seed after heterostyle × heterostyle and homostyle × heterostyle crosses among species in *Primula* section Aleuritia. P = pin, T = thrum, H = homostyle. Female parent first

		Seed set(%)
Heterostyle × heterostyle		
P. farinosa	P × T, T × P	51.1
	P × P	0.4
	T × T	0
P. modesta	P × T, T × P	26.3
	P × P	4.3
	T × T	0
P. farinosa × *modesta*	P × T	10.1
	T × P	0.6
P. modesta × *farinosa*	P × T	0
	T × P	0.1
Heterostyle × homostyle		
P. modesta × *laurentiana*	P × H	20.0
	T × H	0.2
P. laurentiana × *modesta*	H × P	5.6
	H × T	80.8

Occurrence of secondary homostyles

It seems likely that all heterostyle populations have the potential to form new homostyle selfing strains by recombination, although this potential is seldom realized. In *Primula*, about 6% of the species are secondary long homostyles (p. 290), and these are generally polyploid (Wedderburn and Richards, 1992) (p. 292). In *P. vulgaris*, a few populations contain high proportions of long homostyles (p. 287). Homostyles also occur as occasional individuals in other populations. For instance a single short homostyle is reported from a Japanese population of *P. sieboldii* (Washitani *et al.*, 1994), and I have found short homostyle individuals of *Primula* × *tommasinii* and long homostyle individuals of *P. veris* during population searches. Long homostyles are also known in *P. obconica* (Dowrick, 1956). Ernst (1933) grew up many thousands of seedlings of *P. latifolia* and *P.* × *pubescens* in his detailed analysis of the occurrence of the occasional recombinants that he found.

Secondary homostyles also occur in other genera. I once found a self-fertile long-homostyle individual of *Pulmonaria affinis* which had self-sowed from heterostyle stock in my own garden. Fifteen *Limonium* species are recombinational short homostyles, only one being a long homostyle (Baker, 1966), and self-fertile monomorphic populations of *Armeria maritima* in arctic Europe and America are also recombinational in origin. However, crosses between secondarily monomorphic and dimorphic individuals in the Plumbaginaceae show that monomorphic individuals have lost the papillate stigma recognition.

Long homostyle races are also reported for *Turnera ulmifolia* (Shore and Barrett, 1985), and *Pemphis acidula* (Lewis, 1975). In both cases, crosses with heterostyles showed them to be recombinational in origin.

Semi long-homostyles (in which one of the pairs of anthers are in the homostylous position) are reported for all three of the tristylous families, and in each case these are self-fertile for pollen originating in the homostylous anthers (reviewed in Lewis and Jones, 1993). The genetics of these are complicated by the two unlinked *S/s* and *M/m* loci, and it is not clear whether they are recombinational in origin.

Non-recombinational homostyles

Not all homostyles that occur in heterostylous genera are recombinational in origin. Al Wadi and Richards (1993) argue that some homostylous *Primula* are primary homostyles, representing the condition in the genus before heterostyly evolved (p. 265).

Mather (1950) identifies two genes in distylous *P. sinensis*, *A/a* and *M/m*, unlinked to the heterostyly chromosome, the recessive phenotypes of which modify style length and anther position respectively, although they do not affect the incompatibility reaction of the morphs. Pins with the aM phenotype look like short homostyles, and those with the Am phenotype resemble long homostyles. am phenotype plants are 'pseudothrums', but their mating system shows them to be effectively pin for the heterostyly chromosome.

Wedderburn (1988) also finds genes which modify style length in wild populations of *P. veris*. In the hybrid of *P. veris*, *P.* × *tommasinii*, there is strong evidence that at least one of these genes is linked to the heterostyly chromosome (p. 264). Tremayne and Richards (1993) show that considerable variation for style length can also occur in secondary homostyle *Primula* species (Fig. 7.9), and this is also apparently mediated by modifier genes. They argue that such systems which develop herkogamy will be favoured in self-fertile plants with very narrow corolla tubes, as they will encourage some outcrossing to take place.

Also, it seems likely that in *Linum austriacum* (Heitz, 1973) and in *Amsinckia* (Ganders, 1975), variably homostyle conditions have arisen which are mediated by modifiers, rather than by recombination. These groups are basically self-fertile, but in *Anchusa officinalis* (Schou and Philipp, 1983) variably homostyle individuals occur in a group with multiallelic self-incompatibility.

Elizabeth Arnold and I are currently studying a single population of *Primula farinosa* which, uniquely in this species, is acaulescent (flowers borne deep in the rosette). On first examination, half of these plants seemed to be long homostyle, the remainder being thrum. A closer analysis shows that the acaulescent plants have a shorter corolla tube than do neighbouring caulescent plants, and the anthers are borne higher in the tube, so that the flowers resemble long homostyles, but genetically they behave as pins. It seems that this adaptation has occurred in response to crawling, pollen-feeding flower visitors in this extremely exposed station which may not be able to reach conventionally positioned pin anthers. Interestingly, pin pollen of acaulescents is larger than that in coexisting caulescent pins, but it is smaller than in both acaulescent and caulescent thrums. This suggests that as pollen has become more available to the visitor, male selection for large pollen number has become less intense.

In this population of *Primula*, it seems that another coadapted linkage group, unlinked to the heterostyly supergene, has evolved in response to exceptional pollinators.

Other recombinations within the heterostyly supergene

Kurian and Richards (1997) have been working with a strain of *Primula* × *tommasinii* in which a high proportion of the pin plants are very self-fertile, having half their pollen 'pin' sized, and the other half 'thrum' sized (Fig. 7.9(c)). Pin-sized and thrum-sized pollen germinates equally readily

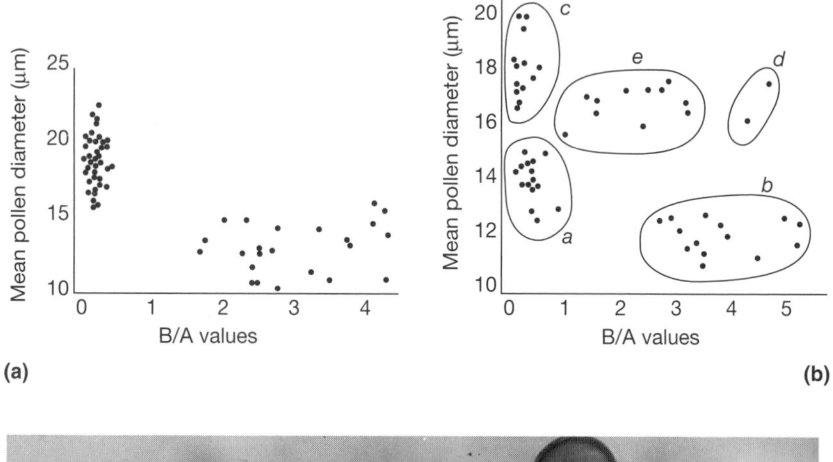

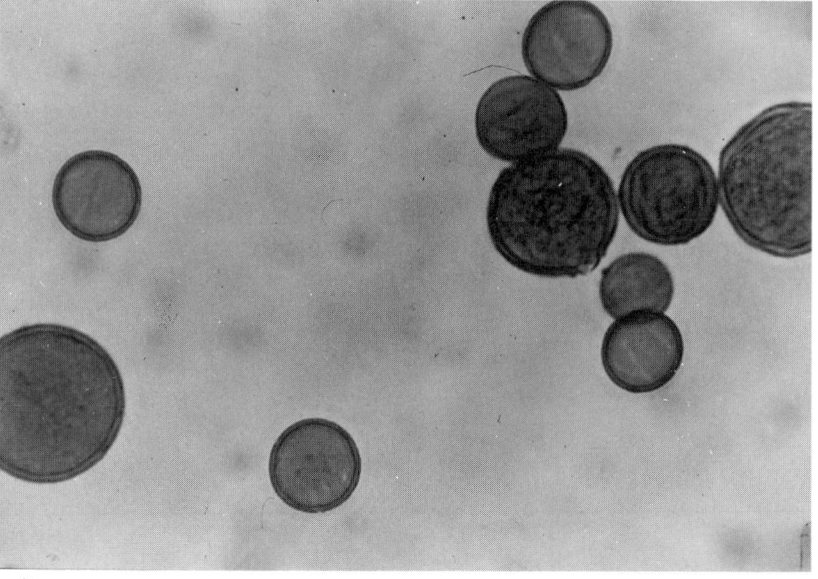

Fig. 7.9 (a) Distribution of the ratio of style length against anther position (B/A), plotted against mean pollen diameter in μm for individuals of ten distylous species: *Primula apoclita, P. cernua, P. deflexa, P. giraldiana, P. gracilenta, P. mairei, P. muscaroides, P. pinnatifida, P. violacea* and *P. vialii*. (After Tremayne and Richards, 1993.) (b) Distribution of the ratio of style length against anther position (B/A) plotted against mean pollen diameter in μm for *P. bellidifolia* and subsp. *hyacinthina*. (After Tremayne and Richards, 1993.) *a*, Diploid thrums, *b*, diploid pins, *c*, tetraploid pins, *d*, tetraploid thrums, *e*, presumptive homostyles. (c) A mixture of pollen of sizes typical of both pins and thrums found in the P* recombinant of *Primula* × *tommasinii*.

on a pin style. These abnormal pins (p*) also differ in that the style and stigma papillae are slightly but significantly shorter than in normal pins. When the abnormal pins are selfed, two-thirds of the offspring are abnormal pins (P*), while the remainder are 'normal' pins (P). When these abnormal pins are crossed with 'normal' pins, in either direction, half the offspring are normal and half are abnormal.

Offspring of selfs and crosses involving abnormal self-fertile pins in *Primula* × *tommasinii*

P* self offspring	P* male × P cross offspring
P* 50 P 31	P* 139 P 133
chi-squared test against 2:1 hypothesis n.s.	chi-squared test against 1:1 hypothesis n.s.

Several interesting conclusions emerge from a consideration of these abnormal pin plants.

1. It seems that they carry the thrum allele *P* controlling pollen size and male incompatibility, so that they have the putative genotype *gaP*/*gap*, but the dominant control for pollen size has been lost, pollen size being expressed by the genotype of the individual grain.
2. However, the male mating-type recognition has not lost its dominance, so that all grains have the thrum mating phenotype P, whether they carry the allele *P* or *p*.
3. This suggests that different, closely linked genes control pollen size and male mating type, and that each is subject to different linked dominance modifiers. Because the male mating type dominance modifier has been recombined, but the pollen size dominance modifier has not, this suggests that the position of the latter is proximal to that of the former.
4. We presume that this syndrome is recombinational rather than mutational in origin as it involves five different characteristics (male mating type, pollen size, mating type dominance modifier, recessive lethality and style length modifiers) which are likely to be coded by different genes.
5. No P* offspring ever have 100% thrum-sized pollen, or ever give rise to offspring which are all abnormal. Thus, *gaP*/*gaP* homozygotes must be lethal, because we would otherwise expect 25% of the offspring of abnormal pin selfs to be of this type. We conclude that a thrum-linked lethal has been recombined onto the pin chromosome.
6. If this recombinant arose from a single chiasma, it follows that *P*/*p* is distal to the supergene and that the correct gene order is in fact *GAP*, not *GPA*. This conflicts with the conclusions of Lewis and Jones (1993),

based on the absence of putative double recombinants in Ernst's work on *P.* × *pubescens* (p. 256).

There are three possible solutions to this conundrum. First, that the P* syndrome arose by a double recombination, between *G/g* and *P/p*, and between *P/p* and *A/a*. Second, that the numbers of Ernst's recombinants are too small to be certain which recombinant types are the rarest. Third, that the heterostyly supergene evolved independently in *P.* × *pubescens* (subgenus *Auriculastrum*) and in *P.* × *tommasinii* (subgenus *Primula*). The first and second explanations are inherently very unlikely. It seems more probable that heterostyly has had a separate origin in these two very distantly related plants (Richards, 1993).

If we assume the latter explanation, we can now suggest a tentative model for part of the structure of the heterostyly supergene in *P.* × *tommasinii*:

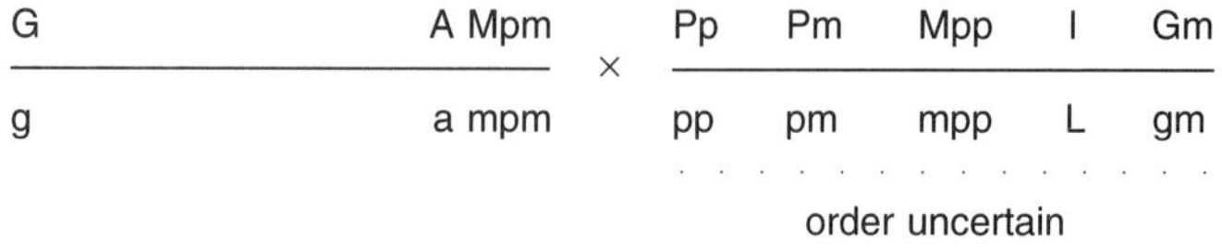

G		A Mpm		Pp	Pm	Mpp	l	Gm
			×					
g		a mpm		pp	pm	mpp	L	gm
				order uncertain				

where

G/g controls all female characters
A/a controls anther position
Mpp/mpp controls pollen size dominance
Pp/pp controls pollen size
Pm/pm controls pollen mating type
Mpm/mpm controls pollen mating type dominance
l/L controls recessive lethality
Gm/gm contributes to stylar shortening
(× is the position of putative recombination in P* plants).

Possible protection against recombination in the Primula heterostyly supergene

Clearly, the heterostyly supergene is very tightly linked, for Lewis and Jones (1993) estimate that it is responsible for only about one two-hundredth of the length of the S chromosome (p. 257). Other S-linked genes, for instance *Ma/ma* and *Mg/mg* which are responsible for anthocyanin petal coloration in *Primula*, are much more loosely linked, being at least 33 map units apart (Kurian, 1996). Consequently, recombinants within the *Primula* supergene seem to be sufficiently unusual that no protection against recombination has evolved in most cases (unlike the situation for the sex chromosome in many dioecious species, p. 307).

Thus, it is surprising that Bruun (1932) reported that in some species in *Primula* section Muscarioides (e.g. *P. vialii*), a heteromorphic pair of chromosomes occurs in thrums, but not in pins. At meiosis, these chromosomes clearly differ for an inversion, suggesting that the *Primula* heterostyly supergene may well be protected from recombination within an inversion loop in these plants. As yet, there is no other indication in any heteromorphic species that the heterostyly supergene is so protected.

PRIMARY HOMOSTYLES

In several heteromorphic genera, related groups of self-compatible or partially s-c species or races occur which include homostyles, heterostyles, and a variety of intermediate conditions, for instance in *Amsinckia* (Ganders, 1975), *Narcissus* (Barrett, 1993) and *Primula* (Al Wadi and Richards, 1993). Intermediate forms are also found in *Anchusa officinalis* (Schou and Philipp, 1983), but these have a multiallelic incompatibility system. Such plants have traditionally been interpreted as representative of stages in the evolutionary breakdown of a full distyly (e.g. Barrett, 1993).

However, the particularly elegant conditions existing in the various isolated species of *Primula* section Sphondylia strongly suggest that in this case at least the various populations may represent stages in the original evolution of distyly (Al Wadi and Richards, 1993). I have called these conditions primary homostyly to differentiate them from secondary, recombinational homostyles. (In Barrett, 1993, the term monomorphy is preferred for pre-heterostyly conditions, but I find this usage confused, as it can refer to secondary homostyles, and to monomorphic plants in dimorphic, non-heterostylous species.)

Eight species are classified in *Primula* section Sphondylia (Richards, 1993). All are relicts with scattered, highly localized distributions beside north-facing waterfalls in subdesert regions (Table 7.7). All are self-compatible diploids.

Thus, the isolated localities of these eight species comprise between them a linear series from north-east Africa in the south-west to the western Himalaya in the north-east. In general, morphological features also follow this linear series; for instance species in the south-west have the longest corolla tubes, and those in the north-east have the shortest.

The features of the distyly also show a remarkable correspondence to this 'step-cline' (Fig. 7.10). For instance, *P. verticillata* shows more herkogamous variation than does *P. simensis*, and the degree of the separation of the reciprocal herkogamy becomes more marked in the distylous species in a north-easterly direction (Figs. 7.10, 7.11). In this context, the very localized and highly endangered *P. boveana* is a particularly interesting case, for this has a fairly marked distyly and pollen size dimorphy, but

Table 7.7 The species in *Primula* section Sphondylia

Species	Distribution	Styles	Anthers	Pollen
simensis	Ethiopia	Homomorphic	Homomorphic	Homomorphic
verticillata	Yemen, Saudi Arabia	Variably homomorphic	Homomorphic	Homomorphic
boveana	Sinai	Dimorphic	Homomorphic	Dimorphic
davisii	Kurdistan	Dimorphic	Dimorphic	Dimorphic
gaubeana	Luristan	Dimorphic	Dimorphic	Dimorphic
edelbergii	Afghanistan	Dimorphic	Dimorphic	Dimorphic
afghanica	Afghanistan	Dimorphic	Dimorphic	Dimorphic
floribunda	NW Himalaya	Homomorphic to dimorphic	Homomorphic to dimorphic	Homomorphic to dimorphic

the anther position is monomorphic (Fig. 7.11). Geographically and morphologically it is also in an intermediate position within the group.

The Himalayan *P. floribunda* is also instructive. In this winter-flowering species, low altitude populations are distylous, but populations at higher levels are homostylous, intermediate populations showing intermediate conditions (Fig. 7.12). This suggests that populations at higher altitudes have selected for automatic self-pollination in the absence of insect visits, but at lower altitudes where insect activity in the winter is more prevalent, reciprocal herkogamy has been favoured.

In a number of morphological features, *Primula* section Sphondylia has been regarded as being among the least derived groups in the genus (Wendelbo, 1961a, discussed in Richards, 1993), so this adds credence to the hypothesis that these isolated populations do in fact represent stages in the evolution of heterostyly. As all species are self-compatible, it is not possible to test this hypothesis by examining the fertility of reciprocal homostyle × heterostyle crosses, as is done for species in section Aleuritia in Table 7.6. However, Al Wadi and Richards (1993) argue that it is highly unlikely that the various homostyle/heterostyle conditions exhibited in this group represent stages in the breakdown of heterostyly by recombination. Not only are all these species self-fertile, but the subtle intergradations between the homostyle and heterostyle conditions that these species demonstrate are most unlikely to have resulted from recombination.

If indeed these species do exhibit geographical stages in the evolution of full heterostyly, we can make the following deductions concerning the evolution of heterostyly in this group.

1. Heterostyly evolved in self-fertile plants.
2. A short style variant (*G*) evolved first, followed by a small pollen variant (*p*), and lastly by a low anther position variant (*a*).

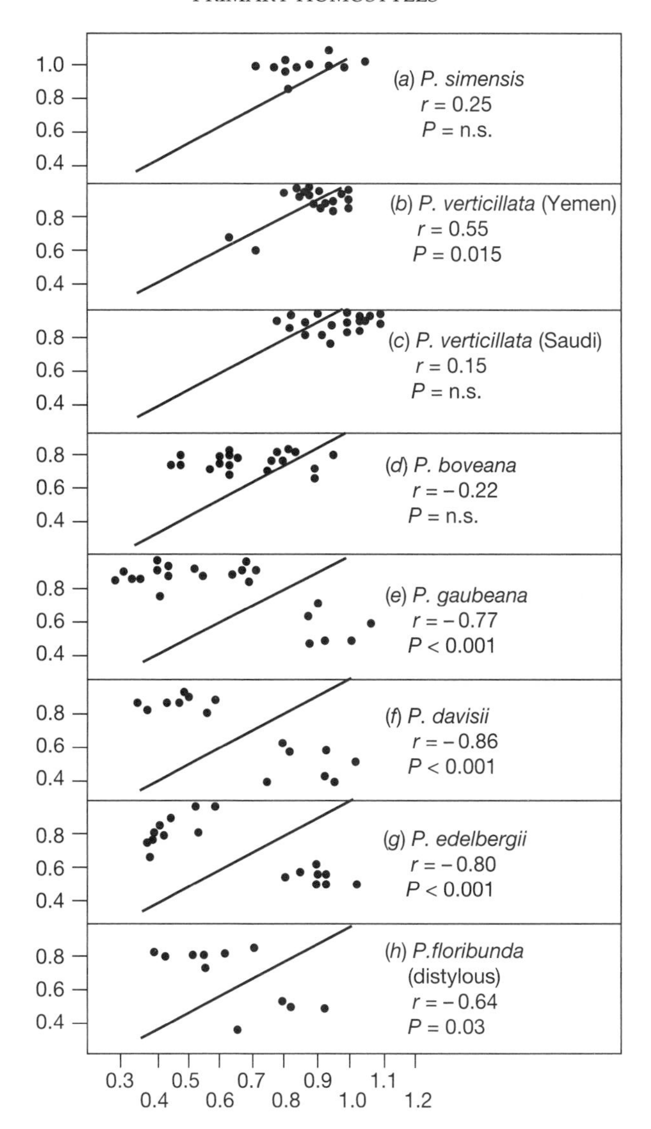

Fig. 7.10 Mean style length relative to corolla-tube length (x axis) and mean anther position relative to corolla-tube length (y axis) in various species of *Primula*. Lines represent homostylous positions. (After Al Wadi and Richards, 1993.)

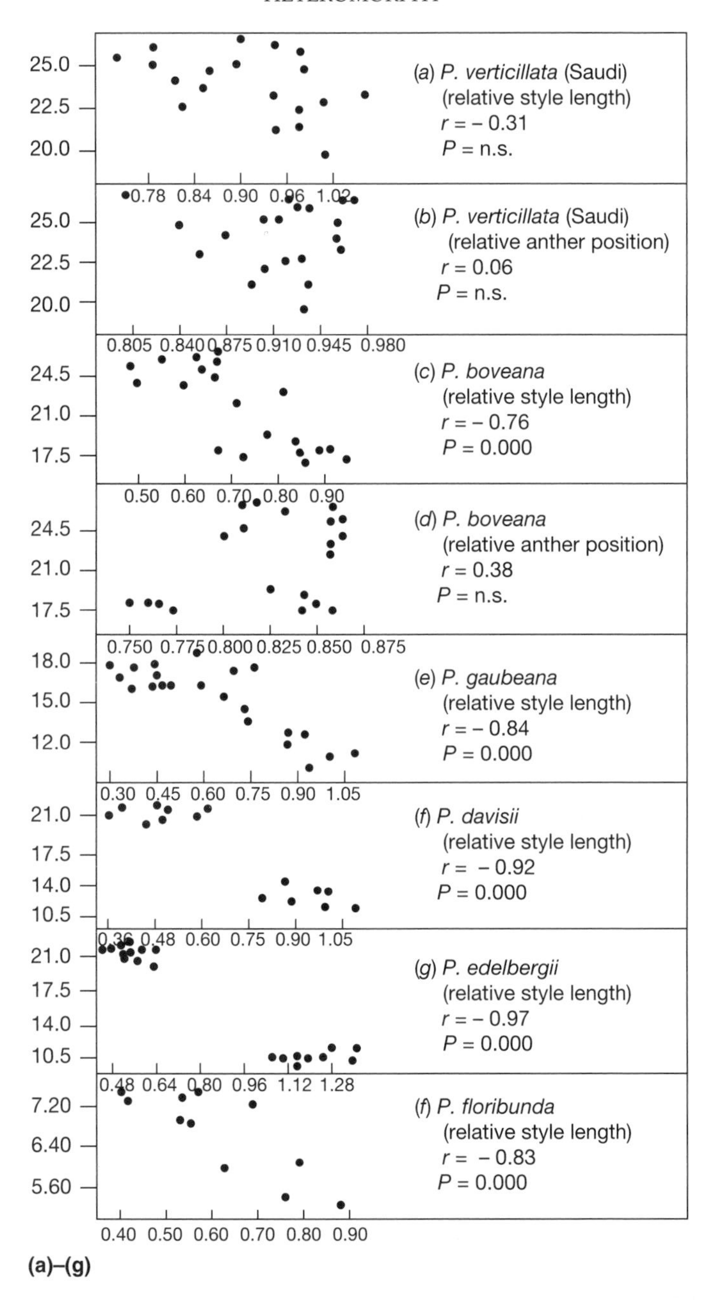

Fig. 7.11 Mean style length relative to corolla-tube length (a,c,e–h). Mean anther position relative to corolla-tube length (b,d) (x axis) and mean pollen diameter in μm (y axis) in wild populations of *Primula verticillata* (Saudi) (a,b) and *P. boveana* (c,d), and in herbarium samples of *P. gaubeana* (e), *P. davisii* (f), *P. edelbergii* (g) and distylous *P. floribunda* (h). (After Al Wadi and Richards, 1993.)

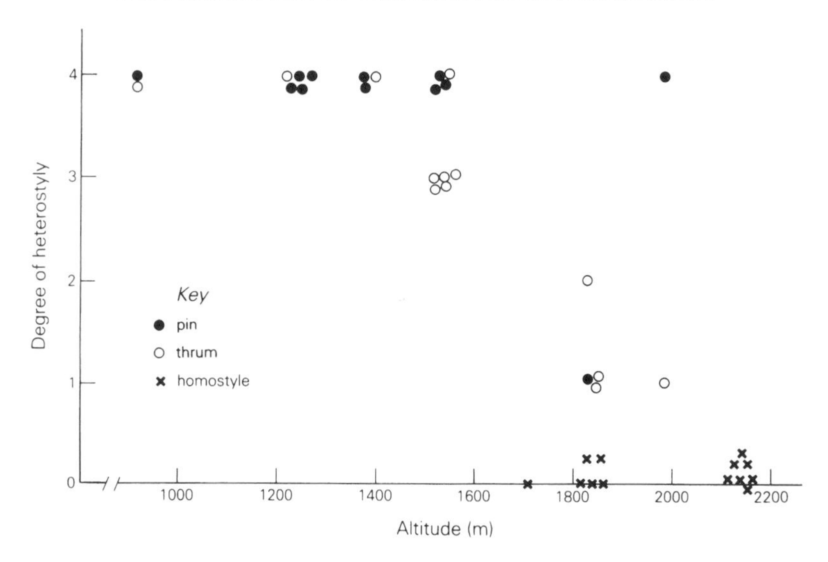

Fig. 7.12 Relationship between altitude (x axis) and the degree of heterostyly (y axis, scale 0 to 4) for individual herbarium specimens of *Primula floribunda* collected in the wild in the western Himalayas. Plants scoring 0 are homostyles, those scoring 4 have anthers and stigma displaced by 7 mm or more; those scoring 1 to 3 are intermediate. Closed circles are pins, open circles are thrums and x denotes homostyles.

3. Although these three conditions are apparently controlled by major genes, later evolving modifiers of minor effect increased the accuracy of the reciprocal herkogamy and pollen size dimorphy.

THE FUNCTIONING OF DIALLELIC SELF-INCOMPATIBILITY

Heteromorphic breeding systems are secondary and polyphyletic in origin, so it is likely that the mating system recognition functions in various ways in different groups. However, the genetic control of mating in most heteromorphic systems shares the following features:

1. two or three mating types which correspond to the flower morphs
2. diallelic control, where one morph (usually the thrum) is heterozygous *Ss* and the other homozygous *ss*
3. sporophytic control, in that the pollen of the heterozygote (usually the thrum) all behave the same way on illegitimate or legitimate stigmas, and the stigma of the heterozygote shows dominance rather than independent reaction with respect to the genotype of incoming grains.

By analogy to multi-allelic sporophytic mating systems (p. 236) we might expect the functioning of diallelic incompatibility to be restricted to the stigmatic surface. However, in *Primula* at least, a good deal of the inhibition of illegitimate grains in many species occurs deep in the style (Wedderburn and Richards, 1990). The same seems to be true of *Pulmonaria* (Table 7.8). As it might be surprising if exine-bound products of tapetal origin are expressed after the pollen tube has grown a considerable distance into the style, it is possible that some of the incompatibility function has a gametophytic control. In this context, it is noteworthy that in those species of *Primula* with total within-thrum incompatibility, as in species of *Primula* section Aleuritia, all the incompatibility is expressed at the stigma surface (Wedderburn and Richards, 1990) and is thus presumably fully sporophytic in its control. In general, within-thrum inhibition more commonly occurs at the stigma surface than does within-pin inhibition suggesting that it is more frequently completely sporophytic in control. (Under gametophytic control, the heterogametic (*S*) (*s*) thrum grains should behave differentially on illegitimate thrum stigmas.)

Wedderburn and Richards (1990) have investigated the site of illegitimate pollen inhibition for 52 species of *Primula*, classified within 18 sections of the genus. From this work the following conclusions can be drawn.

1. Illegitimate and legitimate grains adhere to pin and thrum stigmas equally well.
2. Illegitimate pollen inhibition occurs at three sites, namely:
   (a) failure of germination
   (b) failure to penetrate the stigma
   (c) failure of pollen tube to grow down the stigma and style

Table 7.8 Pollination, pollen trapped beneath coronal hooks, pollen germination and pollen tube growth in *Pulmonaria affinis* after open pollination (%)

				x̄ tubes		
Stigma cross	x̄ pollen on stigmas	x̄ pollen trapped	x̄ % pollen germinating	mid-style	style base	ovary
Pin legitimate	11.60	5.1 (44.0%)	40.2	8.3	4.7	1.5
Pin illegitimate	8.8	3.9 (44.3%)	49.8	12.3	5.9	0.7
Thrum legitimate	11.36	7.4 (65.1%)	75.9	5.1	3.4	1.1
Thrum illegitimate	4.63	0.1 (0.02%)	70.6	2.9	0.6	0

3. Within a species/morph, illegitimate pollen inhibition most commonly occurs at more than one site. Where it occurs at only one site, this is usually germination failure.
4. The site(s) of illegitimate pollen inhibition almost always differs between the two morphs of the same species. This strongly suggests that within-pin and within-thrum inhibitions have different evolutionary origins, and functions.
5. Closely related species tend to show similar patterns of illegitimate pollen inhibition.
6. Total illegitimate pollen inhibition in both morphs was only observed for two species, so that a small amount of selfing is possible in most *Primula* species. In general, the more sites at which illegitimate inhibition occurred, the more efficient was the overall inhibition. Where inhibition occurred at one or two sites, stigmatic inhibition tended to be more efficient than stylar inhibition (Table 7.9).
7. In some taxonomic groups, inhibition site complexity has developed in pins, whereas in others it has developed in thrums. This suggests that inhibition site complexity and efficiency have evolved by at least two separate routes in *Primula*. It is possible to consider those species with three-site inhibition in both thrums and pins to be the most derived. They also tended to have the most efficient incompatibility.

Table 7.9 Average percentages of ovules developing into seeds in *Primula* after illegitimate pollinations, classified according to the position(s) at which most of the incompatibility inhibition is expressed (number of species in each category in brackets) (after Richards, 1993)

Inhibition site	Pins	Thrums	Total
One site			
Pollen germination	6.9 (9)	5.3 (21)	5.8 (30)
Stigma penetration			5.2 (2)
Pollen tube growth	3.9 (3)	33.0 (3)	18.4 (6)
Mean one site			6.3 (38)
Two sites			
Pollen germination and stigma penetration	1.1 (5)	11.0 (7)	6.9 (12)
Pollen germination and pollen tube growth	15.6 (8)	0.6 (2)	12.6 (10)
Stigma penetration and pollen tube growth	0.6 (3)	8.2 (1)	2.5 (4)
Mean two sites			8.4 (26)
Three sites			
Pollen germination, stigma penetration and pollen tube growth	4.3 (6)	0 (2)	3.2 (8)

In other heteromorphic genera, illegitimate pollen recognition has evolved in different ways, and in two cases these involve mechanical systems.

In *Pulmonaria affinis*, Richards and Mitchell (1990) describe how the remarkable coronal hooks of the stigma papillae (Figs. 7.13 and 7.14) act as a filter. Pollen grains can only hydrate and germinate when positioned beneath these hooks. Thrum pollen is larger than pin pollen, and the stigma papillae of thrums create smaller interpapillar spaces, so that they filter out the larger thrum pollen, while pin pollen can fall beneath the thrum hooks into a position where they can germinate. This mechanism reinforces a considerable disassortative pollination (Table 7.8), but it does not prevent pin pollen becoming trapped on illegitimate pin stigmas, where it germinates well. However, both thrum and pin illegitimate pollen is inhibited in the style.

Coronal stigmatic papillar hooks are also found in *Anchusa* (Philipp and Schou, 1981; Schou and Philipp, 1983), but it is not clear whether they function in the (multiallelic) incompatibility reaction in that genus.

Another mechanical system operates in *Armeria maritima* (Matsson, 1983) and presumably in other Plumbaginaceae. Cob ('type A') pollen has large bacculae (Figs. 7.15 and 7.16) and can only become hydrated and germinate if one of these bacculae is penetrated by a long papilla on the stigma of the legitimate papillate morph (Fig. 7.6). This reaction is mediated by tapetal lipoprotein which is only carried within the apertures of the cob pollen. Consequently, cob pollen does not adhere well to cob stigmas, so that if *Armeria* stigmas are collected and examined microscopically, it appears that cob stigmas receive a massively disassortative pollination. However, this is clearly an artefact of the mating system.

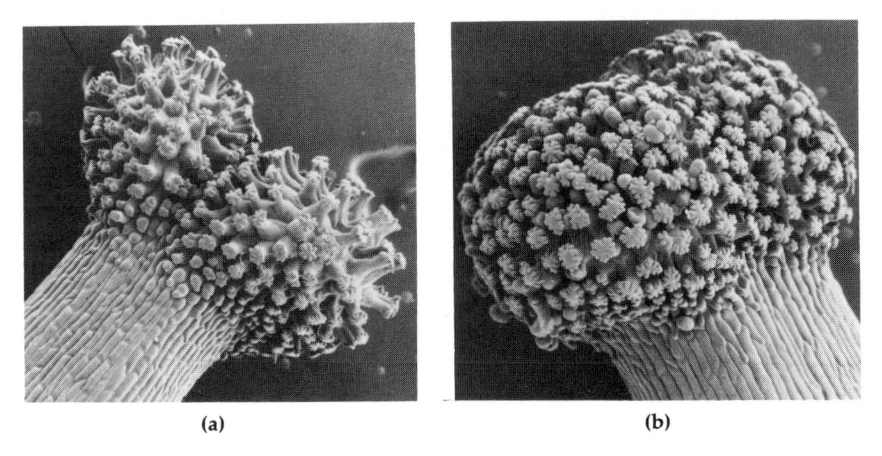

(a) (b)

Fig. 7.13 Scanning electron micrographs (×65) of the stigma of (a) thrum and (b) pin *Pulmonaria affinis* (Boraginaceae).

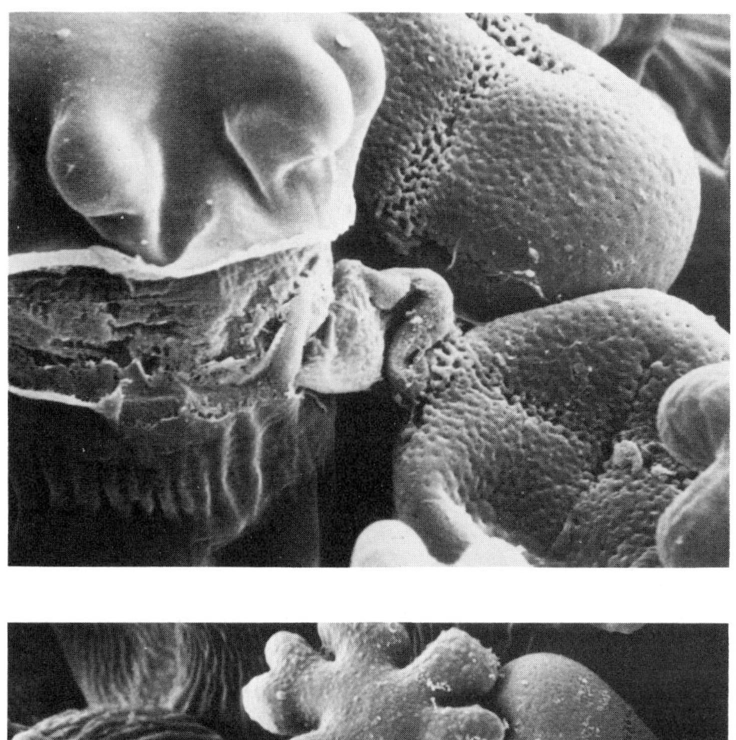

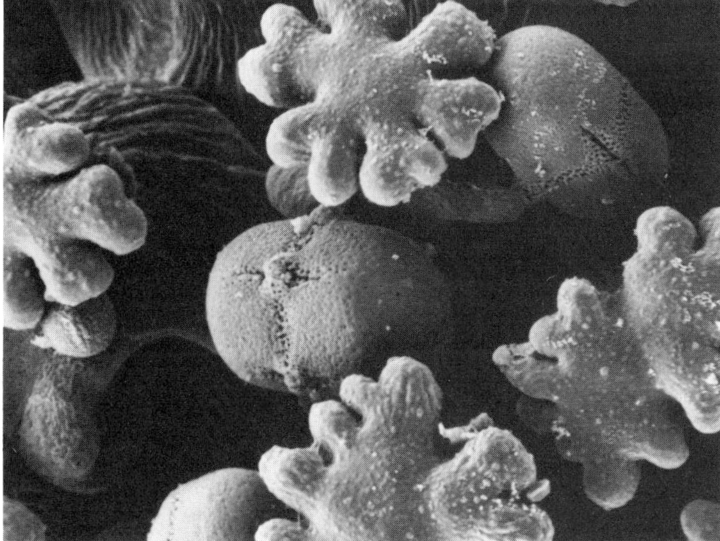

Fig. 7.14 Scanning electron micrographs of pollen tubes from (a) pin (×1500) and (b) thrum (×1000) *Pulmonaria affinis*. In (a) the pin pollen tube is penetrating a stigmatic papilla of a thrum flower beneath the apical 'corona' of hooks. In (b) the thrum pollen tube is penetrating a stigmatic papilla of a pin flower. It is considered that the apical corona of hooks on the papillae, which are more marked in the pin, help the lodgement of the pollen grains, which are larger in the thrum.

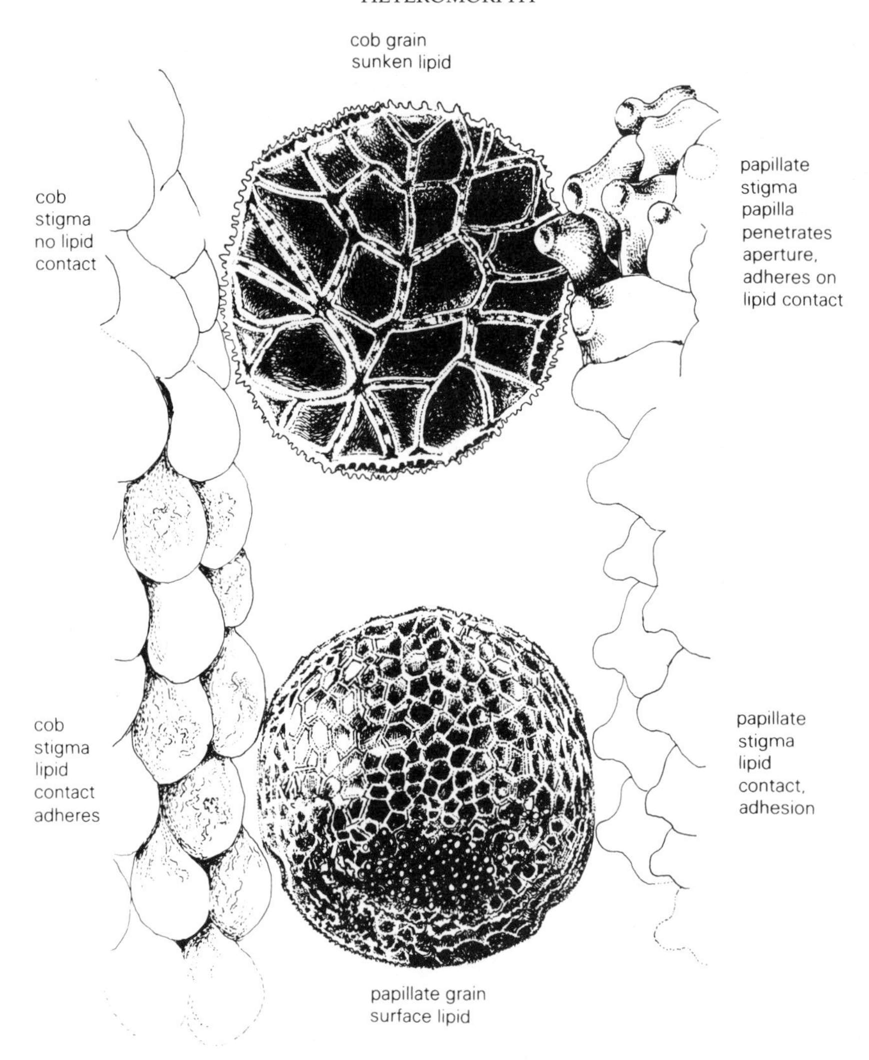

Fig. 7.15 Scheme to illustrate the operation of within-morph incompatibility in the thrift, *Armeria maritima*. Lipid is borne external to the pollen grain in the finely reticulated papillate grain (below), and is able to adhere to the smooth cob stigma, but less well to the papillate stigma. In the coarsely reticulated cob grain, the lipid is sunken in the tectum apertures, where it can only make contact with the papillate stigma.

In contrast, the finely reticulated pollen of papillate plants will stick to most smooth surfaces, even glass, suggesting that lipoprotein of tapetal origin is borne externally on these grains. Consequently papillate pollen adheres to cob and papillate stigmas in roughly equal amounts. Few papillate grains germinate on papillate stigmas (0.8%), but if a pollen-free

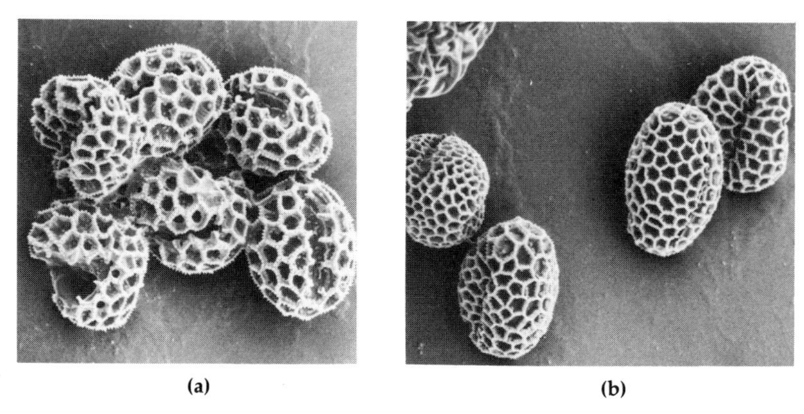

Fig. 7.16 Scanning electron micrographs (×450) of the pollen of (a) cob and (b) papillate *Armeria maritima*.

extract made from cob pollen is added to selfed papillate stigmas, the level of germination is raised to 10.3%, suggesting that exine-borne lipoprotein is involved in pollen recognition and hydration.

The molecular basis of mating-type recognition in Primula

Pollen germination

Primula pollen differs from that of many plants with sporophytic incompatibility, in that it hydrates and germinates readily in the absence of a lectic attachment to the stigma. Eisikowitch and Woodell (1975) show that if the erect flowers of the primrose (*P. vulgaris*) fill with rain, 30% of the pollen in the anthers germinates, although if the anthers are then removed, 90% of the pollen germinates in water, suggesting that the anthers contain a partial germination inhibitor. The anthers of the oxlip, *P. elatior*, with naturally pendant flowers, contains no such inhibitor. Sometimes, cowslip (*P. veris*) pollen can be found germinating where it has fallen into the nectaries at the base of the flower (Richards and Ibrahim, 1982), and at abnormally high temperatures pollen tubes can even penetrate the ovary from this position. Richards and Ibrahim (1982) and Shivanna, Heslop-Harrison and Heslop-Harrison (1981) show that water-soluble extracts of anthers and pollen can elicit incompatibility type responses on same-morph stigmas.

Shivanna, Heslop-Harrison and Heslop-Harrison (1981, 1983) show that primrose pollen is very dependent on humidity for hydration and germination (Table 7.10). Illegitimate pollen germination is greater for thrums at high humidities. On p. 255, I suggest that the long papillae of pin stigmas might create localized conditions of high humidity which would favour the germination of legitimate (thrum) pollen. However,

Table 7.10 The effect of external relative humidity on the germination and penetration of illegitimate pollen tubes in *Primula vulgaris* (after Shivanna *et al.*, 1983)

Relative humidity(%)	Germination of pollen(%)		Penetration of stigma(%)	
	P × P	T × T	P × P	T × T
5–10	0	0	0	0
56–65	5	25	3	0
95–100	47	75	25	13

in some *Primula* species, illegitimate pollen germination never occurs, as for thrums of species in section Aleuritia (Wedderburn and Richards, 1990).

Shivanna, Heslop-Harrison and Heslop-Harrison (1983) removed tapetal lipoprotein from thrum primrose pollen with cyclohexane and showed that such pollen had lost its ability to rehydrate on thrum stigmas, although it was still able to rehydrate and germinate when placed on a neutral (*Silene dioica*) stigma. By using protein extracts from pin and thrum stigmas in an investigation of primrose pollen germination *in vitro*, Shivanna, Heslop-Harrison and Heslop-Harrison (1981) showed that same-morph extracts particularly interfered with the rehydration and germination of thrum pollen, and that this is caused by dialysates of less than 10 000 daltons in size (larger fractions are ineffective at this stage).

From this work I conclude that the inhibition of same-morph pollen germination in *Primula*:

1. is controlled by the inability of pollen grains to rehydrate;
2. involves a recognition interaction between water-soluble pollen exine lipoprotein of tapetal origin and stigma-bound glycoproteins which mediates or interrupts pollen rehydration;
3. is total in some cases, but in others can be overcome by high levels of humidity which may dilute recognition responses based on water-soluble lipoprotein.

Stigma penetration

Heslop-Harrison *et al.* (1981) have shown that the legitimate penetration of the stigma papillae differs in the two morphs of primrose. Thrum grains enter the long pin papillae subapically where the papilla cuticle is relatively thick. However, pin grains enter the shorter thrum papillae basally, and intercellularly, where the papilla cuticle is thin (Fig. 7.17).

These authors also demonstrate a major difference in the behaviour of illegitimate pollinations in this species, similar to those found for the

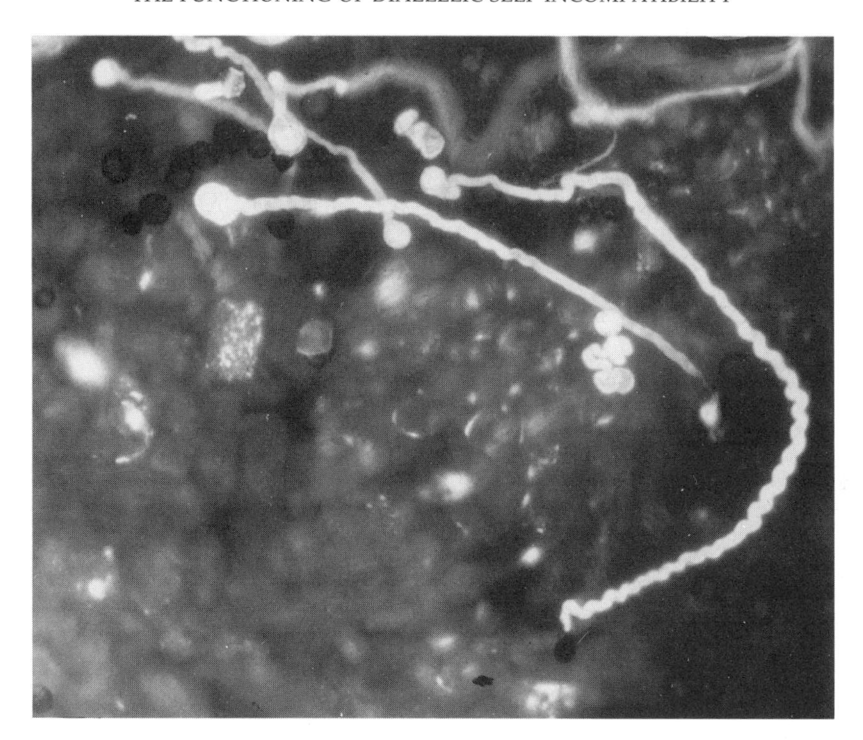

Fig. 7.17 Penetration of a stigmatic papilla of pin *Primula veris* by a germinating pollen tube.

cowslip *P. veris* by Richards and Ibrahim (1982) and by Stevens and Murray (1982) for *P. obconica*. In the case of illegitimate pin pollinations, the majority of pollen tubes that germinate penetrate the pin stigma. However, illegitimate thrum tubes are very abnormal in appearance, being thick and sinuous (Fig. 7.18), and these wander over the surface of the stigma, causing callose to be deposited in their path. Richards and Ibhrahim (1982) show that pollen-free aqueous extracts of thrum pollen elicit similar callose responses on thrum stigmas. It seems likely that the abnormal morphology of these tubes prevents their stigmatic penetration.

Pollen tube growth

Shivanna, Heslop-Harrison and Heslop-Harrison (1981) dialysed stigma protein extracts and showed that larger fractions (over 10 000 daltons) have a marked effect on pollen tube growth *in vitro*, although they do not affect pollen germination (p. 276). Thrum extracts influence thrum pollen tube growth, and pin extracts influence pin pollen tube growth. In both cases, tubes grow slowly and have an abnormal morphology resembling that of thrum self pollen tubes *in vivo*.

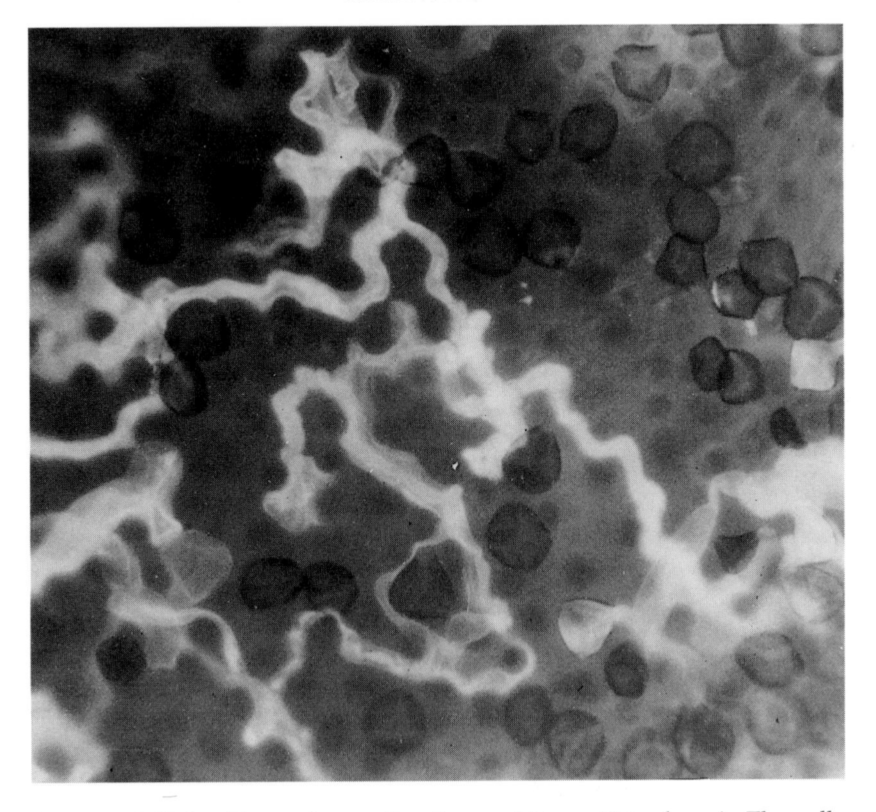

Fig. 7.18 Selfed pollen grains on the stigma of thrum *Primula veris*. The pollen grain has germinated, but the thick and irregularly swollen pollen tube wanders aimlessly over the surface of the stigma, and does not penetrate a stigma papilla. Microphotograph under ultraviolet light (×300). Note stigmatic papillae fluorescing in neighbourhood of the illegitimate grain, apparently due to the deposition of callose.

Working with *P. obconica*, Golynskaya, Bashkirova and Tomchuk (1976) obtained similar results with thrum stigma extracts, showing that the proteins which influence thrum pollen tube growth may be phytohaemagglutinins. However, for illegitimate pin interactions, stylar extracts which are non-agglutinizing have the greatest influence on pin pollen tube growth. This agrees well with the findings of Wedderburn and Richards (1990) for *P. obconica*, a species in which most thrum × thrum pollen inhibition takes place at the stigma surface, but in which most pin × pin inhibition occurred within the style. Working with decapitated styles in *P. obconica*, Stevens and Murray (1982) show that thrum pollen grows well in thrum styles, but that pin pollen grows poorly in pin styles.

This tends to confirm the conclusion that incompatibility inhibition resides chiefly in the thrum stigma but in the pin style in this species.

At least five different gene products seem to be involved in the control of the *Primula* mating system:

1. exine bound lipoproteins of tapetal origin; these presumably differ between pins and thrums and there is some evidence that the protein is more water-soluble for thrum pollen than it is for pin pollen;
2. a thrum stigmatic glycoprotein of less than 10 000 daltons involved in inhibiting the hydration of thrum grains through an interaction with exine-bound lipoprotein;
3. a thrum stigmatic agglutinizing glycoprotein of more than 10 000 daltons which causes thrum pollen tubes to have an abnormal morphology;
4. a pin stylar non-agglutinizing glycoprotein which inhibits the growth of pin pollen tubes within the style.

As within-pin and within-thrum illegitimate pollen can, in various *Primula* species, be inhibited at three distinct sites, there could be at least six distinct gene products mediating self-recognition of pollen on the female side in this genus, although not all will operate in any one species, and not more than three will occur within an individual. This concept has been challenged by Barrett and Cruzan (1994) who state 'the evidence to support this remarkable claim is weak at best'. Here I merely present the evidence and let the reader decide!

However, it is possible that fewer recombinable genes are involved than there are gene products. By analogy to what we know of the control of gametophytic incompatibility (p. 211), it is possible that, for any one morph and incompatibility site, only one recognition gene is involved. This may be differently expressed in the tapetum and the stigma cuticle by pollen-part and stylar-part operators, for instance as lipoproteins and glycoproteins.

THE EVOLUTION OF HETEROMORPHIC SYSTEMS

The Charlesworths' model: incompatibility evolves first from a homostylous ancestor

Charlesworth and Charlesworth (1979) developed a model for the evolution of the typical heterostylous syndrome. This disagreed with Darwin's original ideas (1877), for they proposed that a two mating-type incompatibility should evolve first, and the genetic control for a heterostyly which encourages disassortative pollination between the mating types latterly becomes linked to the genetic control of the mating system. Essentially, their argument presumes that the heterostylous syndrome arose in response to the evolutionary disadvantages of selfing, and assumes that the selfing ancestors of heterostyles were homostylous, without herkogamy. Because of the automatic genetic disadvantages of an outcrossing gene (p. 360), they suggest that ancestral selfers must be highly selfed for an outcrossing gene to be sufficiently advantageous to become established. This encouraged them to suggest that the original outcrossing genes were highly efficient, involving incompatibility.

Richards (1986, 1993) has pointed out that this model did not agree well with the facts, in that there is no certain case of a monomorphic species with a two mating-type incompatibility. However, there are a number of examples of heteromorphic species which are self-fertile (*Amsinckia*, Ganders, 1979; *Primula* section Sphondylia, Al Wadi and Richards, 1993); or which have a multiallelic incompatibility (*Anchusa*, Dulberger, 1970, Schou and Philipp, 1983; *Narcissus*, Lloyd, Webb and Dulberger, 1990).

Charlesworth and Charlesworth (1979) could not explain how reciprocal herkogamy could evolve from a self-fertile long-homostylous ancestor. They point out that if a short-styled dominant mutant (*G*) arose first, for such a mutation to succeed, it should be so successful that it should proceed to fixation to give rise to a species with reverse herkogamy. The male fitness and female fitness of such a mutant would be as great as in coexisting homostyles, it would give rise to short-styled offspring when crossed onto homostyle mothers, and it would have the considerable additional advantage of being outcrossed to a greater extent. However, if a low anther recessive mutant (*a*) arose first, such plants would suffer from reduced male fitness in comparison with long homostyles and this mutant would not spread.

The Lloyd and Webb model: heterostyly evolves first from a herkogamous ancestor

Lloyd and Webb (1993a,b) overcome this objection by proposing that the ancestors of heterostyles already possessed approach herkogamy with reduced within-flower selfing where the anthers are in a mid-flower position (Fig. 7.19). In this case, increased fitness advantages incurred by a further reduction in selfing by novel short-styled mutants would be at best slight. They suggest that Charlesworth and Charlesworth (1979) were mistaken in thinking that the advantage of a short-style mutant was entirely a reduction in selfing. Rather, they suggest that new mutants should benefit by being more efficient male parents to other-type stigmas.

In their scenario, pollinator behavior should ensure that the pollen of a short-styled dominant mutant *G* would outnumber *g* carrying pollen on the stigmas of long-styled plants. Equally, *G* carrying pollen would be less

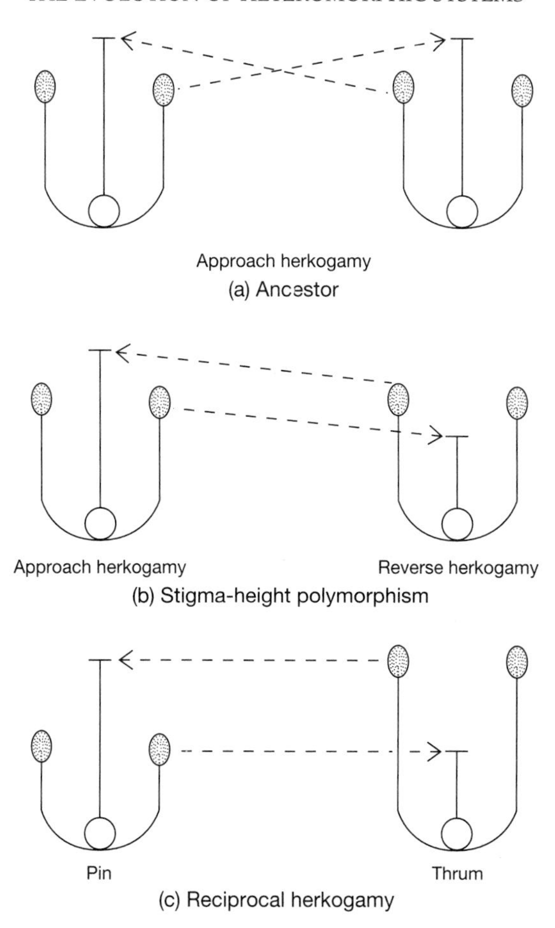

Fig. 7.19 The postulated principal stages in the evolution of reciprocal herkogamy. (a) The ancestral condition, approach herkogamy. (b) A stigma-height polymorphism. (c) Reciprocal herkogamy. The arrows show the direction of most proficient pollen transfer. (After Lloyd and Webb, 1993a.)

common on short-styled stigmas, thus reducing levels of geitonogamous selfing. Consequently, the frequency of *G* should increase in the population, but only to the point where short-styled plants equal the frequency of long-styled plants. Once short-styled plants form the majority of the population, short-styled pollen *G* suffers from a shortage of long-styled stigmas, whereas long-styled pollen *g* benefits from an abundance of short-styled potential mothers (Fig. 7.19). Thus, in the Lloyd and Webb model, frequency-dependent selection should maintain short-styled and long-styled plants at roughly equal frequencies, each morph having the greatest male fitness when it is rarest.

This contrasts with the model of Charlesworth and Charlesworth (1979), where an original homostyle is not advantaged when it becomes scarce, for, unlike long-styled pollen, homostyle pollen will not occur on short-styled stigmas more frequently than does short-styled pollen. Additionally, homostyle mothers will bear the additional penalty of producing offspring more likely to be of a selfed origin than are the offspring of short-styled mothers.

Consequently, Lloyd and Webb (1993a) produce models which show that a stable distyly can evolve before a two mating-type incompatibility system if the original founder has approach herkogamy, but it cannot evolve from self-fertile homostyles.

They point out that once the frequency-dependent distylous polymorphism has become established, a further dominant *G*-linked mutation *A* which moves the position of short-styled anthers up to the mouth of the flower (into the 'thrum' position) would be favoured as it would further increase the fitness of short-styled pollen on long-styled stigmas (Fig. 7.19). In these circumstances, thrum pollen would be fitter on reciprocally placed long-styled stigmas than was long-styled pollen on non-reciprocally placed short-styled stigmas. These conditions should favour long-styled linked mutations which increase the fitness of long-styled pollen on short-styled stigmas. This increased fitness could result from

1. a mutation which decreases pollen size and increases pollen number (*p*);
2. a recessive mutation *a* which lowers the anther position into the 'pin' position, where it is reciprocally placed to thrum stigmas.

This order of evolutionary events (*G*, followed by *p*, followed by *a*) agrees well with those suggested by a study of *Primula* section Sphondylia (p. 266).

Either of these mutations should individually increase long-style pollen fitness and reduce the likelihood of within-flower pin selfing. In some distylous species, only one of these pin male fitness attributes has become established (*Primula boveana* and *Linum grandiflorum* have no anther position dimorphies, whereas *Vitaliana primuliflora* has no pollen size dimorphy). In the majority of cases, it seems that both these recessive mutations have become linked to the pin chromosome.

Lloyd and Webb (1993a) suggest that once a stable, frequency dependent, reciprocal herkogamy evolved in self-fertile plants, any mutations linked to the distyly supergene which allowed an incompatibility self-recognition to occur would subsequently be favoured. Such mutants would further reduce levels of selfing, while interfering relatively little with reproductive efficiency, as for each morph the majority of pollen would travel to legitimate rather than to illegitimate stigmas.

The Lloyd and Webb model for the evolution of distyly is attractive, being plausible and capable of explaining many of the distylous conditions which occur in various genera. For instance, self-fertile heterostyly in *Amsinckia* and *Primula* section Sphondylia should be maintained even in the absence of a later-evolving incompatibility, while in genera such as *Narcissus* and *Anchusa* it would allow for the evolution of a secondary self-incompatibility which is multiallelic and not linked to the distyly supergene.

A modification of the Lloyd and Webb scheme can also cope with conditions such as those found in *Hypericum aegypticum* (Ornduff, 1979) and *Limonium* (Baker, 1966) in which, unusually, it is the pin morph which is heterozygous, pin features being dominant. By analogy to the evolution of dioecy (p. 307), we would expect the dominant morph to have evolved after the recessive morph, so that it is expressed in the heterozygote (this supports the suggestion that the original condition in *Primula* was a 'pin'-like approach herkogamy). We may suppose that in *Hypericum* and *Limonium*, founder species were reverse herkogamous, and that the original style length mutant which became established was a long-style variant. In most genera, such a novel mutant might not be favoured, as it might suffer from reduced male fitness on short-style stigmas. However, it is noteworthy that both these genera are untypical for distylous plants in that they have very open corolla tubes within which pin pollen is probably as accessible to pollinators as is thrum pollen. In such conditions, a long-style variant should not suffer from reduced male fitness.

The Richards model: thrum-linked lethals allow the initial evolution of heterostyly from a homostylous ancestor

Despite the accommodating nature of the Lloyd and Webb hypothesis, there are two features of many distylous genera which suggest that distyly has not always evolved by the Lloyd and Webb route. First, in genera such as *Primula* and *Amsinckia*, where we strongly suspect that monomorphic species ancestral to the evolution of heterostyly survive, these tend to be homostylous rather than approach herkogamous (although they may show both approach and reverse herkogamy to some degree, Al Wadi and Richards, 1993).

Second, and crucially, it seems that thrum-linked lethals commonly occur. It is rare to find complete thrum self-sterility (e.g. Wedderburn and Richards, 1990), so that some thrum selfing commonly occurs in many heterostylous species. Yet, true-breeding thrum homozygotes which should form one quarter of the offspring of an original thrum self seem to occur very rarely (whereas true-breeding pins are frequently encountered). Despite many investigations, true-breeding thrums have only been found in *Primula* in *P. sinensis* (Mather and De Winton, 1941) and these thrum homozygotes only have 70% of the viability of thrum heterozygotes. Although some *Armeria maritima* are somewhat cob (heterozygote) self-fertile (Richards *et al.*, 1989), these selfs never give rise to homozygote cobs, for true-breeding cobs have never been found. Only in self-fertile *Amsinckia* do true-breeding thrums commonly occur (Ganders, 1979), but this genus does not have a stable distyly.

We must conclude from these findings that a recessive lethal or sublethal gene is usually linked to the dominant (thrum) chromosome.

The work of Valsa Kurian (p. 263) clearly demonstrates the presence of this thrum-linked lethal in polyanthus primulas. Moreover, a thrum-linked sublethal in their primrose parents is clearly inferred from studies of recombinational homostyles in this species. Crosby (1949) interpreted Catcheside's results as showing that long homostyle (i.e. 'thrum male') homozygotes were only about 65% as viable as homostyle heterozygotes (p. 288). Despite being more fecund, homostyles at high frequencies are less fit than heterostyles, and Crosby assigned much of this reduction in fitness to homostyle homozygote sublethality. Clearly, if the thrum lethal is linked to *A* and *P*, it should be inherited by recombinational long homostyles.

Remarkably, in the otherwise very thorough review of heterostyly by Barrett (1993), there is no single mention of thrum recessive lethality. Yet, clearly it is a pervasive feature of most heterostylous plants. It would be possible to argue that such recessive mutants might accumulate by 'Müller's ratchet' (p. 23) on the dominant thrum *GPA* chromosome, protected from recombination (p. 264).

However, if a recessive lethal became linked to the original dominant short-styled mutant early in the evolutionary history of heterostyly, this initial thrum mutant could establish from a homostylous ancestor without immediately proceeding to fixation. Thrums would only exist as heterozygotes, would yield one-third homostyles when selfed, and despite their obvious advantages would nevertheless carry a fitness penalty in that one-quarter of their offspring would be lethal. Such conditions overcome the Charlesworths' objections to the evolution of heterostyly by this route. Also, they would favour the establishment of subsequent anther position and pollen size/number mutations, causing homostyles to become 'pins', which would enhance their male fitness by encouraging cross-pollination onto sunken thrum styles (Fig. 7.19).

It seems likely that heterostyly evolved by the Richards model in some genera, and by the Lloyd and Webb model in others. Subsequent research could be usefully targetted towards investigations concerning the occurrence of thrum-linked lethals in various heteromorphic genera.

The subsequent evolution of two mating-type incompatibility

As Lloyd and Webb (1993a) have pointed out, any mutants which arose on the thrum and pin linkage groups which caused the failure of illegitimate pollen on same-type stigmas would be favoured, as they would further discourage residual levels of selfing in a heterostylous plant.

It is not realistic to suggest, as Charlesworth and Charlesworth (1979) do, that single mutations of a 'complementary' type could 'take advantage' of residual features of an 'ancient' (presumably gametophytic) system already present in the self-compatible parents. Indeed (p. 270), there may be elements of the *Primula* incompatibility which are gametophytic in nature, but these may have evolved secondarily to reinforce a primarily sporophytic system.

It is very likely (p. 271) that the original two mating-type incompatibility in *Primula* controlled pollen hydration and germination, as most single-site mechanisms found today are of this type. Later stigmatic and stylar reinforcing mechanisms apparently evolved independently. Any one of these would have been selected for, as they would decrease levels of selfing individually as well as collectively. Furthermore, it is clear that all the features of the *Primula* incompatibility are oppositional, rather than complementary, in nature.

From what we know of the *Primula* incompatibility mechanism (p. 279), its evolution seems to have been complex, involving no less than three different gene products on the female side for each of pins and thrums. As suggested on p. 279, these recognition substances may, however, be controlled by a single recognition gene for the male and female sides, the expression being modified by pollen-part and stylar-part operators.

In the original, self-compatible, heterostyle *Primulas*, a self-incompatibility is most likely to have arisen first in pins. In most heterostylous genera, pins tend to receive more selfed pollen than do thrums (Ganders, 1979), by virtue of their exposed stigma position, and this is also true for *Primula* (Piper and Charlesworth, 1986). Thus, a recessive self-sterility mechanism affecting pollen hydration and germination *s* would be established readily if linked to the pin chromosome. It is simplest to assume that this gene became duplicated in association with a pollen operator (*sm*) linked to *p* and with a stylar operator (*sf*) linked to *g* (Richards, 1993). Presumably, later genes controlling recognition at stylar and stigmatic sites would also have occurred in duplication at or near the *sm* and *sf* sites, possibly involving the same, duplicated, operators *m* and *f*.

Once pin pollen germination incompatibility became established, the outcrossing pin chromosome should succeed at the expense of the partially selfing thrum chromosome and would become more common in the population. However, pins would depend on pollen originating from plants containing the thrum chromosome for reproduction, so limiting pin morph frequency.

Thrums without self-incompatibility would, on present evidence (e.g. Piper and Charlesworth, 1986), already be substantially outcrossed. Thus, if a dominant outcrossing gene controlling thrum pollen germination incompatibility arose secondarily, it would be most likely to be selected for if it was very efficient, thus further reducing levels of thrum selfing. Interestingly, thrum pollen germination incompatibility in *Primula* tends to be stronger than is the corresponding pin reaction (p. 270). However, as thrums give rise to one-third pin offspring when selfed, thrum outcrossing genes can become established if only slightly advantageous in comparison with selfing genes (see below).

As may be the case for the pin outcrossing gene, I assume that the thrum gene is duplicated *SM SF*. Thus, the hypothetical structure of the supergene (p. 264) now reads as:

G	SF	A	Mpp	SM	Pp	Pm	Mpm	I	Gm
g	sf	a	mpp	sm	pp	pm	mpm	L	gm

SF is the thrum stigma recognition gene
SM is the thrum pollen recognition gene
sf is the pin stigma recognition gene
sm is the pin pollen recognition gene

where these genes code for tapetal lipoproteins on the male side and papilla cuticle glycoproteins on the female side. Additional stigmatic and stylar incompatibility factors are not presented here. Other loci are as on p. 264.

Theoretically, outcrossing genes carry a penalty in comparison with selfing genes, as outcrossing mothers give rise to selfers when crossed with selfers pollen (p. 361), whereas selfing mothers should normally produce only selfing offspring. Although the benefits of an outcrossing gene arising among highly self-pollinated pins should outweigh its disadvantages, the benefits of an outcrossing gene arising among thrums which are already relatively highly cross-pollinated are less obvious. However, the selfing advantage of heterozygous thrum *Ss* coexisting with an outcrossing pin is only slight. Thrums when selfed will yield one-third outcrossing pins in their offspring, whereas an outcrossed thrum will yield one-half outcrossing pins in its offspring. Thus, the outcrossing penalty is minimal for thrums, readily allowing a slightly advantageous outcrossing gene to become established.

CASUAL SECONDARY HOMOSTYLY

There are a number of species in mostly heterostylous genera, for instance in *Primula* and *Limonium*, which are stable, self-fertile, long homostyles of recombinational origin (p. 290). Also, single recombinational homostyle individuals are occasionally reported from otherwise heterostyle populations (p. 260). Thus, it is curious that populations polymorphic for heterostyly and recombinational secondary homostyly are rarely reported. A possible explanation is that a thrum-linked sublethal/lethal gene (p. 283) becomes recombined onto the homostyle chromosome, so that true-breeding homostyle homozygotes cannot form.

The most intensively studied cases refer to populations of the primrose (*P. vulgaris*) in Somerset, UK. These were originally reported by Crosby in 1940, but were more intensively studied by Crosby (1949) and later by Curtis and Curtis (1985), Piper and Charlesworth (1986), Piper, Charlesworth and Charlesworth (1986) and Boyd, Silvertown and Tucker (1990). Similar populations also occur in the English Chilterns.

Homostyle advantage

The Somerset long homostyle populations are limited to an area with a diameter of about 30 km, within which almost all populations have some homostyles, but no population has become entirely homostyle. Where homostyles are scarce, pins and thrums occur in roughly equal proportions. However, when homostyles are common, thrums disappear, leaving only pins. Crosby explained this by suggesting that homostyles have an automatic advantage over thrums (p. 361), in that the 'thrum-type' pollen of homostyles competes equally with thrum pollen on pin stigmas, but homostyles can also self, which thrums cannot. Pins are not so disadvantaged because when homostyle heterozygotes self they give rise to one-third pin offspring. Also, pin primroses are relatively self-fertile themselves (p. 271) as well as acting as effective mothers to homostyle (thrum-type) pollen.

Being self-pollinating and self-fertilizing (thrum in male function, but pin in female function), long homostyles have a reproductive advantage as well as a genetic advantage over heterostyles. The average reproductive advantage of homostyles over heterostyles as extra seed set is about 65% overall.

The homostyle (thrum) sublethal

Crosby noted that although the homostyle chromosome with its genetic advantage and extra reproductive fitness should proceed to fixation, it never does. He suggested that homostyle homozygotes *gAP*/*gAP* were less fit than homostyle heterozygotes *gAP*/*gpa*, citing Catcheside's unpublished results which show that after homostyle heterozygotes are selfed, homostyle homozygotes only occur at about two-thirds of the expected frequency. This strongly indicates that the recombinant homostyle chromosome has indeed inherited a sublethal gene from the thrum chromosome (p. 284).

Crosby calculated that if homostyle homozygotes were more than 81.5% as fit as pins, pins as well as thrums should disappear from the population. However, if as estimated, homostyle homozygotes were only about 65% as fit as pins, populations should reach an ESS equilibrium with 80% homostyles:20% pins.

Interestingly, it is clear that homostyles are relatively more disadvantageous at a high frequency. Although some populations do approach the 80:20 ratio predicted by Crosby's model, Curtis and Curtis (1985) showed that in such populations, homostyles subsequently tended to become less frequent, whereas in those with a low initial homostyle frequency, homostyles latterly became more common, over a 30-year

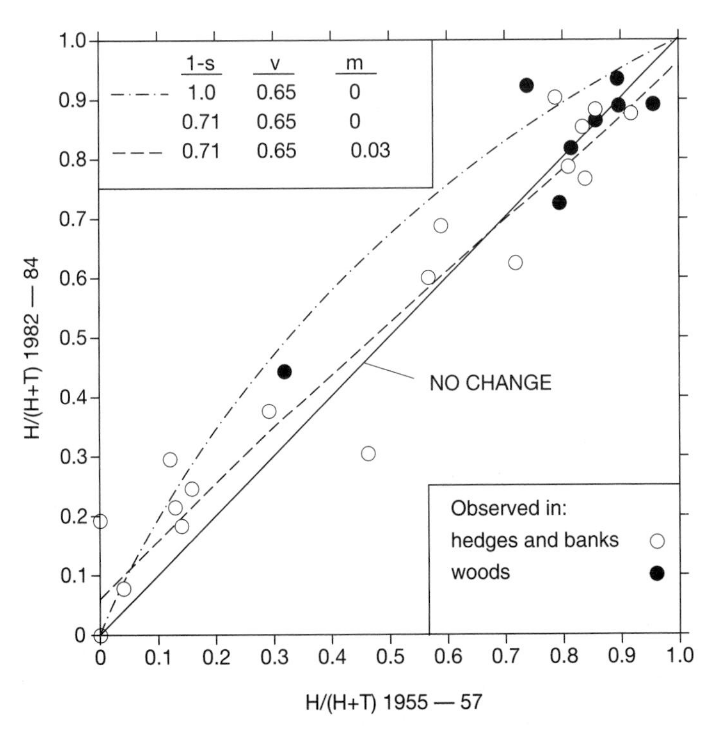

Fig. 7.20 Relative frequency of homostyles (H) against all plants (H + T) of primrose *Primula vulgaris* in Somerset woods in 1955–1957 (x axis) and 1982–1984 (y axis). (After Curtis and Curtis, 1985.)

period. This agrees well with Crosby's model, for at a high homostyle frequency, more of the pollen incoming onto homostyle stigmas will be carrying the homostyle chromosome, and so more of the progeny of heterozygote homostyles will be homozygote homostyles with a low viability.

The Curtises found that only a narrow range of homostyle fitnesses, relative to heterostyles, could explain the frequency-dependent curve for homostyle frequency change that they discovered (Fig. 7.20). If Crosby's suggestion that homostyle homozygotes are about 65% as fit as pins is correct, then homostyle heterozygotes must also be less fit than pins (about 80% as fit as pins).

The reproductive disadvantage of homostyles

Boyd, Silvertown and Tucker (1990) have thoroughly investigated a range of fitness attributes for mature homostyles in Somerset populations, and in almost all cases they found that homostyle performance was equal to, or superior to, that of pins, and where they occurred, that of thrums too. Only in one regard did they find that homostyles suffered. Although homostyles set more seed per capsule, this seed tended to be individually smaller, there being an inverse relationship between seed number and seed weight as is commonly found in plants with multiseeded fruits (p. 31).

As our own work with *Primula farinosa* shows that large seeds from few-seeded capsules give rise to seedlings with greatly superior attributes to those of small seeds (Baker, Richards and Tremayne, 1994), we are inclined to attach some significance to this finding. It seems to us that homostyly has never become fixed in primrose populations for the following reasons.

1. It may pay perennial outcrossers to set fewer seeds than do selfing relatives, because that seed is heavier, better resourced, and gives rise to fitter seedlings.
2. A sublethal gene *l*, possibly involved in the evolution of heterostyly and originally linked to the thrum chromosome, has been recombined onto the homostyle chromosome (being linked to the *P* allele, p. 264), so that homostyle homozygotes have a reduced viability.

Crosby's work on homostyle primroses has attracted a good deal of attention, as it provides a superb field laboratory for investigations into the evolution of outcrossing and selfing. Jack Crosby made a television film for the BBC entitled 'The Wars of the Primroses', in which he highlighted the ongoing competition between heterostyles and recombinant homostyles in the woods of Somerset as a classic example of mating system evolution in progress.

Now that John Piper, Mark Boyd and the Curtises have published their findings, we have acquired important insights about the 'rules' governing these 'wars'. It seems that the high seed quality and avoidance of inbreeding depression inherent in outcrossing effectively counteract the genetic and reproductive advantages of selfing. Our concept of what determines 'fitness' in such populations has become more realistic, and less naive.

Nevertheless, mysteries, thankfully, remain. For instance, we still fail to understand why recombinational homostyly should have become established in primrose in just these two limited areas, and, apparently, nowhere else.

Recombinational short homostyles

No short homostyle species or populations are known, although short homostyles should arise by recombination at the same frequency as do long homostyles (p. 256) and are sometimes reported occurring as casual individuals (p. 260). The short homostyle chromosome *Gap* or *Gpa* carries an automatic genetic disadvantage compared with the long homostyle chromosome *gAP* or *gPA*, because when crossed with a compatible heterostyle, only a quarter, rather than a half, of the offspring are homostyle:

short homostyle male × thrum > thrum: hom.: thrum: pin *Gap*/*gap* × *GAP*/*gap* > *GAP*/*Gap*: *Gap*/*gap*: *GAP*/*gap*: *gap*/*gap*
long homostyle male × pin > hom.: hom: pin: pin *gAP*/*gap* × *gap*/*gap* > *gAP*/*gap*: *gAP*/*gap*: *gap*/*gap*: *gap*/*gap*

Thus, long homostyle pollen competes successfully with thrum pollen on compatible pin stigmas, but short-homostyle pollen does not compete successfully with pin pollen on thrum stigmas. Also, whereas long homostyles receive about 10% outcrossed pollen onto their exposed stigmas (Crosby, 1949; Piper and Charlesworth, 1986), it is unlikely that short homostyles undertake much outcrossing as either mothers or fathers. Consequently, they would be expected to suffer severely from inbreeding depression after establishment.

PERMANENT SECONDARY HOMOSTYLY

Richards (1993) reports that of the 426 species of *Primula*, 38 (9%) are long-homostyle. For reasons given on p. 265, we believe that some of these are primary homostyles, representing the condition before the evolution of heterostyly. In other cases, the behaviour of homostyles when crossed with related heterostyles shows them to be secondary homostyles which have arisen from heterostyles by recombination (p. 259).

Table 7.11 Evidence used to suggest whether *Primula* homostyles are primary or secondary (after Richards, 1993)

Chromosome number	Diploid *primary*	Polyploid *secondary*
Heterostyle individuals also known	Yes *primary*	No *secondary*
Systematic affiliations	'Primitive' *primary*	'Derived' *secondary*
Geographical distribution	Warm temperate *primary*	Arctic or alpine *secondary*

Where this kind of evidence is lacking, other circumstantial features can be used to suggest whether homostyly in a species is primary or secondary (Table 7.11).

On the basis of these considerations, Wedderburn and Richards (1992) believe that 23 *Primula* species are best regarded as secondary homostyles. Of those for which we know the chromosome number, 16/20 (80%) are polyploid, whereas only 8% of the heterostyle species in the genus as a whole are known to be polyploid (Richards, 1993). Clearly, there is a strong relationship in *Primula* between polyploidy and secondary homostyly.

The formation of polyploid homostyles carrying a recessive lethal

It is possible that autopolyploids arose from homozygous homostyles *gPA/gPA* which had lost the recombined lethal/sublethal, perhaps by a terminal deletion (p. 264). In this case, polyploid homostyles would be true-breeding. In contrast, polyploid homostyles resulting from a homostyle heterozygote *gPA/gpa* would segregate out pin offspring. However, it is far more likely that polyploid homostyles are allo-polyploids. If a weakly viable (sublethal) homostyle homozygote *gPA/gPA* formed an allopolyploid as a male parent to a pin *gpa/gpa* of another species, and if there was no homoeologous pairing, the resultant alloploid *gPA/gPA//gpa/gpa* would give rise to true-breeding homostyle heterozygotes with no sublethality. Thus, polyploidy would allow homostyly to become fixed, despite the presence of a homostyle-linked sublethal. The strong relationship between homostyly and polyploidy strongly supports this hypothesis.

Richards (1993) advanced four other secondary explanations for the relationship between polyploidy and homostyly.

1. The genetic control of the heterostyle system does not usually work in polyploids.

2. Most polyploids, and most secondary homostyles, are found in the arctic, i.e. it is an indirect relationship.
3. Polyploidy is most likely to become established in a self-fertile plant which can mate with itself.
4. Polyploids are more likely to resist the inbreeding effects resulting from selfing in homostyles.

All these explanations could be true in part. However, it is certainly the case that heterostyly can operate perfectly well in polyploid *Primulas*. Also, there are some polyploid homostyle *Primulas* which occur in temperate or even subtropical climates.

Polyploid homostyly and latitude

There is also a strong relationship between polyploid secondary homostyles and latitude, and, to a lesser extent, altitude. This is particularly well shown by various species in *Primula* section Aleuritia (Table 7.12).

The most obvious explanation for the relationship between homostyly and latitude is reproductive assurance. Plants from the far north, and from

Table 7.12 Distribution, heterostyly (het) or homostyly (hom) and diploid chromosome number in species of *Primula* section Aleuritia

Species	Distribution	het/hom	$2n$=
modesta	Japan 33–43°N	het	18
specuicola	SW USA 36°N	het	18
algida	W. Asia 36–40°N	het	18
exigua	Bulgaria 42°N	het	18
frondosa	Bulgaria 43°N	het	18
daraliaca	Caucasus 43°N	het	18
alcalina	NE Idaho 45°N	het	18
longiscapa	C. Asia 39–46°N	het	18
farinosa	W. & C. Europe 43–63°N	het	18
mistassinica	N. America 42–60°N	het	18
anvilensis	Alaska 65°N	het	18
borealis	Alaska, E. Siberia 52–70°N (average 46.5°N)	het	36
halleri	Alps-Bulgaria 42–47°N >2000 m	hom	36
yuparensis	Hokkaido, Japan 43°N	hom	36
scotica	N. Scotland 59°N	hom	54
incana	USA 38–61°N	hom	54
scandinavica	Scandinavia 59–70°N	hom	72
laurentiana	USA 45–55°N	hom	72
magellanica	S. America 43–55°S	hom	72
stricta	Europe, America 62–73°N (average 53.4°N)	hom	126

other cold, unreliable habitats, are far less likely to receive insect visits than are those from further south, and it will pay them to adopt a selfing mode. This explanation is strengthened by the occurrence of homostyly in several species from subtropical regions (*P. verticillata*, *P. simensis*, *P. floribunda*, *P. filipes*) which flower in mid-winter, a time when pollinator service is likely to be unreliable even at these latitudes.

It is noteworthy that arctic populations of thrift, *Armeria maritima*, are also homomorphic and self-fertile (Baker, 1966; Vekemans *et al.*, 1990). American populations are also recombinationally self-fertile, suggesting that they colonized southwards from the arctic, but have been unable to re-evolve an outcrossing mode, except by gynodioecy (p. 318).

A supplementary explanation for the relationship between homostyly and the arctic considers the recolonization of arctic regions in the wake of glacial retreat. Single long-distance colonizing events are most successfully undertaken by self-fertile individuals, so that homostyle *Primulas* may have been able to recolonize arctic regions before heterostyle species could reach these sites.

The penalty apparently associated with homostyle selfing in these northern species is well illustrated by Bullard *et al.* (1987) for *Primula scotica* (Fig. 7.21). Most individuals are short-lived and flower poorly, but a few are long-lived and free-flowering, contributing chiefly to the seed-bank. It is suggested that these vigorous individuals arise from cross-fertilizations made during the brief and unusual spells of warm weather. However, heterozygosity of outcrossing origin is very rare in this species,

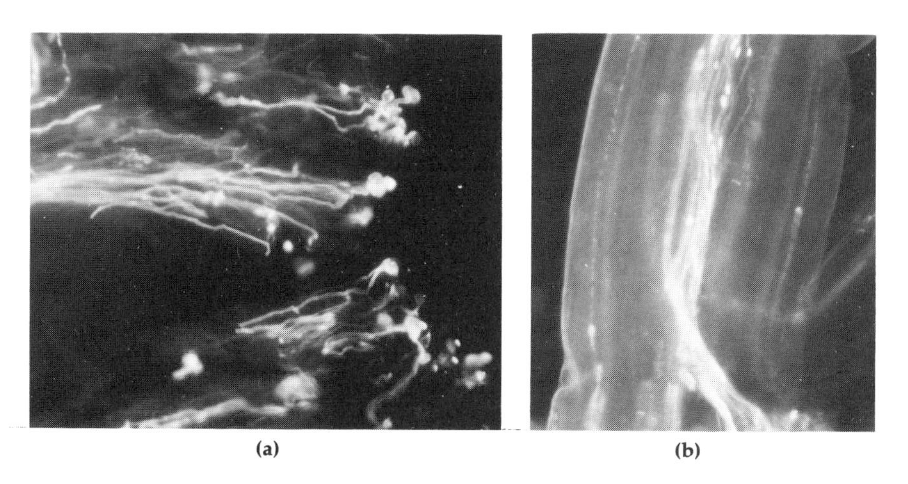

(a) **(b)**

Fig. 7.21 Microphotographs taken under ultraviolet irradiation of (a) good pollen germination and pollen-tube growth into the stigma in a self in the homostylous *Primula japonica* (×100), and (b) good pollen-tube growth in the central style of the homostylous *Primula scotica*, after a self (×40).

occurring in only 4/61 individuals in 1/15 loci examined (Glover and Abbott, 1995). Significantly perhaps, no less than 7/15 loci displayed invariably fixed heterozygosity however, which doubtless reflects the allohexaploid origin of this species.

The development of herkogamy in secondary homostyles

Tremayne and Richards (1993) investigated the floral structure of three tetraploid homostylous species in *Primula* section Muscarioides. In this largely Chinese group, most species develop extremely narrow floral tubes, and it is likely that only lepidopterans can form effective pollinators. In the short-tubed *P. concholoba*, only a slight reverse herkogamy develops, anther–stigma distances rarely exceeding 1 mm. However, in the long-tubed *P. bellidifolia* (which also has heterostylous populations) and in *P. watsonii*, homostyles have developed both approach (to 5 mm) and reverse herkogamy (to 2 mm) to a remarkable degree (Fig. 7.9). In *P. watsonii*, populations from higher altitudes were less herkogamous than those from lower altitudes, and were more likely to be reverse herkogamous, whereas those from lower altitudes were approach herkogamous.

In these lepidopterophilous, narrow-tubed flowers, it is likely that homostylous flowers will be caused to be selfed when probed by a proboscis. Where flowers are frequently visited, at lower altitudes, herkogamous flowers are more likely to be outcrossed, thus promoting the evolution of a non-heterostylous secondary herkogamy in these plants. This represents yet another evolutionary cycle in the outcrossing/selfing 'war'.

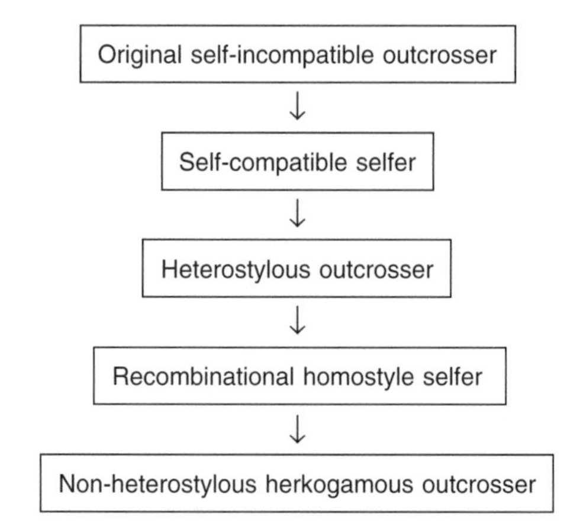

The condition in the secondary homostyle *P. watsonii* resembles that in the primary homostyle *P. verticillata* (Fig. 7.10). Both are self-fertile, but have long, very narrow-tubed lepidopterophilous flowers in which a continuous variation for a non-heterostylous herkogamy has evolved.

OTHER MEANS BY WHICH HETEROSTYLY CAN BREAK DOWN

Apart from *Primula*, the other group in which the secondary evolution of selfing from heterostyly has been intensively studied is the tristylous Pontederiaceae. Most of this work has been undertaken by S. C. H. Barrett and his co-workers, and is comprehensively reviewed by Weller (1993).

In the other tristylous families, Lythraceae and Oxalidaceae, a loss of complexity from a tristylous condition to a secondarily distylous condition has occurred in several examples, and results from the loss of the mid-styled, or the short-styled morphs by mutations to the dominant *M* or *S* alleles (p. 248). The evolution to a secondary distyly has usually been triggered by an initial increase in levels of pollen self-compatibility.

In the Pontederiaceae however, Barrett and Anderson (1985) conclude that a breakdown of the tristylous condition is usually initiated in the mid-styled morph. In this, both sets of anthers are equidistant to the stigma. Mutations which move one set of mid-style anthers to the homostylous position subsequently encourage the evolution of self-compatibility in this morph. These semi-homostyle phenotypes may be variably affected by the environment (Barrett and Harder, 1992). They do not differ greatly in pollen characteristics of size or number from short-anther variants, but show a greater male fitness on mid-style stigmas (Manicacci and Barrett, 1995), which has doubtless favoured self-compatible mid-style mutants for 'long' short-anthered plants. Once the mid-styled morph becomes self-fertilizing, the short-styled morph loses fitness, and the dominant *S* allele is lost from the population. It takes longer for the recessive *m* allele which determines the long-styled morph

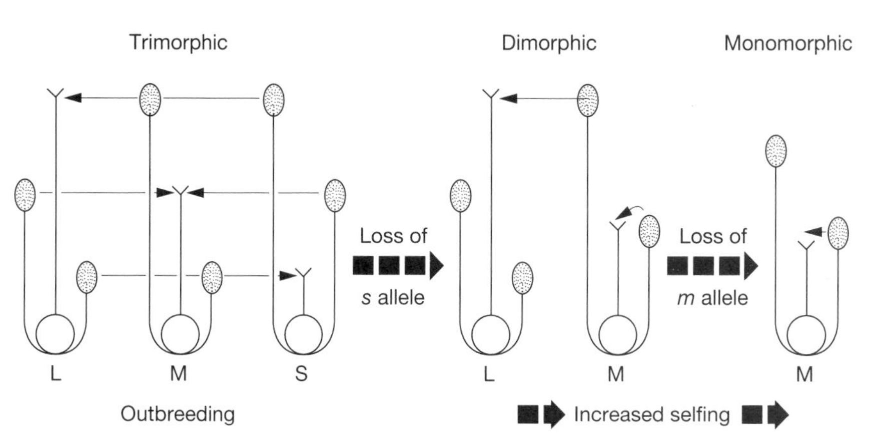

Fig. 7.22 Model of the breakdown of tristyly to semi-homostyly in *Eichhornia paniculata*. Arrows indicate predominant matings. Note the modifications in the short-stamen position of the M morph in dimorphic and monomorphic populations. (After Barrett, 1988.)

to disappear, but eventually, the population becomes monomorphic for the newly self-compatible mid-styled morph ssMM (Fig. 7.22). Of 110 populations surveyed in north-west Brazil and Jamaica, 53% were trimorphic, 25% were dimorphic (although mid-style morphs predominated) and 22% were monomorphic for mid-style morph (Barrett, Morgan and Husband, 1989).

These authors conclude that loss of the tristylous system usually accompanies populational bottlenecks such as founder events and cold spells. Non-tristylous populations tend to be geographically and ecologically marginal. Also, the short-styled morph tends to lack reproductive efficiency when long-tongued pollinators such as sphingids are absent. Pollinator service is particularly important in clonal water plants such as these, in which one clone may easily predominate in small populations, and for which distances between interfertile clones may be considerable.

Unlike dimorphic genera such as *Primula, Pulmonaria, Turnera, Pemphis, Limonium* and *Armeria* where evolution to a self-fertilizing condition is mediated by recombination, breakdown cannot operate this way in tristylous plants where two unlinked *S/s* and *M/m* supergenes control the heteromorphy. A breakdown by the accumulation of mutants is less immediate than a recombinational breakdown, but in circumstances where self-fertilization is favoured, a number of examples demonstrate that such a breakdown can nevertheless evolve.

CHAPTER 8
Dicliny

DEFINITIONS

A flower is considered to be male when it bears an androecium (stamens, the microsporangia) with viable pollen (microspores) capable of forming fertile pollen tubes (male gametophytes). A flower is considered to be female when it bears a gynoecium (= pistils), the megasporophylls, including the stigma(s), style(s) and an ovary containing ovules (the megasporangia with integuments). The ovules when fertile contain viable embryo-sacs (female gametophytes), each with a fertile egg cell (Chapter 3).

Dicliny is said to occur when all members of a population are not regularly hermaphrodite (sometimes called co-sexual or 'perfect', bearing male and female flowers). There are a number of conditions with respect to the distribution of male and female organs on individual plants (genets), and these have rather confusing names, which are explained below. They include both hermaphrodite and diclinous situations.

	Hermaphrodite	% of species[a]
all flowers hermaphrodite	(⚥)	72
all flowers monoecious	(♂ ♀)	5
gynomonoecious	(♀ ⚥)	2.8
andromonoecious	(♂ ⚥)	1.7
	Diclinous	
subgynoecious	(♂) (♀ ⚥)	2?
subandroecious*	(♂ ⚥) (♀)	2?
gynodioecious*	(♀) (⚥)	7
androdioecious	(♂) (⚥)	0?
polygamous (including trioecious and subdioecious)	(♂) (♂ ⚥) (⚥) (⚥ ♀) (♀) (not all may occur)	3.6
dioecious*	(♂) (♀)	4%

[a] After Yampolsky and Yampolsky (1922). *See p. 298.

For each of these categories, the symbols within brackets represent the sexual states that can be represented within an individual genet. In all cases except full hermaphrodity, some flowers have only one gender. For andromonoecy and androdioecy this single gender is male, and all flowers with female function also have male function, so that all fruit-bearing flowers can be selfed.

In all other conditions, there are some flowers which are female, and these have to receive pollen from another flower in order to set seed. Thus, they can be outcrossed, but they may also be geitonogamously selfed by pollen from another flower on the same plant. Only in the three diclinous conditions marked with an asterisk (*) are there at least some individual genets which are entirely female, which must be outcrossed for seed to be set.

Two of these three conditions, gynodioecy and dioecy, are the commonest forms of dicliny, and these conditions most commonly lead to outcrossing. The other conditions are mostly of interest in so far as they represent intermediate evolutionary steps towards dioecy, or result from environmentally triggered variations in the expression of genetically controlled dioecy or gynodioecy.

It should be noted that dioecy in seed plants is not homologous with so-called dioecy (sometimes distinguished as 'dioicy') in bryophytes, nor does it have the same genetic consequences. In bryophytes, it is the gametophyte (haploid) generation which may be unisexual ('dioicious'); this generation is always unisexual in seed plants. Variations in gender expression in seed plants (above) occur in the sporophyte generation. In bryophytes however, the sporophyte generation is always uniform, producing only one type of sporangium and spore.

Thus, bryophytes lack a mechanism for preventing mating between meiotic products of the same sporophyte (selfing), which is achieved by dioecy in seed plants. The term dioecy should be reserved for those systems in which selfing is rendered impossible by the unisexuality of the sporophyte. This condition is restricted to seed plants, and bryologists should seek another term.

DIOECY

Distribution of dioecy

In the flowering plants, dioecy is a rather unusual condition which is found in about 4% of all species. It is scattered through many plant families at a low frequency, and few families are entirely dioecious. The willows and poplars (Salicaceae) form a familiar example of an entirely dioecious family (Fig. 8.1). This distribution pattern strongly suggests that dioecy has arisen secondarily from hermaphrodite breeding systems on many occasions (polyphyletically), and that it is reversible, or frequently

Fig. 8.1 Male (left) and female (right) inflorescences of the dioecious goat willow, *Salix caprea* (×0.25).

becomes extinct. Dioecy rarely seems to last long enough in evolutionary time, or to be successful enough, to establish a dynasty, i.e. to dominate a higher taxonomic category such as a family, tribe or even a genus.

It must be noted here that a minority view persists that a dioecious, wind-pollinated condition is primitive to the Angiosperms (A. D. J. Meeuse, 1973, 1978). I prefer to follow the majority view that the original Angiosperm condition was a hermaphrodite, beetle-pollinated flower with gametophytic self-incompatibility (p. 68).

Dioecy is a mechanism that ensures total outbreeding, but it is also inefficient in that only about half of the genets in the population bear seeds.

Anderson and Stebbins (1984) present an alternative view. They suggest that dioecy is often more reproductively efficient than is hermaphrodity, when gsi hermaphrodites share rather few *S* alleles in a population. This can occur when few genets undergo extensive vegetative reproduction, which might explain how dioecy evolved in *Fragaria*, or in *Spartina patens*. Gsi allele starvation in hermaphrodites could also result from a founder effect, as might occur in island colonization, which could also encourage the evolution of dioecy. Unisexual flowers also set more fruits than do bisexual flowers (Table 2.4), perhaps as a result of shedding male load.

The distribution of dioecy within the Angiosperms strongly suggests that dioecy has usually arisen secondarily from self-compatible inbreeders (p. 13), in response to renewed selection pressures for outcrossing. Thomson and Barrett (1981) point out that dioecy has mostly arisen within taxonomic groupings which have lost the original gsi outcrossing system.

The explanation that dioecy has arisen from selfing progenitors in response to selection pressures which favour outcrossing has a long and respectable history, originating with Darwin (1877). More recently Willson (1979), Bawa (1980) and others argue rather that dioecy has been successful because it promotes the separation of male and female function. A familiar general explanation is that sexual selection may favour dioecy, as dioecy tends to minimize competitive interactions between gender functions within a species (e.g. Kay, 1987b). Geber and Charnov (1986) show that dioecy is only likely to evolve successfully when hermaphrodite founders do not separate gender function in space, or more particularly, in time, i.e. when they are homogamous rather than dichogamous.

Certainly, dioecy is associated with certain life forms and habitats. Bawa (1980) has gathered information as to the proportion of dioecious species in the floras of 12 areas (Table 8.1). Yampolsky and Yampolsky (1922) discovered that 3–4% of species in many parts of the world are dioecious (the figure for the British Isles has been estimated at 4.3% of species, Kay and Stevens, 1986). Bawa's figures agree with this assess-

Table 8.1 Incidence of dioecy in 11 different floras (after Bawa, 1981)

Area	Type of flora	Dioecious species (%)
Barro Colorado Island, Panama	Tropical forest	9.0
Costa Rica	Lowland rainforest	23
Central Chile	Montane sclerophyllous forest	9.0 (Arroyo and Uslar, 1990)
India	Tropical, various	6.7
Ecuador	Tropical, various	3.0
Hawaii	Subtropical oceanic	27.7
New Zealand	Various, oceanic	14.5
South-west Australia	Mediterranean	4.4
South Australia	Various	3.9
North Carolina	Warm temperate	3.5
South California	Mediterranean	2.5
British Isles	Cool temperate	4.3

ment, but with two notable exceptions. These are tropical forests and oceanic islands.

Tropical forests

The figure of 9% dioecious species in the wet tropical forest of Barro Colorado Island, Panama, may be typical of many tropical forests. A high proportion of species in this habitat are trees, and trees tend to be dioecious, especially in tropical forest (Table 8.2). Bawa's own work in a Costa Rican forest (Bawa, 1979) gives similar figures, as many as 22% of the tree species being dioecious, while other figures for the proportion of dioecy among lowland rain forest trees are similar (Sarawak 26%, Ashton, 1969; Panama 21%, Croat, 1979).

A breakdown by growth habit of the data for a Chilean sclerophyllous montane flora (Arroyo and Uslar, 1990) is even more dramatic.

annual herbs	0% dioecious
perennial herbs	2% dioecious
shrubs	17% dioecious
trees	57% dioecious

Associations between dioecy and a woody habit have also been found in such diverse floras as those of California (Freeman, Harper and Ostler, 1980), Puerto Rico and the Virgin Islands (Flores and Schemske, 1984) and Alaska (Fox, 1985).

Lahav-Ginott and Cronk (1993) show that in the Malesian tropical nettle genus *Elatostemma*, dioecy is correlated with large size and a woody habit (liane species); this condition is unusual among herbaceous perennials in this genus.

Bawa (1980) has argued that the association between dioecy and tropical trees results from problems faced by the seedling tropical tree. This must germinate and grow at very low light levels on the forest floor. If it is to succeed, and become a sexually reproducing forest giant, it must survive until a gap appears in the canopy caused by the death of an older tree. Therefore, there is a premium on large seeds with large maternal

Table 8.2 Frequency of dioecy in different growth forms in three floras (after Bawa, 1980)

	% dioecious species		
Life form	North Carolina	Barro Colorado Island	California
Trees	12	21	20–33
Shrubs	14	11	0–23
Lianes	16	11	–
Herbs	1	2	4–9

investments of energy; trees producing such seeds are most likely to pass their genes on to the next generation as these offspring are most likely to persist and survive.

Bawa argues that a high maternal investment in the fruit will also be selectively advantageous. Nutritious fruits which are attractive to birds, bats and monkeys are more likely to be dispersed away from the competitive environment of the parent tree, perhaps to conditions that have been made favourably open by the animals' roost or nest. Bawa demonstrates a striking association between dioecy and animal-dispersed fruits in two areas of Costa Rica (Table 8.3) although, as animal-dispersed fruits preponderate in this flora, it would be more accurate to state that the fruits of dioecious species are rarely wind dispersed. Fox (1985) and Flores and Schemske (1984) are other authors who have found an association between dioecy and fleshy fruits in several parts of the USA, and in the West Indies, respectively.

More recently, Muenchow (1987) has also tested this relationship, finding a significant three-way correlation between dioecy, small inconspicuous flowers and fleshy fruits. However, Muenchow argues that the relationship between dioecy and fleshy fruits may be indirect, as both attributes may be related to the understorey habitat.

Nevertheless, large seeds and nutritious fruits will cause a considerable drain on maternal resources. Bawa considers that such plants may be more successful when these heavy maternal loads are separated from male loads, in the dioecious condition. Givnish (1980) has provided a similar argument to explain the high levels of dioecy among tropical forest Gymnosperms, which also tend to have very large, animal-dispersed seeds (e.g. cycads).

There is another line of argument concerning the occurrence of dioecy among tropical trees. Within a tropical tree species, individuals usually occur at a very low density, and are distant from one another, with many other species in between. As a result, self-compatible hermaphrodites will be very largely selfed, with an accompanying loss of vigour and variabil-

Table 8.3 Numbers of animal- and wind-dispersed species that are dioecious or hermaphrodite in two Costa Rican floras (after Bawa, 1980)

		Fruit dispersal	
Type of forest	Breeding system	Animal	Wind
Dry deciduous	Dioecious	30	3
	Hermaphrodite	60	26
Wet evergreen	Dioecious	66	0
	Hermaphrodite	222	29

ity in their offspring (p. 384). If unisexual outcrossing variants arise among such populations, cross-pollinated by 'trap-lining' solitary bees (p. 165), they may be favoured by the quality of their offspring, even though they are very much less fecund than are self-fertilizing hermaphrodites (p. 358).

Interestingly, work on patterns of both pollen and pathogen spore dispersal in white campion (Alexander and Antonovics, 1993) shows them to be frequency rather than density dependent. Specialist pollinators apparently adjust flight distances to accommodate to very different levels of host density.

A long-lived forest tree only has to produce one successful seedling from many millions of seeds if it is to perpetuate itself, so offspring quality is likely to be more important than offspring quantity. This premium on offspring quality is further enhanced for plants that invest heavily in female reproduction (large seeds, nutritious fruits); the law of the jungle should favour the production of few high-quality seeds, rather than many low-quality seeds (i.e. 'K' rather than 'r' strategy, p. 145).

Dioecious species are inconvenienced in another way, when pollinated by animals which collect pollen. Although males and females both produce nectar, only males produce pollen. Kevan and Lack (1986) show that the only reward to pollinators provided by the Indonesian tree *Decaspermum parviflorum* is pollen; females provide sterile pollen as a reward to visiting bees.

If cross-pollination is to be effective among dioecious tropical trees, the plant will be most successful when it attracts visitors which are primarily nectar feeders. It is to be expected that pollen feeders will be poorly represented among visitors to tropical dioecious trees, and indeed beetles and large social bees are much rarer visitors than they are to comparable hermaphrodite flowers (Table 8.4). The dominant visitors to dioecious

Table 8.4 Most important pollinators of dioecious and hermaphrodite tree species in a Costa Rican dry deciduous forest (after Bawa, 1980)

	% of tree species	
Main pollinators of species	Hermaphrodite ($n = 94$)	Dioecious ($n = 28$)
Medium or large bee	25	1
Small bee	26	80
Beetle	14	3
Moth	19	9
Other	16	7
Total	100	100

flowers are small, non-social bees which tend to be entirely nectar feeders, and which fly long distances between nectar sources ('trap-line', p. 165).

Some dioecious tropical trees produce stigmatic nectar. Richards (1990b) shows for several *Garcinia* species that males produce sterile gynoecia which attract *Trigona* bees to their nectariferous stigmata.

Wind pollination, a common condition among dioecious woody species in temperate latitudes (p. 88), is rare in tropical forests, where the density of the vegetation and the sparsity of individuals is likely to impede wind-borne pollen flow between plants. It is perhaps surprising that more tropical dioecious trees are not visited by vertebrate nectar feeders such as birds and bats; maybe the heavy energetic load of producing large, nectar-rich flowers suitable for vertebrate flower feeders is unsuccessful in combination with expensive fruit production.

Nevertheless, Cox (1982) has suggested that large vertebrate pollinators such as fruit bats, parrots and monkeys may indeed select for dioecy in the tropical forest in some circumstances. He notes for the liane *Freycinetia reineckei* that bats frequently destroy male flowers while visiting them, although at the same time they do pick up much pollen externally. Cox argues that in these conditions it will pay the plant to separate 'disposable' male flowers from female flowers which, lacking pollen, are less attractive food targets. In this way, female flowers are allowed to mature their fruit.

Oceanic islands

The highest levels of dioecy among species of flowering plants are to be found on oceanic islands of volcanic origin, such as New Zealand and Hawaii (see Table 8.1) (Carlquist, 1974; Godley, 1975, 1979). It is assumed that such islands have been colonized from neighbouring continents, and that intense speciation has subsequently occurred into different niches, and onto different islands, as on the Galapagos, made famous by Darwin.

Baker (1955) suggests that most successful colonizers will be hermaphrodite and self-fertile. Most founders will arrive as single individuals, which could not set seed if self-sterile or unisexual (dioecious). It is thus remarkable that such areas should today be notable for the high levels of dioecious and gynodioecious species.

It is reasonable to suppose that oceanic islands have a low biological diversity and incomplete vegetational cover (Carlquist, 1974). Ample opportunities exist for genetic diversification of founders into different ecological niches to occur and for speciation to result.

Such diversification will depend on high levels of genetic variability in populations and offspring families, which will result from outcrossing (p. 14). If founders to island populations are self-fertile inbreeders, secondarily arising outbred mutants are likely to be favoured (Baker, 1967). Thus male-sterile mutants should be favoured after a self-fertile hermaphrodite founder has arrived on an oceanic island. Although reproductively inefficient, such diclinous mutants should be more outcrossed than their progenitors. Their variable offspring should be able to diversify into a wider variety of niches than the less variable offspring of the inbred progenitors, and thus outbred parents will be fitter than selfing parents (an extension of the 'tangled bank' hypothesis, p. 18).

One test for this hypothesis is to examine the frequency of oceanic island species which have the 'original' gametophytic self-incompatibility. As discussed by Thomson and Barrett (1981), the floras of New Zealand, Hawaii and the Galapagos Islands do indeed have a very low (considerably less than 10%) proportion of species with gsi. This indeed suggests that such islands were originally mostly colonized by s-c immigrants.

THE CONTROL OF DIOECY

In all dioecious species, the expression of gender is basically controlled by special chromosomes, the sex chromosomes. One gender is usually heterogametic XY with two types of sex chromosome, and the other is always homogametic XX. As will be seen later, it is important to differentiate between the gender genotype, based on the sex chromosomes, and the gender phenotype, in which gender expression may be modified by environmental influences on growth substances which transmit genetic messages. In most plants, like most animals, the male is heterogametic XY and the female homogametic XX, but there are some exceptions.

Because males are bound to cross with females, and the heterogametic gender should generate X gametes and Y gametes in equal numbers, genders might be expected to arise from seed at a ratio of 1 : 1 (a frequency of 0.50 for each). This parallels heteromorphy (p. 242) where a heterozygote *Ss* usually crosses with a homozygote *ss* so that frequencies of each morph from seed are also 0.50. There is, however, an important distinction between dioecy and heteromorphy. Heteromorphs are hermaphrodite, and the morphs do not usually differ significantly in their gender effort. Consequently, it is very unusual to find heteromorphic populations with a significant anisoplethy (but see Richards *et al.*, 1989). However, for dioecy, gender ratios which depart from unity are commonplace (p. 351).

The evolution of dioecy

Dioecy is secondary in origin, and in many species the presence of vestigial sex organs in unisexual individuals plainly indicates a hermaphrodite origin (Fig. 8.2). Unisexuality is thus caused by the suppression of one gender. A female is male sterile, and a male is female sterile.

Fig. 8.2 Male flower (above) and female flower (below) of the dioecious tetraploid form of the shrubby cinquefoil, *Potentilla fruticosa*. Female flowers have sterile anthers, whereas in the male flowers the gynoecia are replaced by tufts of hairs. In this species, dioecy is clearly of secondary origin.

In most cases, male sterility is expressed through a reduction of anthers to empty (often very small) staminodes, whereas female sterility is caused by a failure of (again, often very small) gynoecia to contain fertile ovules. In many cases, particularly where gynodioecious and androdioecious species are examined (p. 318), it is clear that these male sterile and female sterile mutations are controlled, or at least triggered, by single genes.

Cryptic dioecy

In some cases however, dioecy can be highly cryptic, and such examples often have a functional explanation. For instance, the genus *Solanum* was long thought to be entirely hermaphrodite, as all species have 'perfect' flowers which bear abundant pollen, and bear gynoecia which are usually normal in appearence. However, in the American *S. appendiculatum*, and in no less than nine Australian species, female-fertile individuals bear pollen which is non-porate, so that it cannot hydrate and germinate. Equally, porate pollen cannot germinate on the stigmas of individuals which themselves have fertile, porate pollen. In the Australian species the gynoecia are highly reduced, so that the species initially appear to be androdioecious (p. 332). Thus, these two cryptic morphs are essentially female and male respectively (Levine and Anderson, 1986; Anderson and Symon, 1988, 1989). These authors suggest that this condition is successful as visits by solitary, pollen-feeding bees to both genders are encouraged by the provision of pollen in females, and of stigmatic nectar in males. A similar argument is evinced by Kevan *et al.* (1990) for *Rosa setigera* which has a similarly cryptic dioecy.

The evolution of sex chromosomes

In a dioecious species, the primary gender characters are controlled by gender sterility genes which are located on the sex chromosomes. One gender (most commonly female) is homogametic XX. In this, male sterility is homozygous. The other gender (most commonly male) is heterogametic XY. Female sterility must be carried on the Y chromosome. However, the heterogametic gender also has an X chromosome, which carries male sterility. To function correctly, the Y-carried female sterility must be dominant to the X-carried male sterility.

Thus, it is reasonable to suppose that in most dioecious plants which have heterogametic males, that femaleness evolved before maleness. Maleness could only be expressed secondarily if it was dominant to the femaleness already present. Gynodioecy, with females and hermaphrodites, is much more common than androdioecy, and probably formed the main evolutionary pathway to full dioecy.

In some plants the sex chromosomes are visually distinct from the other chromosomes (autosomes). Typically, they are smaller, with less active DNA (more heterochromatic), and although they may associate with each other at meiosis, they fail to form chiasmata, or chiasmata are limited to specific regions of the chromosomes. Often, the X and Y chromosomes can also be distinguished visually. The Y chromosome is frequently even smaller and more heterochromatic than is the X chromosome.

However, in many dioecious plants (Table 8.5) it is difficult or impossible to identify the sex chromosomes, especially when all the chromosomes are very small, and segments of the sex chromosomes regularly form chiasmata at meiosis. If dioecy is to be maintained, crossing over cannot occur at meiosis in the heterogametic gender between those segments of the X and Y chromosomes that carry the primary and secondary sexual character loci. It is probable that these sex character linkage groups are protected from crossing over by one being inverted with respect to the other (e.g. Darlington, 1939). Often, much of the length of the chromosomes differs by a large pericentric inversion. Crossovers that do occur within the inversion loop at meiosis will lead to non-viable products.

Table 8.5 A list of some dioecious species indicating differentiation of sex chromosomes, and sex of the heterogamete (after Westergaard, 1958; Williams, 1964)

Species	Differentiation of sex chromosomes		Heterogametic sex
Cannabis sativa	XY	XX	male
Humulus lupulus	XY	XX	male
H. japonicus	XYY	XX	male
H. lupulus var. *cordifolium*	XXYY	XXXXX	male
Rumex acetosella	XY	XX	male
aggregate	3XY	4X	
	5XY	6X	
	7XY	8X	
R. acetosa	XYY	XX	male
R. hastatulus	XYY	XX	male
R. paucifolius	3XY	4X	male
Silene latifolia (alba)	XY	XX	male
S. dioica	XY	XX	male
Asparagus officinalis	XY	XX	male
Elodea canadensis	XY	XX	male
Salix spp.	XY	XX	male
Populus spp.	XY	XX	male
Urtica dioica	XY	XX	male
Spinacia oleracea	XY	XX	male
Coccinea indica	XY	XX	male
Dioscorea spp.	XO XY	XX	male
Fragaria elatior	XX	XX	female
Silene otites	XX	XX	male and female?
Valeriana dioica	XX	XX	male and female?
Sedum rosea			?
Thalictrum fendleri			male
Mercurialis annua			male
M. perennis			?
Vitis vinifera			male
Carica papaya			male
Bryonia dioica			male
Empetrum nigrum			?
Ecballium elaterium			male
Potentilla fruticosa			female
Cotula spp.			female

Note where XY and XX are both given, the X and Y chromosomes can be distinguished visually (are heteromorphic). Where XX XX is given, the sex chromosomes can be distinguished from autosomes, but not told apart.

Heterogametic females

In three genera of flowering plants, females of dioecious species, rather than males, have been shown to be heterogametic. These are *Fragaria* (Staudt, 1952), *Potentilla fruticosa* (see Fig. 8.2; Grewal and Ellis, 1972) and

Cotula (Lloyd, 1975b). In these cases, the sex chromosomes cannot be differentiated visually. However, when females are crossed with hermaphrodite relatives, the offspring vary in gender. The reciprocal crosses, using males, yield offspring that are uniform in gender. This non-reciprocity is explained by the donation of either X or Y chromosomes to the offspring from the heterogametic female.

In *Silene otites*, which is sometimes subandroecious (p. 297), Correns (1928) reported that the offspring of selfed hermaphrodites were all male, suggesting that female heterogamety may also occur in this species.

Heterogametic males

There are three plausible explanations why most flowering plants have heterogametic males.

1. Haldane (1922) has suggested that where one gender is weaker or less viable than the other, especially when hybrid, it is always the heterogametic gender that is so. When disadvantageous recessive mutations arise in the homogametic (XX) gender, they are protected from selection, and they can be recombined into a 'disposable' state on 'redundant' X chromosomes. However, Y-linked sublethals are exposed to selection, and they cannot be recombined, if they arise in a non-chiasmatic region. Thus, Y chromosomes are liable to 'Mueller's Ratchet', whereas X chromosomes are not. There is a greater premium on vigour and longevity in the seed-bearing female, so that heterogamety may be more successfully associated with 'disposable' males than with seed-bearing females.
2. Lloyd (1974b) suggests that when females are lost from gynodioecious populations, hermaphrodites (which may later evolve to maleness can only regenerate new females if the hermaphrodites are heterogametic.
3. If, as it seems, dioecy usually evolves from gynodioecy (p. 320), male sterility evolves first (XX). Thus, female sterility must be dominant to male sterility, and the heterogametic sex (XY) must be female sterile, i.e. male (p. 305).

Evolution of the Y chromosome

The Y chromosome has two functions:

1. to carry female-sterile (male) genes, and
2. to suppress male-sterile (female) genes.

However, primary male characteristics (female sterility) and secondary male factors can also function if they are encoded on the other chromosomes (autosomes), as long as male-sterile (female) genes are suppressed by the Y chromosome. Even the female suppression function of the Y chromosome can be replaced by gene dosage effects between the X chromosome and the autosomes.

In some dioecious genera a clear evolutionary progression can be observed in the reduction in size and function of the Y chromosome. As the sex chromosomes evolved from autosomes, both are initially large and rich in active DNA. The X-linked factors, which can recombine harmful mutations, remain linked to the X chromosome, which remains relatively large and rich in DNA.

However, the Y-linked factors can become associated with harmful mutations which accumulate by 'Müller's ratchet' as these mutations cannot be shed by recombination (p. 309). Thus, the translocation of male-determining genes to the autosomes may be a successful strategy, as these genes are removed from close linkage with disadvantageous Y-linked mutants. Autosomal male factors, once translocated, now behave as one large Y unit, balanced against the X chromosome. In time the Y chromosome may disappear completely (XO), or lose all function. In the absence of a functional and dominant Y chromosome, changes in balance between the X chromosomes and the autosomes may allow hermaphrodity to re-establish, particularly in polyploids.

This evolutionary progression is elegantly observed in the sorrels (*Rumex*). In the *R. acetosella* polyploid complex, diploids ($2n = 14$), tetraploids ($2n = 28$), hexaploids ($2n = 42$) and octoploids ($2n = 56$) occur. In females, the X chromosome occurs 2, 4, 6 and 8 times respectively; in males the X chromosome occurs 1, 3, 5 and 7 times, respectively, but the Y chromosome occurs only once (Løve, 1944). This mechanism gives rise to some puzzling and unexplained questions about the formation of these polyploids. However, not until the decaploid level ($2n = 70$) is reached do plants with a single Y chromosome exhibit any female tendency. Here is an overdominant Y chromosome with very strong X suppressant features, which may be regarded as rather primitive.

In the related field sorrel, *R. acetosa*, wild-occurring males have $2n = 15$, with two Y and one X chromosomes, whereas females have $2n = 14$ with two X chromosomes. At male meiosis, a trivalent forms with the X chromosome in a central position. The trivalent orientates on the spindle so that the X chromosome passes to one pole at anaphase I, and the two Y chromosomes pass to the other pole. In artificially induced higher polyploids, in hybrids and in chromosome mutants, the number of Y chromosomes may vary in the male. This variation has no effect on the sex of the plant, which remains male, and is entirely controlled by the ratio of X chromosomes to autosomes (Ono, 1935; Yamamota, 1938; Zuk, 1963). The Y chromosome is inert, and the autosomes act collectively as the male factor.

	X	XX	XXX	XXXX
autosomes 2n	male	female		
3n	male	hermaphrodite	female	
4n		male	hermaphrodite	female

The genus *Rumex* exhibits the full range of Y chromosome control in plants. An intermediate condition has been found in the American *R. hastulatus* by Smith (1963). Here the diploid sex control is by X chromosomes and autosomes, as in *R. acetosa*, but in artificial polyploids the Y chromosome modifies intersexes towards maleness. This condition has been interpreted as an intermediate stage in the progression towards total Y inertness. It is very similar to the control described for *Silene dioica* (Fig. 8.3) by Winge (1931), Westergaard (1940, 1946, 1948) and Warmke (1946). It contrasts with the report of Løve and Sarker (1956) for the octoploid ($2n = 56$) *R. paucifolius* in which males are XY or XO, and females XX. In this species Y chromosomes are inert, but sex chromosomes are only present disomically (twice) rather than polysomically as is the case in other polyploid *Rumex*.

Dioecy and polyploidy

Dioecy is most common in diploids. Where a primitive dominant Y occurs, as in *Rumex acetosella* (above), dioecy also persists in polyploids. Where male control has passed to the autosomes, it is rare for dioecy to remain stable in polyploids unless the sex chromosomes are disomic (only present twice) as in *Rumex paucifolius*. Gametes from the tetrasomic male XXYY or XXOO will tend to contain one X chromosome if there is regular disjunction, and thus trisomic intersexes XXXY or XXXO will commonly occur in the next generation. Hermaphrodites commonly occur among the offspring of artificially polyploid sorrel *Rumex acetosa*, or white campion *Silene latifolia* (*S. alba*).

It is often observed in diploid/tetraploid 'species pairs' that the diploid is dioecious, but the tetraploid is hermaphrodite, as in the crowberries *E. nigrum* (diploid) and *E. hermaphroditum* (tetraploid).

In other genera, the relationship between polyploidy and sex expression can be more complex. In the annual mercuries (*Mercurialis annua*), the following is found (Durand, 1963):

2x	4x	6x	8x 10x 12x 14x
dioecious	dioecious androdioecious	dioecious androdioecious hermaphrodite	hermaphrodite

However, even in diploid *M. annua*, the genetic control of gender is complex, for it depends not only on one male and two female determining genes, but also on complementary interactions between these loci. For instance the genotype A/A bi/bi b2/b2 is female, but A/A bi/bi B2/b2 is male (Durand and Durand, 1991).

In *Salix* (willows) all the species in a highly polyploid complex are dioecious (see Fig. 8.1), and we must presume that all still possess a highly dominant Y chromosome, of the *Rumex acetosella* type.

In two of the genera with heterogametic females, the usual situation is reversed, that is diploids are hermaphrodite and polyploids are dioecious. In *Potentilla fruticosa* (see Fig. 8.2), diploids ($2n = 14$) from the Alps, Pyrenees and North America are hermaphrodite, whereas tetraploids and hexaploids are dioecious, including the tetraploid British race (Elkington, 1969). In the strawberries (*Fragaria*), all the diploid species ($2n = 14$) are hermaphrodite, and all the polyploids ($2n = 28, 42, 56$) are dioecious. The sex chromosomes cannot be identified in these genera. However, it seems likely that the sex chromosomes are disomic, with only a single pair occurring, even in high polyploids.

THE STABILITY OF SEX EXPRESSION

Obligately dioecious species, where intersexes or hermaphrodites are never found, seem to be uncommon. In a survey of the 60 dioecious species in the British flora, Kay and Stevens (1986) report that certain or probable exceptions to a strict dioecy can occur in 35 species, and many of the others have not been fully investigated. However, in some cases, such as the genus *Salix*, gender-determining genes seem usually to override environmental influences, so that intersexes are very rare among willows (nevertheless, occasional hermaphrodity has been reported among five of the 18 British willow species).

Other dioecious groups are much more fickle in their gender expression, which tends to be subject to environmental influence. The gender genotype, as evidenced by the chromosomes and inheritance of gender in the offspring, may not be the same as the gender expressed. This has often allowed gender control to be analysed genetically, as individuals which are genetically unisexual can be selfed (p. 318).

Genetic gender seems often to be expressed by means of relative concentrations of plant growth substances (reviewed by Irish and Nelson, 1989). These growth substances have manifold functions in the plant, many of which control plant behaviour with respect to environmental influences.

Freeman, Harper and Charnov (1980) review gender lability, identifying no less than 21 distinct parameters involved in the change of gender in over 60 species (Table 8.6).

Table 8.6 Some factors known to modify the sexual expression of vascular plants under controlled conditions. (After Freeman, Harper and Charnov, 1980)

Factor	Direction	Species
Abscisic acid	♂ → ♀	*Cannabis sativa* L.*
	♂ → ♀	*Cucumis sativa* L.
	♂ → ♀	*Cucurbita pepo* L.
Age (size)	♂ → ♀	*Arisaema japonica* Bl.
	♂ → ♀	*A. triphyllum* (L.) Torr.
	♂ → ♀	*Castillea elastica* Sesse*
	♂ → ♀	*Catasetum macrocarpum* L.C. Rich. ex. Knuth
	♂ → ♀	*Cynoches densiflorum* Rolfe
	♂ → ♀	*Eucommia ulmoides* D. Oliver*
	♂ → ♀	*Ilex opaca* Ait.
	♂ → ♀	*Metasequoia glyptostroboides*
Auxin	♂ → ♀	*Cannabis sativa* L.*
	♂ → ♀	*Cleome iberidella* Welw. ex Oliv.
	♂ → ♀	*Cucumis sativa* L.
	♂ → ♀	*Cucurbita pepo* L.
High boron	♂ ← ♀	*Cannabis sativa* L.*
Carbon monoxide	♂ → ♀	*Cannabis sativa* L.*
	♂ → ♀	*Cucumis sativa* L.
	♂ → ♀	*Mercurialis annua* L.
Cold weather	♂ ← ♀	*Atriplex canescens* Pursh.*
	♂ ← ♀	*Cycas circinalis* L.*
Dry soil	♂ ← ♀	*Acer grandidentatum* Nutt.
	♂ ← ♀	*Arisaema triphyllum* L. Schott.
	♂ ← ♀	*Cucumis sativa* L.
	♂ ← ♀	*Juniperus osteosperma* (Torrey) Little
	♂ ← ♀	*Quercus gambelii* Nutt.
	♂ ← ♀	*Sarcobatus vermiculatus* (Hook.) Torr.
	♂ ← ♀	*Triticum aestivum* L.
Ethylene	♂ → ♀	*Cucumis sativa* L.
		C. melo L.
		Cucurbita pepo L.
Gibberellins	♂ ← ♀	*Cannabis sativa* L.*
	♂ → ♀	*Cleome iberidella* Welw. ex Oliv.
	♂ → ♀	*Cleome spinosa* Jacq.
	♂ → ♀	*Cucumis sativa* L.
	♂ → ♀	*C. melo* L.
	♂ → ♀	*Ricinus communis* L.
	♂ ← ♀	*Spinacea oleracea* L.*

Table 8.6 *Continued*

Factor	Direction	Species
Growth retardant	♂ → ♀	*Cucumis melo* L.
Eriophid mites	♂ → ♀	*Salix andersoniana* Sm.*
	♂ → ♀	*S. aurita* L.*
	♂ → ♀	*S. caprea* L.*
	♂ → ♀	*S. cinerea* L.*
	♂ → ♀	*S. grandifolia* Seringe*
	♂ → ♀	*S. silesiaca* Willd.*
Cytokinins	♂ → ♀	*Cannabis sativa* L.*
	♂ → ♀	*Kalanchoe integra* Medic. (*K. crenata*)
	♂ → ♀	*Cleome spinosa* Jacq.
	♂ → ♀	*Spinacia oleracea* L.*
High light intensity	♂ → ♀	*Catasetum expansum* Rchbif.*
	♂ → ♀	*C. macrocarpum* (L.) Rich. ex Kuth*
	♂ → ♀	*C. macroglossum* Rchbif.
	♂ → ♀	*C. platyglossum* Schltr.
	♂ → ♀	*C. ventricosum* Batem.
	♂ → ♀	*C. densiflorum* Rolfe
	♂ → ♀	*Cynoches dianae* Rchbif.
	♂ → ♀	*C. lehmannii* Rchb. f.
	♂ → ♀	*C. stenodactylon* Schltr.
	♂ → ♀	*Cucumis sativa* L.
	♂ → ♀	*Elaeis guinanensis* Jacq.
	♂ ← ♀	*Kalanchoe (Bryophyllum) daigremontiam* Hamet & Perr. A. Berger
	♂ → ♀	*Mormodes buccinator* Lindl.
Manure	♂ → ♀	*Arisaema triphyllum* (L.) Schott
	♂ → ♀	*Humulus japonicus* Siebold & Zucc.*
Methylene blue	♂ → ♀	*Cucurbita sativa* L.
High nitrogen	♂ → ♀	*Begonia semperflorens* Link & Otto
	♂ → ♀	*Ceratopteris thalictroides* (L.) Brongn. (gametophyte)
	♂ → ♀	*Cleome spinosa* Jacq.
	♂ → ♀	*Cucumis anguira* L.
	♂ → ♀	*C. melo* L.
	♂ → ♀	*C. sativa* L.
	♂ → ♀	*Lycopersicon esculentum* Mills
	♂ → ♀	*Osmunda regalis* L. (gametophyte)
	♂ → ♀	*Zea mays* L.
	♂ → ♀	*Spinacia oleracea* L.*
Photoperiod long day	♂ ← ♀	*Cannabis sativa* L.*
	♂ ← ♀	*Cucumis anguira* L.

Table 8.6 *Continued*

Factor	Direction	Species
	♂ ← ♀	*C. sativa* L.
	♂ ← ♀	*Heteropogon contortus* (L.) Beauv. ex R. & S.
	♂ ← ♀	*Humulus japonicus* Siebold & Zucc.*
	♂ ← ♀	*Xanthium strumarium*
	♂ ← ♀	*Zea mays* L.
Photoperiod short day	♂ → ♀	*Ambrosia artemisifolia* L. (*A. elatior* L.)
	♂ → ♀	*A. trifida* L.
	♂ → ♀	*Cannabis sativa* L.*
	♂ → ♀	*Cucurbita pepo* L.
	♂ → ♀	*Cucumis sativa* L.
	♂ → ♀	*Heteropogon contortus* L. Beauv. ex R. & S.
	♂ → ♀	*Humulus japonicus* Siebold & Zucc.*
	♂ → ♀	*Silene pedula* L.
	♂ → ♀	*Spinacia oleracea* L.*
	♂ → ♀	*Xanthium pennsylvanicum* Wally.
	♂ → ♀	*X. strumarium* L.
	♂ → ♀	*Zea mays* L.
Potassium	♂ → ♀	*Ricinus communis* L.
'Rich soil'	♂ → ♀	*Arisaema triphyllum* (L.) Torr.
	♂ → ♀	*A. dracontium* (L.) Schott.
	♂ → ♀	*Begonia semperflorens* Link & Otto
	♂ → ♀	*Cucumis sativa* L.
	♂ → ♀	*Equisetum* (gametophyte)*
High temperature	♂ ← ♀	*Ambrosia trifida* L.
	♂ ← ♀	*Carica papaya* L.
	♂ ← ♀	*Citrullus lanatus* (Thunb.) Matsum & Nakai
	♂ ← ♀	*Cucurbita pepo* L.
	♂ ← ♀	*Cucumis sativa* L.
	♂ ← ♀	*Spinacia oleracea* L.*
Trauma	♂ → ♀	*Acer negundo* L.*
	♂ → ♀	*Cannabis sativa* L.*
removal of leaves or flowers or crown pruning	♂ → ♀	*Carica papaya* L.
	♂ → ♀	*Cleome spinosa* Jacq.
	♂ → ♀	*Cucumis sativa* L.
	♂ → ♀	*Mercurialis annua* L.*
	♂ → ♀	*Morus alba* L.*
	♂ → ♀	*Musa paradisiaca* L.
removal of storage tissue	♂ ← ♀	*Arisaema japonica* Bl.
	♂ ← ♀	*A. triphyllum* (L.) Torr.

For instance, in *Cannabis sativa*, spring-sown well-grown material expresses the genotypic gender, with males and females in stable 1:1 ratios. However, material grown at low light intensities or in conditions of water stress is very unstable for gender expression, and hermaphrodite, gynomonoecious and andromonoecious individuals all occur. Indeed, the gender genotype may even be reversed in expression, with genotypic females becoming male and vice versa (Schaffner, 1921, 1923). Heslop Harrison (1957) showed that gender expression in this species is governed by auxin concentrations in shoots. Auxin (indole acetic acid, or IAA) is an important plant growth substance, which responds to a number of environmental influences, such as the quantity and quality of light.

In *Mercurialis annua*, it has been found that gender is under the control of auxin/cytokinin ratios. Using gas-chromatography, Dauphin-Guerin, Teller and Durand (1980) were able to resolve this relationship more satisfactorily. They discovered that the cytokinin metabolite zeatin (in the absence of its nucleotide) was particularly associated with the development of floral initials into gynoecia. Various steps in cytokinin metabolism are mediated by the *A* and *B* genes which determine sex in this species (p. 312); for plants in which the gene products of both *A* and *B* are absent, cytokinin metabolism proceeds as far as zeatin, which triggers development of initials past the male stage, towards gynoecial development. Other late cytokinin metabolites induce the production of IAA oxidases, which reduce auxin levels, so that cytokinin/auxin ratios are higher for female initial development than for male initial development.

Cytokinins are metabolized in the root, and various forms of stress, such as drought, tend to inhibit the transport of cytokinin from the root to the developing shoot. This is the physiological expanation for the general finding that stressed plants tend to form male flowers rather than female flowers. As stressed plants are less capable of supporting expensive female reproduction, this functional relationship between stress and male gender has an adaptive significance.

Other dioecious plants in which genotypic gender can be modified by day length and temperature include the hop *Humulus lupulus* and *Rumex hastatulus* (Westergaard, 1958).

An opposite trend is found in many Cucurbitaceae (melons, squashes, cucumbers, etc.). These are typically monoecious, with both male and female flowers on an individual, and they are genotypically hermaphrodite. However, environmental conditions can result in only male flowers or only female flowers being produced, and thus individuals become effectively unisexual. Nitsch *et al.* (1952) showed that high temperatures and long days encourage the production of male flowers, whereas low temperatures and short days promote female flowers. Other plant growth substances such as maleic hydrazide, which encourages male sterility, are known to influence the expression of gender in addition to auxin and cytokinins. No doubt, day length and temperature influence gender expression by causing variations in levels of several growth substances.

Freeman *et al.* (1981) have suggested that those monoecious (hermaphrodite) species which may become phenotypically (but not genotypically) dioecious, such as the Cucurbitaceae (see above) are also more likely to become male under stressed (e.g. xeric) conditions. Under mesic conditions, they tend to become female. This may form a general rule for monoecious 'quasi-dioecious' species (p. 338). Dodson's (1962) classic experiments with epiphytic *Catasetum* orchids, repeated and extended by Zimmerman (1980) reinforce this expectation. Males growing at low light levels become female when moved to a higher light level where they were better able to resource expensive fruits. Moved back into gloomy surroundings they once again reverted to the male condition.

Sex expression can also be altered by disease. It had been known for many years that infection of females of the red and white campions (*Silene dioica* and *S. latifolia* (= *S. alba*) by the smut *Ustilago violacea* induces the formation of anthers. These may produce some pollen, but they are mostly filled with smut spores, thus promoting dispersal of the smut (Fig. 8.3) (Westergaard, 1953, 1958). The smut, which is very common in the British Isles, causes species that are usually fully dioecious to become androdioecious.

Fig. 8.3 Male (left) and female (right) flowers of the dioecious red campion, *Silene dioica*. The central flower, genetically female, has been infected by the smut fungus *Ustilago violacea* which has caused it to produce stamens, the anthers of which are filled with smut spores. Photo by M. Wilson (×2).

The occurrence of homogametic males (YY)

In cases (listed above) in which the phenotypic expression of genotypic sex is unstable, it is possible for crosses to occur between heterogametic (XY) hermaphrodites. This will result in a proportion of the offspring being YY homogametes. In various species of meadow-rue, *Thalictrum* (Kuhn, 1939), and in cultivated *Asparagus officinalis* (Rick and Hanna, 1943) such plants are male, and do not differ in appearance from XY males. In such cases, genes that are vital to the well-being of males do not occur on the X chromosome, which must be primarily concerned with female determination (male sterility).

In other dioecious species, such as *Mercurialis annua* (Kuhn, 1939; Gabe, 1939; Durand, 1963), YY plants are weak and rather sterile; in these the X chromosome probably contains genes which are important to the metabolic efficiency of the male, and indeed to its sexual function. An alternative (or identical) explanation is that disadvantageous recessive genes have become linked to the Y chromosome (p. 309).

GYNODIOECY

Gynodioecy, where females and hermaphrodites coexist, can result from unstable gender expression in a genetically dioecious plant. However, gynodioecy more frequently results from the polymorphic occurrence of male sterile genes in otherwise hermaphrodite populations. It can often also form an intermediate stage in the evolution of full dioecy.

The only evolutionarily stable diclinous conditions are dioecy and gynodioecy, because these are the only conditions usually encountered where females, which can only reproduce by outcrossing, coexist with males (as explained on p. 319, subandroecy (♂ ⚥) (♀) tends to be evolutionarily unstable, not attaining to an ESS equilibrium).

Distribution of gynodioecy

Gynodioecy has been regarded as an important breeding system since Darwin (1877). It has a rather different distribution from dioecy. In temperate floras, such as those in Europe, it is roughly twice as frequent as dioecy. Delannay (1978) shows that 7.5% of the species native to Belgium and Luxemburg are gynodioecious, and lists 223 European gynodioecious species, belonging to 89 genera and 25 families. Similar surveys do not seem to have been made for other temperate floras, such as those of north America. Gynodioecy is a less significant outbreeding mechanism than dioecy in the well-studied oceanic island floras of New Zealand and Hawaii, but it is found in at least seven genera in Hawaii and eleven in

New Zealand. In New Zealand, several important genera have many gynodioecious species, and *Hebe, Pimelea, Cortaderia* and *Fuchsia* among others have attracted attention (Ross, 1978; Godley, 1979). It may be that the evolution of gynodioecy on oceanic islands can be explained by similar arguments to those adduced for dioecy, i.e. selection pressures favoured the development of outcrossers among self-fertile colonizers.

However, in strict contrast to dioecy, gynodioecy appears to be very rare in tropical forests. No examples were discovered by Bawa (1979) in a sample of 309 species of forest tree in Costa Rica. This is consistent with the arguments of Bawa (1980) concerning the high frequency of dioecy among forest trees. If hermaphrodites have their female function disadvantaged by male load, and by self-pollination, hermaphrodites should disappear from gynodioecious populations, leading to a full dioecy.

Stable and unstable gynodioecy

Ross (1978) and Bawa (1980) distinguish between stable and unstable gynodioecy. In stable gynodioecy, male sterility is normally controlled by at least two unlinked genetic factors, one of which is usually controlled cytoplasmically; it seems never controlled by a single nuclear locus (Lewis, 1941), although Stevens and van Damme (1988) show that nuclear-controlled stable gynodioecy can establish when females are better vegetative reproducers than are hermaphrodites.

Stable gynodioecy is widespread within a species, and often constant for that species; within a population, frequencies of hermaphrodites and females show little variation with time. As Lewis points out, stable gynodioecy is a feature of certain families in which dioecy is rare. Stable gynodioecy can be considered as a well-established breeding system, and not an intermediate step in the evolution of dioecy, in such families as the Lamiaceae (mints and relatives), Asteraceae (daisies and relatives) and Dipsacaceae (scabiouses and relatives).

In unstable gynodioecy, male sterility is controlled by a single factor, which may be under nuclear or cytoplasmic control. When the control is nuclear, gynodioecy can represent an intermediate condition towards the evolution of dioecy. If unsuccessful, nuclear male-sterile genes will not become permanently established (unless they are associated with better vegetative reproduction). However, if females are successful, they become better mothers than do hermaphrodites, so that mutations which encourage male function to predominate in hermaphrodites will be favoured, leading to the evolution of subandroecy, and, ultimately, to full dioecy (Fig. 8.4).

There is another reason why gynodioecy with dominant nuclear male sterility is unlikely to become widely established within a species, and this seems not to have been stated before. When new populations are founded

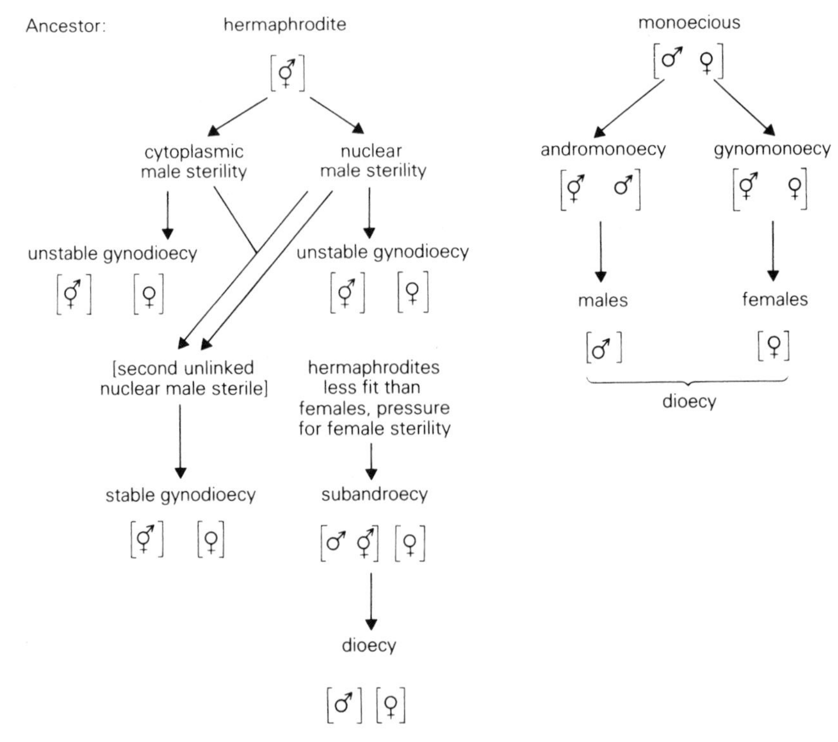

Fig. 8.4 The most important suggested pathways leading to the evolution of gynodioecy and dioecy. (Based on Ross, 1978 and Bawa, 1980.)

by a single seed, this founder must be a self-fertile hermaphrodite, lacking a male sterile gene, so that the new population will not be gynodioecious, at least initially.

Cytoplasmic male-sterile genes cannot become linked to a sex chromosome, and thus such genes cannot lead to the evolution of full dioecy. However, they may lead to stable gynodioecy if associated with nuclear genes which restore male-fertility (Fig. 8.4).

The origin and genetic control of gynodioecy

Genetically controlled male sterility, whether cytoplasmic or nuclear, is a common phenomenon among flowering plants, and thus there is no shortage of variation on which evolution can act to give rise to gynodioecious populations. Note however that male sterility is not always under genetic control; flowers that are produced abnormally early or late in a hermaphrodite flowering season are frequently male sterile, but this condition is not inherited. Also, male sterility can be environmentally induced by chemicals such as methanoproline, synthesized from the fruits of *Aesculus parviflorus*.

Most male-sterile mutants do not persist in populations, so that gynodioecy does not become established in these cases. However, any male sterile mutants can be selected for plant breeding purposes, for they offer ideal opportunities to obtain outcrossed (hybrid) seed which gives rise to vigorous lines. The development of such seed has transformed modern agriculture. In many crops such as maize, rice, barley and tomatoes most seeds sold today are F_1 hybrid lines obtained from true-bred male-sterile maternal lines grown together with different true-bred hermaphrodite lines which act as fathers (Simmonds, 1976). Many male-sterile genes can be collected by breeders, from which a suitable strain can be chosen. For instance, at least 37 separate male-sterile nuclear genes are recognized in tomatoes (Clayberg *et al.*, 1966) and more than 100 have been catalogued for maize. One of these (TS2) is now known to code for a short-chain alcohol dehydrogenase which mediates gynoecial abortion during the early stages of male flower development. When the gene is missing, gynoecial development interferes with anther differentiation, causing male sterility.

Male-sterile genes are inevitably passed to the next generation by the mother. If the inheritance is cytoplasmic, all the offspring of the male sterile mother will be male sterile (female). If the inheritance is nuclear, and male sterility is dominant, the mother is bound to be heterozygous at this locus; half the offspring will be hermaphrodite, and half will be female. If the inheritance is nuclear and recessive, the offspring will either be half hermaphrodite and half female, or all hermaphrodite, depending on whether the father is homozygous or heterozygous (Fig. 8.5). Unlike cytoplasmically inherited male sterility, the offspring of a nuclear-sterile female are never all male sterile.

It is frequently found that interspecific hybrids differ between reciprocal crosses with respect to male sterility, as in *Epilobium* (Michaelis, 1954). Such non-reciprocity is explained by interactions between hybrid nuclei and non-hybrid (maternal) cytoplasm. These findings indicate that cytoplasmic male sterility is frequently modified by nuclear male fertility restorer genes.

Early examples of stable gynodioecy which were studied by Darwin (1877) and Correns (1928) seemed to be controlled entirely by cytoplasmic factors, in such a way that female mothers always gave female offspring, and hermaphrodite mothers always gave hermaphrodite offspring. However, Ross (1978) suggests that a reappraisal of the original data in the light of modern knowledge indicates that interactions between cytoplasmic and nuclear factors always operate in stable gynodioecy. (Nevertheless, Stevens and van Damme (1988) suggest that when cytoplasmically

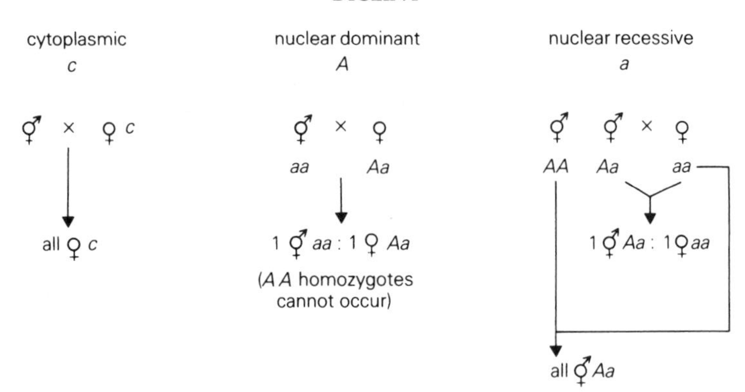

Fig. 8.5 Inheritance of single-factor male sterility.

controlled females are less fit than hermaphrodites, but are better vegetative reproducers, then a stable gynodioecy can establish.)

Ross argues that gynodioecy will only be favoured when females make fitter mothers than do hermaphrodites, for reasons that will be examined below. As argued on p. 319, nuclear-controlled gynodioecy is an inherently unstable condition which favours the evolution of dioecy. However, dioecy can only evolve when female (male sterile) genes form the basis of a sex (X) chromosome. If female factors are cytoplasmic, dioecy cannot evolve.

Where the control of gender is solely cytoplasmic, females will only give rise to females, and if females form fitter mothers than do hermaphrodites, then females will predominate in populations to the point at which their fitness is limited by male (hermaphrodite) frequency. In such a case, the overall fitness of the population is no greater than if the population were entirely hermaphrodite.

However, if cytoplasmic male sterility is modified by polymorphic nuclear restorer genes for male fertility, the cytoplasmic gene can proceed to fixation in the population. Females will mother some hermaphrodite offspring, and female frequency can stabilize at an ESS which will maximize the overall fitness of the population. Ross (1978) predicts that stable gynodioecy should always involve an interaction between a cytoplasmic gene for male sterility and one or more nuclear genes which restore male fertility for females carrying the male-sterile cytoplasm.

Charlesworth and Ganders (1979) show that a stable gynodioecy should not establish when a single gender phenotype (e.g. hermaphrodite) with a constant fitness is controlled by more than one gender genotype. However, these conditions will not apply when the cytoplasmic gene for male sterility is fixed in the population, as will usually be the case.

One-locus nuclear male-restorer systems which interact with a cytoplasmic gene for male sterility have occasionally been reported, as for *Chenopodium quinoa* (Simmonds, 1971). However, two-locus nuclear male-restorer systems with cytoplasmic interactions are probably more common. A number of examples are quoted by Ross (1978), several of which are herbs in the Lamiaceae, e.g. marjoram, *Origanum vulgare*. One case has been analysed in my laboratory by David Stevens (Stevens and Richards, 1985) for the meadow saxifrage, *Saxifraga granulata*. This species is usually hermaphrodite, but Stevens has discovered three gynodioecious populations in upland English hill pastures. This complex system of control of male sterility probably involves a cytoplasmic gene, and two unlinked nuclear loci showing complementation:

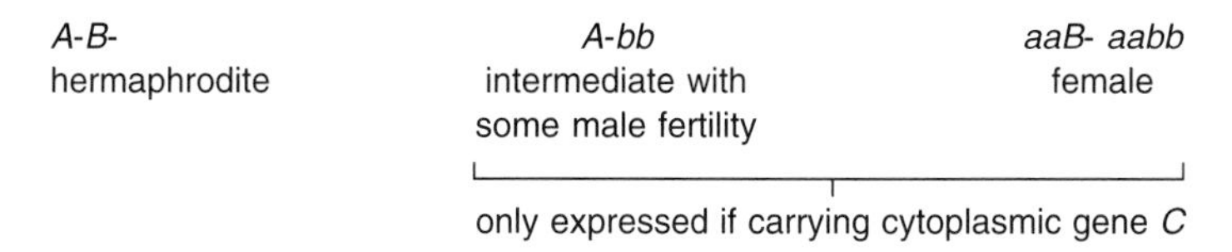

Thus, *A* and *B* interact to override the cytoplasmic gene, but if only *A* or only *B* or neither dominant is present, there is some, or total, male sterility in the presence of *C*, but not in its absence. Offspring sex ratios of controlled crosses of females and hermaphrodites, and of controlled selfs of hermaphrodites fit this model very well.

This complex form of gender control for stable gynodioecy seems to be rather typical. Possibly, such systems have succeeded, for they allow a single hermaphrodite founder to new populations to segregate out female descendants in subsequent generations. Thus, the commonest system reported seems to be a cytoplasmic male sterility/ two unlinked locus nuclear restorer, where some complementation between the restorers characteristically occurs. Typically, selfed hermaphrodites yield 13:3 hermaphrodite:female ratios. This mode of inheritance is explained in Table 8.7.

In the ribwort plantain, *Plantago lanceolata*, dominant complementation occurs, so that 15:1 ratios are obtained from crosses between hermaphrodites (van Damme, 1984). Both alleles controlling male sterility are recessive. This mechanism may be uncommon, as it will tend to result in low female frequencies. For *P. lanceolata*, Krohne, Baker and Baker (1980) reported female frequencies of only 0–6% from inland Californian populations, although in wetter coastal areas females are more common at 2–31% of populations. Various populations around Newcastle upon Tyne, UK, which our classes analyse annually, average at 16.2% females (range 9.2–24.7% females). Female frequency may be density dependent (p. 332), denser populations having more than twice as many females than do sparse populations. In this species, male-sterility is exhibited in

Table 8.7 Explanation of 13:3 ratio of hermaphrodites: females in families from hermaphrodite selfs in stable gynodioecious populations

		Dihybrid *SsMm* selfed		
Gametes	*SM*	*Sm*	*sM*	*sm*
SM	*SSMM*	*SSMm*	*SsMM*	*SsMm*
Sm	*SSMm*	*SSmm*	*SsMm*	*Ssmm*
sM	*SsMM*	*SsMm*	*ssMM*	*ssMm*
sm	*SsMm*	*Ssmm*	*ssMm*	*ssmm*

To explain 13:3 ratios, it must be assumed that two unlinked loci control male sterility in which a dominant allele *M* and a recessive allele *s* interact to give sterility. The other phenotypes MS, mS and ms give male fertility, and will be hermaphrodite. To arrive at this result, two true breeding lines are obtained: (a) which only yields hermaphrodites by removing all female s from the line until only alleles *S* and *m* are left, and (b) which yields only females by crossing females with hermaphrodite siblings until only alleles s and *M* are left. These lines are then crossed to yield the dihybrid of known genotype *SsMm* which is then selfed. Stable gynodioecious systems which nearly always give this result are unable to progress to subandroecy and full dioecy as the female determinants are not linked on a single sex chromosome.

several ways. For instance, males bearing sterile pollen are about twice as frequent as those bearing no pollen in Newcastle populations (Fig. 8.6). Van Damme and van Delden (1984) also reported intersexes in which some anthers are sterile and others fertile. As yet, the developmental genetics of these different male-sterile phenotypes has not been worked out.

The evolution of gynodioecy

Gynodioecy is favoured when females are fitter mothers than are hermaphrodites. Females acquire extra fitness by producing more seed, or by producing seed of better quality (for instance heavier seed), or by mothering fitter (e.g. more vigorous) offspring. Either way, they should mother more viable offspring than do hermaphrodites.

Selection for females will be frequency dependent; if hermaphrodites become too infrequent, female fitness will be limited by a scarcity of pollen, so that hermaphrodite selfing genes should be favoured. However, when females are scarce, pollen supply should be adequate, and the extra fitness of female determining outcrossing genes should allow them to become more frequent. The ESS for female frequency will be deter-

Fig. 8.6 Gynodioecy in *Plantago lanceolata* (ribwort plantain). Inflorescences in this species are protogynous, the flowers opening from the bottom of the spike. The central inflorescence is in the female stage. The other two inflorescences are in the male stage (although with some flowers still in the female stage at the top). That on the right has yellow, narrow, collapsed anthers, which have no pollen, and is female. The inflorescence on the left has plump creamy-white anthers full of pollen, and is a hermaphrodite (×1.5).

mined by the disadvantages of hermaphrodite mothers resulting from selfing and male load on one hand, and the frequency-dependent reproductive disadvantage of females on the other (p. 329).

Selfing hermaphrodites will also possess an automatic genetic advantage compared with crossing females (p. 361), in that selfing hermaphrodite mothers will produce more selfing offspring than outcrossing females produce outcrossing female offspring (Lewis, 1941; Ross, 1978). However, in a stable gynodioecy, where gender is controlled by a cytoplasmic/nuclear system of inheritance, and cytoplasmic male sterility is fixed in the population, this automatic advantage may not be very great. The advantage will depend on

1. the frequency of various male-restorer genes in the population;
2. the modes of interaction between male-restorer genes (e.g. complementation), and between male-restorer genes and the cytoplasmic gene (e.g. dominance).

and is independent of the amount of selfing that takes place in hermaphrodites.

In general, where male-restorer genes are dominant, females will give rise to offspring at least 50% of which are hermaphrodite, whereas hermaphrodites will rarely yield less than 75% hermaphrodites. However, this hermaphrodite advantage is also frequency dependent. When male-restorer genes are very common in the population, and females are rare, the genetic advantage of hermaphrodites becomes disappearingly low (Lloyd, 1975a). In these conditions, very few female offspring are produced by either hermaphrodites or females, but the extra fitness over hermaphrodites for those few females which are produced should be correspondingly high.

The fitness advantage of females over hermaphrodites in a gynodioecious population can be simply stated as

1. their offspring are more outcrossed, and thus they should be more vigorous (heterotic), and more variable (p. 36)
2. the allocation of female resource to fruits and seeds is spared the cost of allocation to male resource

Thus, we would expect gynodioecy to normally evolve in self-compatible, partially selfing species, for in such species with mixed-mating strategies (p. 175) the outcrossing advantage of females will be greater, the more selfed are hermaphrodites. However, if the progenitor species is mostly selfed, gynodioecy is unlikely to succeed, because females would receive little outcrossed pollen. Also, mostly selfed species show little inbreeding depression (p. 390), so that the outcrossing advantage of females becomes very low in such cases. Ross (1978) predicts that gynodioecy should evolve in partially selfed species where the level of outcrossing is nevertheless above 50%.

Estimates of outcrossing onto gynodioecious hermaphrodites do in fact vary from 83% in *Cirsium palustre* (Correns, 1916b) to only 25% in *Cortaderia richardii* (Connor, 1973). Valdeyron, Dommee and Vernet (1977) have shown from 51 to 90% outcrossing in hermaphrodites of different populations of the culinary thyme, *Thymus vulgaris*, in France.

Female/hermaphrodite fitness estimates

It is often found that females set more seed than do hermaphrodites, or that seed produced by females is heavier than is seed produced by hermaphrodites (Table 8.9). Seed size and seed number seem to be entirely under maternal control; apparently, no mechanism exists whereby fitness attributes of the embryo or endosperm can be transferred to the seed (p. 36). Thus, superior female fitness for seed attributes must be prezygotic (p. 330) and largely result from savings made on male resource expenditure.

(It is possible to argue that at high female frequencies, female mothers are more likely to be the heterotic offspring of outcrossed females than are hermaphrodite mothers, but this 'generation effect', if it exists at all, is probably slight.)

Most models for gynodioecy presuppose that all hermaphrodites are equally fit as mothers and as fathers. Moret *et al.* (1992) describe an unusual situation in the gynodioecious bulb *Romulea bulbocodium*, where homogamous hermaphrodites are mostly selfed, but herkogamous hermaphrodites are poor (although mostly outcrossed) female parents, but form much better male parents to females. This condition has yet to be modelled. However, as it fulfils many of the gender fitness requirements of subandroecy, while maintaining a morphological and genetic gynodioecy, it may represent an ESS stable equilibrium.

Postzygotic fitness

Evidence for superior postzygotic female fitness requires a detailed study of the offspring of females and of hermaphrodites at all stages of the life cycle, so Stevens' (1988) findings for *Saxifraga granulata* are of particular interest (Table 8.8). In this species, hermaphrodites are maternally superior to females for all prezygotic attributes, except, interestingly, seed volume. For this attribute, females may be enabled to make larger seeds because they only set about half the number of seeds, on average, that hermaphrodites do (p. 31). There is no other indication that male savings affect prezygotic fitness in this species.

In stark contrast, for all postzygotic parameters, the females of meadow saxifrage produce much fitter offspring than do hermaphrodite mothers. In this case, it seems that nearly all of the superior fitness of females results from heterotic effects, or from an avoidance of inbreeding depression (p. 384). Another clear example of superior female postzygotic heterotic fitness is reported for seed germination in central American *Fuchsias* (Arroyo and Raven, 1975).

The enhanced postzygotic fitness of the offspring of females may be expressed throughout the life cycle of the offspring, and such attributes are rarely studied. The work of Stevens (1988) is unusual in this context, as he shows that the offspring of females, whatever their own gender, produce heavier aestivating bulbils than do the offspring of hermaphrodites. Heavier bulbils are more likely to survive, and are more likely to flower during the following spring.

The vegetative attributes of the offspring of different genders have largely been ignored, but they can be very important. Stevens and van Damme (1988) produce models which show that if the offspring of females have a superior capacity for vegetative reproduction in comparison with those of hermaphrodites, then ESS stable female frequencies can

Table 8.8 Reproductive attributes of a gynodioecious population of *Saxifraga granulata* (only characters that showed statistically significant differences between hermaphrodite and female mothers are listed) (from Stevens, 1985, 1988)

Character means	Hermaphrodite (parent)	Female (parent)	Female: hermaphrodite ratio
Parental characters in the field			
anther (mm)	1.25	0.97	0.776
anther width (mm)	1.05	0.70	0.666
filament length (mm)	12.91	11.83	0.916
petal length (mm)	11.09	10.61	0.957
petal width (mm)	5.91	5.50	0.931
seed/ovule ratio	0.381	0.217	0.570
seed/capsule ratio	203.68	118.85	0.583 (= *f*)
seed volume ($mm^3 \times 10^{-3}$)	4.24	4.96	1.170
style length (mm)	4.27	4.83	1.131
Offspring characters in cultivation			
final seed germination (%)	44.0	57.6	1.309[a]
survival of seedlings to flowering (%)	59.3 ($n = 123$)	76.3 ($n = 156$)	1.287
proportion of seedlings flowering after 1 year (%)	75.3 ($n = 73$)	84.0 ($n = 119$)	1.116
weight of oversummering bulbils of seedlings from different mothers (g)	1.49	1.85	1.242

[a]This difference was not statistically significant at 5%.

develop even when females are less fecund than are hermaphrodites. In such conditions, ESS stable female frequencies can even develop where the control of male sterility is nuclear only, or cytoplasmic only.

As stated on p. 325, the ESS frequency equilibrium for females in a cytoplasmic/nuclear stable gynodioecy should be determined by hermaphrodite fitness disadvantages on one hand, and frequency-dependent female reproductive disadvantages on the other. Unfortunately, gynodioecious systems which are under cytoplasmic/nuclear control have proved difficult to model, as ESS equilibria will depend on the precise mode of cytoplasmic/nuclear genetic interaction involved (p. 325). Thus, regrettably, the models of Charlesworth and Charlesworth (1978), for systems which are under nuclear control only, or under cytoplasmic control only, are hardly relevant to stable gynodioecy.

For systems under nuclear control only, Charlesworth and Charlesworth predict equilibrium proportions (p) of females by:

$$p = \frac{(f + 2sd - 2)}{2(f + sd - 1)}$$

where f is seed number per genet in females compared with that in hermaphrodities, s is the proportion of seeds that result from self-fertilization (on a scale from 0 to 1) and d is the inbreeding depression, expressed as:

$$1 - \frac{(\text{the fitness of the offspring of hermaphrodites})}{(\text{the fitness of the offspring of females})}$$

For a male-sterile gene to spread, Charlesworth and Charlesworth predict that $f > 2 - 2sd$ is necessary for nuclear-inherited male sterility, but that $f > 1 - sd$ is required for cytoplasmically inherited male sterility.

For the commonly encountered stable cytoplasmic/nuclear systems, f should lie between these values (in a position determined by the genetic nature of the cytoplasmic/nuclear interaction) for an ESS female frequency to become established.

If, as it seems, values of s (amount of selfing in hermaphrodites) commonly approximate to 0.5, where f values (prezygotic advantage to females resulting from freedom of male load) are less than 1 (Table 8.8), then values for d (relative postzygotic inbreeding depression of the offspring of hermaphrodites that arise from selfing) should lie between 0.5 and 1.2. That is hermaphrodite offspring should be somewhat less fit (or possibly almost as fit) as those of females, for a female frequency equilibrium to become established.

When prezygotic f values are greater than 1, so that freedom of male load allows females to set more seed than do hermaphrodites, an ESS for female frequency can establish when values for d (postzygotic inbreeding depression of hermaphrodite offspring) are as low as 0.2.

There are few instances for which we have adequate estimates for f and for sd so that attempts can be made to fit these to the Charlesworths' equations (which do not, however, deal with cytoplasmic/nuclear systems). Such a case is the meadow saxifrage, *Saxifraga granulata* (Stevens, 1988, Table 8.8).

For prezygotic fitness estimates of f

female seed-set relative to hermaphrodites = 0.583
female seed-weight relative to hermaphrodites = 1.17

estimate of f = 0.682

For postzygotic estimates of sd

seed germination from female mothers relative to that of hermaphrodite mothers = 1.31
seedlings from female mothers surviving relative to those from hermaphrodite mothers = 1.29

Reciprocal of product of female postzygotic fitness = sd = 0.59

Thus, if these estimates for f and for sd are substituted into the Charlesworths' equation

$$p = \frac{(f + 2sd - 2)}{2(f + sd - 1)}$$

$p = 0.25$, or 25% of the population might be expected to be female.

In fact, for the population from which these estimates were made, exactly 25% of the population were found to be female (Stevens, 1988). This is surprising, for it suggests that the genetic control of this complex cytoplasmic/nuclear system (p. 323) operates in a manner very similar to that with a simple nuclear control, for which the Charlesworths' model was designed. However, for the other two gynodioecious populations known in this species, the proportion of females was found to be 20% and 4%. There may be frequency dependent seed-set, so that where females are rare in this species (e.g. 4%), estimates for f may be higher, leading to lower expectancies for the frequency of females p.

Prezygotic fitness

Prezygotic advantages to mothers result from freedom from male load, and should be expressed in reproductive performance only. Such fitness attributes are most readily studied with respect to seed number and seed weight (Table 8.9). As van Damme and van Delden (1984) point out, such detailed investigations often show little if any fitness advantage to female mothers. However, as their own work shows, prezygotic female advantage may not only be expressed through such simple reproductive characters.

Although nearly all species in which gynodioecy becomes established are partial selfers when they are hermaphrodite (as is the case for *S. granulata*), this is not the case for *Plantago lanceolata* which is fully self-incompatible (Ross, 1973). In such cases, when there is no selfing and no inbreeding depression, $sd = 1$ and female advantage can only be expressed prezygotically, as freedom from male load (f).

Van Damme and van Delden (1984) thoroughly examined the fitness attributes of females and hermaphrodites at all stages of the life history of this species. They show that females are no fitter than are hermaphrodites until they start to flower. However, adult flowering females have no less than six times the life expectancy of that of hermaphrodites, and they set both more seeds, and in some cases, heavier seeds than do hermaphro-

Table 8.9 Reproductive performance of females in gynodioecious species, expressed as a ratio to hermaphrodites (based on Godley, 1979 and Stevens, 1985)

Species		Flowers/ plant	Ovules/ flower	Seed/ ovule	Seed/ fruit	Seed/ plant	Weight/ seed
Thymus vulgaris (Assouad *et al.*, 1978)		1	1	2.24–3.25[c]	2.24–3.25[a,b]	3.14	–
Plantago lanceolata (Van Damme, 1984)	MS1	1.69[a]	1	1[c]	1	1.69[c]	1.15[a]
	MS2	1.69[a]	1	0.70[c]	0.70[a]	1.18[c]	0.90[a]
Stellaria longipes (Phillip, 1980)		1.90[b]	1	0.92[c]	0.84	(1.60)	–
Geranium sylvaticum (Vaarama and Jaaskelainen, 1967)		0.69	–	–	1	(0.67)	>
Hirschfeldia incana (Horovitz and Beiles, 1980)		–	1.08	1.03[c]	1.11	–	1.17
Leucopogon melaleucoides (McCusker, 1962)		–	1	1	1	–	–
Origanum vulgare (Lewis and Crowe, 1956)		–	1	1.19[c]	1.19[a]	–	–
Iris douglasiana (Uno, 1982)		–	1	1[c]	1	–	0.95
Saxifraga granulata (Stevens, 1985)		1	1	0.57[a]	0.58[a]	0.54[c]	1.17[c]

[a]Stated to be significant at $p < 0.05$ (results stated to be insignificant recorded as 1, otherwise reported as given by the authors).
[b]From cultivated plants.
[c]Calculated directly from other values in this table.
– No data.

dites. However, the extra seed set by females over that set by hermaphrodites is solely a function of the greater number of flowers produced by females. A very similar proportion of ovules are fertilized and develop in both females and hermaphrodites (Table 8.10).

These findings are precisely what would be expected in a self-incompatible gynodioecious species in which female advantage is only prezygotic. Seeds, seedlings and young plants should have no advantage if their mother was female, as the offspring of females and of hermaphrodites would be equally of outcrossed origin. However, a flowering adult female, freed of male load, should be able to transfer more reproductive function to female effort, thus being able to produce more flowers and

Table 8.10 Prezygotic and postzygotic fitness attributes in *Plantago lanceolata* (after van Damme and van Delden, 1984)

	Female	Hermaphrodite	
Prezygotic			
Half-life of adults (years)	13.7	2.2	***
Flowering spike length (mm)	8.59	7.32	***
Seeds/mm spike	4.15	4.91	ns
Seed production/plant	77.7	64.7	***
Postzygotic			
Number of seedling leaves	2.49	2.57	ns
Leaf length (mm)	26.2	24.6	ns

seeds. In this case, it seems that it is also enabled to survive considerably longer. In this wind-pollinated species, the male component of reproductive effort must be very considerable.

Where hermaphrodites are self-incompatible, as for *P. lanceolata*, the ESS equilibria for female frequency will depend entirely on how f values vary at different female frequencies (when females become common, they will suffer from a pollen shortage, and f values will drop). This relationship is also density-dependent, for pollen supply in this largely wind-pollinated species will become more reliable at a high density, at which female ESS frequencies should be higher (p. 323).

ANDRODIOECY

In contrast to gynodioecy, androdioecy, where males and hermaphrodites coexist, seems to be a very rare phenomenon, although reported long ago by Correns (1928) in *Pulsatilla* and *Geum*. It has been suggested by Lloyd (1975a) that androdioecy is unlikely to be successful, as all females, being male as well, are not protected from selfing, and 'spare males' suffer from having no female fitness. Only in conditions of pollen shortage, where an increased reproductive fitness of females is commensurate with the 'wasteful' production of 'spare males', is androdioecy likely to be favoured. However, in a wind-pollinated species which does not self automatically, if males produce far more pollen than do hermaphrodites, their pollen will dominate the pollen pool, thus also increasing levels of outcrossing.

The first proven examples of androdioecy are very recent (Liston *et al.*, 1990; Molau and Prentice, 1992). The best-known example is for the

American cannabis-like herb *Datisca glomerata* (Fritsch and Riesenberg, 1992; Riesenberg *et al.*, 1992, 1993), whose nearest relative, the Eurasian *D. cannabina*, is dioecious. Hermaphrodity results from the presence of one or both of two linked dominant alleles; males are double recessive homozygotes. This accords with the proposal that androdioecy in *D. glomerata* has evolved from dioecious *D. cannabina*, rather than from some hermaphrodite common ancestor, hermaphrodites rather than males being heterogametic having 'evolved last' (p. 309). Possibly *D. cannabina* has homogametic males (p. 308) although this does not match an old report.

As might be predicted, male *D. glomerata*, which gain fitness through male function only, but have no maternal costs, produce more than three times as much pollen on average as do hermaphrodites. The offspring of hermaphrodites experience inbreeding depression after selfing, and outcrossing rates onto females are high (in excess of 62%). Doubtless, fecund 'spare males' contribute very significantly to the pollen pool in this wind-pollinated species and boost levels of outcrossing. In these conditions recessive male genes will be favoured over pollen-borne hermaphrodite genes as their offspring will tend to be fitter and will maintain males in the population at an ESS equilibrium frequency dependent on the proportion of male pollen in the pollen pool, selfing rate and the level of inbreeding depression experienced after selfs.

The second recently proven case of androdioecy was even more unexpected. The widespread arctic drooping saxifrage *S. cernua* chiefly reproduces apomictically, by means of inflorescence bulbils, and many populations lack flowers. However, in Abisko, northern Sweden, floriferous populations are androdioecious (Molau and Prentice, 1992). Males predominate, often to the exclusion of any hermaphrodites in subpopulations, and are rendered female-sterile by the absence of stigma papillae. Males and hermaphrodites do not differ in ovule number, pollen number or pollen fertility (typically for a high polyploid apomict, pollen is rather sterile, averaging less than 50% stainable). Males produce more bulbils than do hermaphrodites, perhaps as a result of a saving on female load. Remarkably, hermaphrodites are completely self-incompatible.*

It is clear that in this species, androdioecy does not promote outcrossing. Rather, it seems that vegetative apomicts have increased asexual fitness by shedding their reproductive female load, while still being able to fertilize hermaphrodites. Although the genetic control of androdioecy is not known in this case, males are once again likely to be recessive; if dominant the female-sterile mutant should rapidly overtake the entire population. This example provides an androdioecious parallel to the prediction of Stevens and van Damme (1988) (p. 327) that vegetative repro-

* Brenda Jones (unpublished) has recently demonstrated that androdioecy is widespread in another pan-boreal herb with vegetative reproduction, *Lloydia serotina*.

duction should aid the persistence of gynodioecy. It also gives us a vegetative example of the role played by male fertility among apomicts (Mogie, 1992, p. 439).

There are a few tropical fruit trees, for instance the rambutan, *Nephelium lappaceum*, for which androdioecy has also been reported (Simmonds, 1976). It is possible that this and other apparently androdioecious species have sterile pollen in 'hermaphrodites' and are thus functionally dioecious, as Kevan and Lack (1986) have shown for *Decaspermum parviflorum* in Indonesia. Anderson and Symon (1988) show that nine apparently androdioecious Australian *Solanum* species are in fact dioecious, because morphological hermaphrodites have non-functional, non-porate, pollen. Related Australian *Solanums* are andromonoecious, bearing one hermaphrodite flower and many male flowers, suggesting that where pollination is effected by pollen feeders, diclinous mechanisms are only likely to succeed in cases where pollen supply is ensured, and even heavily promoted, in functional females.

Working with the Mediterranean shrub, *Phillyrea angustifolia*, Lepart and Dommee (1992), show that hermaphrodite pollen does not function on any stigmas, although male pollen does, so that one morphologically androdioecious population is in fact functionally dioecious. (However, in a second, introduced population in the Camargue, there is some hermaphrodite pollen function, so that this population is truly androdioecious, possibly because some plants have been introduced from wholly hermaphrodite populations.)

Otherwise, androdioecy probably occurs only as a cytogenetic accident, as in some strains of *Mercurialis annua* (p. 311), or as the result of pathogen attack, as in *Silene dioica* (p. 317).

SUBANDROECY, TRIOECY, POLYGAMY AND SUBDIOECY (WHERE SOME FLOWERS ARE HERMAPHRODITE)

Subandroecy

As shown on p. 319, an instability in the expression of genotypic gender results in females which produce some flowers with male function (subgynoecious (♂) (⚥ ♀)), or males which produce some flowers with female function (subandroecious, (♂ ⚥) (♀)) (Thomson and Brunet, 1990).

Subandroecy seems often to represent an intermediate position between unstable monofactoral gynodioecy, and full dioecy (Fig. 8.4). Females occur together with hermaphrodites with reduced female function, so that some flowers on hermaphrodites are entirely male.

In Western Australian populations of the tiny lily relative, *Wurmbea dioica*, Barrett (1992) shows that mesic populations tend to be hermaphrodite, but populations in stressed environments are subandroecious. In these, there was a significant relationship between the frequency of females and the frequency of pure males. These populations, which seem clearly to show an evolutionary sequence from hermaphrodity to dioecy via the subandroecious route, would well repay a detailed analysis as to the environmental influences which favour full dioecy.

Another cause of subandroecy is the unstable expression of genetically (X/Y) controlled full dioecy due to environmental modifications (p. 312). Subandroecy will occur in a dioecious species, particularly when flowering during unusual day-lengths, causing male plants to produce some gynoecia. Similarly, abnormal environmental stimuli may cause genetically female plants to produce some stamens (subgynoecy), or both males and females may show unstable sex expression (polygamy).

Male sterility almost always becomes established before female sterility, because when females mutate from a hermaphrodite selfer they produce advantageously outcrossed offspring, but newly arising males have no such advantage, and are unproductive. A newly arising gynodioecy will be unstable (p. 319), favouring the development of subandroecy, with reduced female function in hermaphrodites.

Unlike subandroecy, subgynoecy rarely if ever forms an intermediate stage between hermaphrodity and dioecy, for it would evolve from androdioecy, which itself rarely if ever occurs (see Fig. 8.4). Subgynoecy and polygamy usually result from unstable sex expression in a genetically dioecious species.

Subandroecy may be obvious, as in the New Zealand umbellifer *Gingidia montana*, in which primarily male plants often produce a few fruits (Webb, 1979a; Lloyd, 1980a) or it may be cryptic. Arroyo and Raven (1975) showed that two American *Fuchsias*, *F. thymifolia* and *F. microphylla*, are not gynodioecious as they superficially appear to be. In both species, about 90% of the apparently hermaphrodite flowers are female sterile (although bearing full-formed gynoecia), and are thus male. Cryptic subandroecy may also occur in apparently dioecious species. In the familiar pest creeping thistle, *Cirsium arvense*, Kay (1985b) finds that occasional male-fertile florets bear achenes, although most are male only. Fully female florets never show sporadic male fertility, but in South Wales, Kay has encountered occasional clones which appear to be fully hermaphrodite. This latter case is probably best described as partially polygamous, although variably male females are absent.

Trioecy, polygamy and subdioecy

Polygamy is a condition where both male and female expression is unstable, so that variably female males and variably male females coexist. In some cases, full males, full females, and balanced hermaphrodites may also occur in the population, but this is very unusual. A good example is described by El-Keblawy *et al.* (1995) for the Mediterranean subshrub *Thymelaea hirsuta*. Here, males, females, protandrous hermaphrodites (in which male flowers are produced first) and protogynous hermaphrodites (in which female flowers are produced first) all coexist. In fact, a total of no less than nine gender conditions are reported for this species, but the others are scarce, and possibly less stable. El-Keblawy *et al.* (1996) show that gender phenotype frequencies change on a cline from Egyptian coastal localities to those inland, and that the commonest phenotypes in a locality tend to produce seed which is fittest among the gender phenotypes for that locality.

More usually, polygamy represents a condition where subandroecy and subgynoecy can be said to coexist. It may occasionally represent evolutionary progressions towards a full dioecy, where male and female suppressors are not fully operational. More commonly, it probably results from environmentally labile gender expression (p. 312), although there is not a clear distinction between these two concepts.

Subdioecy differs from polygamy only in degree. Plants are mostly male, or mostly female, but one or both genders occasionally produces hermaphrodite flowers, or flowers of the other sex. A number of examples of what is best described as subdioecy are reported by Kay and Stevens (1986), as for instance in several species of willow (*Salix*).

In trioecy (which is also rare), only pure males, pure females and balanced hermaphrodites coexist. Trioecy probably operates when both male sterile and female sterile mutants are polymorphic in a population, but unlinked, so that when both are missing, hermaphrodites occur. Trioecy may represent a stage in the evolution of full dioecy (see Fig. 8.4), but it is unlikely to persist once male-sterile and female-sterile factors become linked. In contrast, subandroecy and polygamy typically show unstable sex expression, with both male and hermaphrodite, or less commonly, hermaphrodite and female flowers on the same plant.

Darwin (1877) claimed that both trioecy and polygamy occur in the European spindle, *Euonymus europaeus*. Webb (1979b) concludes that populations of this species introduced into New Zealand are in fact subandroecious, for most males are variably female fertile. However, this species should be re-examined in its native area.

Probably the best example of true trioecy, in some populations, is the moss campion, *Silene acaulis*. This widespread arctic-alpine varies in gender expression throughout its range. Many plants in the Alps have uniformly hermaphrodite flowers, but in the high arctic (as in north Greenland) populations are usually fully dioecious. In the British Isles, almost every population seems to differ in the gender spectrum. Some, for instance in Shetland, are reported as being truly trioecious, and this seems to be the case for some mainland populations, such as those on Beinn Eighe. However, in most Scottish populations females produce a little pollen, and some males produce occasional fertile capsules, so that the description by Willis and Burkill reported in Kay and Stevens (1986) as 'polygamotrioecious' is as accurate as any.

Another species which may occasionally be trioecious is the rose-root, *Rhodiola rosea*, as in some Greenland populations, quoted in Kay and Stevens (1986).

MONOECY

Monoecy describes the condition where flowers are of a single gender, but individuals bear flowers of both genders. Thus plants, but not flowers, are hermaphrodite. Monoecious plants are more susceptible to changes in gender than are those with hermaphrodite flowers. Gender change in plants with hermaphrodite flowers frequently involves genetic modification (male-sterile or female-sterile mutations). However, where flowers are unisexual, a modification or change in the gender of the individual can readily result from environmental stimuli which influence chemical messages, or merely from the size or age of the plant.

In supposing that the original flowering plant bore hermaphrodite flowers (p. 68), we must consider monoecy to be a derived condition. Some conditions, such as gynomonoecy, where plants bear both hermaphrodite and female flowers, and andromonoecy, where plants bear both hermaphrodite and male flowers, presumably represent evolutionary stages in the progression from floral hermaphrodity to monoecy. These conditions are often found in plants with umbellate or capitulate flowers (p. 137) and can be interpreted as adaptations which allow gender specialization to be expressed within an insect-pollinated hermaphrodite through a division of labour (p. 300).

In the case of andromonoecious species in diverse forest habitats, it may pay functional hermaphrodites to advertise strongly for pollen-feeding insects by the provision of many 'spare' male flowers (Anderson and Symon, 1989). However, Primack and Lloyd (1980) note that for the insect-pollinated New Zealand andromonoecious shrub *Leptospermum scoparium*, poorly resourced plants produce a greater proportion of male flowers. They suggest that in this case andromonoecy may have become successful as a means by which plants can readily escape from excessive female costs at times of stress.

The evolution of monoecy from floral hermaphrodity presumably involved mutations which allowed male sterility and female sterility to be co-expressed developmentally and spatially within the same individual. Thus, it is usually found that there is a distinct pattern in the arrangement of male and female flowers within the plant. For instance, in the majority of sedges (*Carex*) female spikes or flowers are borne proximally and male flowers or spikes are borne distally (Fig. 4.41). It is scarcely surprising that such developmentally triggered gender expression is labile, responding to age or the environment.

In fact, the evolutionary genetics of monoecy, unlike those of dioecy and gynodioecy, seem to be largely unstudied, and would seem to form a potentially fruitful field for investigation. Those rare examples where hermaphrodity can be expressed either within the flower, or between flowers, should form potent tools for this investigation. The mostly dioecious nettle *Urtica dioica* provides such a system (Pollard and Briggs, 1984). In the mostly monoecious cucumber (*Cucumis sativa*), Malepszy and Niemirowicz-Szczytt (1991) identify four different genes which influence the expression of hermaphrodite flowers (i.e. causing andromonoecy or polygamy), so the genetic control of monoecy is not necessarily simple.

Although some plants with compound inflorescences have developed a full monoecy (below), monoecy is most usually found in plants with an innately inefficient and expensive pollination system such as those which use the wind or water, e.g. trees, sedges and waterweeds (p. 88). As Sutherland and Delph (1984) show, monoecy is actually a more reproductively efficient system for outcrossers than is the hermaphrodite flower, presumably because male and female resource allocation are separated in space and time within the plant (p. 41). Mazer (1992) points out that floral hermaphrodity is a more stable condition, however. Monoecy is most likely to have become evolutionarily successful in those plants where the innate lability of gender expression actually becomes advantageous to the plant.

For instance, in many monoecious trees, young plants mostly or entirely bear male flowers, whereas the preponderance of flowers borne by large old trees are female (Freeman *et al.*, 1981; Goldman and Willson, 1986). Large trees are better resourced to carry the load of a large fruit crop, and so the reproductive efficiency of a related guild is increased (Charnov, 1984). Equally, stressed plants, for instance those in more xeric sites, may have more male flowers than those in mesic sites, as for *Acer grandidentatum* (Barker, Freeman and Harper, 1982). Similar results were obtained for three other North American trees by Freeman *et al.* (1981). Possibly, the adjustment of male/female function ratios to environmental circumstances may be controlled by cytokinin/auxin ratios, as is the case for the mostly dioecious *Mercurialis annua* (p. 316).

Trees are not the only plants to register gender diphasy with size, age or stress. In the Araceae, the specialized trap-flower pollination system (p. 83) requires monoecy in order to function. In the much-studied *Arisaema triphyllum* (review in Schlessman, 1988), plants change from male to female as they age, but after a successful (and expensive) fruiting season, females are likely to revert to a male state, or to become vegetative. Approximately three times as much resource is allocated to reproduction by a fruiting female compared with a male; in the former case almost half its biomass. Even epiphytic orchids such as *Catasetum viridiflavum* tend to be male when small and in low light conditions, but become female when higher light levels allow them to grow larger (Zimmerman, 1980).

The evolution of dioecy from monoecy

In New Zealand species of *Cotula* (now *Leptinella*), in the Compositae, florets are usually unisexual; in hermaphrodites the outer florets in a head are generally male and the inner female (monoecious). When bisexual (hermaphrodite) florets occur, they do so peripherally; the inner florets are always female (gynomonoecious) (Lloyd, 1972a,b, 1975b).

All gradations between gynomonoecy and full dioecy are reported in *Cotula*. In some plants the number of peripheral male florets becomes very reduced and inconstant; in others the number of central female florets becomes reduced and inconstant. Plants with few or no female florets tend to have longer florets, perhaps as an aid to pollen transfer. A great deal of variation in gender distribution occurs between populations, even within a species. However, in general, the degree of gender specialization within a population is symmetrical between the genders. Thus, gynodioecy and androdioecy do not usually occur, nor do subgynoecy or subandroecy. The intermediate states between monoecy and dioecy that occur are best termed subdioecious or polygamous.

Inconstancy of sex expression is not under environmental influence in these plants, and often as much inconstancy can be found at the same time within one individual as in the whole population. Some populations which appear to be fully dioecious in one year may prove to have an inconstant gender control in the next. This has allowed Lloyd (1980a) to suggest that dioecious *Cotula* have reverted to monoecy on at least three occasions, perhaps in response to the rarity or extinction of one gender. Populations of alpine scree *Cotula* are frequently very small and subject to catastrophe, and individual clones can spread over a large area and live to a considerable age. In these circumstances, an enforced unisexuality would encourage a sexually inconstant system to evolve.

In *Cotula* there is a conventional XY genetic control of sex, so that crosses between males and females give male and female offspring at ratios approaching unity. However, females rather than males are heterogametic. Lloyd (1975b) suggests that unisexuality in a head is achieved by control of timing of the sequential production of male and then female florets, rather than by genetically controlled sterility. Doubtless this provides an explanation that satisfactorily accounts for the ready reversion to monoecy in this genus. It also explains why femaleness does not precede maleness in the evolution of dioecy in this case; pressures favouring outbreeding do not lead to gynodioecy, with females and hermaphrodites, but to gynomonoecy in which females still have some male function. Sexual selection should encourage maleness to develop in andromonoecious lines in direct correspondence to the development of femaleness in gynomonoecious lines. Sexual selection for gender function should result in the eventual evolution of full dioecy, as long as dioecy remains reproductively efficient.

Bawa (1980) and Ross (1978, 1982) have recognized the symmetrical development of dioecy from monoecy, as exemplified by *Cotula*, as one of five mechanisms by which dioecy is potentially able to evolve. Frequently, hermaphrodites are symmetrically protandrous (male flowers develop before female flowers) and protogynous (female flowers develop before the male on an individual, as for *Thymelaea hirsuta* (El-Keblawy *et al.*, 1995). Earlier authors such as Cruden and Hermann-Parker (1977) and Cruden (1988) have identified such 'temporal dioecism' as a potential evolutionary mechanism whereby dioecy can evolve from monoecy.

However, with respect to the evolution of dioecy from hermaphrodite flowers, by far the most important is the asymmetrical mechanism involving male and then female sterile mutants, via gynodioecy and subandroecy (see Fig. 8.4).

Artificial selection for monoecy from dioecy

When the environmental modification of gender allows genetic unisexuals to become monoecious hermaphrodites (p. 312), the selfing of genetically male hermaphrodites sometimes results in progeny with females and 'males' at the ratio 1:3 (as in the economically important *Asparagus officinalis*). In this case, this strongly suggests that 'males' are XY which give 1 XX (females): 2 XY: 1 YY (males) when selfed. The assumption that YY 'males' are viable and fertile is confirmed as some 'males' when crossed to females (XX) yield only 'males' (XY). The identification of YY 'males' (actually andromonoecious hermaphrodites) in *Asparagus* has allowed commercial breeders to raise offspring consisting entirely of the higher-yielding 'males' (p. 347).

In the grape, *Vitis vinifera*, complex patterns of inheritance occur in what was probably a fully dioecious species before cultivation. Monoecious mutants have been favoured by man, for males also hear grapes, and do not take up valuable space non-productively. The following types of offspring result from the selfing of monoecious hermaphrodites in different grapes:

1. all hermaphrodite	?hermaphrodites are XX due to breakdown of male sterility in females
2. 3 hermaphrodites: 1 female	XY and YY hermaphrodites, XX females

3. 9 hermaphrodites: 3 females: 4 males	two-locus unlinked dominant control with complementation

In grapes, monoecy seems to have arisen from dioecy on at least three occasions, and by three different mechanisms:

1. loss of male sterility in females (XX hermaphrodites);
2. loss of female sterility in males (XY and YY hermaphrodites);
3. loss of linkage of male and female sterility (? translocation on primary sex control onto an autosome).

Selection for monoecy by man has also occurred in another important crop that was originally dioecious, *Cannabis sativa*. The genetic gender of individuals in this species can be judged by a number of secondary sexual characters, or by chromosome karyotype, for Menzel (1964) has distinguished XY from XX karyologically. At least some monoecious hermaphrodites are genetically female XX, and thus there may have been a reversion of male sterility in some strains. The picture is, however, confused by the influence of the environment on the expression of genotypic gender. Selfing of monoecious or andromonoecious males can result in many types of sex expression in the offspring, as is the case for grapes; translocation of the primary male and female factors onto unlinked autosomes has clearly occurred, and some secondary sex characters also show segregation, suggesting that they have also been translocated onto autosomes in some hermaphrodite hemps.

Yet another crop plant, spinach (*Spinacia oleracea*) is normally dioecious with equal proportions of male and female plants. Many gradations of monoecism and hermaphroditism are known in this species, which has enabled man to select 'highly male' and 'highly female' lines. These lines can subsequently be used to produce 'hybrid' varieties (Simmonds, 1976, see also Janick and Stevenson, 1955).

Clearly, man has tended to find dioecy an inconvenient trait in crop plants, and has selected for hermaphrodites of several different types of origin, even within one species.

SECONDARY SEX CHARACTERS

We are accustomed to the idea that male and female animals not only differ by possessing testes or ovaries, but also often have very different external appearances. Such secondary sex characters are not directly concerned with the primary functions of gender, mating and the bearing of offspring although they may often mediate mate choice.

In the case of the bright plumage assumed by male birds such as ducks, pheasants, finches, chats, birds of paradise, etc., mating is achieved after females select among displaying males. Brilliant plumage, and/or vigorous and complex display, song etc. have evolved as cues whereby females can assess the genetic and environmental potential of a mate and his territory. In this way, sexual selection may directionally select for fantastic features that are selectively disadvantageous in other ways.

Among unisexual animals, the relative 'disposability' of males is also commonly observed. Males do not bear young. As a result they usually have lighter reproductive loads than do females, and often they are evolutionarily irrelevant once mating has occurred. Thus, where the evolutionary dichotomies of gender result in differential survivorships, successful strategies should normally assign the lower survivorship to the male, as follows.

1. Males are usually hemizygous for the X chromosome so that X-linked recessive mutants are exposed to selection.
2. Males usually bear the Y chromosome which acts as a non-recombinable target for harmful mutations through 'Müllers ratchet' (p. 23) (it is noteworthy that in birds and butterflies where the female is the heterogametic gender, the Y chromosome is either missing, or is completely inert).
3. Males commonly bear conspicuous (non-cryptic) colours involved in mating rituals, so that they are more liable to predation. It follows that choice for mate quality is usually made by females, rather than by males.
4. Where genders are diphasic for Batesian mimicry, females rather than males tend to be mimics.
5. Where prey-specialization has encouraged the development of gender-based size-dimorphism, and the smaller gender has the poorer survivorship, that gender is usually male (as in birds of prey).
6. In extreme cases (mantis, many spiders) the very much smaller male is eaten by the female after mating.

Unlike many animals, plants do not consciously choose their mates. Thus, conspicuous secondary sex characters relating to territoriality or mate choice are lacking.

However, secondary sexual characters in plants do occur and have been recognized since the time of von Mohl (1863) and Darwin (1877). Lloyd and Webb (1977) have produced a useful later review. In general, such characters should have evolved in response to one of the following strategies:

1. A reduction in competition between genders, allowing genders with different reproductive loads to coexist;
2. A differentiation in the distribution of reproductive load, so that the gender with the higher reproductive load (usually the female) can compete successfully with the other gender;

3. A differentiation in the supply or distribution of rewards or signals to pollinators, so that female plants, which lack pollen, are nevertheless visited by pollinators.

Not all morphological features by which males and females differ are evolutionary features which are under genetic control however. Some may well be phenotypic responses to different reproductive loads. The distinction between these non-genetic phenotypes and those which are controlled by genes linked to the sex chromosomes is a continuing problem. It would be possible to test whether secondary sex characters are under sex chromosome-linked genetic control in some cases by making crosses between dioecious plants and related hermaphrodites, but to my knowledge these tests seem never to have been made for any plant.

Secondary sex characters in plants are usually expressed more subtly than are those in animals, and usually concern patterns of growth, resource allocation, timing or longevity. There are exceptions, for instance in *Cannabis*, for which male and female plants have a different overall shape. Other such cases are reviewed by Ornduff (1969). According to Lloyd and Webb (1977), secondary sex characters have been found in all dioecious species which have been carefully investigated. Most of these can be interpreted as adaptations to, or effects of, the different reproductive loads of the sexes. Only in a few cases do morphological differences between genders become obviously apparent before flowering commences.

Flower size, number and reward

One of the commonest expressions of secondary sexual characters is in flower size, for which males and females commonly differ (Delph, 1996). For this character, there is an interesting general distinction between gynodioecious and dioecious systems.

Darwin (1877) showed that in many gynodioecious species, females tend to have smaller flowers, and Baker (1948) provides a list of no less than 73 species of the north-west European flora for which female flowers are smaller than hermaphrodite flowers (some are gynomonoecious). In contrast, he only provides seven examples of fully dioecious species with smaller female flowers. Bawa and Opler (1975) found that 14 out of 20 dioecious Costa Rican species studied have larger female flowers. More recently however, Delph (1996) has analysed 436 dioecious and monoecious species, and has shown that there is a gender-linked flower size difference in no less than 85% of cases, but that there was a slight majority of cases in which male flowers were larger than female flowers. However, in temperate localities, almost three times as many cases had larger male flowers, rather than larger female flowers, but in tropical localities, cases with larger male flowers were equal in number to those with larger female flowers. Nevertheless, it does seem, even in temperate localities, that there is a greater chance of gynodioecious, rather than dioecious, species having larger male (hermaphrodite) flowers.

One possible explanation for this distinction depends on the expectation that gynodioecious females, being freed from male load, should have a lighter reproductive load than do (male) hermaphrodites of the same species. In contrast, dioecious females commonly have a heavier reproductive load than do males.

Thus, it will pay gynodioecious females to produce more flowers than do hermaphrodites, as is usually the case (e.g. Philipp, 1980; Table 8.9), but dioecious females usually produce fewer flowers than do males (Bawa, 1980). This may allow females to maximize fruit quality by transferring resource from flower number into flower size. In 16 species of diclinous New Zealand umbellifer investigated by Lloyd and Webb (1977), all produce more flowers per male inflorescence than per female inflorescence, and in all species except one, more inflorescences are also produced by male genets than by female genets. Similar results were obtained by Kay *et al.* (1984) for the red campion, *Silene dioica.*

An alternative (although by no means exclusive) explanation is that the (fewer) pollenless flowers of dioecious females will be more successful if they advertise strongly to pollen-feeding pollinators by use of a larger flower size. In monoecious *Begonia,* female flowers offer no reward, so that Schemske, Agren and Le Corff (1996) consider these to mimic pollen-bearing male flowers (p. 123). Rather unexpectedly, male flowers are noticeably larger than female flowers, and also tend to be more symmetrical in shape. Not surprisingly, male flower visitation rates are more than four times those of females, and pollinators spend very much longer on male flowers. Nevertheless, females show almost no reproductive disadvantage, setting almost as many seeds as pollinated controls. We can speculate that stabilizing selection acts on flower size in females: if the flowers are too small, they would attract too few pollinators which 'expect' them to be male and rewarding; if they are too large, costs in flower production would outweigh reproductive benefits in female fitness.

It should also follow that female flowers should provide more nectar than do male flowers, if, as is usually the case, female flowers are fewer than male flowers, and if the pollinator feeds on both pollen and nectar. Working with red campion, Kay *et al.* (1984) show that females produce nectar over a much longer period than is the case for male flowers, and pollinator visit rates closely track the amount of nectar available. We have obtained results very similar to those of Kay and co-workers from a very different part of the British Isles. In Costa Rica, Bawa and Opler (1975) found that five out of six dioecious species investigated produced more nectar overall in female flowers.

However, gynodioecious females, with numerous flowers, have no need for this extra advertisement or reward, and may be able to maximize their reproductive output by transferring resource from flower size to flower number.

One test to distinguish between these hypotheses would be to compare flower sizes between males and pollen-bearing females, as for Australian *Solanum* (Anderson and Symon, 1989). In all five dioecious species examined, females have significantly larger flowers. In these species females have only one flower per inflorescence, whereas males have between 6 and 60, so females are able to transfer resource from flower number to flower size so that expensive berries can be resoured.

Of course it is possible that flower size is controlled by a growth substance involved in the development of gender, so that differences in flower size between gender are 'accidental' by-products of these substances. Placke (1958) shows that gibberellins affect flower size, but these have not been implicated in the control of gender (p. 316).

Differences in flower number between the genders may be expressed within or between inflorescences. In the rose-root, *Rhodiola rosea*, males produce more stems than do females, but the number of flowers per inflorescence does not differ between genders.

With respect to the commonly found differences in flower number between males and females, Bawa (1980) contrasts causal and adaptive explanations. By expending less on each flower, males may be able to afford to make more flowers than do females, so that this distinction would not necessarily be under genetic control. Conversely, Bawa suggests that superior male flower production has resulted from sexual selection. He argues that female fitness should be maximized by expenditure on resource allocation to individual seeds and fruits. However, male fitness is maximized only by the production and dissemination of pollen grains. There will be a direct relationship between flower number and pollen production.

Sexual selection on males and females is disharmonious, and indeed may act in directly opposing ways. Put simply, males should select for quantity, and females for quality of reproductive effort (see also p. 2). In some conditions, such as a tropical forest trees in which this disharmony becomes great, dioecy may become advantageous, however inefficient a mechanism it may be.

Vegetative characters

In a dioecious or a gynodioecious species, males and females will compete with each other. Despite having different flower numbers, the average total reproductive energy budgets of the two genders almost inevitably differs, and is usually greater for females.

Vegetative differences between males and females may represent phenotypic side effects of these different reproductive loads, or they may be under genetic control and adaptive. In some cases, adaptive differences may directly relate to reproductive resourcing, as for the shoots of *Asparagus* which are always heavier in females (Table 8.11) In other instances, it seems that males and females may be adapted to different environmental niches, thereby reducing intergender competition. However, it can be very difficult to distinguish between these conditions.

Phenotypic vegetative differences resulting from reproductive resource expenditure are usually expressed as follows. First, more male shoots than female shoots are produced per genet. This relationship is found for most herbaceous and shrubby dioecious species examined, but is liable to be modified by

1. density; male dog's mercury (*Mercurialis perennis*) is more vigorous in competition with females at an intermediate density, but is less so at high and low densities (Wade, 1981b);
2. frequency; in the sheep's sorrel, *Rumex acetosella*, the rarer sex shows a better productivity at a range of densities (Putwain and Harper, 1972);
3. overall vigour; in the bog myrtle, *Myrica gale*, size difference between the genders is maximized in the most vigorous populations (Lloyd and Webb, 1977).

Certainly, it is not possible to state that males always produce more stems than females after examining one or a few populations. In a study of *Rumex acetosella* and *R. acetosa* on the Scottish island of Great Cumbrae, Ken Craggs (unpublished) found a significant excess of male stem production in one of four populations in both species, but in the other six populations the genders did not differ in the number of stems produced. Equally, there is no regular difference in shoot production between the genders for *Silene dioica* on this island. In general, males tended to show a greater shoot production than females in open, well-resourced localities than in heavily shaded sites.

Secondly, male predominant ratios among dioecious species are frequently explained by the greater longevity of males (e.g. Richards, 1988 for *Ilex aquifolium* and *Rhodiola rosea*). However, relationships between gender and longevity can be complex. Where individuals of *Potentilla fruticosa* by the River Tees, England could actually be aged, it was found that the average age of coexisting genders did not differ, but that males tended to predominate in short-lived stands, and females in long-lived stands (Richards, 1975). This topic is further explored on p. 356 and by Lloyd (1974a,b).

Thirdly, there is production of larger male shoots (unusual).

Table 8.11 Performance of male and female *Asparagus officinalis* (after Lloyd and Webb, 1977)

	Year	Male	Female	Male : Female ratio
Mean numbers of spears per plant	1925	2.98	1.95	1.53
	1926	15.68	8.61	1.82
Mean weight of spears (g)	1925	18.74	21.90	0.86
	1926	23.77	27.73	0.86
Mean weight of spears per plant	1925	55.90	42.70	1.31
	1926	372.80	238.80	1.56

1. In trees such as yew (*Taxus baccata*), *Ginkgo biloba* and poplars (*Populus*), males tend to be taller.
2. Vegetative rosettes of the red and white campions, *Silene dioica* and *S. latifolia* are larger in males.

A representative example of secondary sex characters is found in the cultivated *Asparagus* (Table 8.11). Males consistently show a heavier yield, largely due to the greater number of male spears (ramets) produced per plant (genet). In fact, female spears are consistently slightly heavier than males. The greater production by males makes them agronomically more desirable.

Dioecious species of strawberry (*Fragaria*) show a similar relationship between the sexes, with males producing far more runners, as originally recorded by Darwin (1877).

COMPETITIVE INTERACTIONS BETWEEN GENDERS

In diclinous conditions such as dioecy and gynodioecy, one gender may very often be at a competitive disadvantage with respect to the other. Differential reproductive loads, or disadvantageous sex chromosome-linked genes, may cause one gender to be less vigorous and long-lived than the other, and the resulting anisoplethy may seriously affect the fecundity of the population.

In long-lived perennials with efficient vegetative reproduction and dispersal, such a low fertility may not be disadvantageous. The Canadian pondweed, *Elodea canadensis*; and various species of butterbur, *Petasites*, are among many examples of dioecious species which may be very successful when only one gender is present. In contrast, dioecy is unusual in annual and other monocarpic life forms, perhaps as a result of the effects of intergender competition on reproductive efficiency in plants in which fitness is maximized by optimal seed-set.

Gender-linked ecological differentiation

Bierzychudek and Eckhart (1988) suggest that '. . . SSS (spatial segregation of the sexes) would be favoured only if the deleterious effects of competition between males and females are more severe than those of competition between individuals of the same sex. This seems unlikely, given the ecological similarity between individuals of the same sex.' From this argument, these authors deduce '. . . that a reduction of competition between males and females is unlikely to be an evolutionary cause of SSS.'

I wholly disagree with this view. The authors have totally missed the point that intragender competition does not lower fecundity, but intergender competition will usually do so. Because males and females usually have very disparate reproductive loads, intergender competition will normally be more severe in its effects on one gender, and hence on sexual reproduction in the population, than would be the case for intragender competition.

Consequently, within the environmental constraints of a single microsite, one gender will usually predominate by outcompeting the other. Thus, any sex-linked attributes which cause relative gender fitnesses to change across microsite gradients are likely to be selected for. Such attributes will tend to increase fecundity by bringing males and females within reproductive range.

Bierzychudek and Eckhart (1988) in fact agree that 'SSS can be favoured if male and female fitness respond differently across environments'. They note that this does in fact seem to have occurred in 66% of the 32 cases they list, for which male/female niche separation has been reported. However, they assign such differences to 'nonadaptive SSS' resulting from 'sex-differential mortality by habitat'. As mortality commonly results from intraspecific competition between the genders, it is difficult to understand how niche differentiation between the genders can be considered anything but adaptive.

Niche differentiation between genders can be expected to respond to small-scale and constant environmental heterogeneities, such as small gaps in forests. Also, such differentiation might be most successful in wind-pollinated plants with distant pollen travel. Just such an example is dog's mercury, *Mercurialis perennis*, for which Mukerji (1936) claimed that males predominate in better illuminated areas of woodland, and females in more shaded areas. Some confirmation is provided by Wade, Armstrong and Woodell (1981). Wade (1981a,b) shows that males grow better than females at high density, under high illumination, and in competition with the creeping soft-grass, *Holcus mollis* (its frequent companion in British woods). Competition experiments show that competitive effects are lower between genders than within a gender at a range of frequencies, densities and illuminations. In this species, some niche differentiation has occurred between the genders, although the relationship between gender and niche is not a simple one.

Dioecious *Aralias* are another group where genders show differential compartmentation for levels of shade on the woodland floor (Fig. 8.7) (Barrett, 1984). In this case however, males perform relatively well in poor light.

In another forest-floor herb, the North American liliaceous *Chamaelirium luteum*, Meagher (1980, 1981) showed that male: female ratios were heterogeneous for comparisons between quadrats of areas between 1 m^2 and 5 m^2. This spatial heterogeneity was presumed to exist with respect to different niches on the forest floor which were not identified.

It may be typical for males, which often have the lighter reproductive load, to predominate in habitats where the species performs well. This seems to occur in the culinary thyme, *Thymus vulgaris* (Dommée, 1976), which is gynodioecious. Hermaphrodites predominate in the driest and most open habitats, and Dommée finds a positive correlation between the frequency of male fertiles and the total performance of the species per area of ground. In bog myrtle, *Myrica gale*, plants perform with the greatest vigour in wetter habitats, and in these males also predominate (Davey and Gibson, 1917; Lloyd and Webb, 1977). Males, with a lower reproductive load and correspondingly stronger growth, should be able to colonize a

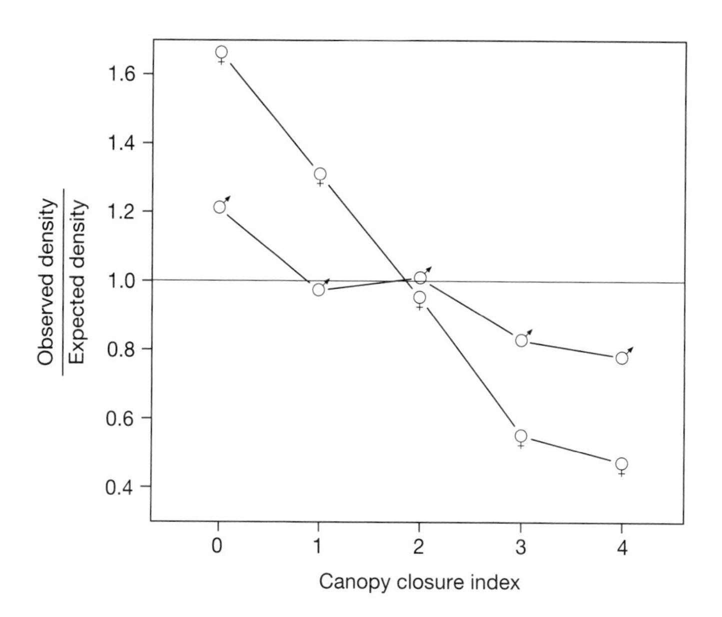

Fig. 8.7 Density response of staminate and pistillate flowering ramets of *Aralia nudicaulis* to canopy closure. Closure index: 0 = most open canopy; 4 = most closed canopy. The female response is significantly stronger than the male response (G = 18.61, df = 4, $P < 0.001$). (After Barrett, 1984.)

habitat in which the species is successful more rapidly than can females. This advantage will be less apparent in more marginal habitats.

Frequency-dependent competitive interactions between genders

Male advantage may be minimized by frequency-dependent effects. In three species for which detailed within- and between-gender competition experiments have been conducted, the rarer gender in a competitive environment tends to perform better both vegetatively and sexually than it does at more equal gender frequencies (Putwain and Harper, 1972 for *Rumex acetosella* and *R. acetosa* and Wade, 1981b for *Mercurialis perennis*).

Typically, genders differ in both productivity and timing. In *R. acetosa*, males grow faster, flower earlier and produce more, smaller shoots than do females. However, in *Aralia nudicaulis*, it is the females which flower first, producing fewer flowers than do males (Flanagan and Moser, 1985). Despite the difference in flowering time between genders in this species, which reduces overlap between gender functions for both flowering time and resource usage, it seems that seed-set tends to be resource-limited rather than pollen-limited.

Time–niche differentiation between the genders may be both cryptic and subtle. For white campion (*S. latifolia*), Purrington (1993) shows that male seeds germinate about 10 hours before females do (remarkably, this difference is statistically significant). This effect may favour males, but only where seeds germinate very thickly. Later, females may nevertheless flower a day or so earlier than do males, but only under conditions of nutrient stress.

Temporal differentiation may reduce intergender competition in species where vegetative growth patterns lead to an intermingling of the ramets of both genders. If differences in timing and productivity do minimize competitive effects between genders (and there seems to be no direct proof of this), a rare gender will compete with its own gender rarely, and should do well. However, when frequencies of the genders within a microniche are equal, intragender contacts will be disadvantageously common, and each gender will perform relatively poorly. This frequency-dependent selection will tend to favour the more vigorous gender to a point where the less vigorous gender becomes advantaged by its scarcity. Thus, ESS-stable gender frequencies for ramets should become established within any microniche.

Adaptive gender-linked niche differentiation is not 'group selectionist', invoking 'voluntary death' as suggested by Bierzychudek and Eckhart (1988). Rather, it will pay the gender with the heaviest reproductive load, which is therefore the poorer competitor, usually the female, to adopt sex-linked genes which reduce competition with vigorous males, so that females perform well in minority situations. Such environmentally related gender differentiation is likely to lead to the evolution of gender-related niche differentiation which should enhance both female and male fitness.

GENDER RATIOS

For dioecious species, the heterogametic gender (XY) must mate with the homogametic gender (XX). At the simplest level, one should expect that males and females will be produced at equal frequencies (1:1). However, there are a number of points in the sexual cycle at which differential selection between X and Y chromosomes can occur:

1. competition between meiotic products in the heterogametic gender (i.e. usually in pollen formation);
2. competition between gametophytes in the heterogametic gender (i.e. usually between pollen grains or pollen tubes at the stigma, or in the style and ovary);
3. competition between zygotes (i.e. between heterogametic XY and homogametic XX seed embryos in the developing fruit or between XY and XX seedlings or adults).

It is very important to distinguish between the gender ratio of genets, and the gender ratio of ramets. In order to do so, it is necessary to be able to identify the limits of genets. At times, marker genes, for instance for inflorescence colour in *Urtica dioica* or *Rumex acetosella*, makes this easy. In *Silene dioica* or *Rumex acetosa*, each genet only makes a few ramets, so that the identification of the individual is usually simple. In other cases (notoriously so in *Myrica gale*), it can be very difficult to delimit the individual, except by the use of isozymes.

Males commonly produce more ramets per individual than do females (p. 346), so that they appear to be more common, whereas the ratio of genets may not in fact differ from equality.

Gametophytic competition

Early studies (Correns, 1928; Lewis, 1942a) considered that female-predominant ratios were in fact more common among dioecious species. In four species (*Silene latifolia, Rumex acetosa, Humulus lupulus* and *Cannabis sativa*) it was found that gender ratios from seed differed from equality when there was a high density of pollen on stigmas. At low pollen densities, males and females were produced from seed at a ratio of 1:1, whereas at high densities, an excess of females was produced. It was suggested that X-carrying pollen grains outcompeted Y-carrying pollen grains at the stigma or in the style, resulting in an excess of homogametic (female) offspring. A microscopic examination of pollen tubes indicated that one-half germinated and grew more quickly than the other half. These, it was considered, were from the X-carrying grains.

This model is very attractive, for it carries an elegant negative feedback element. If superior male vegetative reproduction creates an excess of male ramets, there will be surplus pollen available for cross pollination, and stigmas will receive pollen at a high density. As a result, an excess of females will occur in the next seed generation, to a point when males are sufficiently scarce for pollen grains received by stigmas to drop below the level where all ovules are fertilized. Gender ratios from seed will return to equality, allowing for a corresponding build up of male-ramet frequency to develop. Frequency-dependent selection should result in an ESS frequency of ramet gender, which will depend on the ramet gender frequency at which the saturation of stigmas by pollen occurs. Interestingly, this ESS point will allow for more male ramets to occur than would give the maximum seed production per unit area for the population.

This system may not, in fact, operate at all. For *Silene latifolia*, Purrington (1993) found a surplus of females was produced from seed, irrespective of the nutrient status of the parents, or the pollination position. If competitive effects do occur between X-carrying and Y-carrying pollen, these should give rise to a negative relationship between male:female ramet ratios in the field, and male:female genet ratios from seed. Mulcahy (1967, 1968) was unable to discover such a relationship in *S. latifolia* (*S. alba*). Working with *S. dioica* on several occasions, students and I have never discovered any relationship between ramet density and gender ratio (if Correns was correct, dense populations should have relatively high proportions of female genets). For the rare relative of these campions, *S. diclinis*, Prentice (1984) consistently finds that an excess of females comes from seed, suggesting that the superior performance of X-carrying pollen, or of XX zygotes, is density independent in this case.

Gender ratios are non-adaptive

As we have seen, such negative feedback systems for the control of gender frequency would not necessarily result in gender frequencies which are reproductively the most efficient for the population as a whole. Dioecy is a reproductively inefficient system, not so much because pollen must be transmitted from one genet to another, but rather because the products of X/Y competition at the gametic and zygotic levels result in inefficiently low densities of female ramets in comparison to male ramets.

However, Lloyd (1974b) challenges the assumption that fitness in a dioecious population will necessarily be directly related to the reproductive output of seed per unit area in any case. He points out that as seed-set in females increases, so does the disparity in reproductive load between males and females. High levels of seed-set among females may later give rise to inefficiently high male: female ramet ratios.

Moreover, Lloyd (1974a,b) makes it clear that plants cannot select for gender ratios anyway, for there is no mechanism by which natural selection, which works on individuals, can do so. Rather, the gender ratio of genets will be an accidental by-product of male/female competition, depending on:

1. competition between and X and Y-carrying male gametophytes at the stigma and the resulting gender ratios in seedlings (he views the poor performance of Y-carrying gametophytes as an unavoidable by-product of dioecious systems in which half the male gametophytes lack X-linked genes);
2. differential survival and growth of male and female genets;
3. efficiency of pollen-flow from male to female flowers.

All three of these attributes are themselves dependent on male: female ramet (not genet) frequencies. Genet frequencies (gender ratios) will occur at ESS equilibria dependent on all three ramet-ratio attributes. These attributes differ from species to species (and to a certain extent from population to population), so it is not surprising to find that genet gender ratios are female predominant in some species and male predominant in others.

Female predominant ratios

One uncanny correspondence in gender ratios between two species has been found for the arctic willows, *Salix polaris* and *S. herbacea* (Crawford and Balfour, 1983). In both species, the former in Svalbard and the latter in Iceland, the frequency of female genets throughout a total of 60 populations is invariably very close to 0.59 (Fig. 8.8). The authors show that females are more variable for leaf resistance with respect to water loss than are males, and conclude from this that a wider variety of microsite habitats can probably be colonized by females, compared to males.

Quoting the model of Wright (1931), Crawford and Balfour suggest that female frequencies do not exceed 0.62, because effective population sizes drop rapidly when this proportion of females is exceeded (Fig. 8.9). This explanation is clearly incorrect. When as many as 95% of the population is female, the effective population size N_e is still 20. Furthermore, the effective population size suggestion is group-selectionist, as pointed out by Prentice (1984), for it assumes that the population can evolve to adjust its female frequency should this become inefficiently high.

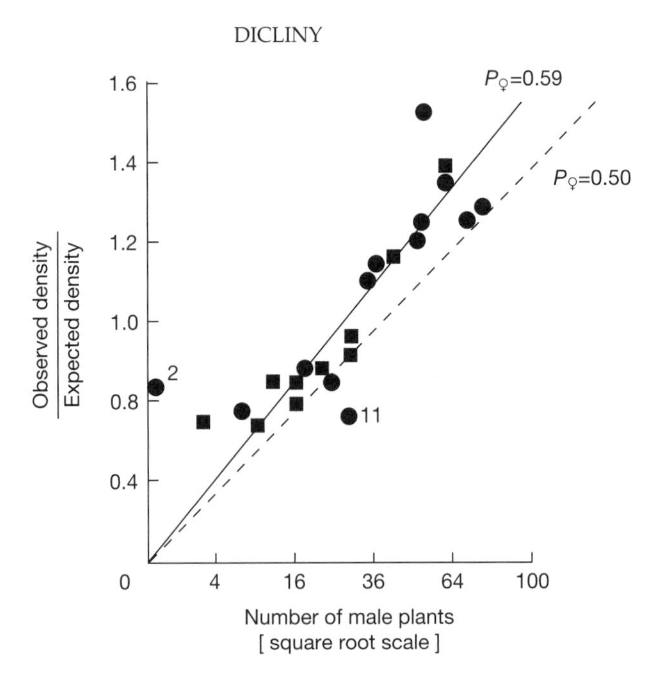

Fig. 8.8 A binomial plot of (•) *Salix polaris* at each of 14 sites in Spitsbergen and (■) *S. herbacea* at each of nine sites in Iceland. The solid line represents the mean proportion of female plants (0.59) and the broken line equal numbers of each sex. The two small island sites are indicated by their numbers, 2 and 11. (After Crawford and Balfour, 1983.)

Rather, if we assume that there is no pollen shortage and no gametic competition in these willows, so that males and females come from seed at an equal frequency (there is no evidence either way), the equilibrium frequency of females at 0.59 can only result from a slightly higher female survivorship. This greater overall male mortality will itself result from the product of the differential ability of the genders to become established (apparently greater in females), and the different reproductive load of the genders (probably on average a greater disadvantage to females). The extraordinary thing seems to be that the long-term difference in survivorship between the genders should be so similar between sites and between species.

Not all willows have this particular gender ratio however. J. W. Heslop-Harrison (1924) reports 1 : 1 gender ratios for three diploid species of *Salix*, but consistent 3 : 1 ratios of females to males in four polyploid species. This relationship between 'ploidy and gender' ratio is certainly not consistent within the genus. *S. herbacea* is a diploid and *S. polaris* a tetraploid, yet both have 59% females. We can only conclude that willows commonly have female-predominant gender ratios.

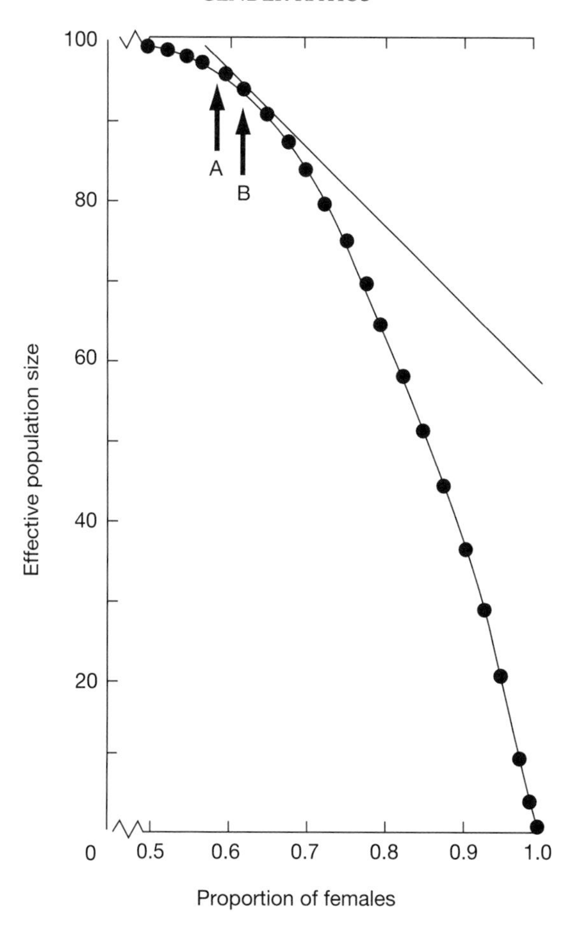

Fig. 8.9 Relationship between the variation of sex distribution from equality and the effective breeding size of any population (Wright, 1931). The arrow A represents the observed proportion of female plants of *Salix polaris* in Spitsbergen and of *S. herbacea* in Iceland. The arrow B is the hypothetical equilibrium point between increase in seed-bearing population and decrease in effective population breeding size. (After Crawford and Balfour, 1983.)

Male-predominant ratios

Although Correns (1928) and Lewis (1942a) considered that such female-predominant gender ratios were typical among dioecious species, a survey by Lloyd (1974a) points out that male-predominant ratios are in fact rather more common. For 16 New Zealand dioecious species, Godley (1964) found male-predominant ratios in 10, and a female-predominant ratio in only one, whereas Webb and Lloyd (1980) surveyed 53 populations of New Zealand Umbelliferae, which are dioecious or gynodioecious, and found male-predominant ratios in 43. Male-predominant ratios are expected for gynodioecious species (p. 329), but in 18 populations of 11 fully dioecious species in their survey, 78% have male-predominant ratios which significantly differ from equality. The authors explain these male-predominant ratios by the superior survival of male genets resulting from their lower reproductive loads.

Most reports of gender ratios among populations result from 'snapshot' surveys, and such surveys can be very misleading for the species as a whole (Kay and Stevens, 1986; Richards, 1988). In *Potentilla fruticosa* (Richards, 1975), both male-predominant and female-predominant subpopulations have been reported from three distinct areas: Upper Teesdale, England; the Burren, Ireland; and Öland, Sweden (Elkington and Woodell, 1963). In this species it seems that the gender ratio found in a subpopulation is a function of the longevity of that subpopulation. Short-lived populations tend to be male predominant, and long-lived populations tend to have predominantly female genets (and ramets)

Opler and Bawa (1978) reached similar conclusions about the gender ratio of 23 species of tropical forest tree from Costa Rica. Of these, eight had male-predominant sex ratios and two had female-predominant ratios. Males tend to flower earlier, and to live longer. Sex ratios were thought to be the product of differential times to maturity and life span of the sexes, which themselves probably reflect the different reproductive loads of the sexes.

Certainly, a wealth of circumstantial evidence confirms the position of Lloyd (1974a), that sex ratios cannot in themselves be adaptive, but are rather the result of differential selection on individual X and Y chromosome-carrying gametophytes and sporophytes.

CHAPTER 9

Self-fertilization and inbreeding

INTRODUCTION

Hermaphrodity seems to have been the original condition in the flowering plants, and 95% of the species are still hermaphrodite (p. 10). Therefore, unlike unisexual higher animals, most Angiosperms have the potential for self-fertilization (or 'selfing'). Although many flowers avoid selfing by a chemical self-recognition (gsi p. 207), or by other outcrossing mechanisms such as dioecy, some 40% of flowering plant species can and do self at various times, whereas as many as 20% may habitually do so (Fryxell, 1957).

Such self-fertilization is unknown in complex animals such as vertebrates and most arthropods which are unisexual, although it commonly occurs in other groups such as molluscs and many parasites (Chapter 1). It seems possible that flowering plants can tolerate selfing more readily than do many animals, because many disadvantageous recessive genes that would lead to inbreeding depression in plants may have been 'screened out' when exposed to selection in the haploid gametophyte generation (p. 1).

It seems that plants originally evolved to the selfing condition by losing gsi and becoming self-compatible. In many circumstances, such secondary selfers may have proved fitter than their outcrossing ancestors for a variety of reasons.

1. Colonizers can establish from single disseminules when self-fertile ('Baker's (1995) rule', p. 304).
2. Cross-pollination (especially when animals are pollen vectors) may be unreliable.
3. Selfing saves on male expenditure.
4. Selfers have control over the identity of the father of their offspring.
5. Monocarpic plants have only a single opportunity to reproduce.

Once selfing became established as a successful mode of reproduction, mutants that arose in selfing lines which secondarily enhanced the efficiency of selfing (e.g. homogamy, autogamy, even cleistogamy), or that saved on male expenditure (small flower size, savings on pollen and nectar production) would have been favoured.

It seems that the life style of many plants has encouraged the evolution of a mixed strategy mating system in which outcrossing and selfing coexist (Chapter 1). Only in relatively few has selfing become habitual, and obligate selfing is almost unknown. Habitual selfing seems to confer severe penalties, which are discussed in greater detail later in this chapter. These mostly depend on the situation whereby the proportion of heterozygote loci in the population is halved after each selfing generation. The penalties which result from this loss of heterozygosity include:

1. high levels of homozygosity, so that recessive lethals and other disadvantageous genes that accumulate in outcrossers become phenotypically expressed after a bout of selfing, resulting in 'inbreeding depression' (repeated selfing may cleanse a line of some of this 'genetic load', but a completely homozygous line may by chance have fixed some relatively unfit alleles);
2. 'overdominance' effects resulting from genes which display vigour effects in the heterozygous condition will be lost. This alternative cause of inbreeding depression will not be lost after repeated selfing, and is the probable reason why some loci often remain in a heterozygous phase after many generations of selfing;
3. homozygosity results in a loss of segregational and recombinational variability among gametes; also, if variability is lost between individuals in a population, then almost all zygotes produced by that population will be genetically uniform. The penalties incurred by this genetic uniformity are similar to those encountered by asexual populations (Chapter 1).

However, it is noteworthy that segregation and recombination do still occur in habitual selfers. Thus, disadvantageous mutants can be 'recombined away' from a selfing line, so that Müller's ratchet (p. 23) does not apply. In this regard, habitual selfing differs from and is evolutionarily superior to asexuality.

Reproductive assurance

Outcrossing will usually succeed when reproductive fitness is not limited by a paucity of animal visits. Where such limitations occur, features that maximize within-flower selfing (autogamy) may be favoured. Fitness constraints resulting from limited visiting may occur in inimical environments in which animal visitors are few or uncertain, such as arctic, alpine or arid environments. Within-flower selfing is more frequently encountered in these habitats, but rarely predominates, perhaps because most denizens of harsh conditions are long-lived perennials for which constraints on reproductive fitness by seed are low.

Nevertheless, it is clear that in species for which fitness is highly reliant on sexual reproductive efficiency, selfing must usually be favoured. It is a common experience to discover little if any seed-set in the wild in fully self-incompatible species. In a wild population of the New Zealand alpine daisy, *Celmisia hectori*, I discovered one seed set in 120 heads (approximately 4000 ovules; 0.025%). The best field indication of sexuality in the mostly apomictic genus *Taraxacum* is poor seed-set. Many heads in sexuals, which are mostly fully self-incompatible, set no seed at all. The average percentage seed-set in a population of the diploid sexual *T. erythrospermum* in Slovakia was only 20%; this coexisted with several apomictic species of *Taraxacum* in which seed-set rarely fell below 90% (author's unpublished work).

In these Asteraceae, it is easy to estimate seed-set, and to demonstrate by experimental cross-pollinations that failure of cross-pollination is the cause of low seed-set. However, in many plants, relatively few ovules set seed even when all have been efficiently cross-pollinated. Only about 1% of the single ovule flowers of Australian *Banksia* usually set seed, although most may be cross-pollinated. The fruits are very large (Fig. 2.11) and the plant is adapted to develop very few according to energetic constraints. Reproductive assurance selection only operates when fewer ovules are fertilized than the mother can succour.

Perhaps the most thoroughly studied example of comparative fecundity in related selfers and outcrossers is in the American cruciferous genus *Leavenworthia*. Solbrig and Rollins (1977) show that reproductive efficiency, in terms of ovule set, varies from 52 to 62% in self-incompatible populations, and from 73 to 94% in self-compatible populations, thus giving approximately a 3:2 reproductive advantage to the selfers. It is noted, however, that seed-set in outcrossers varies noticeably with the weather, which presumably affects pollinator activity (see also Table 2.4).

Life style of selfers

Reproductive constraints are much greater in short-lived plants, and especially in monocarpic plants which flower only once. They only have one opportunity to contribute to the next generation, and thus there will be very great directional selection pressures, which are likely to outweigh any disadvantages of selfing, for reproductive input to be maximized. Most monocarpic plants are at least partially autogamous (Fig. 9.1), thereby maximizing reproductive output in the absence of the unreliable agents of cross-pollination.

By channelling all reproductive energy into seed production, a monocarpic plant forgoes the chance of producing any vegetative propagule or perennating organ which would continue that genet to another flowering.

Seeds have capabilities for travel, dormancy and establishment rarely found in vegetative organs. Thus, colonizing species adapted to take advantage of open ground will achieve greater success by maximizing

Fig. 9.1 *Saxifraga longifolia*, a monocarpic species of the Pyrenees and other mountains of Spain and North Africa. Rosettes grow for many years (often ten or more) until they are perhaps 25 cm in diameter before producing a huge inflorescence, which may contain in excess of 1000 flowers. After fruiting, the individual invariably dies.

seed production through autogamous selfing. Where arctic and alpine environments have been recolonized after glacial epochs, selfers would have been more likely to have become established in isolated nunataks than would outcrossers for which two disseminules must travel together if colonization by seed is to succeed. The same argument holds for oceanic islands (p. 304), or indeed any kind of 'biological island' recolonization, where high proportions of potentially selfing species are likely to be found.

Polyploids, which buffer to some extent the genetical effects of selfing (p. 44), are much more likely to be partial selfers than are diploids (p. 382). It is also the case that polyploidy is more likely to arise in selfers, as such plants are immediately self-fertile despite the 'minority effect' (Fowler and Levin, 1984).

The automatic advantage of selfing

Selfers have been considered to have an automatic advantage over outcrossers. It has been frequently suggested that selfers should donate more selfing genes to the next generation than coexisting outcrossers donate outcrossing genes to that generation. This is partly because selfers, being more reproductively efficient, should be more fecund than are outcrossers. However, it is also because the pollen of selfers can introduce selfing genes into outcrossing lines, but selfers, by selfing early, usually prevent any incursion of outcrossing genes into selfing lines.

Undoubtedly, this argument holds for gynodioecy (p. 325), wherein outcrossing females have no pollen to donate to the hermaphrodite selfers. It also holds for the advantage of selfing homostyles over thrum heterostyles (p. 287), for the reason, particular to the heteromorphic system, that homostyles produce fewer recessive pin offspring than do thrums.

Where we are considering advantages of a selfing gene arising among otherwise uniform outcrossing hermaphrodites, I am less certain that an automatic selfing advantage obtains. For it to do so, we must assume that the selfing gene inhibits the male outcrossing function less than it does the female outcrossing function. But, features which predispose an outcrosser to selfing, such as homogamy, small flowers, lack of attraction or reward to pollinators, equally inhibit the outcrossing male function of that flower.

SELF-POLLINATION

An essential feature of habitual self-fertilization is autogamy, that is to say within-flower self-pollination. It is common to find complex, animal-pollinated flowers that are incapable of such autogamy, but are nevertheless self-compatible (most Fabaceae, Scrophulariaceae, Lamiaceae, Orchidaceae or Araceae for instance). As these flowers commonly occur in a raceme, with several flowers open and receptive together, insect-mediated pollen transfer between flowers results in some geitonogamous selfing. Such plants maintain a balanced strategy of mixed selfing and crossing, depending on proportions of pollen transferred within and between genets.

In others, within-flower selfing can occur, but may only be mediated by an animal visit. This strategy is useful for both self-incompatible and self-compatible species, in that stigmatic sites will not suffer from prior blockage by automatically selfed pollen, so that such crossed pollen as may be carried into the flower by the visitor is given a chance to succeed.

Even more frequently, it is found for many species that the individual flower is initially female in function (protogynous) so that opportunities for outcrossing occur before the flower enters a homogamous phase and become automatically selfed. A similar strategy is for the flower to open only in conditions favourable for cross-pollination (*Crocus* and many other bulbous plants); in less suitable weather the flower becomes automatically selfed.

Because habitually selfing plants are not normally pollinated by insects or wind, they must have efficient within-flower selfing (autogamy). This is achieved by removing those features of structure and timing that have evolved in insect-visited flowers to minimize stigmatic site blocking by selfed pollen. Mutations which decrease the herkogamous separation within the flower of pollen issue (stamens) and pollen receipt (stigma) are likely to be favoured in a selfing plant. Similarly, mutations which affect the timing of pollen dehiscence or stigmatic receptivity are likely to be favoured in a selfing plant if flowers that were dichogamous become homogamous. In this case, if anthers shed pollen as the stigma matures within the same flower, successful within-flower pollination is likely to occur.

Such shifts in floral timing are under polygenic control, and natural heritable variation in floral timing, which can respond to selection for increased within-flower selfing, probably occurs in all protogynous or protandrous species, as for instance in *Gilia* (see Table 9.1). Wyatt (1984) shows that *Arenaria uniflora,* widespread on granitic outcrops in the southeastern United States, shows considerable variation in protandry, which is correlated with flower size. Extreme autogamous populations, homogamous, and with many small flowers, have arisen on at least two occasions, and have been called *A. alabamica.*

In more complex flowers, special mechanisms may achieve within-flower selfing. As many as 50 species of the insect mimic orchids, *Ophrys,* are cross-pollinated by male Hymenoptera which pseudocopulate with the flowers. However, in the bee orchid, O. *apifera,* the pollinia develop abnormally long stalks and droop into the stigmatic cavities (Fig. 4.38). In northern populations, in which the pollinator seems to be missing, automatic self-pollination is thus achieved.

In another orchid genus, *Epipactis* (helleborines), outcrossing species such as *E. palustris, E. helleborine, E. purpurata, E. atrorubens, E. microphylla* and *E. gigantea* have a sticky cap (viscidium) to the projection of the stigma (rostellum), above which the pollinia are situated (Fig. 4.37). The wasp (*Vespa* spp.) visitors are attracted to the coloured, nectar-filled hypochile, and to reach this, brush against the viscidium, which adheres to the wasp, taking the pollinia with it. In contrast, all autogamous species (*E. leptochila, E. phyllanthes, E. youngiana, E. albensis, E. pontica*) lack the viscidium, an absence that may be controlled by only a single gene, and in most the rostellum is slight or absent. In these, the pollinia are released from the anther early (in bud in *E. phyllanthes*) and automatically self-pollinate (Fig. 4.37). However, despite assertions in the literature that such species are invariably selfed, these species have nectar and also receive visits from wasps, and cross-pollination of loose massulae and pollen tetrads is likely to occur. Only in the scarce forms, *E. leptochila* var *cleistogama* and *E. phyllanthes* var *vectensis,* in which the flower

Table 9.1 Breeding system and genetic structure of *Gilia achilleifolia* (after Schoen, 1982,b)

Population	Protandry Index	Stigma exertion (mm)	Flower weight (mg)	Seed-set in absence of pollinators (%)	F_0[a]	Number of loci	t[b]
Eagle Ridge	0.00	–0.6 ± 0.2	1.4 ± 0.3	91	0.47	4	0.15
Arroyo Mocho	0.13	1.7 ± 0.4	2.5 ± 0.2	–	0.05	6	0.29
Metcalf Road	0.37	1.2 ± 0.4	2.8 ± 0.3	84	0.10	5	0.42
Arroyo Seco	0.69	2.1 ± 0.3	–	42	0.07	9	0.58 *Adh-1* 0.92 *Pgi-2*
San Ardo	0.69	–	3.5 ± 0.3	–	0.20	6	0.80
Poly Canyon	0.86	2.3 ± 0.3	2.7 ± 0.2	48	0.22	5	0.64
Hastings	0.88	1.2 ± 0.2	2.6 ± 0.1	–	0.04	8	0.85 *Adh-2* 1.06 *Pgi-1*

[a] F_0, the fixation index, is based on deficiencies in heterozygote frequency from that expected by panmixis, averaged for the number of loci stated for each population, and has a maximum value (heterozygotes absent) of one.

[b] t the estimated outcrossing rate, is based on deficiencies in heterozygote frequencies from that expected by panmixis, and is estimated from whichever of the loci *Adh-1, 2* and *Pgi-1, 2* show the highest level of polymorphism for each population.

For Arroyo Seco and Hastings, readings from two loci are given.

usually fails to open at all, is autogamy likely to be absolute (Richards, 1982).

Not only are the autogamous *Epipactis* less variable than the outcrossing species, but they tend to have smaller flowers as well. Thus, the primary cause of selfing, the absence of the viscidium, which has allowed outcrossing species to evolve into autogamous species, perhaps as the result of a single mutation, is reinforced by other, less vital characters, which improve the efficiency of autogamy. The autogamous species tend to have smaller, more drooping, less open and less brightly coloured flowers with a smaller rostellum. Such features involve a lower expenditure of energy, and may well be favoured in populations that have become largely autogamous (p. 27).

In other plant families, it is also common to find that autogamous races have smaller flowers than their relatives. Thus, Schoen (1982a) showed that races of the north American *Gilia achilleifolia*, which are mostly autogamous by virtue of their near homogamy and juxtaposition of their anthers and stigma, also have the smallest flowers (see Table 9.1). Likewise, Strid (1970) demonstrated that the autogamous love-in-a-mist *Nigella doerfleri* that grows on small, dry, hot Aegean islands has much smaller flowers than its more mesic relative *N. arvensis* agg. which also tends to be more outcrossed. Derived small-flowered inbreeders frequently inhabit dry areas. Moore and Lewis (1965) described the origin of a small-flowered, autogamous segregate of *Clarkia xantiana*, *C. franciscana*, after a dry period extended the desert margin near San Francisco into the range of the former, thus favouring the inbred mutant. Lewis (1962) has termed such a phenomenon 'catastrophe speciation'.

Cleistogamy

The ultimate reduction in floral display is the flower that never opens, and thus shows obligate autogamy. Such closed 'cleistogamous' flowers are not uncommon, but chiefly occur in outcrossing species, usually after the 'conventional' outcrossed flowers have failed to set seed. Failure of seed-set is perhaps especially common in vernal species, adapted to flower in deciduous woodland before the leaf canopy develops, and which are thus subject to the vicissitudes of unreliable spring weather. The development of cleistogamous flowers later in the season, after the canopy has closed over and pollinators are few, is found in such typical herbs of the woodland floor of temperate Eurasia and North America as wood violets (*Viola* section Rostrates) (Fig. 9.2) and wood sorrel (*Oxalis acetosella* agg.). Such cleistogamous flowers may be regarded as 'fail-safe' devices, which ensure seed-set in the absence of outcrossed seed earlier on, but may also provide 'mixed strategy mating' with some variable (outbred) and some invariable (inbred) offspring. In some years, plentiful outcrossed seed is

Fig. 9.2 Cleistogamy in the European wood violet, *Viola riviniana*. Large chasmogamous flowers in spring are often followed by small, bud-like flowers which automatically self-fertilize without opening. Cleistogamous flowers can be seen to the right of the photo.

produced, thus avoiding the genetic problems resulting from repeated selfing. Other woodland herbs showing late-season cleistogamy are touch-me-not, *Impatiens noli-tangere*, *I. capensis* (Waller, 1984) and ground ivy, *Glechoma hederacea*. Annual weeds such as *Lamium amplexicaule* (henbit) can also have some cleistogamous flowers, as do some plants from seasonally wet areas, which produce cleistogamous flowers in the dry season (*Hesperolirion* spp.). In the Arabian *Commelina forskahlei*, the dry-season cleistogamous flowers are subterranean.

As suggested by Lord (1981) in his review on cleistogamy, totally cleistogamous species are apparently very rare. Certain aquatics are usually cleistogamous, such as awl weed, *Subularia aquatica* and mud weed, *Limosella aquatica*. Even the relatively showy flowers of water lobelia, *Lobelia dortmanna*, are self-pollinated in bud and are thus effectively cleistogamous. Casual observations can be misleading, however. Populations of *Subularia* flowering in hot weather on a dried-up lake margin do open their flowers, and are visited by small flies, although this species more usually flowers under the water and is automatically selfed.

Similarly, observations on certain annual grasses in which the stigmas and anthers are usually included within the palea can be misleading. Most British populations of the wall barley, *Hordeum murinum*, appear to be cleistogamous (Fig. 9.3), yet both chromosomal and seed-protein evidence show the origin of diversification in the group to be due to allopolyploidy resulting from between-populational crossing, and in warmer areas of southern Europe chasmogamy is frequently observed in this group (Booth

Fig. 9.3 Spikelet of the annual wall barley grass, *Hordeum murinum*. The two lateral spikelets are sterile; the central floret contains stigmas and anthers which are rarely if ever released from the palea, at least in the UK. The floret is therefore cleistogamous and automatic self-pollination almost invariably occurs. Photo by T. Booth (×4).

and Richards, 1978). Even in the apparently fully cleistogamous grass *Festuca microstachys* (Adams and Allard, 1982), bursts of genetic variability are still observed, suggesting the occurrence of occasional outcrossing. Perhaps the least outcrossed species yet reported is the familiar genetic tool, *Arabidopsis thaliana*. In their survey of 16 UK populations, Abbott and Gomes (1989) mostly find no evidence for outcrossing whatever, and outcrossing never occurs at a level above an estimated 0.3% of matings. Like the corn spurrey *Spergula arvensis* for which New (1959) also found outcrossing events to be very rare, *Arabidopsis* is not strictly cleistogamous, although the flowers are very small and may usually be selfed before the bud opens.

LEVELS OF SELFING

Early work on the population genetics of selfing plants relied either on patterns of pollen transfer (e.g. Levin, 1979), or on the phenotypic effects of major genes. By the time of the seminal review by Allard, Jain and Workman (1968), the dynamics of genetic polymorphisms under various levels of selfing were quite well characterized, but had scarcely been studied in wild populations. However, the necessity to accurately delimit levels of outcrossing (or its reciprocal, selfing) in order to understand evolutionary processes and breeding strategies for selfers was well understood. For the study of natural populations, the breakthrough came in 1970, when isozymes were first used to indirectly estimate mating systems (Brown and Allard, 1970) in wild oats (Marshall and Allard, 1970). Perhaps the best example where outcrossing estimates were compared using both the traditional methods of examining floral structure and the amount of self-fertilization that occurs in the absence of pollinators, and indirect estimates from isozymes comes from the outcrosser *Limnanthes alba* and the selfer *L. flocculosa* (de Arroyo, 1975; Table 9.2).

Since about 1980, isozymes have been used routinely to estimate levels of selfing (or outcrossing). Such techniques depend on the fact that selfing reduces heterozygote frequencies below those expected in panmictic conditions. They have the advantage that numbers of polymorphic loci can be readily sampled for numbers of individuals, and that electrophoretic phenotypes rarely demonstrate dominance, so that heterozygotes are readily identified.

However, a number of other conditions can cause departures from genotype frequencies predicted for panmixis by the Hardy–Weinberg equilibrium (p. 42). These include small population size or restricted gene-flow within population ('subgrouping', p. 184), so that non-panmictic genotype distributions frequently occur among obligate outcrossers which cannot possibly be caused by selfing.

Table 9.2 Variation in breeding system and enzyme polymorphism in *Limnanthes* spp. (after de Arroyo, 1975)

Taxon	Within-flower protandry (days)	Seed-set in absence of pollinators (%)	Estimated selfing from floral timing (%)	Estimated selfing from deficiency of heterozygotes (%)	PLP[a] (%)	k	H (%)
L. alba var *alba*	3	3–7	0	0.3	41.7	1.5	16.3
var *versicolor*	2	19–32	0	2.6	36.1	1.3	15.1
L. flocculosa							
ssp. *californica*	1–2	100	50	26.2	33.3	1.4	23.5
ssp. *grandiflora*	1–2	93	50	21.1	41.7	1.4	9.7
ssp. *pumila*	1–2	89	50	21.2	41.7	1.4	5.3
ssp. *flocculosa*	0.1	95–100	100	79.4	31.2	1.3	25.6
ssp. *bellingeriana*	0	69–100	100	68.9	16.7	1.2	2.0

[a] Abbreviation: PLP, the proportion of loci polymorphic for a species or population; k, the mean number of alleles per locus for a species or population; H, the mean proportion of heterozygous loci.

Selective differentials between genotypes, and gene migration, are other phenomena which might bias estimates of selfing made from heterozygote frequencies using isozymes. Brown (1979) wrote of the 'heterozygosity paradox', whereby inbreeders tend to be heterozygous at more loci than predicted (perhaps due to overdominance at these loci), but outcrossers can be homozygous at more polymorphic loci than expected (maybe due to subgrouping effects). Barrett and Eckert (1990) review the various models that have been devised to estimate outcrossing and selfing rates using isozyme surveys.

Fixation indices

There are two main procedures for estimating the degree of outcrossing (t) or selfing ($s = (1 - t)$) in wild populations. Fixation indices (F-statistics, based on Wright, 1951; Schaal, 1975; Nei, 1973, 1975) take a 'snap-shot' view of the genetic structure of the population, and are based solely on deviations from levels of heterozygosity expected by panmixis. They are more accurate when employing a number of variable loci (Ritland and Jain, 1981), but can only provide overviews of populations or parts of populations. To see how mating system varies between individuals, mixed mating models are required. By sampling the isozyme (or DNA) genotypes of seedlings in comparison with that of the mother, maximum-likelihood techniques estimate selfing rates (Clegg, 1980; Ritland, 1984).

For mixed mating models, it is assumed that the genotypes of mothers, offspring and the allele frequencies of the population as a whole (pollen genotype frequencies) are known. Naturally, mixed mating models are more accurate, but are more time-consuming to construct as they require that seedlings should be germinated.

The basic fixation index statistics are:

F_{IS} the fixation index for a given polymorphic locus for a (sub)population
F_{ST} the mean fixation index over all polymorphic loci for a (sub)population
F_{IT} the fixation index over all (sub)populations

where

$$F = \frac{1-(\text{observed frequency of heterozygotes})}{\left(\text{frequency of heterozygotes expected by } (p+q)^2\right)}$$

where p and q are the frequency of the two alleles of a single locus.

The frequency of heterozygotes expected in a panmictic (sub)population $(p + q)^2$ is termed H_S (otherwise known as PI or the polymorphic index). The average H_S across all subpopulations is $H_{S,}$ whereas H_T refers to the PI for all subpopulations, treated as a single population. D_{ST} is the difference between H_S and H_T, and D_{ST}/H_T gives G_{ST}, a measure of the degree of genetic differentiation between populations.

Likewise, F_{ST}, the degree of genetic fixation due to population subgrouping, is directly related to G_{ST} which is merely the weighted average of F_{ST} estimates.

Mixed-mating models

These are well explained in Brown (1990). For diallelic loci AlA2

1. expected maternal progeny of A1A1 = $(m_1 + m_2)(1 - tq)$
2. expected non-maternal progeny of A1A1 = $(m_1 + m_2)tq$
3. expected maternal progeny of A2A2 = $(m_3 + m_4)(1 - tp)$
4. expected non-maternal progeny of A2A2 = $(m_3 + m_4)tp$
5. expected A1A1 progeny of A1A2 = $(m_5 + m_6)(1 - tq + tp)/2$
6. expected A2A2 progeny of A1A2 = $(m_5 + m_6)(1 + tq - tp)/2$

where $m_1 \ldots m_6$ are the observed numbers in each of the classes 1 . . . 6, p is the frequency of A1 and $q = 1 - p$, the frequency of A2, and t is the proportion of outcrossed matings. From these, the average figure for t can readily be estimated.

Estimates of outcrossing (selfing) rate

Several reviews provide a range of estimates for outcrossing rate among many species. The majority of these studies have involved a single polymorphic locus, a procedure which may lead to bias through the possibility of linkage to non-neutral loci. Brown (1990) usefully compares single locus estimates of outcrossing rates with those based on a number of loci for the same species; estimates obtained are in fact remarkably similar in most cases.

In support of theoretical studies, selfing rate estimates have been used in conjunction with data on inbreeding depression for 29 plant species by Charlesworth and Charlesworth (1987). Schemske and Lande (1985) accumulated data for 55 species in support of their contention (Lande and Schemske, 1985) that mixed mating systems are inherently unstable, and do not result in ESS.

Selfing rate and reproductive assurance

Lande and Schemske (1985) suggest that selfing only normally evolves when extrinsic forces such as founder effects and population size bottlenecks cause an outcrossing population to inbreed. Such unavoidable inbreeding, they argue, will reduce genetic loads in populations by exposing more than half of disadvantageous recessive alleles to selection. The resulting populations, thus 'bled' of much of their potential for inbreeding depression (p. 385), will be able to tolerate selfing mutants, which may otherwise be evolutionarily favoured by their reproductive efficiency and automatic selfing advantage (p. 361).

Consequently, Schemske and Lande (1985) expected the distribution of selfing rate in plants to be bimodally distributed, with the majority of plants either mating by outcrossing, or by being mostly selfed. In fact, these authors did find the distribution of selfing rate between species to be bimodal, with only about 30% of the species they studied showing selfing rates which lay between 20% and 80% of matings. A reanalysis by Barrett and Eckert (1990) using a larger data set (129 species) also found such a bimodal distribution (Fig. 9.4), although with some 36% of species with selfing rates between 20% and 80%.

Whatever the reality of Lande and Schemske's original model (and Charlesworth and Charlesworth, 1987, 1990; Charlesworth, Morgan and Charlesworth, 1990 are among those who find it genetically over-simplistic), we can agree with Barrett and Eckert (1990) that 'at the present time. . . . theoretical ideas on inbreeding depression and mating system evolution have far outstripped empirical data from natural populations.'

Barrett and Eckert (1990) further break down the distribution of selfing rate between species (Fig. 9.4). Most woody perennials are outcrossers, as are most gymnosperms, not surprisingly as they are mostly woody perennials.

Although most selfers are annuals, the converse is untrue, perhaps surprisingly, for the majority of annuals are at least partially outcrossed. This suggests that it pays many annuals to invest in attributes which attract cross-pollinating insects. For many, it seems, outcrossing fecundity is sufficient to prevent the selection of an autogamous life style.

In fact, bimodality in mating system is most clearly expressed in wind-pollinated species, virtually all of which are less than 20% or more than 80% outcrossed. There is no reason why wind-pollinated plants should be more subject to small population-type inbreeding than are animal-pollinated species. However, wind-pollinated plants are more likely to encounter poor reproductive assurance when obligately outcrossed than are animal-pollinated species, for pollen distribution is untargeted and reproductive assurance is correspondingly expensive to achieve (p. 88). Possibly, selfing mutants are more likely to be favoured in conditions of poor reproductive assurance, rather than after circumstances which have fortuitously resulted in a bout of selfing.

Intraspecific variability in outcrossing (selfing) rate

Most estimates of outcrossing rates involve 'snapshots' in space and time. Consequently, there is an understandable tendency to 'label' a species with a selfing rate which has been estimated for only one population in

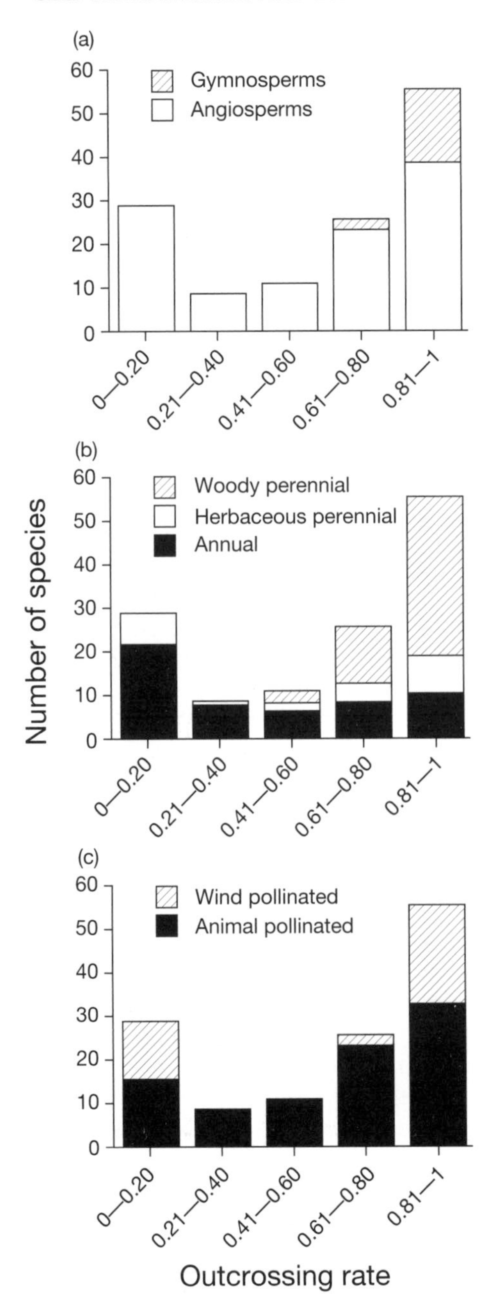

Fig. 9.4 The distribution of outcrossing rates in populations of 129 species of seed plants subdivided into (a) angiosperms and gymnosperms, (b) life forms, and (c) animal- and wind-pollinated species. (After Barrett and Eckert, 1990.)

only one season. However, it is important that we understand how outcrossing or selfing rates differ within a species. This allows us to understand genetic variation for features which influence selfing rate that exist within and between populations and on which natural selection can act (p. 362). It also indicates how selection pressures for outcrossing rate resulting from attributes such as fecundity, population size, and the vigour and variability of offspring, can change from season to season and locality to locality.

Some species vary between individuals from being totally outcrossed to largely selfed. Indeed, such variation is inherent in breeding systems such as gynodioecy where females are by definition totally outcrossed, but hermaphrodites may be largely selfed (p. 326). In Chapter 7, I have discussed populations of the outcrossing heterostylous primose (*Primula vulgaris*) which contain selfing homostyles of recombinational origin. Such cases can give us exceptional insights into the disruptive evolutionary tensions which in part favour both selfing and outcrossing mating systems simultaneously.

Barrett and Eckert (1990) review intraspecific variation in outcrossing rate in 24 species (Table 9.3). They subdivide these into those which vary as a result of genetically controlled factors; those which are influenced by demographic factors; and those which vary according to external, environmental causes.

Barrett and Eckert identify a variety of genetically controlled factors which can influence intraspecific outcrossing rate, not only gynodioecy and heterostyly, but also herkogamy (e.g. the heterostylous but self-fertile *Turnera ulmifolia*), dichogamy as in *Gilia achilleifolia* (Table 9.1) and *Limnanthes* (Table 9.2), flower size and even flower colour. Typically such factors have profound effects on the outcrossing rate, although the effect of flower colour is not surprisingly not great. It is interesting to observe how intraspecific variability in flower size can strongly influence outcrossing rate in the cherry tomato *Lycopersicon pimpinellifolium*. Where pairs of related species occur, one of which is an outcrosser and the other a selfer, characteristically the outcrosser has larger flowers (p. 364). Here, cause and effect may become inextricably interlinked. Smaller flowers attract fewer visitors and so may cause fewer outcrossing events to occur. Those selfing flowers are likely to be favoured which 'spend' less on male resource, which includes flower display (p. 172).

Of the so-called demographic effects, pollination system has relatively little impact on outcrossing rate (although it can certainly affect effective population number, Schmitt, 1980). However, pollinator abundance might possibly influence outcrossing rate, as has been suggested for *Lupinus nanus* which has extremely variable outcrossing rates between populations and between seasons (Horovitz and Harding, 1972). However, as this self-fertile plant nevertheless requires an insect visit for pol-

Table 9.3 Genetic, demographic and environmental factors that contribute to intraspecific variation of outcrossing rate (t) in natural plant populations

Factor	Species	Mean t	Range[a] of t	Relation[b] with t
GENETIC FACTORS				
A. Quantitative floral traits				
Flower size	*Lycopersicon pimpinellifolium*	0.13	0.00–0.40†	+
Herkogamy	*Turnera ulmifolia*	0.34	0.04–0.79†	+NS
Dichogamy	*Clarkia tembloriensis*	0.55	0.08–0.83	+
	Gilia achilleifolia	0.57	0.15–0.95	+
B. Polymorphic floral traits				
Flower colour				
pigmentation	*Ipomoea purpurea*	0.65	0.60–0.86†	+
blue reflectance	*Lupinus nanus*	0.43	0.27–0.34	+
Heterostyly				
frequency of homostyles	*Amsinckia spectabilis*	0.29	0.03–0.53	–
	Eichhornia paniculata	0.55	0.00–0.96	–
	Primula vulgaris	0.85	0.08–1.1†	–
Gynodioecy				
frequency of females	*Bidens menziesii*	0.62	0.58–0.65	+
	Plantago coronopus	0.86	0.62–0.98	+
Capitulum polymorphism				
frequency of radiate morph	*Senecio vulgaris*	0.04	0.03–0.04†	+
DEMOGRAPHIC FACTORS				
Plant size	*Malva moschata*	0.65	0.22–1.0†	–
Plant density				
animal pollination	*Cavanillesia plantanifolia*	0.46	0.35–0.57†	+
	Echium plantagineum	0.88	0.73–0.97	+
	Helianthus annuus	0.75	0.65–0.86	–
wind pollination	*Pinus ponderosa*	0.92	0.85–0.96	+
	Plantago coronopus	0.70	0.62–0.92	+
Population size	*Eichhornia paniculata*	0.55	0.00–0.96	+
ENVIRONMENTAL FACTORS				
Moisture	*Abies lasiocarpa*	0.89	0.65–0.99†	+
	Bromus mollis	0.09	0.04–0.16†	+NS
	Hordeum spontaneum	0.02	0.00–0.02	+
	Picea englemanni	0.86	0.85–0.93†	+
Altitude	*Abies balsamea*	0.89	0.78–0.99	–

[a] Ranges marked with '†' are from within-population studies; all others are among-population ranges.
[b] The relationship of each factor with t is indicated as '+' = positive; and '–' = negative ('NS' denotes a non-significant but consistent trend).

lination to occur, the relationship between pollinator abundance and mating system is somewhat obscure. Such a relationship is much more likely to occur in a species where flowers show 'fail-safe' autogamy after an outcrossing phase.

Population size will chiefly affect outcrossing rate when, as in *Eichhornia paniculata*, the plant has vigorous vegetative reproduction and mostly reproduces clonally. In such cases, apparently large populations may in fact consist of only a single individual. Further colonization of new sites by seed can only result from self-fertile mutants. Where a species with vigorous vegetative spread has specialized disjunct 'island' habitats, this life style almost inevitably selects for self-fertility, which in the case of one-clone sites will be obligate.

On p. 40, I have discussed how stabilizing selection for size may operate for self-fertile bee-pollinated racemose inflorescences, because large rewarding inflorescences may preferentially attract pollinators to the extent that they may become mostly geitonogamously selfed. The results for musk mallow, *Malva moschata* (Crawford, 1984b) provide another example of this phenomenon. Although large flower size promotes outcrossing, large plant size may promote selfing. Harder and Barrett (1995) later presented a similar report for *Eichhornia paniculata*.

In contrast to these large effects on the mating system that intrinsic and extrinsic biotic factors can wield, abiotic (environmental) factors appear to have little influence on the mating system, intraspecifically.

Within-individual variability in selfing rate

Perhaps the most interesting examples of between-flower variation in mating system come from division of labour between flowers. Where early season chasmogamous flowers are followed by late season cleistogamous flowers, seedlings from the latter result by definition from selfs, whereas early season seedlings, resulting mostly from crosses, tend to be both more variable and more vigorous (Waller, 1984; Clay and Antonovics, 1985).

Division of labour between florets in the capitulum of many members of the families Compositae or Dipsacaceae may also result in a bimodality of mating system within an individual, albeit rather more subtly. In many UK populations, the groundsel *Senecio vulgaris* is polymorphic for capitula which bear, or lack, ligulate ('ray') female florets (Fig. 9.5). As disk florets are hermaphrodite, radiate plants are gynomonoecious.

Remarkably, genes which control the 'radiate' groundsel phenotype in the UK are derived from the introduced species *S. squalidus*, which is an obligate outcrosser (Abbott and Forbes, 1993). Thus, radiate groundsels form excellent examples of introgressive hybridization across a 'ploidy barrier' (Abbott, Ashton and Forbes, 1992), which has been resynthesized

Fig. 9.5 Capitula of (from right) two unrayed groundsels, *Senecio vulgaris* (*rr*), a heterozygote rayed groundsel (*Rr*) and two homozygous rayed groundsels (*RR*). The large capitulum on the left is of the Oxford ragwort *Senecio squalidus* from which the rayed gene in groundel arose through introgressive hybridization (×1.5).

in the laboratory (Ingram, Weir and Abbott, 1980). It is of particular interest that one parent to these introgressive populations is an outcrosser, and the other is usually selfed.

Marshall and Abbott (1984a,b) have shown that radiate morphs are more outcrossed than are the rayless morphs, and they attribute this largely to the 10–13 female ligulate (ray) florets distributed around the capitulum. These comprise about 20% of all florets (Fig. 9.5). Capitula bearing ray florets appear to be more conspicuous to insects than are the non-radiate capitula (Fig. 9.6) and Marshall and Abbott (1984b) have shown that radiate capitula receive more insect visits, which may account in part for this increased level of outcrossing (Table 9.4). Nevertheless, seed resulting from the female ray florets is more likely to be outcrossed than those from unrayed florets in the same capitulum, as estimated by the use of marker genes.

Abbott (1985) points out that where the radiate morph occurs, it always does so polymorphically. This suggests that balanced selection pressures for radiate and non-radiate morphs have resulted in ESS equilibrium frequencies. It might be thought that such equilibria might have resulted as a result of the contrasting advantages of reproductive assurance for all florets of the hermaphrodite non-radiate morph, compared with the vigour and variability of the offspring of the female ray florets of the gynomonoecious radiate morph which have an outcrossed origin. However, Abbott finds no direct evidence for enhanced vigour among the offspring of ray florets. Rather he suggests that for a Cardiff population,

Table 9.4 Outcrossing frequencies of (a) the 'outer' and 'inner' floret fractions of capitula from radiate plants, and (b) the radiate and non-radiate morphs, in populations of *Senecio vulgaris* sampled from Cardiff and Leeds, together with heterogeneity χ^2 comparisons (after Marshall and Abbott, 1984b)

Population	Outcrossing frequencies	
	Cardiff	Leeds
(a) Floret fractions within radiate capitula		
outer (ray) florets	0.225 (436)	0.063 (627)
inner (disc) florets	0.098 (559)	0.036 (565)
χ^2 (1)	48.16***	7.26**
(b) Morphs[a] in populations		
radiate	0.137 (999)	0.060 (1316)
non-radiate	0.007 (1046)	0.031 (1302)
χ^2 (1)	230.6***	1.54 (N.S.)

Numbers in brackets represent the progeny scored.
[a] Morph values are from Marshall and Abbott (1984a).
N.S., not significant.
** Significant at $P = 0.01$.
*** Significant at $P = 0.001$.

Fig. 9.6 A small hover-fly visits the capitulum of a rayed groundsel (×3).

the radiate plants, containing as they do genes of the larger parent *S. squalidus*, are more fecund, and that this outweighs the poor reproductive assurance of the female florets. However, for an Edinburgh population, Ross and Abbott (1987) report the opposite finding, that is that the non-radiate morph is the more fecund. Yet in both cases similar equilibrium frequencies of the radiate morph have resulted. These populations are known to have originated from different introgressive events (Irwin and Abbott, 1992), and are likely to have incorporated different parts of the *S. squalidus* genome. Although this is likely to explain the different innate fecundities of the radiate morph, it is remarkable that a stable polymorphism obtains in both conditions.

In a way, these findings parallel those for heterostylous and homostylous primrose *Primula vulgaris* (p. 288) where similar ESS polymorphic equilibrium frequencies for outcrossers and selfers are also found between different populations. In this case as well, outcrossing advantage seems to have accrued not from inbreeding depression or low variability among the offspring of the selfer, but from an intrinsic feature of the outcrosser.

THE EFFECT OF SELFING

Theoretical considerations

Darwin said 'nature abhors perpetual self-fertilization' (Darwin 1862, 1876), although he was well aware of the apparently greater reproductive fitness of many selfers. Repeated self-fertilization rapidly results in a reduction in genetic variation within populations, and in a decrease in the proportion of loci in populations and individuals that are heterozygous: most loci become fixed in the homozygous state. How this occurs is most simply examined by following the fate of a heterozygous locus *A/a* in a line experiencing repeated selfing. The proportion of heterozygotes is halved in each successive generation, for a selfed heterozygote yields 50% heterozygotes and 50% homozygotes, which can subsequently only form homozygotes. The proportion $f^{(n)}$ after n generations selfing is thus $(0.5)^n f^0$, where f^0 is the original proportion of heterozygotes (Table 9.5).

Thus within relatively few generations, often less than ten, one may presume that an obligate selfer would become homozygous (fixed) at most loci, in the absence of heterozygous advantage. It does not follow, however, that an obligate selfing population will become genetically invariable. Different lines within the population may show fixation of different homozygotes from originally heterozygous loci either by chance, if the alleles are nearly neutral with respect to each other, or by disruptive selection. If the alleles are neutral, the proportion of the two homozygotes *AA* and *aa* will depend on the initial frequency of *A* and *a* before selfing.

We have seen, however, that obligately selfing populations are probably very rare, and some outcrossing occurs onto nearly all autogams. In such a partially autogamous population, the proportion of heterozygotes $f^{(n)}$ in the population after n generations is calculated by:

Table 9.5 The fate of a heterozygous locus A/a during repeated self-fertilization

	Fate of locus	Generation	Proportion of offspring of original parent heterozygous
Homozygotes	*Aa* selfed		
AA and *aa*	1 *AA* / 2*Aa* \ 1 *aa*	F_1	0.5
can only	1 *AA* / 2*Aa* \ 1 *aa*	F_2	0.25
form	1 *AA* / 2*Aa* \ 1 *aa*	F_3	0.125
homozygote	1 *AA* / 2*Aa* \ 1 *aa*	F_4	0.062
offspring	1 *AA* / 2*Aa* \ 1 *aa*	F_5	0.031
	1 *AA* / 2*Aa* \ 1 *aa*	F_6	0.015
	etc.		

$$(s/2)^n f^0 + 2pq(2t/2 - s)\left(1 - (s/2)^n\right)$$

where s is the proportion of selfing, t is the proportion of random outcrossing $(1 - s)$, f^0 is the proportion of heterozygotes at generation 0, p is the frequency of A in generation 0, and q is the frequency of a in generation 0. For further reading on the modelling of the genetic structure of inbred populations, refer to Allard (1960), Gale (1980) or Cook (1971).

As n increases, so $(s/2)^n$ tends towards zero, so that at equilibrium, the frequency of heterozygotes f becomes $2pq\ (2t/2 - s)$ and is thus dependent on the initial allele frequencies, and the proportion of selfing. These equilibrium frequencies of heterozygotes will vary from 0.017 (less than 2%) where one allele is initially rare (p = 0.05) and selfing is high (90% t = 0.1) to 0.50 where p initially = 0.5 and the populations are fully outbreeding (t = 1) (Table 9.6).

Consideration of the effect of selfing on many originally heterozygous loci is simple when they are unlinked on different chromosomes, as each locus will obey the same rules independently. The number of generations needed for an individual to become fully homozygous increases with the number of loci considered (Allard, Jain and Workman, 1968, Fig. 9.7) but is still remarkably small. After 12 generations of repeated total selfing, 98% of individuals originally heterozygous for 100 loci (assuming so many could be unlinked!) will have become totally homozygous for these loci. However, even low frequencies of outcrossing, of the order of 1%, will allow few loci to fix, and fixation equilibria are rarely reached, because crossing will occur between differentially fixed homozygotes (*AA* and *aa*) in the same population.

Table 9.6 Expected equilibrium proportions of heterozygotes under mixed random mating and selfing (without selection) for various assumptions about t and P (initial frequency of allele A_1)

Probability of outcrossing (t)	Initial frequency of A_1 (P)					
	0.05	0.10	0.20	0.30	0.40	0.50
0.1	0.0173	0.0327	0.0582	0.0764	0.0783	0.0909
0.2	0.0317	0.0600	0.1067	0.1400	0.1600	0.1667
0.3	0.0439	0.0831	0.1477	0.1939	0.2216	0.2308
0.4	0.543	0.1029	0.1829	0.2610	0.2743	0.2857
0.5	0.0633	0.1200	0.2133	0.2800	0.3200	0.3333
0.6	0.713	0.1350	0.2400	0.3150	0.3600	0.3750
0.7	0.0783	0.1482	0.2636	0.3459	0.3953	0.4118
0.8	0.0844	0.1600	0.2844	0.3733	0.4266	0.4444
0.9	0.0900	0.1705	0.3022	0.3969	0.4548	0.4737
1.0	0.0950	0.1800	0.3200	0.4200	0.4800	0.5000

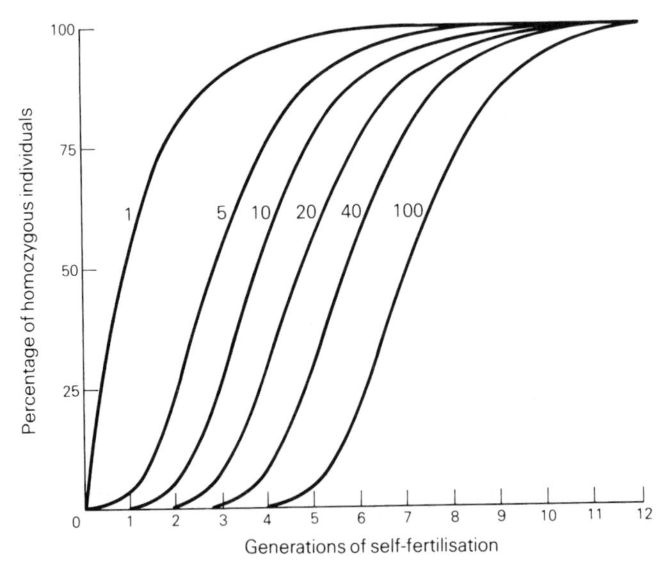

Fig. 9.7 Percentage of homozygous individuals after various generations of self-fertilization, when the number of independently inherited heterozygous gene-pairs is 1, 5, 10, 20, 40 and 100. (After Allard *et al.*, 1968)

Linkage disequilibria

Where loci are linked on the same chromosome, considerations of increased homozygosity and genetic fixation become more complex (usefully summarized in Allard, Jain and Workman, 1968). Linkage of two loci *A/a* and *B/b* on the same chromosome will tend to result in linkage disequilibria, so that the frequency of the four possible gamete genotypes *AB*, *Ab*, *aB* and *ab* differs from that predicted by the products of the frequencies of the alleles *A*, *a*, *B* and *b* in the population. In the absence of differential selection in an outbreeding population, linkage disequilibria will disappear through recombination in time, although the more closely linked the loci (low recombination values), the more slowly will this occur. Total selfing of a population carrying two heterozygous linked loci in disequilibrium will probably result in homozygous fixation at disequilibrium, with some genotypes (e.g. *AAbb*) being more common and some (e.g. *AABB*) less common than predicted by allele frequencies.

In a partly inbred situation, linkage can cause linkage disequilibria to be maintained, and fixation to occur very slowly. In Fig. 9.8, where $s = 0.9$ ($t = 0.1$), random segregation ($c = 0.50$) causes linkage disequilibria to almost halve in ten generations and zygotic associations, which monitor genetic

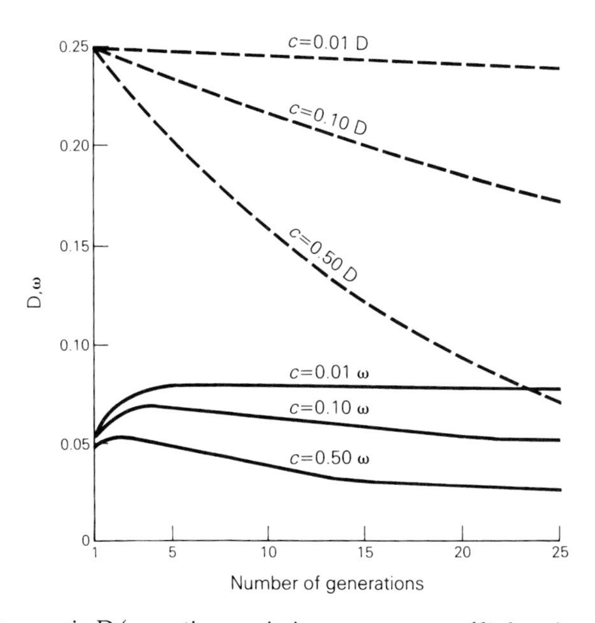

Fig. 9.8 Changes in D (gametic associations, a measure of linkage) and ω (zygotic associations, a measure of genetic fixation) in a 90% selfed ($s = 0.90$) model population for three different levels of linkage between *A/a* and *B/b*. Where $c = 0.50$ there is random segregation; where $c = 0.10$ and 0.01, *A/a* and *b/b* are recombined in 20% and 2% of meioses, respectively. The experiment runs for 25 generations (*x* axis). The original population started with the linkage groups *AB* and *ab* equally frequent. (After Allard *et al.*, 1968.)

fixation, increase. However, tight linkage between two loci ($c = 0.01$) causes almost no diminution of disequilibrium, and fixation proceeds more slowly. As one allele proceeds to homozygous fixation through the force of inbreeding, it 'drags' closely linked alleles of different loci with it, irrespective of their type, a process sometimes known as 'hitchhiking' (p. 16). Linkage, especially close linkage in which the 'normalizing' effects of recombination are minimized, removes the independence of a locus proceeding to fixation. However, in totally autogamous populations, recombination will eventually ensure that chromosomes as well as genes will become entirely homozygous (Gale, 1980, pp. 22–33).

Polyploidy

Just as linkage tends to slow down the process of homozygous fixation of polymorphic loci in inbreeders, so does polyploidy. At polysomic loci, the proportion of heterozygotes remaining in the population H^n after total selfing is determined by:

$$[(4K-3)/2(2K-1)]^n \times H^0$$

where H^0 is the original frequency of heterozygotes, n is the number of generations of obligate selfing, and K is half the 'ploidy level (2x = 1, 4x = 2, 6x = 3, etc). After one generation of selfing in a diploid, where $H^0 = 0.5$, H^n becomes 0.25, i.e. the proportion of heterozygotes is halved. Thus, in a tetraploid, H^n is 0.416 and the proportion of heterozygotes is reduced by 1/6, and in a hexaploid, H^n is 0.45 and the proportion of heterozygotes is reduced by 1/10. At a tetrasomic locus, heterozygotes can exist in simplex (*Aaaa*), duplex (*AAaa*) and triplex (*AAAa*) forms. Thus, polyploids buffer polymorphic loci against the effects of selfing more effectively than diploids. Of course, most polyploids are allopolyploid, and are not necessarily polysomic for any one locus; however, resulting as they do from hybrids between related parents, allopolyploids may be more often effectively polysomic than appears from the homology and pairing relationships of chromosomes from the parental genomes. Tetrasomic inheritance is well known in many loci of allotetraploids such as dutch clover *Trifolium repens* or birds-foot trefoil *Lotus corniculatus*.

It is frequently suggested that inbreeders are more often polyploid than are outbreeders (p. 360). Outbreeding systems such as dioecy (p. 311), heteromorphy (p. 291) or gametophytic self-incompatibility (p. 220) often fail to work efficiently at the polyploid level. However, polyploids are much more likely to originate among selfers due to minority effects on the fertility of newly arising polyploids (Fowler and Levin, 1984). Furthermore, polyploids are better able to maintain relatively high levels of variability and heterozygosity when partially selfed, or through temporary bouts of total selfing than are diploids; in high polyploids, the accumulation of homozygosity may be very slow indeed. A good example is seen in the local Scottish endemic homostyle *Primula scotica* for which no variation for DNA after RAPD analysis could be found between individuals from four populations, and only four out of 61 individuals displayed a heterozygous phenotype for one isozyme locus; no variation was discovered for another 14 loci. However, no less than 7 of the 15 isozyme loci displayed fixed heterozygous phenotypes, which no doubt reflect the allohexaploid origin of this largely selfed species (Glover and Abbott, 1995). An exception to the general association between polyploidy and selfing is found in *Draba* (Brochman, 1993), in which all the outcrossing species are polyploid.

Disadvantageous alleles

It is likely that a high proportion of the recessive alleles carried heterozygously will be disadvantageous if expressed phenotypically in the homozygous condition. In a randomly outbred, panmictic condition, such homozygotes will occur, and be selected against, but as selection renders the allele increasingly scarce, the homozygote becomes still scarcer, at the square of the frequency of the allele. In a population in which the frequency of a recessive allele is 1 in 100 (0.01), it will be expressed in only 1 in every 10000 individuals (0.0001). At low frequencies, selection against harmful recessive alleles proceeds very slowly indeed in panmictic populations. This allows high levels of residual variability and heterozygosity to be maintained.

A selfing population is quite different, for heterozygotes that protect disadvantageous recessive alleles from expression rapidly disappear, and these alleles are exposed to selection in a homozygous condition. Instead of being protected from the rigours of selection at low frequencies, as in a panmictic condition, selection will cause disadvantageous alleles to disappear rapidly from fully selfing populations, and such genes will be unusual even in less completely selfed plants. Consequently, selfing populations carry little if any 'genetic load' of deleterious recessives, and are genetically 'clean'.

The early phases of inbreeding may expose deleterious recessives in phenotypically obvious ways. Experimentally selfed material from naturally outcrossing populations may produce some progenies with a proportion (usually 1:3) of non-viable offspring that may be albino (chlorophyll-less), very slow growing, amino acid requiring or which may simply fail to germinate (lethal). Continued selfing will lead to genetically 'clean' stocks in which such lethals or sublethals no longer occur.

It is worth pointing out briefly that for those many bryophytes and pteridophytes which commonly or habitually self-fertilize in the haploid generation, the offspring are completely homozygous. Presumably such plants carry virtually no genetic load, particularly in the bryophytes where the dominant generation, in which most genes would be expressed, is haploid anyway.

INBREEDING DEPRESSION AND HYBRID VIGOUR

Inbreeding depression has probably been recognized as long as humans have domesticated plants and animals; it was certainly discussed by Virgil two millenia ago. It can be defined as a loss of vigour in offspring which result from bouts of inbreeding, i.e. selfing or near-relative matings, for instance in small and genetically depleted populations, compared with equivalent outcrossed families. In the absence of selfing, inbreeding is usually quantified by Wright's (1921) coefficient of inbreeding f (p. 369). For at least the last century, plant and animal breeders have focused their efforts on means whereby the effects of inbreeding depression can be avoided among highly selected, genetically depleted, stocks. However, cruder measures to this end have been employed for much longer; the mule is one of many examples that could be quoted.

The terms hybrid vigour and heterosis are used interchangeably here; the term hybrid begs the question as to what is a species. However, hybrid vigour is more often used with respect to interspecific crosses. Hybrid vigour is frequently thought of as the straight corollary of inbreeding depression; that is the less two parents are related genetically, the less likely it is that the offspring of their matings will show inbreeding depression. In fact, the hybrid offspring of two different species may prove to be much more vigorous than either of the parental species ever is, as any gardener will testify. However, such a level of hybrid vigour may never be manifested in the wild, as the hybrids may never occur naturally, or may be maladapted to natural habitats, or they may be sterile.

There is little doubt that inbreeding depression and hybrid vigour are strongly associated with the proportion of loci that occur within an organism in a heterozygous condition. Relatively homozygous individuals tend to be less vigorous than are relatively heterozygous ones. As selfing and near-relative mating increase the proportion of loci that are homozygous within an individual (p. 379), these mating systems tend to result in a loss of vigour, or inbreeding depression.

There are remarkably few studies which specifically relate inbreeding depression to levels of homozygosity. One of the best known, for the lima bean, *Phaseolus lunatus*, is depicted in Fig. 9.9 (Harding, Allard and Smeltzer, 1966). The classical study of Schaal and Levin (1976) on *Liatris cylindracea*, which demonstrates a strong relationship between levels of heterozygosity and survivorship in this species, is of a kind that deserves repeating more often (Fig. 9.10).

The basis of inbreeding depression

It is a remarkable fact that although we have known about inbreeding depression and hybrid vigour for so long, although they dominate our breeding strategies, and although these strategies have been largely responsible for a doubling in global food productivity between 1960 and 1990, we are still some way from a full understanding of why inbreeding depression occurs.

In fact there are no less than four hypotheses which may explain in part the relationship between inbreeding depression and homozygosis. However, it should be pointed out that these are not self-exclusive, and to some extent they may all be describing the same phenomenon from different perspectives:

1. Inbreeding (including selfing) causes deleterious recessive alleles to be phenotypically expressed more frequently by occurring in the homozygous phase, as detailed in the previous section (the 'partial dominance' theory).
2. Where heterozygous loci result in phenotypes superior to either homozygote (overdominance), inbreeding results in the inferior

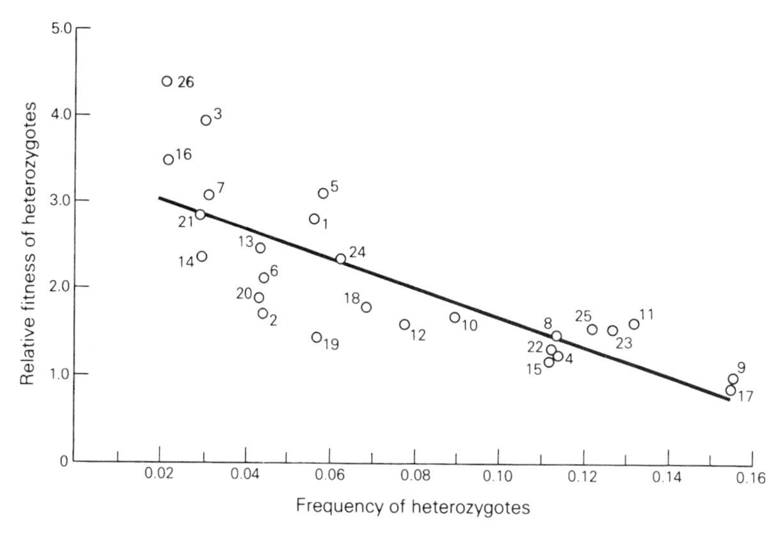

Fig. 9.9 Relationship between the frequency of heterozygotes and their fitness relative to homozygotes for the *S/s* locus in lima bean (*Phaseolus lunatus*) populations. Homozygotes were assigned a relative fitness of 1.0. (After Harding *et al.*, 1966, quoted in Allard *et al.*, 1968.)

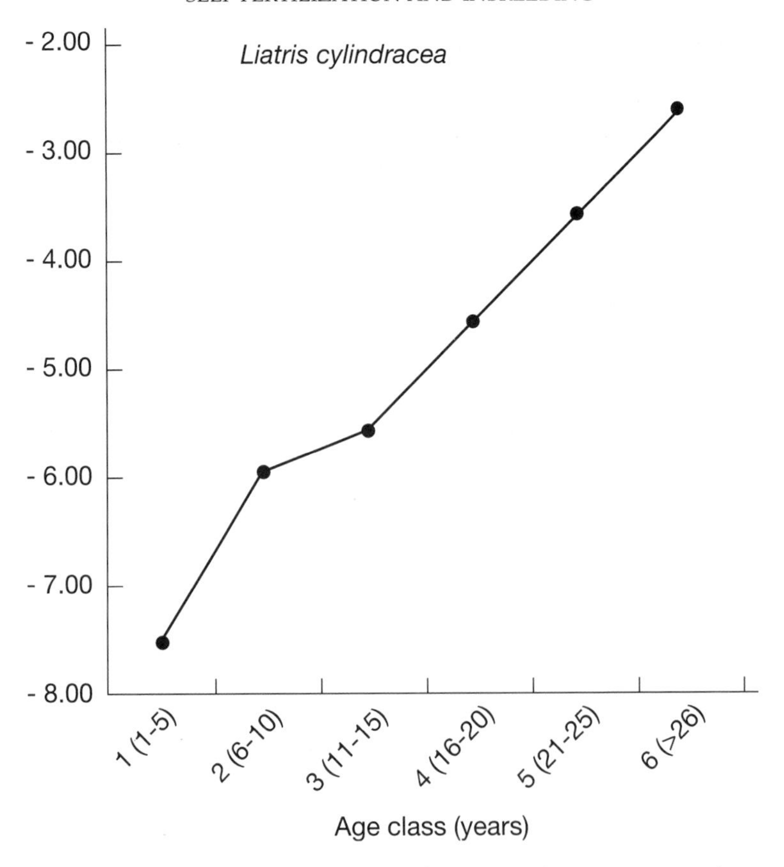

Fig. 9.10 The decrease in heterozygote deficiency with age in a natural population of *Liatris cylindracea*. Ordinate is deviation from expectation assuming random mating. (After Schaal and Levin, 1976.)

homozygous phenotypes being expressed more frequently (the 'overdominance' theory).

3. Where dominance effects contribute substantially to phenotypic variance for characters which are controlled by additive polygenic inheritance, inbreeding removes this dominance-controlled input into the variance.
4. Gene products are enzymes, so inbreeding reduces the numbers of enzymes produced. As the efficiency of metabolic systems is enzyme reaction rate-limited, a restriction in the number of alternative metabolic pathways ('shunts') may cause metabolic 'bottlenecks' to occur, thus slowing metabolic systems and growth rates.

Inbreeding depression has been subject to a number of reviews, most recently by Charlesworth and Charlesworth (1987). These authors show that genetic dominance and overdominance can undoubtedly explain inbreeding depression fully at a genetic level, although not on a molecular basis. Any instance where the phenotype resulting from the expression of a recessive allele is less fit than that caused by the dominant allele at the same locus, or where the heterozygote is more fit than either homozygote, will result in inbreeding depression (e.g. Crow and Kimura, 1970). This is because inbreeding increases the proportion of homozygous loci within the individual through which less fit phenotypes thus tend to be expressed more frequently.

In general, alleles whose phenotypic expressions lower individual fitness tend to be recessive to more commonly occurring 'wild type' alleles. Most new mutants will be genetically recessive by failing to produce a gene product at all, whereas in other cases the gene product may fail in the absence of appropriate operons. In contrast, the genetic and chromosomal environment of phenotypically successful alleles will have evolved through evolutionary time to ensure that their gene products are expressed (dominant). Consequently, the association of recessive homozygosity with low fitness (inbreeding depression) is inevitable.

Partial dominance versus overdominance

If phenotypic expression of the 'genetic load' was the sole explanation for inbreeding depression, we would expect:

1. that total selfing would lead to total homozygosity;
2. that as an originally outcrossing strain continued to self, later generations would show fewer signs of inbreeding depression as the genetic load decreased, until finally the homozygous strain was as vigorous as the original outcrossers (Lande and Schemske, 1985);
3. that if two totally homozygous, true-breeding strains purged of all genetic load are crossed, their offspring should not show hybrid vigour (heterosis), nor should the secondary selfing of these offspring lead once again to a reduction of vigour;
4. that interspecific crosses between two fully panmictic species should show no hybrid vigour.

As discussed by Lande and Schemske (1985) and Charlesworth and Charlesworth (1987) none of these propositions is altogether true for nearly every instance examined, although it is certainly the case that levels of vigour usually become somewhat restored as a bout of selfing continues through several generations (Barrett and Charlesworth, 1991). It is usually assumed that the fittest allele will always proceed to fixation, but stochastic effects and changes in selection pressures through space and time will often cause the invariable genotypes which fix to be less than optimal. The consequences of such localized differentiation between alleles fixed by selfing in different populations may resemble those caused by overdominance.

Nevertheless, it is difficult to escape the conclusion that some key fitness loci show overdominance effects, where the heterozygote is fitter than either homozygote, in most organisms and are usually at least in part responsible for inbreeding depression.

Charlesworth and Charlesworth (1987) use a modification of Mather's (1949) classical breeding equations for polygenic inheritance to assess what proportion of inbreeding depression is caused by partial dominance, and what proportion by overdominance effects (assuming interacting loci to be unlinked).

1. Two pure-breeding lines are obtained by continued selfing, and these are used to estimate the variance in later generations caused by environmental heterogeneity E.
2. The F_1 generation between these lines should also be genetically uniform and vary only by E.
3. F_2 and the two backcross generations B_1 and B_2 are next created, and the variance of each generation is measured.
4. Variance of the F_2 is accounted for by $1/2D + 1/4H + E$.
   Variance of the B_1 and B_2 together results in $1/2D + 1/2H + E$.

where D is the sum of the variance resulting from additive gene effects and H is the sum of the variance resulting from dominance interactions between genes. As E can be estimated, H, and thus D, can also be estimated by substitution.

Where the ratio a = root (H/D) is substantially greater than one, most inbreeding depression should result from overdominance effects, but when a is less than one, most results from partial dominance effects.

Molecular basis of inbreeding depression

The partial dominance and the overdominance models for inbreeding depression deal with the problem at a genetical rather than a molecular level. As yet, little evidence has accrued as to the molecular basis of inbreeding depression, but we may suppose that these two explanations are likely to have rather different molecular causes.

I assume that the enzyme reaction rate limitation hypothesis (p. 386) underlies both the partial dominance and the overdominance models for inbreeding depression. In this case, it is reasonable to expect that where a locus shows heterozygous advantage (overdominance), both alleles code for metabolically active gene products (enzymes) which provide alternative metabolic pathways, so that the metabolic reaction mediated by these enzymes is most efficient when both enzymes are present. Thus, either homozygote phenotype will be inferior to that of the heterozygote.

However, where inbreeding depression results from a partial dominance effect, we may assume that the recessive allele is null, coding for no gene product, or it codes for a product which is inferior at mediating a metabolic reaction, compared to that coded for by the dominant allele. In this case, the recessive homozygote phenotype will be metabolically inferior to that of the heterozygote, or of the dominant homozygote.

Evidence for inbreeding depression

The first major study of inbreeding depression was made by Darwin (1876). This extensive series of experiments would not stand up too well to modern scrutiny, particularly with respect to the origin and sampling of the strains selected, but nevertheless the scale and scope of his work is still impressive. The performance, assessed as weight of seed produced per plant (and other attributes), was compared for selfed and crossed lines over ten generations for more than 40 species of flowering plants.

The results of these experiments are summarized in Table 9.7. Darwin's major conclusion is that although the majority of species selfed showed some inbreeding depression in the first seedling generation, and that this usually continued to increase throughout all 10 generations of selfing, habitual selfers were less likely to show inbreeding depression than were habitual outcrossers.

Darwin also showed that only one crossing episode between different selfed lines will often nullify the effects of inbreeding depression over

Table 9.7 Lists of species that showed, and did not show, inbreeding depression after (usually ten) generations of artificial selfing when compared with artificially crossed lines of the same species (after Darwin, 1876)

Species showing inbreeding depression		Species showing no inbreeding depression	
Outbreeders	Inbreeders	Outbreeders	Inbreeders
Ipomaea purpurea	*Lactuca sativa*	*Scabiosa atropurpurea*	*Borago officinalis*
Primula veris (heterostyle)	*Iberis umbellata*	*Passiflora gracilis*	*Nolana prostrata*
Cyclamen persicum	*Papaver dubium*	*Hibiscus africanus*	*Legousia hybrida*
Nicotiana tabacum	*Reseda lutea*	*Dianthus caryophylleus*	*Bartonia aurea*
Lobelia fulgens	*R. odorata*	*Primula sinensis*	*Adonis aestivalis*
Nemophila insignis	*Viola tricolor*	*Petunia violacea*	*Phaseolus multiflorus*
Mimulus luteus	*Tropaeolum minus*		*Pisum sativum*
Digitalis purpurea			*Ononis minutissimum*
Verbascum thapsus			*Phalaris canariensis*
Origanum vulgare			*Primula veris* (homostyle)
Brassica oleracea			
Escholtzia californica			
Pelargonium zonatum			
Limnanthes douglasii			
Lupinus luteus			
Lathyrus odoratus			
Cytisus scoparius			

several generations. Another finding of his was that it was possible to select fully selfed, true-breeding lines of habitual outbreeders such as *Petunia violacea* which were at least as vigorous as any outcrossed line. Today, we would suggest that in this instance no (over) dominance variance for vigour occurred, and that careful selection had caused the optimal concurrence of genes with additive effects (in other words that a would approach zero in this case).

In fact, inbreeding depression seems to be almost universal among non-haploid organisms (Jarne and Charlesworth, 1993). The few cases where it is claimed not to occur seem always to be habitual selfers (e.g. Wyatt, 1990). Nevertheless, inbreeding depression has rarely been quantified accurately, nor its effects compartmented through a whole reproductive cycle. Stevens' (Stevens and Richards, 1985) work with gynodioecious *Saxifraga granulata* forms an example, illustrating well how evidence for inbreeding depression can persist into another (vegetative) generation (Table 8.8). Tremayne and Richards (1997) have also shown how the effects of inbreeding depression can persist throughout a generation in the field, whereas phenotypic effects on vigour resulting from different seed sizes are soon lost after transplantation (Table 9.8). Interestingly, in this case the study species, *Primula scotica*, is a homostyle selfer. It may be relevant that it is an allohexaploid which characteristically displays fixed heterozygosity at many loci.

Charlesworth and Charlesworth (1987) compartment evidence for inbreeding depression for some 20 species of flowering plant into seed set,

Table 9.8 Germination rate, seedling parameters and survival and flowering performance one year after transplantation into wild populations for seeds >60 mg after selfs and crosses for the selfing homostyle *Primula scotica*

	Cross	Self	$P =$
Days to germination + SE	8.02 ± 0.29	9.09 ± 0.33	<0.05
$\bar{x}$ seedling dry wt (g) at 4 weeks + SE			ns
at 8 weeks			ns
$\bar{x}$ leaf number at transplantation	20.2 ± 1.1	15.3 ± 0.2	<0.001
$\bar{x}$ leaf number after one year's transplantation	14.2 ± 0.8	11.8 ± 0.7	<0.01
$\bar{x}$ rosette diameter (mm) at transplantation	44.8 ± 1.5	37.9 ± 0.5	<0.001
$\bar{x}$ rosette diameter (mm) after one year's transpl.	28.0 ± 0.9	24.5 ± 0.2	<0.001
Proportion of transplants surviving one year	45/50	41/50	ns
Proportion of transplants flowering after one year	8/45	0/41	<0.01

seed germination and seedling performance (Table 9.9). Overall, there is a remarkable correspondence in the relative performance of selfs and their offspring compared with that of crosses and their offspring, both between species, and between stages in the reproductive cycle.

Nevertheless, examples exist where the effects of inbreeding depression seem to be limited to only some features of the reproductive cycle. For instance, Karron (1989) found that selfing in *Astragalus linifolius* massively reduced seedling biomass, but inbreeding depression was not expressed through seed-set or seed germination at all. Similar results were reported for *Lupinus texensis* (Schaal, 1984) and *Impatiens capensis* (Waller, 1984) (Table 9.10).

Working with a fully self-fertile recombinant strain of *Primula* × *tommasinii* (p. 263), Valsa Kurian has similarly compartmented the effects of inbreeding depression over four generations of selfing (Table 9.11), and has found similar results. The interesting extra feature of this experiment, which is confined to a single strain of horticultural origin, is that selfs and their offspring are compared with both illegitimate and legitimate crosses and their offspring. Any inbreeding depression found in the offspring of illegitimate crosses must be due to fitness-related genes that are linked to the heterostyly *S*/*s* chromosome.

Table 9.9 Inbreeding depression estimates in angiosperms calculated as ratio of value for selfed seed or progeny to value for outcrossed seed or progeny (After Charlesworth and Charlesworth, 1987)

	Character affected			
Species	Number of non-aborted seed per pollination	Germination rate	Size[(a)], fertility[(b)], or survival[(c)]	Selfing rate
Gilia achilleifolia	ns	ns	0.56[c]	0.04
Papaver dubium	–	–	0.80[b]	0.75
Costus allenii	0.67	0.88	0.75[a]	~0
Costus laevis	0.63	1.07	0.80[a]	0
Costus guanaiensis	0.66	0.94	0.81[a]	0
Impatiens capensis	–	ns	0.59[a]–0.93[c]	High
Impatiens pallida	–	–	1.5–0.6[b]	High
Thlaspi alpestre	–	0.82	–	0.95
Limnanthes alba	–	–	0.40–1.00[b]	0.03–0.57
Delphinium nelsoni	0.57	–	0.156[c]	–
Limnanthes douglasii	–	–	0.78[b]	0.23
	–	–	0.93[b]	0.06
Silene vulgaris	0.66–0.82	0.57–0.73	0.81[b]–0.04[b]	~0.1–0.3
Phlox drummondii	0.83	–	0.83[c]	0[SI]
Thymus vulgaris	0.69	0.66	0.70	0.1–0.70
Erythronium americanum	0.25	–	–	0.38[2]

Table 9.10 Germinability and survivorship of cleistogamous (CL) and chasmogamous (CH) seeds and seedlings in *Impatiens capensis* (after Waller, 1984)

Study	Seed type	No. of seeds	Number germinated	Prob. of germination	Number surviving	Prob. of survival	Prob. of germ. and survival
1978	CL	322	225	0.70	218	0.969	0.677
	CH	322	232	0.72	231	0.996	0.718
1979	CL	209	189	0.904	118/185	0.638	0.577
	CH	211	205	0.972	168/186 *	0.903	0.877
1980	CL	450	445	0.989	425	0.955	0.944
	CH	450	449	0.998	434	0.967	0.965

*Proportions surviving differ significantly: Chi-squared = 4.47, $P < 0.05$.

Table 9.11 Performance of illegitimate crosses (P × P) and selfs and their offspring relative to that of legitimate crosses (T × P and P × T) and their offspring through several generations for *Primula* × *tommasinii*, compartmented into three stages of the life cycle. In all cases, differences between crosses/selfs within a generation are significant at $P < 0.01$

Generation/P × P or self	% seed set	% seed germination	% survival to flowering
1. P × P	0.656	0.435	0.902
1. self	0.411	0.326	0.423
2. P × P	0.651	0.584	0.896
2. self	0.367	0.348	0.355
3. P × P	0.622	0.583	0.712
3. self	0.421	0.311	0.307
4. P × P	0.625		
4. self	0.332	not tested	not tested

From the results in Table 9.11 we can deduce:

1. that inbreeding depression is expressed at all stages of the reproductive cycle;
2. that in this case, inbreeding depression for seed-set and seed germination increased little if at all over successive generations, although it did appear to increase over successive generations for seedling survival to flowering;
3. that on average, at each stage of the reproductive cycle, selfing resulted in reproductive and offspring fitnesses only approximately one third of those from legitimate crosses; thus the product of fitness estimates from selfs may not exceed one ninth of that from legitimate outcrosses;
4. for seed set and seed germination, illegitimate crosses are less than two-thirds as fit as are legitimate crosses; however, the offspring of

illegitimate crosses are only slightly less fit than are those of legitimate crosses, at least for the first two generations of illegitimate crossing.

From this information, it seems that as many as half the genes which influence vigour and fertility in this primula may be linked to the heterostyly (*S*/*s*) chromosome, although this is only one of eleven chromosomes in the haploid genome.

Inbreeding depression among outcrossers

All of the above information results from inbreeding caused by selfing. In higher plants there is little information about the effects of small population size on inbreeding depression (reviewed in Barrett and Kohn, 1991). In contrast, such information is very well documented for higher animals, which cannot self. For instance Soulé (1976) has demonstrated a remarkably close link between population size and levels of heterozygosity for a large number of animal populations.

In their review on this topic, Ellstrand and Elam (1993) point out that small populations of flowering plants may inbreed for several reasons. The gene pool of such populations is limited, so that many matings may occur between near relatives. However, where potential mating partners are few, and scattered, most matings for self-fertile species will in fact be self. Furthermore, the 'Baker effect' is likely to select for self-fertility among isolated colonies, which may have become 'conditioned' to inbreeding after passing through 'bottlenecks' of small population size (p. 46).

Inbreeding depression in small populations is not only of evolutionary interest (p. 47), but may also be of considerable conservational importance where restricted populations are threatened.

One of the few examples which seems to show that small population size can result in inbreeding depression in plants comes from the work of Menges (1991) who worked with fragmented populations of the prairie catchfly *Silene regia*. Menges showed that both the rate and the reliability of seed germination was greater for large populations than for small ones.

Often, the potential for inbreeding depression in small populations can be estimated by comparing the fecundity and offspring vigour of selfs in comparison with crosses. For instance, Karron (1989, 1991) worked with widespread and geographically limited montane *Astragalus* in dry-land regions of Colorado, USA. He found that for very rare and self-fertile *A. linifolius*, for which as few as 1500 flowering individuals may remain in three isolated populations, artificially selfed plants nevertheless show considerable signs of inbreeding depression. Seedlings resulting from selfs only produced about one-third of the dry weight of those arising from outcrossing. However, significant variation for inbreeding depression was found between maternal lines, suggesting that some individuals carry a heavier genetic load than do others.

The implication is that despite the precarious state of this partially selfed species, many individuals in small populations remain heterozygous enough to carry a considerable genetic load. Only when artificial selfs in rare species fail to elicit inbreeding depression can it be assumed that the population is already substantially inbred.

Such information is perhaps more reliable than general surveys of genetic variability with respect to population size as reviewed by Karron (1987). Karron found that geographically restricted populations often appeared to lack genetic variability. However, such surveys do not target that part of the genetic load which specifically affects vigour and causes inbreeding depression. Selection may favour heterozygosity in overdominant genes, even in very small and genetically invariable populations.

PHENOTYPIC (MORPHOLOGICAL) VARIATION AMONG INBREEDERS

Much attention has been focused on the penalties imposed on the selfing plant by inbreeding depression, while relatively little emphasis has been given to the evolutionary effects of the reduction in genetic and phenotypic variability among siblings that inbreeding will cause.

If selfing populations are genetically less variable than outcrossed relatives, one would expect them to be phenotypically less variable as well. However, the phenotype results from interactions between the genotype and the environment. The amount of this plasticity is itself under genetic control. Thus, genotypically invariable populations should respond to selection so that a high level of potential for phenotypic plasticity should evolve among selfers (Scheiner, 1993).

Table 9.12 Coefficients of variation in fruit length and fruit width in self-compatible (s-c) and self-incompatible (s-i) populations of *Leavenworthia* (after Solbrig and Rollins, 1977)

	Parents (field)	Offspring (culture)
Fruit length		
s-c	0.091	0.118
s-i	0.133	0.127
Fruit width		
s-c	0.066	0.118
s-i	0.098	0.119

A typical study which compares morphological variability between selfers and outcrossers can be found for *Leavenworthia* (Solbrig and Rollins, 1977). This study compared the variability of fruit length and width for self-compatible and self-incompatible parental populations (in the field), and for their offspring raised in relatively standard conditions (Table 9.12). In all cases, self-compatible populations are on average less variable than self-incompatible populations, although these differences were not always statistically significant. Interestingly, variability in offspring was greater than in parents, and less distinct between the two breeding systems, although growing conditions were more uniform. This may be a product of the relatively relaxed stabilizing selection operating in artificial growing conditions, allowing a greater range of genotypes to survive to fruiting.

As the authors predict, between-family variance is significantly greater than within-family variance in nearly all populations of the self-compatible *L. exigua*, whereas most families of self-incompatible populations gave between-family variances that were not significantly greater than within-family variances. In addition, between-family variances were greater on average in self-compatible than in self-incompatible populations. Here we see a clear example of the phenotypic effect of homozygous gene fixation in inbreeders; the differential fixation of neutral or nearly neutral alleles has apparently increased between-family differences, but decreased within-family differences in inbreeders compared with related outbreeders.

The phenotypic effects of selfing can be investigated by comparing variability of the offspring of a single individual in cases where chasmogamous and cleistogamous flowers are both produced. Both Clay and Antonovics (1985) for the grass *Danthonia spicata*, and Waller (1984) for *Impatiens capensis* demonstrate that offspring from cleistogamous flowers are less variable than are those from chasmogamous flowers.

Waller (1984) also showed that the offspring of allogamous flowers outcompete the offspring of cleistogamous flowers from the same parent plant of the balsam *Impatiens capensis*. The variability of the outcrossed offspring was considered to form an important component in their competitive success.

CHAPTER 10

Agamospermy

INTRODUCTION

Asexual reproduction in seed plants (apomixis) can be divided into two main classes; vegetative reproduction and agamospermy. Vegetative reproduction was dealt with in Chapter 10 of the first edition of this book (1986). However, vegetative reproduction is scarcely a 'breeding system', and so this subject has been omitted from the current volume. In contrast, agamospermy is derived from elements of sexual reproduction, and is frequently a modification of sexual reproduction, so it forms a valid subject in the present context.

Agamospermy can be defined as the production of fertile seeds in the absence of sexual fusion between gametes or 'seeds without sex' (Brown, 1978). Sexual fusion presupposes a reductional meiosis if the 'ploidy level is to remain stable. In agamospermy, a full reductional meiosis is usually absent, and thus chromosomes do not segregate. In some cases (*Citrus*), the gametophytic generation is avoided entirely, and the new sporophyte embryo is budded directly from the old sporophyte ovular tissue, usually the nucellus (**adventitious embryony**, or **sporophytic apomixis**) (Fig. 10.1).

More commonly, a female gametophyte (embryo-sac) is produced with the sporophytic chromosome number (**gametophytic apomixis**). This non-reduction of chromosome number is achieved either by a complete avoidance of female meiosis (**apospory** and **mitotic diplospory**), in which case the embryo-sac is usually budded directly from the nucellus; or by failure and non-reduction (restitution) of the female meiosis (**meiotic diplospory**).

One of the most striking features of gametophytic apomixis is its extremely strong association with polyploidy (Asker and Jerling, 1992). This association is not yet understood, although we must take seriously the suggestion of Nogler (1984b) that dominant genes for apomixis may usually be linked to recessive lethals which are expressed only in the gametophyte generation. Thus, these recessives are protected from expression in most polyploids, but not in the reductional gametophytes of any diploids. Such an association does not occur for sporophytic agamospermy in which a gametophyte does not occur.

If a diplosporous female meiosis occurs, there is usually an absence of chiasmata, and hence no recombination. Thus, genes remain inextricably

they arose from sexually sterile hybrid polyploids ('apomixis is an escape from sterility'). Being hybrid and polyploid, agamosperms can be expected to be highly heterozygous, and therefore very vigorous (Chapter 2), the heterozygosity being 'fixed' by agamospermy (see Table 10.1, Fig. 10.2).

Disadvantages

There are four main disadvantages to agamospermy.

1. Inability to escape from accumulating disadvantageous (but non-lethal) mutants ('Müller's ratchet'), for the fittest genotypes would be inevitably 'saddled' with the mutants that became linked to them, unable to escape by the processes of recombination and segregation.
2. Inability to recombine novel advantageous mutants that allow current evolution to proceed in the face of environmental change, such as evolution by sexual competitors (the 'Red Queen hypothesis', Chapter 2).
3. A very narrow population niche width; I have assumed that one of the main advantages of panmictic sexuality is that genetic variation be-

Table 10.1 Comparison of variation between an asexual and a sexual population of *Taraxacum*

	Phenotype			
	Sexual		Asexual	
Enzyme locus	Ho	He	Ho	He
MDH-1	0.84	0.16	–	1.00
ME	0.98	0.02	1.00	–
6PGDH-1	0.98	0.02	1.00	–
6PGDH-2	0.70	0.30	0.06	0.94
TYR	0.74	0.26	–	1.00
PER	0.61	0.39	–	1.00
SOD-1	0.78	0.22	–	1.00
ACPH	0.53	0.47	–	1.00
MDH-2	1.00	–	1.00	–
IDH	1.00	–	1.00	–
GDH	1.00	–	1.00	–
CAT	1.00	–	–	1.00
SOD-2	1.00	–	1.00	–
SOD-3	1.00	–	1.00	–
GOT-2	1.00	–	1.00	–
GOT-3	1.00	–	1.00	–

Ho, homozygous; He, heterozygous.
Mean heterozygosity per individual: sexual = 0.12; asexual = 0.43.

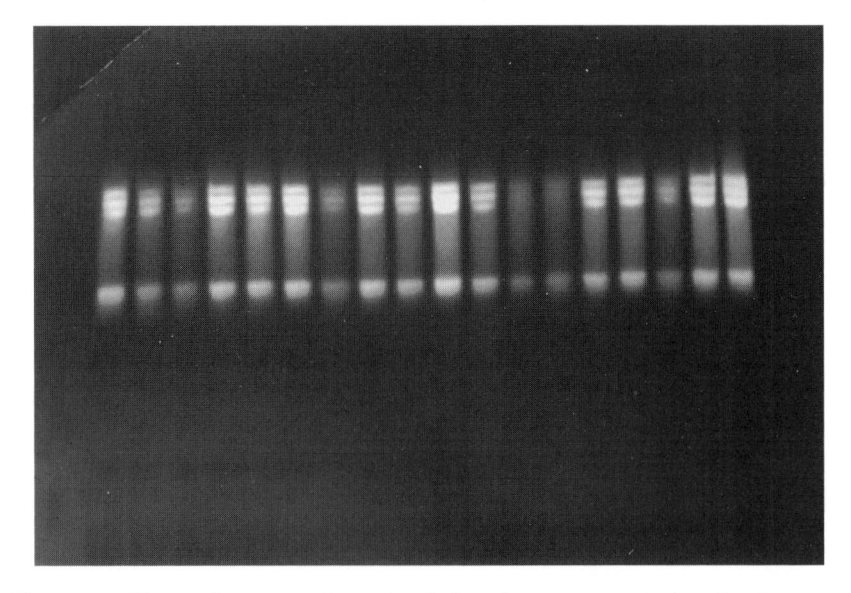

Fig. 10.2 Electrophoretic gels, stained for the enzyme 6-phosphogluconate dehydrogenase. Each gel is taken from a different individual of the apomictic dandelion *Taraxacum pesudohamatum*. For the fast-running (upper) locus, individuals are invariably triple-banded, representing a fixed heterozygous gene giving rise to dimeric products.

tween individuals in the gene pool remains in intermittent contact with itself in time and space. This maximizes the fitness and abundance of DNA by allowing it to fill the maximum number of niches in the environment (Dawkins, 1976).

4. As hybrids, most agamosperms might be expected to lack the adaptive 'fine-tuning' to a particular environmental niche that would be expected of its parental sexual species. In general, an empirical assessment of many agamospecies would agree with this expectation. Many are weeds, or plants of open, transitory or unstable habitats.

However, if point (3) is considered further, the agamospermous gene pool should consist of only a single genotype, and genetic contacts should only be in direct, vertical lineage from mother to daughter in time, rather than being reticulate (Fig. 10.3). As a result, the niche width of the gene pool, and thus the number of opportunities that an evolutionary unit of DNA has to exploit the environment, should be severely limited. This also severely limits the fitness of agamospermous mothers.

In practice, it seems that at least some of this potential niche limitation is offset by unusually high levels of phenotypic plasticity in many agamospecies (as in dandelions, *Taraxacum*, Richards, 1972). This stratagem is also typical of many relatively invariable inbreeders (Chapter 9).

It should also be pointed out that for so-called 'populations' of related agamosperms, for instance belonging to the same genus, many genotypes (agamospecies) may coexist. To take one example, Ollgaard (personal

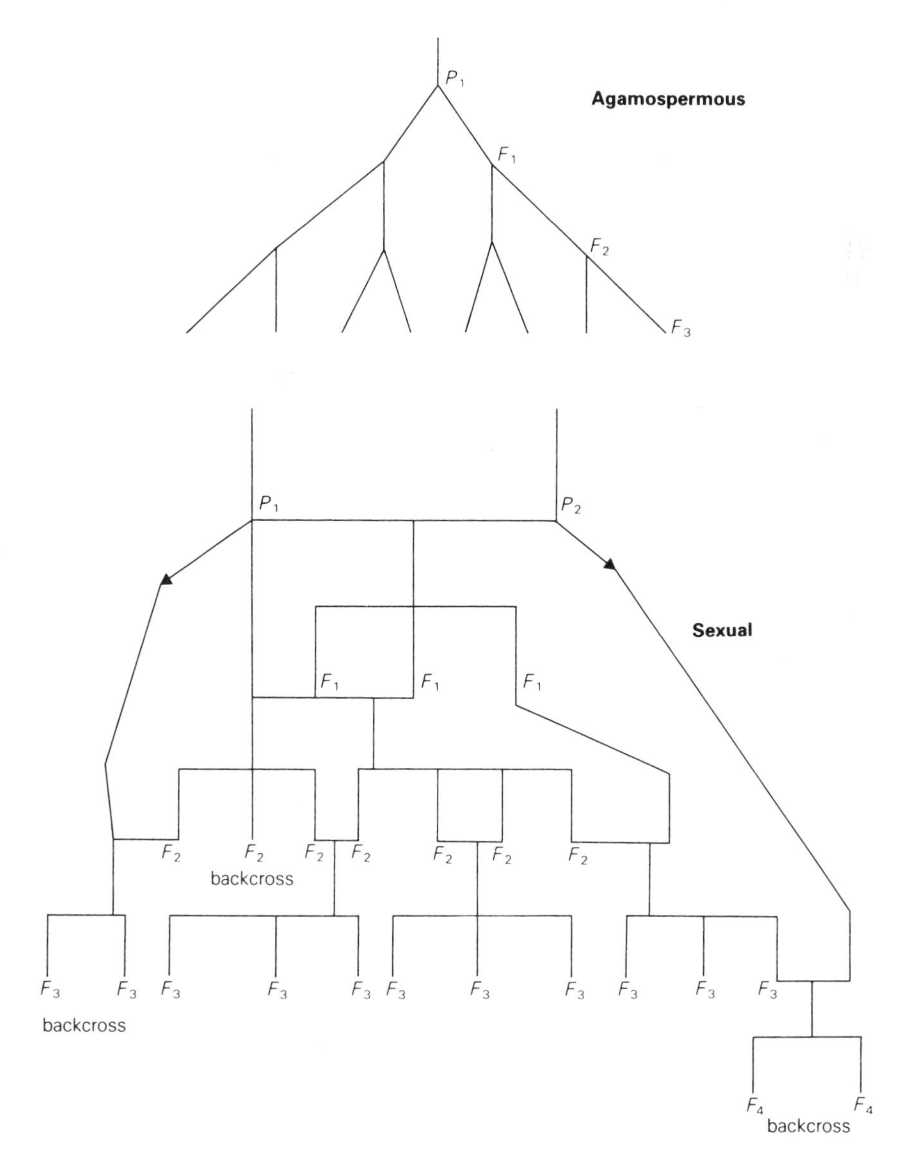

Fig. 10.3 Genetic contact in sexual and agamospermous lineages.

communication) has identified over 100 *Taraxacum* agamospecies inhabiting one hectare of waste ground near Viborg, Denmark. In the UK, it is commonplace to find 20–30 *Taraxacum* agamospecies coexisting. Such levels of diversity can also be found in some other agamic complexes (e.g. *Rubus* and *Ranunculus auricomus*), but in others (e.g. *Alchemilla* and *Sorbus*) it is unusual to find more than two agamospecies coexisting.

The definition of a population in an agamic complex cannot be the same as in a sexual gene pool. One could point out that ten species of grass can coexist, entirely by apomictic (vegetative) reproduction, in a lawn that is regularly mown. We would not consider these to belong to the same population, for the species are not in genetic contact with each other. Similarly, coexisting *Taraxacum* agamospecies are not in genetic contact with each other. Thus, although an agamic complex can show considerable diversity within a location, the diversity within an evolutionary unit (gene pool of DNA) should be very low (Hughes and Richards, 1989).

MECHANISMS OF AGAMOSPERMY

For obligate sporophytic agamospermy to function correctly, it is necessary for three or four changes to be made to the sexual process; these may be independent of each other, or they may be connected.

1. Avoidance of reductional meiosis.
2. Avoidance of sexual fusion.
3. Spontaneous embryony.
4. Spontaneous development of the endosperm (not necessary in pseudogamous species), and adjustment to different genetic balances between the embryo and endosperm.

In addition, it is notable that almost all apomicts are polyploid, for reasons further discussed on p. 417.

There are many and complex mechanisms by which agamosperms have bypassed the basic sexual functions of reductional meiosis and sexual fusion. These mechanisms have suffered detailed and at times confusing nomenclatures which have been comprehensively reviewed by Gustafsson (1946–7) and Battaglia (1963). There is little point in covering all these, with their manifold variations, in the present volume. The main mechanisms are presented as a flow diagram (Fig. 10.4), and are discussed briefly in the following sections. For references to source material, the reader is referred to Gustafsson (1946–7).

Adventitious embryony (sporophytic agamospermy)

In *Citrus* and in most other genera showing adventitious embryony, agamospermy usually only occurs in the presence of normal sexual

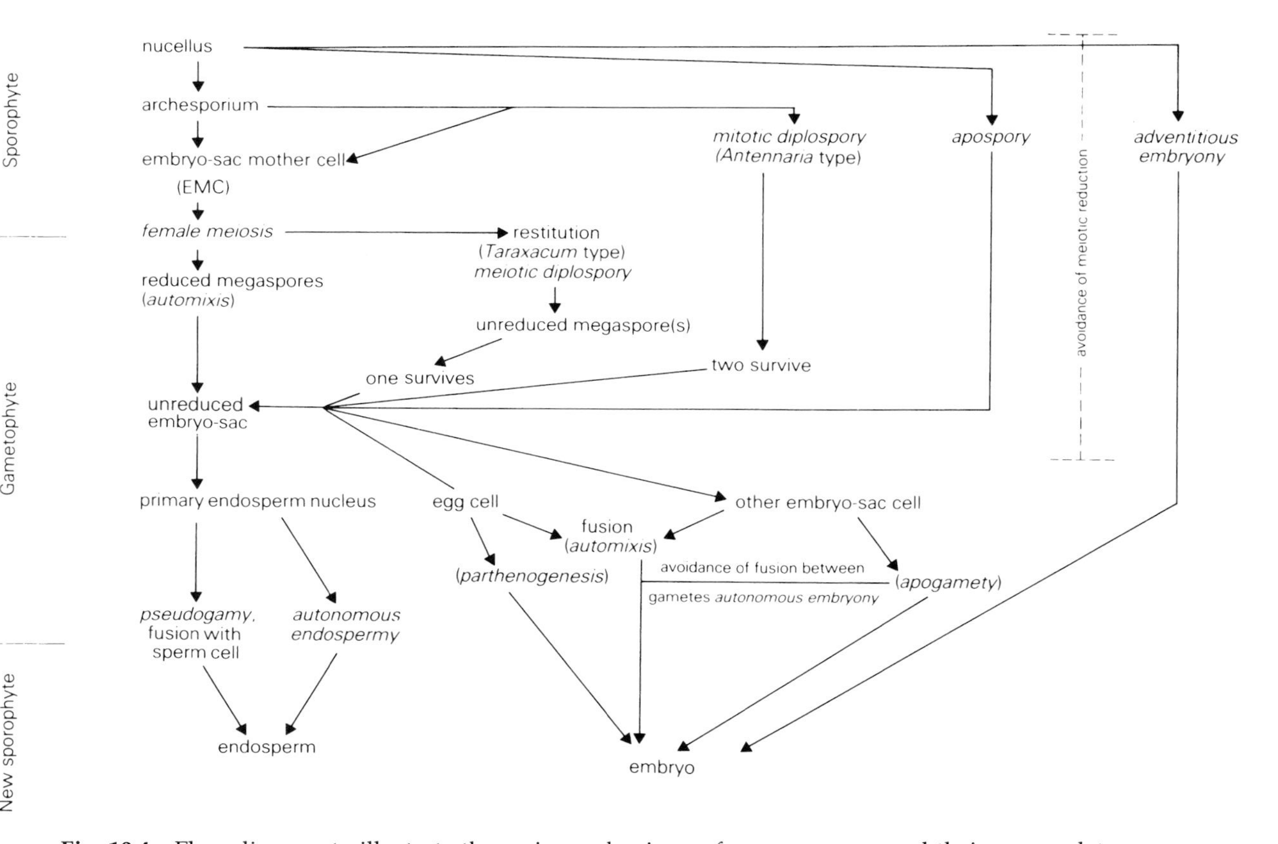

Fig. 10.4 Flow diagram to illustrate the main mechanisms of agamospermy and their nomenclature.

reproduction (Richards, 1990a). Typically, pollination is followed by the double fertilization of a conventional reduced sexual embryo-sac, and the embryo and endosperm start to develop. The stimulus of sexual embryo development often then results in the growth of apomictic proembryos in the nucellus or integument (pollination by itself may also act as a stimulus to proembryony, whereas in some cases the autonomous development of apomictic proembryos can initiate in the bud (p. 443)).

Various fates of the sexual embryo and the apomictic proembryos are now possible, but typically one of the apomictic proembryos invades the embryo-sac, outcompeting the other apomictic and sexual embryos, and commandeers the sexually produced endosperm. Alternatively, the sexual embryo and one of the apomictic embryos coexist, sharing the same endosperm. Thus, the mature seed may have one sexual embryo, one apomictic embryo, or two (or rarely, more) embryos (polyembryony), one of which is sexual and the other apomictic in origin. In common with many aposporous plants, nearly all sporophytic agamosperms are pseudogamous, and thus maintain a need for the production of fertile pollen and pollination. In sporophytic apomixis, pseudogamy seems to have favoured diploid, non-hybrid apomicts in which sexual reproduction persists, but this tends not to be so for apospory (p. 405).

Thus, sporophytic agamospermy is almost always facultative (that is agamospermous and sexual reproduction coexist, in the same population, individual mother, or even seed). Genetic fixation is tempered by sexual variability. Also, unlike other agamospermous mechanisms, only one mutation may be necessary for adventitious embryony to evolve. Thus, sporophytic agamospermy may respond readily to selection, and it is perhaps curious that adventitious embryony is not a more common mechanism.

Most species showing adventitious embryony are unusual apomicts, in that they tend to be tropical trees. In contrast, most species with gametophytic apomixis are perennial herbs from cold climates. Also, many sporophytic apomicts have large fleshy, animal-dispersed fruits. Often, the sexual embryos are large, and of an untypical morphology, being cylindrical or tuberoid in shape. It is possible that such sexual embryos preadapt their parents to the evolution of sporophytic apomixis, in that the asexually produced proembryo, which has a quasi-tumerous initiation, may more easily differentiate into such morphologies (p. 442).

Gametophytic agamospermy

Unlike adventitious embryony, all other forms of agamospermy involve an embryo-sac (female gametophyte) (see Fig. 10.4). A distinction is generally made between apospory, in which the gametophyte originates from the nucellus, and diplospory, where the gametophyte originates in the archesporium. However, this is not always a very clear demarcation, as pointed out by Gustafsson (1946–7). The archesporium itself originates in the nucellus. In the case of so-called mitotic diplospory, a single distinctive embryo-sac mother cell (EMC), which undergoes at least the earlier stages of female meiosis, is replaced by multiple cells which resemble an archesporium and several EMCs (which do not undergo meiosis) to different extents. It is important to distinguish between mechanisms that undergo meiotic diplospory and those that do not. The former have at least the potential for recombination (through crossing over) and segregation so that offspring may differ, whereas this is impossible among the products of apospory and mitotic diplospory.

Apospory

The distinction between apospory and diplospory is also important. Aposporous agamospermy is usually facultative, and diplosporous agamospermy is usually obligate. Just as the nucellar embryos of *Citrus* allow the sexual embryo-sac to undergo regular sexual fertilization and embryony, so the nucellar embryo-sacs of apospory allow the archesporium to undergo a reductional meiosis which can result in a sexual embryo-sac as well. If this is fertilized, embryos of sexual and agamospermous origin may coexist in the ovule; the mature seed may have both (as is often the case in *Poa*), or one only (most frequently the agamospermous one) may survive. The point is, that in aposporous mechanisms the archesporium, by definition, remains intact, and thus there is always the potential for sexuality. In such plants, agamospermy must always be considered at least potentially facultative rather than obligate.

In three of the aposporous groups most thoroughly investigated, the *Poa pratensis* and *P. alpina* complexes, the *Potentilla verna* and *P. argentea* complexes, and the *Ranunculus auricomus* aggregate, agamospermy is very often facultative. Polyembryony is well known in all three groups, and nucellar agamospermy coexists with archesporial sexuality within the population, the individual, or even the ovule. However, individuals which, by virtue of their hybridity, triploidy or other genetic disfunction, have highly irregular female meioses may show poor or quite sterile sexual function, and these will approach obligate agamospermy.

Endosperm development and pseudogamy in apospory

For an embryo to develop normally, it is also necessary that the endosperm should develop. In sexual plants the endosperm only develops after the fusion of the diploid polar nucleus of the embryo-sac with a sperm cell to form a triploid primary endosperm cell (Chapter 3).

The balance of 'ploidy between the diploid embryo and the triploid endosperm, and the balance in genome contributions between the embryo and endosperm are probably critical for the successful development of the sexual endosperm, and embryo (p. 66).

In adventitious embryony, diploid, maternal apomictic embryos commandeer triploid, sexual endosperm. This is a form of pseudogamy, in which pollination and fertilization of the endosperm is required for the successful agamospermous development of the embryo. Pseudogamy is also usual in aposporous agamosperms; in these the balance between the embryo and the endosperm in 'ploidy, and in genome constitution, will differ from that in sexual relatives, and will need to be accommodated by the new agamosperm. The successful agamospermous embryo is accompanied by an endosperm from the same asexual and unreduced embryo-sac. This endosperm is achieved in most cases by the pseudogamous fertilization of the tetraploid polar nucleus (PEN) by a haploid sperm cell. (There are exceptions, for instance four-celled embryo-sacs in which the PEN is diploid, and not the product of autofusion; and diploid sperm cells which result from unreduced pollen.)

As a result, the 'ploidy level of the endosperm will normally differ from that of the embryo in an aposporous pseudogam by 5:2 (3:2 in sexuals) and the male genome contribution balance between the endosperm and embryo will be 0.2:0 (0.33:0.5 in sexuals). The evolution of pseudogamy has been modelled by Strenseth, Kirkendall and Moran (1985).

Apospory is closely associated with pseudogamy; nearly all aposporous agamosperms are pseudogamous, requiring fusion of the sperm cell with the polar nucleus for endospermy, and hence successful seed-set to occur. This association may reflect the facultative nature of aposporous agamospermy. Because the archesporium often persists alongside the aposporous embryo-sac, the potential for sexual reproduction remains. In these cases, no benefit accrues from the loss of pollen or pollination. Instead, the regular presence of fertile pollen tubes at the ovule will facilitate sexual reproduction where it is possible. In apospory, mutants that confer autonomous development of the endosperm in the absence of polar nucleus fertilization may not be at a selective advantage, and male sterile mutations will be disadvantageous.

Nybom (1988) shows that in *Rubus*, apomicts with a highly disturbed meiosis have a reduced seed fertility which she assigns to the failure of genetically abnormal male gametophytes to fertilize the PEN (polar nucleus). This should select for relatively male fertile genotypes, which would also thus be more likely to undertake sexual reproduction successfully. Thus, there is a feedback between apomixis with pseudogamy and sexuality.

Gustafsson (1946–7) shows that when the polar nucleus remains unfertilized in pseudogams, the endosperm fails to develop. Whereas the embryo of *Poa alpina* or *Potentilla collina* may develop as far as the 256-cell stage in the absence of the endosperm, its form is abnormal, and abortion eventually results. In most pseudogams, fertilization of the endosperm seems to be the only function of pollination, for embryo development precedes pollination (it is precocious). Only in *Arabis holboellii* does it seem that pollination alone (not pollen-tube growth or fertilization) is necessary for embryony to proceed (Böcher, 1951). We have already seen (Chapter 3) that embryony in a conventional sexual plant is often dependent on the stimulus of pollination, in addition to fertilization, perhaps through a hormonal message. Thus, many pseudogams seem to have escaped the need for the stimulus of pollination for successful embryony to occur. Indeed, this may be how the parthenogenesis gene operates (p. 412). In this way, they escape fertilization of the egg cell, by developing their embryos early, before pollination. Ironically perhaps, they have not escaped the need for the fertilization of the primary endosperm cell (fused polar nucleus).

Diplospory: autonomous endospermy

In contrast to agamosperms with adventitious embryony and apospory, nearly all those that are diplosporous are non-pseudogamous, i.e. they have autonomous endosperm development. Little seems to be known about the physiological or genetic control of autonomous endospermy. It is possible that it can derive from the same processes that result in autonomous embryony. In any case, neither the endosperm nor the embryo will have a paternal genome contribution. In all known cases, the endosperm usually has twice the 'ploidy level of the embryo, having resulted from the fusion of unreduced polar nuclei. Occasionally, high polyploid endosperm cells, which presumably result from endoduplication, are found as well. The adjustment of non-pseudogamous diplosporous agamosperms to an endosperm: embryo 'ploidy ratio of 2:1 (3:2 in sexuals) may be less difficult than the adjustment of pseudogamous aposporous agamosperms to a 5:2 ratio. In so far as the male: female genome balance is critical in controlling relative rates of endosperm and embryo growth (Chapter 3) this should not trouble the non-pseudogam, with its entirely maternal endosperm and embryo.

Diplospory, avoidance of meiosis

As they are derived from the archesporium, diplosporous unreduced embryo-sacs have forfeited the use of the archesporium for reduced sexual embryo-sacs. Any sexuality in diplosporous agamosperms must therefore depend on inconstancy in the avoidance of sexual fusion, which may be accompanied by inconstancy in spontaneous embryony and by some reduction in the female meiosis. Most commonly, diplosporous plants show obligate agamospermy, and so pseudogamy is no longer favoured, for reasons argued above.

In diplosporous species, the archesporium either undergoes a female meiosis, which fails in some way so that it is restitutional ('*Taraxacum* type', Gustafsson 1946–7) or it divides to form the embryo-sac mitotically ('*Antennaria*-type').

A regular reductional meiosis in sexuals seems to be dependent on regular chromosome pairing (synapsis) and chiasma formation between homologues; synaptic bivalents auto-orientate on the spindle in such a manner that regular reduction and segregation of half-bivalents to the poles of anaphase I is achieved. In the absence of synapsis, the univalents

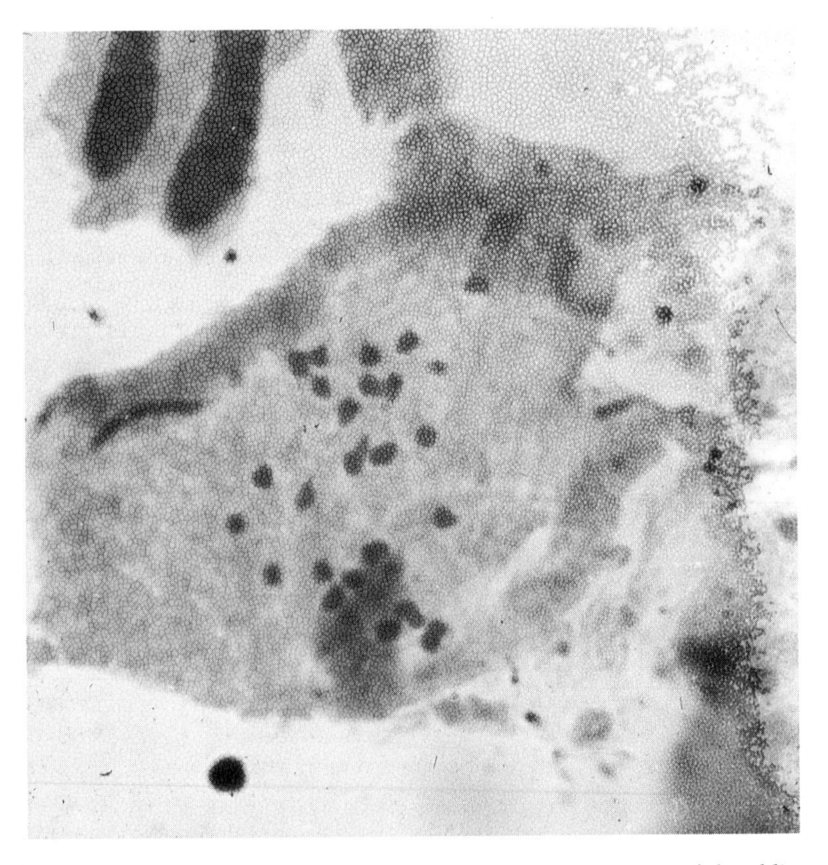

Fig. 10.5 Microphotograph (×2000) of diakinesis in male meiosis of the obligate agamosperm, *Taraxacum hamatum*, with 24 univalents. This meiosis is fully restitutional.

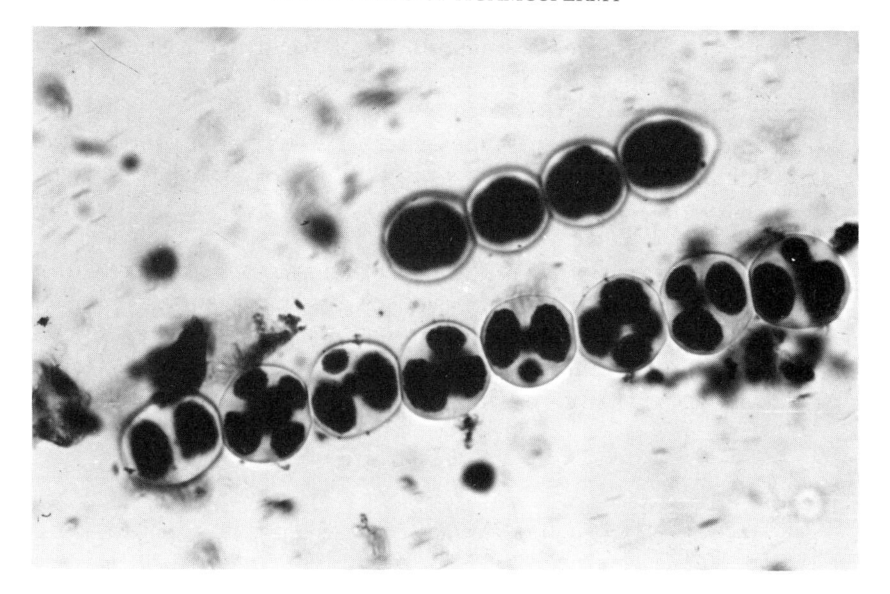

Fig. 10.6 Microphotograph (×200) of pollen 'tetrad' formation in *Taraxacum subcyanolepis*. From the left is a diad (meiosis restitutional), a pentad, a diad with two micronuclei (partial reduction), a tetrad (meiosis reduction), a diad with two micronuclei, a tetrad, and two unequal tetrads caused by a partial reduction. The parent has $2n = 24$, and pollen grains will contain from 2 to 24 chromosomes, with mostly intermediate counts. Only grains with n = 8–12 usually penetrate the stigma.

fail to line up on the spindle at metaphase I, and in the absence of bivalent auto-orientation, disjunction fails to occur. Thus, in *Taraxacum* at least (Richards, 1973), restitution (non-reduction) appears to be a function of asynapsis, through the failure of chromosome homologues to form chiasmata (and thus bivalents at diakinesis and metaphase I) (Fig. 10.5). Asynapsis can be shown to be under simple Mendelian dominant control, unlinked to the control for precocious embryony (also a simple Mendelian dominant).

However, it is not necessary for a female meiosis to be totally asynaptic for it to be restitutional. In certain synaptic triploid species of *Taraxacum*, many meioses are so disturbed (apparently as a result of the allotriploidy itself) that disjunction fails, although neighbouring meioses may, by chance, be sufficiently regular for disjunction and reduction to proceed (Figs. 10.6 and 10.7) (Richards, 1970a; Muller, 1972) (p. 423). Malecka (1965, 1967, 1971, 1973) shows that in some obligately agamospermous *Taraxacum*, asynapsis in female meiosis may be only partial, several bivalents forming in addition to many univalents. These meioses are restitutional.

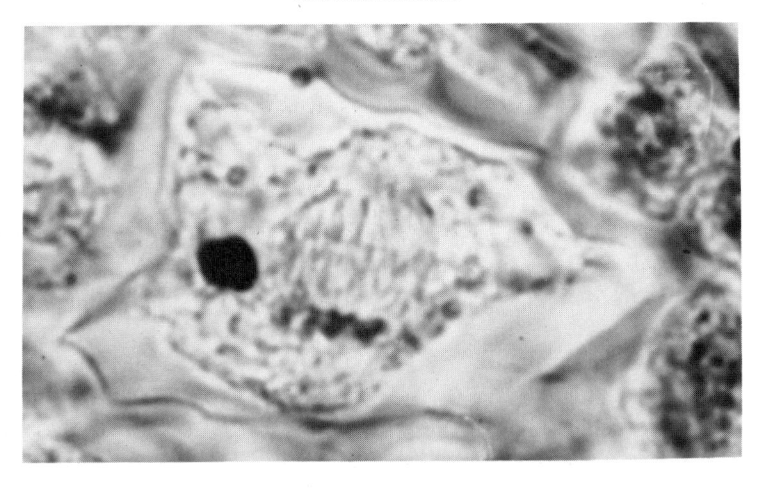

Fig. 10.7 Microphotograph (×1500) of anaphase I at female meiosis in *Taraxacum brachyglossum* ($2n = 24$). This is a reductional meiosis (the spindle and one segregating set of chromosomes only visible in this plane of focus) in a triploid facultative agamosperm. It will give rise to a sexual embryo-sac. In the same plant, restitutional meioses giving rise to asexual embryo-sacs are also found.

Recent work has established that EMC about to undertake a diplosporous megasporogenesis and embryo-sac development lack the callose sheath which seems to be typical of reductional embryo-sac development, at least in grasses (Carman, Crane and Riera-Lizarazu, 1991; Naumova, den Nijs and Willemse, 1993). As yet the significance of this observation seems not to be fully evaluated, although it does suggest that diplospory might be triggered by a single gene product which is expressed at an early stage. However, this observation is already proving to have considerable implications with respect to the screening of progeny for diplosporous behaviour.

Unlike diplosporous species, the EMC of aposporous grasses are callosic, but of course they do not engender apomictic embryo-sacs.

Diplospory: embryony

Most forms of agamospermy pass through an embryo-sac, and this is most usually eight celled (but is four celled in panicoid grasses). These cells show much less constancy in form than is usual in a sexual embryo-sac. Multiple egg cells, an absence of synergids, poor differentiation between egg cells and antipodals, etc. are commonplace. However, in some genera (e.g. *Taraxacum*) the embryo-sac is conventional in appearance, and can apparently function sexually on occasions. Variation in embryo-sac cells can mean that distinctions between development of the egg cell (parthenogenesis) and another cell (apogamety) into the embryo, apparently straightforward, can in practice be difficult to draw. However, in the vast majority of cases, it is a single egg cell that forms the embryo.

The agamospermous attributes of failure of fertilization, and autonomous embryony, are theoretically quite separate, but may in practice be closely correlated. In very many cases, both aposporous and diplosporous (but not usually in adventitious embryony) the egg cell develops into the embryo before the flower opens (precocious embryony), so that fertilization of the agamospermous egg is impossible. Thus, in *Taraxacum* (Murbeck, 1904; Richards, 1973) embryony proceeds 48 hours before anthesis. Even in pseudogamous aposporous species, it is commonly found that the embryo has developed before the flower opens, so that later fertilization of the polar nuclei takes place in the presence of a sizeable embryo. An exception appears to be in *Ranunculus auricomus* (goldilocks), in which autonomous embryony only proceeds after the fertilization of the polar nuclei (Nogler, 1972).

Proving agamospermy

Czapik (1994) has reviewed ways by which plants can be screened for apomixis. Where precocious embryony occurs, apomixis is easily demonstrated, even when the plant is pseudogamous. Microdissection can show the early development of embryos. More difficult are cases such as *Ranunculus* and some *Rubus*, where precocious embryony is absent and pseudogamous agamospermy also occurs. Apomixis may be inferred by the maternal inheritance of chromosome numbers and of other genetic characters (and by an avoidance of female meiosis). Only very careful and painstaking cytology can prove that fusion of a sperm cell with the polar nuclei occurs in the absence of fusion with an egg cell. Non-precocious embryony is also recorded in the diplosporous, non-pseudogamous *Antennaria*. In such a plant, agamospermy is readily proven by emasculation followed by bagging of the flowers to prevent cross-pollination, or by a removal of the stigmas. This is not possible for pseudogamy where pollination is required.

The well-known 'emasculation' procedure for showing agamospermy in *Taraxacum* (Fig. 10.8) removes not only the anthers before anthesis, but also the stigmas, thus making cross-pollination quite impossible in any circumstances. In agamospecies, this traumatic procedure is generally followed by perfect fruit-set.

Two recent methods for the rapid assessment of aposporous behaviour are ovular clearing (Nakagawa, 1990) and the use of auxin to stimulate embryogenesis in the absence of pollination for some pseudogams which have not overcome this need (Matzk, 1991).

Parthenogenesis

It is clear that in *Taraxacum*, and other species with precocious embryony, failure of fertilization is primarily mediated by precocious development of the embryo before the flower opens and pollination becomes possible (Fig. 10.9). Thus, at least one of the functions of parthenogenesis is to overcome the need for the stimulus of pollination for embryony to occur (p. 407). Other mechanisms that prevent fertilization might include full male sterility, failure of the pollen grain to germinate on the stigma or grow down the style, or failure of the sperm cell to fuse with the egg cell. All of these, if they occur (and it is only certain that the last mechanism operates in *Ranunculus* and *Antennaria*) may result from modifications to a self-incompatibility system. However, the physiological basis of failure of fertilization in *Ranunculus* and *Antennaria* seem to be unknown for certain.

As yet, we also seem to be largely ignorant of the physiological processes that lead to autonomous embryony and, in non-pseudogams, autonomous endospermy. Many complex mysteries remain to be discovered by modern physiological and biochemical techniques. Much of the basic work on agamospermy was completed before 1950, and the study of agamospermous processes has not been fashionable in the last three

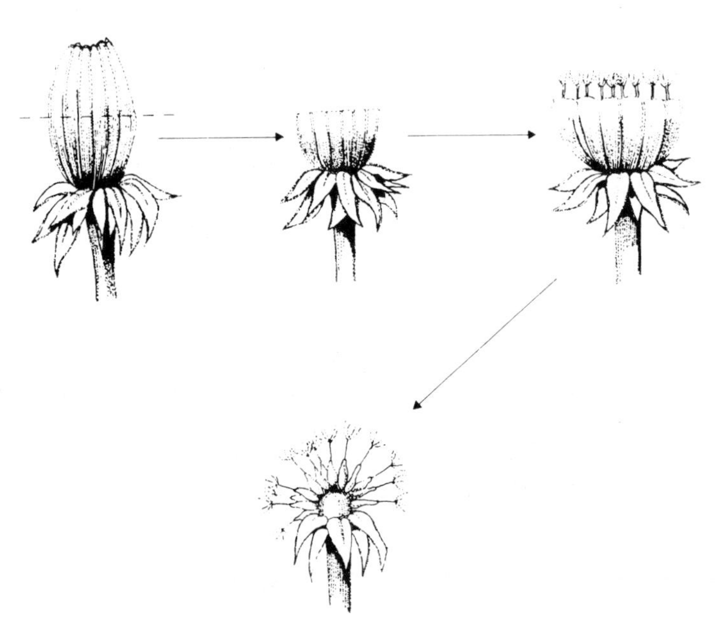

Fig. 10.8 Emasculation of an agamospermous *Taraxacum*. The top portion of the bud, containing anthers and stigma, are sliced off with a razor, but fertile seed is set asexually (×1).

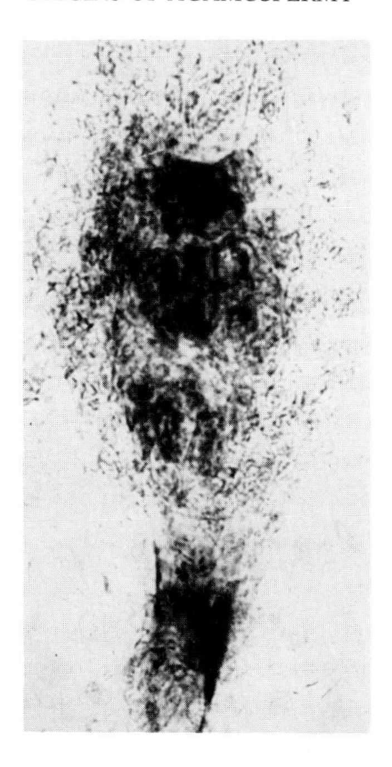

Fig. 10.9 Microphotograph (×60) of an eight-cell embryo (centre), with endosperm (above), surrounding nucellus and neck and basal cell (below) dissected from the bud of *Taraxacum hamatum*, an obligate agamosperm, 24 hours before the inflorescence opens (precocious embryony).

decades, although theoretical and populational studies have grown in popularity. Thus, we have no idea how, for instance, in facultatively agamospermous *Taraxacum*, with irregular partially reductional meioses, the unreduced egg cells develop into embryos precociously in the agamospermous manner, but the unreduced eggs in the same capitulum do not so develop, but wait until anthesis when they are available for sexual fusion (Richards, 1970a). Apparently the message that controls spontaneous embryony (parthenogenesis) is in this case dependent on the chromosome number of the embryo-sac, which is very difficult to credit. However, Asker (1979) quotes many other examples in which this also appears to operate (see Fig. 10.10).

ORIGINS OF AGAMOSPERMY

Modern thinking about the origin of agamospermy, and its control, is reviewed in two influential papers by Asker (1979, 1980), and by Asker and Jerling (1992). The theme of Asker's work is that genes that control

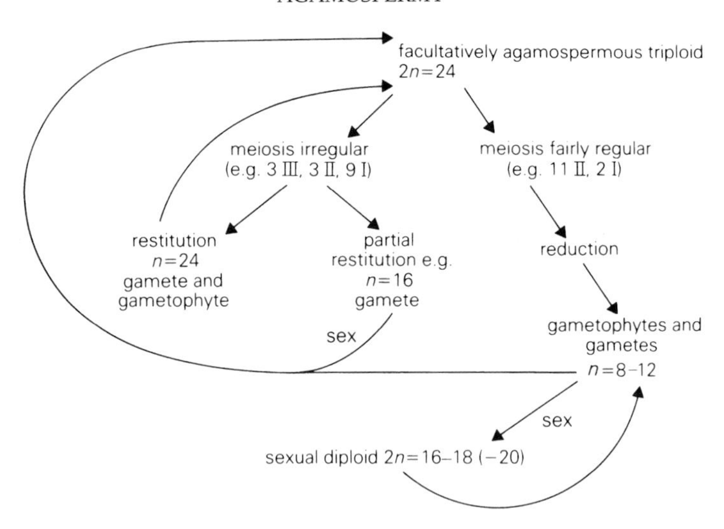

Fig. 10.10 Diploid/triploid cycle in facultatively agamospermous *Taraxacum*. (After Richards, 1970a.)

successful agamospermy when they occur in combination, can be found singly as aberrations in many conventionally sexual populations, where they are likely to be unsuccessful. Examples of such aberrations are listed in Table 10.2. It will be seen that all the functional features of various forms of agamospermy can be found exceptionally in sexual populations (adventitious embryony is excluded as it leads to facultative agamospermy *per se*). Thus, Asker argues that agamospermy is best viewed as an extension of the sexual process, whereas Nogler (1984a) sees apomixis and sexuality not as alternatives, but as 'independent modes of reproduction which can occur side by side'. These ideas are supported by the contention that truly obligate agamospermy, in which all possibility of sexuality has been lost, is a rare phenomenon. These authors argue that some sexuality continues to be a feature of most agamospermous plants. This thesis, which supports that of Gustafsson (1946–7) and de Wet and Stalker (1974) is discussed below.

Clones that carry only one agamospermous mutant are likely to be unsuccessful, since they will reproduce in one of the following ways:

1. Production of parthenogenetic haploids (mutation for autonomous embryony alone) (Chase, 1969);
2. Production of unbalanced high polyploids (mutation for avoidance of reductional meiosis alone);
3. Non-formation of embryo (mutation for avoidance of sexual fusion alone);
4. Non-formation of endosperm, leading to seed sterility.

Table 10.2 Occurrence of independent elements typical of agamospermy amongst habitually sexual species

Autonomous embryony	Endosperm development	Avoidance of reductional meiosis
haploid parthenogenesis e.g. *Zea* (Chase, 1969)	autonomous endospermy *Anemone nemorosa* (Trela, 1963) *Triticum aestivum* (Kandelaki, 1976)	total inhibition of meiosis *Zea* (Palmer, 1971)
progressive subhaploidy 4x *Solanum tuberosum* (Breukelen *et al.*, 1975) ↓ 2x (dihaploid) → (autodiploid) ↓ (chromosome doubling) 1x (haploid)	lack of fusion of polar nuclei in four-cell aposporous embryo-sac followed by sperm-cell (pseudogamous) fusion	total asynapsis (rare) *Zea* (Beadle, 1930)
		partial asynapsis (desynapsis)
synergid apogamety (*Zea*)	Panicoid grasses	*Datura*, *Zea* (Catcheside, 1977) non-disjunction at meiosis I or II e.g. *Zea*, *Datura* (Rhoades, 1956)
diploid parthenogenesis (from endomitotic tetraploid embryo-sac mother cell followed by reductional female meiosis) *Brassica* (Eenink, 1974)		aposporous development of embryo-sac × *Raphanobrassica* (Ellerstrom and Zagorcheva, 1977) *Sanguisorba* (Nordborg, 1967)

However, if two particular mutants come together in a single genet, agamospermy may be successfully achieved at a stroke. For example, an aposporous mutant would generate unreduced embryo-sacs which, after fertilization, would give rise to sterile autotriploids. A separate parthenogenetic mutant in the same species would give rise to sterile, and very possibly non-viable, haploids. If, however, the parthenogenetic mutant (male) is crossed with the aposporous mutant, some offspring might well prove to be aposporous, parthenogenetic triploids which were fertile agamosperms.

If the original parthenogenetic mutant is hybrid and polyploid, with a disturbed meiosis that is partially non-disjunctional, it may be able to reproduce agamospermously for non-reduced ovules. This seems to be the case for some triploid dandelions (*Taraxacum*) which still have synaptic meioses (Richards, 1970a, 1973). Obligate agamospermy in *Taraxacum* only develops after a second mutation for asynapsis (resulting in a totally non-reductional female meiosis). This would presumably be favoured in such facultative triploids with relatively poor seed-set.

Thus, at least two, and often three or more, independent mutations are required before most forms of agamospermy can function. Because reductional meiosis is bypassed by most primary agamospermous mutants, the power to recombine different mutations onto a single chromosome (linkage group) is missing. Segregation of agamospermous alleles will occur in agamosperm × sexual backcrosses. Therefore, it is not often possible for agamosperms to form a single coadapted linkage group in the way that dioecious or heteromorphic (Chapters 7 and 8) plants can (although this has apparently occurred for aposporous *Ranunculus*, p. 420). Agamospermy is only likely to become stable if obligate agamospermy arises, and the whole genotype becomes a single linkage group. Because the evolution of obligate agamospermy, and the loss of recombinational variation, may not be favoured in many subsexual populations, the maintenance of subsexuality is reinforced by the independent segregation of different alleles which control agamospermous functions.

Preadaptation for apomixis

We find that agamospermy is closely associated with certain life styles and genetic conditions in flowering plants. We can assume that such attributes predispose plants to the establishment of agamospermy. These attributes include the following.

1. Perenniality, with limited vegetative spread, but good vegetative persistence. Very efficient vegetative dispersers have their own form of apomictic dissemination. Gadella (1991) notes that sexual *Hieracium pilosella* produce much longer and more abundant stolons than do apomictic strains.

2. Hybridity; hybrids may favour agamospermy by being partially sterile and such hybrids will be vigorous, thus promoting good vegetative persistence until mutants come together. Hybrids may have disturbed metabolic systems which promote abnormal functions such as apospory or autonomous embryony and may also be more likely to combine different mutants from independent genetic lines. The disturbed female meioses of hybrids are likely to be non-disjunctional (restitutional).
3. We can also assume that pseudogamous apomicts will be most successful in a pollen-limited environment when autogamous, as are many *Rubus* (Nybom, 1985), as apomixis then becomes independent of pollinators.

Polyploidy and apomixis

Polyploidy may encourage development of a higher level of parthenogenetic behaviour *per se* as in maize (Yudin, 1970) and it may also lead to partial sterility, which should promote agamospermy; polyploids (especially triploids which will occur commonly when meiosis fails) will have disturbed, and at least partially unreduced, meioses. Additionally, mutants that bypass meiosis (e.g. apospory) will produce unreduced embryo-sacs, so that new agamosperms may originate from crosses between 2x and x gametes and thus also be triploid (3x). Fowler and Levin (1984) have discussed the 'minority type disadvantage' problems which a newly formed polyploid encounters. Most of its outcrossed matings will be with diploids, giving rise to sterile triploids.

The very strong association between gametophytic apomixis and polyploidy has been noted for many years, and has been stressed by Nogler (1984a), Mogie (1992) and Asker and Jerling (1992). Very few cases of diploid gametophytic apomicts are known in nature (single strains of *Potentilla argentea* and *Arabis holboelii*) and diploid gametophytic apomicts are rarely found even after experimental manipulation.

Less frequently, it has been emphasized that this association does not apply for sporophytic agamospermy; indeed the majority of species showing adventitious embryony are diploids (e.g. Richards, 1986). Mogie (1986) has attempted to explain these relationships by a gene dosage model, which assumes that apomixis is controlled by a single gene. As many authors have recently discussed (e.g. Savidan, Grimanelli and Leblanc, 1995; Grimanelli *et al.*, 1995) it never seems possible to explain fully the inheritance of gametophytic apomixis by single gene control.

Theoreticians who discuss selection for sexual reproduction stress the role played by a 'normal' meiosis in the breakage of disadvantageous linkage disequilibrium (e.g. Maynard Smith, 1978) (p. 23). They point out that apomictic lines in contrast act as 'targets' for disadvantageous recessive mutants which accumulate in the lines in which they arise (so-called 'Müller's ratchet'). Undoubtedly, apomicts are depositories for far more mutational 'junk' (the 'genetic load') than are sexuals, and some of this mutational load will become linked to genes for apomixis. Although it remains associated with apomixis, this recessive mutational load will never be expressed.

Facultative apomicts receive genes for apomixis from apomicts, and genes for sexuality from sexuals. The present model depends on the likelihood that dominant genes for apomixis will be linked to recessive lethals, but that genes for sexuality will have recently experienced the genetically 'cleansing' effect of sexual reproduction, and so will be much less likely to be linked to a genetic load. This relationship has been proved in *Ranunculus auricomus* agg., in which Nogler (1984b) has shown that the gene for apospory is closely associated with a recessive lethal.

It has been suggested that as many as 60% of the genes that are expressed in the sporophyte generation of flowering plants are also expressed in the gametophyte (Ottaviano and Mulcahy, 1989). Although this estimate was made for the pollen tube, there is no reason why a similar figure should not apply to the female gametophyte, the embryo-sac, as well. As harmful recessives cannot 'hide' in the haploid generation, diploid flowering plants which reproduce sexually thus have powerful tools which enable them to 'screen out' some of their genetic load of recessive lethals and sublethals. This mechanism is not available to polyploid sexuals which have diploid (or higher 'ploid) gametophytes. In part this explains the higher levels of heterozygosity and genetic load typically associated with polyploids (e.g. Hamrick, Linhart and Mitton, 1979) although there are of course other valid explanations for this association as well (p. 382).

To sum up, most apomicts will normally carry many recessive lethal genes, some of which will be linked to dominant gene(s) which code for apomixis (recessive lethals would not become linked to recessive genes for apomixis). Many of these lethals would be expressed in a haploid embryo-sac or pollen tube, but not of course in the unreduced gametophytes of the gametophytic apomict, where they are protected from expression in a heterozygous condition.

Polyploid apomicts can give rise to diploid offspring in two main ways. In several genera, triploids commonly produce haploid or near haploid gametophytes which mate among themselves, or with gametophytes resulting from sexual diploids, to give rise to diploid offspring (pp. 422–3). These diploid offspring are never apomictic. We can explain this by supposing that haploid gametophytes which carry dominant gene(s) for apomixis linked to a gametophyte-expressed lethals will abort, so that genes for apomixis never enter the diploid offspring. This has been proved for *Ranunculus auricomus* agg. by Nogler (1984b). However, genes for sexuality should have been 'cleansed' of such a linkage load, and can parent diploids which will thus be sexuals.

However, when tetraploid apomicts produce polyhaploid gametophytes which reproduce parthenogenetically to give rise to diploid offspring (p. 421), there is no true haploid phase, so diploid offspring may successfully acquire some dominant genes for apomixis which are linked to lethals. This may explain why in some sexual/apomictic cycles with polyhaploid parthenogenesis such as those in *Potentilla* and *Panicum*, apospory can be expressed in diploids.

The gametophyte-expressed lethal model can successfully explain why diploids in triploid apomict cycles are always sexual. It can also show why gametophytic agamospermy rarely seems to arise in a diploid. Many apomicts still depend on the fertilization of the primary endosperm nucleus for successful seed development, and this pseudogamy (p. 406) must have been a requirement for nearly all newly arising apomicts. If lethal recessives commonly become linked to genes for apomixis, these lethals would be expressed in haploid pollen tubes, so that pseudogamous fertilization of the endosperm would normally fail.

The model also explains why species with sporophytic agamospermy, in contrast, are frequently diploid. Because this mechanism of agamospermy does not involve a gametophyte generation, lethal recessive genes cannot be expressed at this stage.

The rarity of apomixis

Parents of agamosperms carrying mutants that avoid meiosis or promote parthenogenesis, will be sexually unsuccessful, and thus the combination of these mutants in a single genet through sexual reproduction will occur very infrequently. Thus, it is not surprising that agamospermy is a relatively uncommon phenomenon.

The converse view has been expressed by Stebbins (1950), that the rarity of agamospermy is a measure of its long-term failure. However, there is little doubt that agamospermy is currently very successful in genera such as *Taraxacum*, *Hieracium* or *Rubus*, in which some agamospecies may be of a great age, dating back 10 000 years (Wendelbo, 1959), or even to the early Pleistocene (Richards, 1973). Modern work tends rather to emphasise the fitness and evolutionary potential in most agamic complexes (p. 430). Current workers suggest that the scarcity of agamospermy results from the low likelihood of agamospermy arising in a group, or spreading among a genus once it has evolved. The continuing sexuality of many groups of agamosperms, espoused by Gustafsson (1946–7) and emphasized by Asker (1980), suggests that active evolution is still continuing in agamic complexes. It now seems likely that even obligately agamospermous groups are capable of genetic variation and evolution (p. 438). This contrasts markedly with the earlier 'gloom and doom' view of the evolutionary future of obligate agamospermy propounded by Darlington (1939) and Stebbins (1950).

GENETIC CONTROL OF AGAMOSPERMY

This subject is reviewed by Asker (1980) and Nogler (1984a, 1994). For apospory, it seems that the aposporous production of adventitious embryo-sacs is usually essentially controlled by a single gene which is often dominant. For *Ranunculus auricomus* and its relatives however, Nogler (1984b) shows that the control of apospory is not completely dominant in polyploids, but has a dosage effect, so that when only one allele for apospory is present in a pentaploid, aposporous embryo-sacs only occur in a small percentage of ovules. Nogler further demonstrates that the precocious development of aposporous embryo-sacs inhibits the development of sexual embryo-sacs. When the genes for apospory are present at low dosage in polyploids, aposporous embryo-sacs tend to be formed later, so that sexual reproduction is more likely to accompany apospory in the same ovule in these cases.

Probably, genes which cause parthenogenesis are always independent from those which cause apospory (see Mogie (1986, 1992) for another opinion). However, genes for parthenogenesis are sometimes linked to aposporous genes, so that the control of apomixis may appear to be caused by a single factor. In *Ranunculus auricomus* agg., Nogler (1984b) only found one recombinant in which aposporous embryo-sacs behaved sexually.

For diplospory, agamospermy is most commonly controlled by two unlinked loci, between which there may or may not be epistatic reactions, and alleles controlling agamospermy are most commonly recessive. However, several diplosporous apomicts certainly have dominant genes for apomixis, as in *Eragrostis curvula* (Voight and Burson, 1981) and *Taraxacum* (p. 422). Some well known examples of the two-locus genetic control of gametophytic apomixis are given in Table 10.3.

Diplosporous systems and some aposporous systems tend to be complicated by the role played by polyploidy, especially triploidy, in influencing the fate of the meiosis. Thus in *Potentilla*, tetraploids tend to arise by sexual apospory from *aaBB* or *aaBb* diploid genotypes, which are usually the product of sexual × agamosperm crosses. These tetraploids have a reductional meiosis and are usually sexual or parthenogenetic. In the latter case, they may generate aposporous dihaploids *aabb*. Thus, there is

Table 10.3 Some examples of the genetic control of agamospermy (after Asker, 1980)

Sorbus (aposporous)	*A* genomes for *S. aria* (Whitebeam), *B* genomes from *S. aucuparia* (Rowan). *A* genomes apparently influence apospory in polyploids, thus *AA* (*S. aria*), *BB* (*S. aucuparia*) are sexual and *AAAA* (*S. rupicola*), *AAAB*, *AAB* (*S. minima*, *S. arranensis*) are obligate agamosperms. *AABB* (*S. intermedia*, *S. anglica*) are facultative agamosperms with *ABB* (*S.* × *pinnatifida*) being mostly sexual (Liljefors, 1955; Richards, 1975)
Pennisetum (aposporous)	two loci *A*/*a* and *B*/*b*, *A* causes agamospermy, *B* overrides it to give sexuality, e.g. *Aabb* is agmospermous and AaBb is sexual (Taliaferro and Bashaw, 1966)
Panicum (aposporous)	two loci control agamospermy recessively with complementation; thus either recessive homozygote with a heterozygote is agamospermous; *aabb*, *aaBb*, *Aabb* are agamospermous and *AABB*, *AABb*, *AaBB*, *AaBb*, *aaBB*, *AAbb* are sexual (Hanna *et al.*, 1973)
Potentilla (aposporous)	complex, and influenced by 'ploidy, but in some, two-locus recessive control is apparent: *AABB*, *AaBb*, *AABb*, *AaBB* are sexual; *aaBB*, *aaBb* are aposporous and sexual, giving tetraploid sexuals; *AAbb*, *Aabb* are parthenogenetic and meiotic, giving haploids or diploids; *aabb* is agamospermous (Muntzing, 1945)
Taraxacum (diplosporous)	diploids always sexual: parthenogenesis and precocious embryony dominant for unreduced egg cells (on chromosome *H*?) in polyploids with irregular meiosis (facultative); asynapsis, leading to totally unreduced egg-cell dominant (on chromosome *D*?) (obligate)

a tendency for a 2x aposporous/4x reductional/2x aposporous cycle to arise, although the results are in fact much more complex.

Similar cycles are also reported in the African grasses *Panicum*, *Bothriochloa* and *Dichanthium*, which form dominant tussocks over much of the savannah, and probably in the southern hemisphere *Cordateria* spp., including some of the 'pampas grasses' (Philipson, 1978; Asker, 1979; Connor, 1979). These cycles, exemplified by *Panicum maximum* (Savidan and Pernes, 1982), depend on sexual diploids crossing with apomictic tetraploids to give rise to sexual tetraploids which, when crossed with aposporous tetraploids, generate some hybrids with parthenogenesis but no apospory, which themselves give rise to some sexual and some aposporous (diploid) dihaploids (Fig. 10.11). In this way, a genetic continuum is maintained between sexual diploids and apomictic tetraploids.

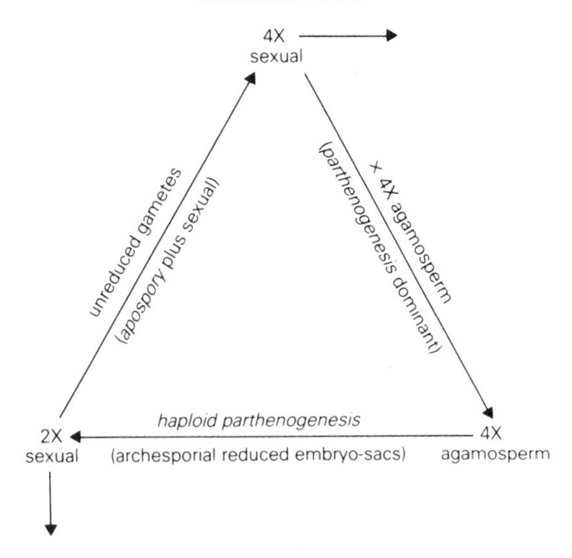

Fig. 10.11 Diploid/tetraploid cycle in *Panicum* and *Dichanthium*. (After Asker, 1979.)

Taraxacum

The great majority of polyploid *Taraxacum* species are obligate agamosperms. However, some sexuality, which can lead to sexual/agamosperm cycles depending on chromosome number, has been reported to arise from two distinct phenomena.

First, in a much quoted piece of work, Sorensen and Gudjonsson (1946) and Sorensen (1958) apparently show that certain disomic aberrants ($2n$ = 23) from triploid species ($2n$ = 24) which are obligate agamosperms, demonstrate some sexuality. They claim that the eutriploid progenitors are autotriploid, having the eight distinct chromosomes of the genome each present three times. Disomic aberrants lacking a chromosome D give rise to sexual diploid offspring, whereas aberrants lacking a chromosome H give rise to higher polyploids. The inference is (Richards, 1973) that the dominant control for synapsis (giving unreduced eggs) is on one of the D chromosomes, whereas the dominant control for parthenogenesis and autonomous embryony is on one of the H chromosomes. Thus, if the asynapsis gene is lost by loss of chromosome D, a reductional meiosis (leading to some reduced sexual eggs) takes place. If the parthenogenetic gene is lost by loss of a chromosome H, parthenogenesis is missing, and unreduced eggs are fertilized.

Unfortunately, it has to be said that the karyology of triploid *Taraxacum* species (including the species investigated by these authors) is nothing like they state it to be. Chromosome types are fairly readily distinguishable, but never occur in regular sets of three (several authors, summarized in Richards, 1972; Mogie, 1982). Nor would one expect them to, for all triploid dandelions are almost certainly hybrid allotriploids arising from species with very diverse karyology. I have occasionally encountered weakly sexual forms with $2n = 23$, which are accompanied by apparent hybrids, and I have little doubt of the basic truth of Sorensen and Gudjonsson's findings. However, we must treat the claimed identification of eight distinct morphological aberrants, each resulting from a different missing chromosome, and the chromosomal location of the agamospermous genes with grave suspicion.

Second, both Richards (1970a) and Muller (1972) have noted that facultative agamospermy is found in a few *Taraxacum* species from central Europe, and in the more widespread *T. brachyglossum* (which occurs in the UK, and adventively in North America). In such plants, emasculation of triploids leads to 40–70% seed-set, and gives rise to triploid offspring agamospermously. However, cross-pollination of these heads gives a higher percentage seed-set, and some of the offspring are diploid or subdiploid ($2n = 16–18$) sexuals. The female meiosis appears to be synaptic, and it is presumed that some female meioses fail, leading to unreduced, agamospermous eggs (which develop parthenogenetically). Other meioses are apparently reductional, leading to haploid, or near-haploid, sexual eggs (Fig. 10.10). Diploid sexuals may cross with reductional pollen, to give rise to more sexual diploids. However, some of the pollen from triploids is non-reductional, or partially reductional, and triploid or near triploid ($2n = 22–29$) agamospermous offspring can arise from agamosperm (male) and sexual diploid crosses (parthenogenesis being dominant and operating only on unreduced eggs).

This remarkable mechanism poses more questions than it solves, but it may account for the high level of sexual diploidy found among the dandelions of some parts of Europe (Fürnkranz, 1960, 1961, 1965; den Nijs and Sterk, 1980, 1984). I have suggested (Richards, 1973) that such plants represent an intermediate condition in the evolution of agamospermy in *Taraxacum*, in which the gene for the parthenogenetic precocious development of unreduced eggs is present, and functions in triploids with an irregular meiosis. However, meiotic asynapsis, leading to total non-reduction of female meiosis and hence obligate agamospermy (as is found in most *Taraxacum*) is not present.

Rosa canina

Before leaving this section, I feel bound to remark on a very interesting mechanism that is apparently unique in its nature and operates in most polyploid roses such as the dog roses (*Rosa canina* agg.) (Gustafsson, 1944;

Gustafsson and Hakansson, 1942; Fagerlind, 1945). These may be tetraploid ($2n = 28$) to hexaploid ($2n = 42$); most have $2n = 35$. In all, there is unequal synapsis in both male and female meiosis. Seven bivalents form and disjoin regularly. The remaining chromosomes (from 14 to 28 depending on the 'ploidy level) remain as univalents. They all become included in one of the daughter nuclei at telophase I of the female meiosis. Thus, one daughter nucleus has $n = 7$, and the other from $n = 21$ to $n = 35$ depending on the 'ploidy level.

After male meiosis, only the $n = 7$ microspores survive meiosis II, the others aborting. Thus, all pollen is $n = 7$, those chromosomes having passed through a regular disjunctional meiosis with recombination and segregation. In contrast, after female meiosis, the $n = 7$ megaspores abort, and one of the others develops an embryo-sac, which will have an egg cell with from $n = 21$ to $n = 35$. Although seven of the chromosomes in this egg have undergone a disjunctional meiosis with recombination and segregation, the remainder have not. Instead, they have been inherited maternally as an invariable, quasi-agamospermous package.

Regular sexual fusion now occurs between the $n = 7$ sperm cell and the $n = 21$–35 egg cell to give an embryo of $2n = 28$–42. There is also fusion between the other sperm cell and the fused polar nucleus. As a result, the endosperm has a female:male ratio of between 6 and 12 to one, much higher than is normal.

In a sense, the *Rosa* system is equivalent to facultative agamospermy, in that sexual and asexual reproduction by seed coexist. What is unique is that features of sexual and asexual reproduction coexist within every individual meiosis. Thus, those genes that are carried on the seven pairs of homologous, synaptic chromosomes undergo the regular sexual cycles of recombination, segregation and random fusion. The remaining genes are protected from the sexual process by asynaptic maternal inheritance. Thus, it should pay a rose to translocate genes that donate fitness and vigour onto the asexual chromosomes, and to transfer disadvantageous mutants onto the sexual chromosomes, where they can be recombined successfully, or lost by segregation. Genes that convey variability in niche specificity might also be most successfully linked to sexual chromosomes.

To my mind, there is a theoretical elegance about the rose system which is totally delightful. If one was to 'play God' and design an ideal breeding system, evolutionarily, this would be it. However, the likelihood of such a system evolving by the processes of mutation and natural selection would seem to be very small, which may explain why *Rosa* seems to be the only example of this mechanism. In particular, the regular inclusion of the asynaptic chromosomes in one daughter nucleus, and the differential abortion of high and low chromosome number spores after male and female meioses, must be very rare phenomena in regular sexual populations from which the rose system must have evolved.

DISTRIBUTION OF AGAMOSPERMY

Systematic distribution

Although agamospory is widespread in Pteridophytes (Manton, 1950), agamospermy is limited to Angiosperms among seed plants, and does not occur in the Gymnosperms. The most comprehensive early review on the occurrence of agamospermy in the Angiosperms is to be found in the appendix of Nygren (1967). He includes species with floral proliferation (so-called 'vivipary'), but I prefer to treat these as a form of vegetative reproduction. I also exclude cases of haploid parthenogenesis (as in *Dactylorhiza*, Hagerup, 1944, 1947 and *Epipactis*, Hagerup, 1945). Of the remainder, Nygren lists about 300 taxa which have been shown to be agamospermous.

Later estimates by Carman (1995) list 406 apomictic taxa. Both estimates are wildly unreliable and must be treated with great caution, due to taxonomic difficulties as to the nature of a species in agamic complexes. Both Nygren and Carman have been very inconsistent as to the taxonomic rank of their units. For instance Nygren treats many *Sorbus* and *Alchemilla* agamospecies as independent units, whereas the genera *Crataegus* (over 100 agamospecies), *Rubus* and *Hieracium* (well over 2000 agamospecies each) are each treated as a single unit. According to Carman, nine *Taraxacum* taxa are apomictic. In fact, the true figure probably approaches 2000, and even the number in which apomixis has been definitely proved runs well into three figures.

With these major reservations in mind, it is interesting to compare numbers of taxonomic units which possess each of the three main classes of agamospermy (Table 10.4). (Note that in deriving the numbers in this table the categories queried by Nygren are not queried here, and species which have not been determined but occur in genera with a consistent

Table 10.4 Numbers of agamospermous taxonomic units showing various types of agamospermy (after Nygren, 1967)

Family	Adventitious embryony	Apospory	Diplospory	Total
Compositae		18	51	69
Gramineae		68	27	95
Liliaceae	6		1	7
Rosaceae		65	3	68
Rutaceae	5	2		7
Urticaceae		2	7	9
28 other families	33	5	15	53
	44	160	104	308

system are 'rounded up' to that system for the sake of simplicity; a further problem arises in distinguishing between plants showing apospory and those showing mitotic diplospory (some *Potentilla, Hieracium*); it can be difficult to decide whether initials in the nucellus form an archesporium and an EMC, or not (Gustafsson, 1947); these figures must not be treated as more than rough estimates.)

Some interesting points arise from a consideration of Table 10.4. Whereas agamospermy is recorded from about 15% of plant families (34 in all), no less than 75% of agamospermous taxa belong to only three families, the Compositae, Gramineae and Rosaceae.

Preadaptation to agamospermy

Although the Compositae and Gramineae comprise two of the three largest plant families; nevertheless, these three families only contain about 10% of Angiosperm species. Thus, it might be supposed that plants in these families are in some sense 'preadapted' to agamospermy, showing for instance unusually high levels of polyploidy, hybridization or haploid parthenogenesis among their sexual members (p. 416).

An alternative hypothesis, that agamospermy, having arisen in a family, leads to the extensive genetic migration of agamospermous traits, or extensive radiation of agamospermous genera through that family, must be discounted. Although facultative agamospermy may encourage the spread of agamospermy within part of a genus, there is no indication that a single origin of agamospermy has ever given rise to more than one genus. Every agamospermous genus has some sexual taxa, which must be considered primitive to it, and agamospermy must be thought of as being essentially highly polyphyletic in nature, arising *de novo* on many occasions, even in closely related genera. Thus in the family tribe Cichorieae (Compositae) which have yellow, dandelion-like flowers, three massive agamic complexes have arisen in the hawksbeards (*Crepis*), hawkweeds (*Hieracium*) and dandelions (*Taraxacum*), as well as in smaller genera (*Chondrilla, Ixeris*). Yet each genus has its primitive sexual progenitors, and in each the mechanism of agamospermy differs (*Crepis*, apospory with parthenogenesis; *Hieracium* subgenus *Hieracium*, non-meiotic diplospory with apogamety; *Hieracium* subgenus *Pilosella*, apospory with apogamety; *Taraxacum*, meiotic diplospory with parthenogenesis; Table 10.5).

An examination of Table 10.4 lends some credence to the 'preadaptation' hypothesis. Thus, apospory is a much commoner method of circumventing meiosis in the Gramineae, and especially in the Rosaceae, than it is in the Compositae, in which diplospory prevails. The pattern of distribution is certainly consistent with a hypothesis that suggests some 'preadaptation' to a certain sort of agamospermous mechanism within a family.

Table 10.5 Some of the more important agamospermous genera, and their major agamospermous mechanisms (after Nygren, 1967)

Adventitious embryony (with some sexual fusion and with pseudogamy)
Citrus (most species and hybrids to varying extents), *Mangifera* (Mango), *Spathiphyllum*, *Opuntia*, *Capparis* (caper), some *Euonymus* spp. (spindles), *Pachira*, *Garcinia mangostana* (Mangosteen), some *Eugenia* spp. (rose-apples), *Ochna serratula*, *Nigritella*
Apospory (with pseudogamy and occasional sexual polyembryony)
Poa pratensis/*alpina*, *Panicum maximum* agg., *Potentilla argentea* agg., *P. verna* agg., *Dichanthium*/*Bothriochloa*, *Cortaderia* sp., *Paspalum* spp., *Pennisetum* spp., *Urochloa* spp., *Rubus fruticosus* agg., *Malus* spp., *Alchemilla* spp., *Cotoneaster* spp.,? *Sorbus* spp.,? *Crataegus* spp., *Ranunculus auricomus* agg., *Crepis* spp., *Hieracium* subgenus *Pilosella*, *Skimmia* spp.
Mitotic diplospory (sexuality very rare, no pseudogamy)
Antennaria, *Ixeris*, *Hieracium* subgenus *Hieracium*, *Arnica* spp., *Erigeron* spp., *Elatostema* spp., *Balanophora* spp., *Calamagrostis* spp.
Meiotic diplospory (sexuality usually absent except for *Rubus*, no pseudogamy)
Taraxacum, *Chondrilla* spp., *Rudbeckia* spp., *Nardus stricta*,? *Limonium* spp., automictic *Rubus*.

Predispositions to agamospermy, if they exist, are not necessarily confined to chromosomal or physiological phenomena. The predominant growth habits and ecology of the various families are similar. Thus the Gramineae and Compositae are mostly composed of perennial herbs of temperate areas, and indeed are the dominant families in these biomes. The Rosaceae, although including some agamospermous perennial herbs (*Alchemilla*, *Potentilla*) or even annuals (*Aphanes*), also has many agamospermous trees (*Crataegus*, *Sorbus*). Common factors in these three great families are non-specific pollination mechanisms by wind (Gramineae) or insects, temperate or grassland distributions, and single-seeded fruits. Non-specific pollination may be associated with hybridization; and temperate distributions seem to have encouraged polyploidy (associated with glacial conditions which may also have favoured agamospermy).

In a single-seeded fruit, the selective fitness of an asexual embryo (of a maternal genotype) may be optimized with respect to sexually arising half-siblings. In comparison, asexually and sexually arising half-siblings which compete for maternal resource in a multiseeded fruit may be compelled to share that resource more equally. There may also be greater competition between half-sibling seedlings after dispersal from certain sorts of multiseeded fruits. As yet there seems to be no information to support these very tentative suggestions.

We do not seem to understand what has apparently predisposed the Compositae to displospory, and the Gramineae and Rosaceae to apospory. If there is a logical basis to these apparent preadaptations, we may have to seek it in currently little-understood areas of developmental control and differentiation.

However, our confidence that preadaptive mechanisms for the origin of agamospermy repay consideration is strengthened by an examination of the distribution of adventitious embryony. This is totally lacking in the temperate, often herbaceous, predominantly polyploid and single-seed fruited Compositae, Gramineae and Rosaceae. Yet 58% of the remaining 76 agamospermous taxa have adventitious embryony.

These sporophytic apomicts are typically subtropical or tropical, woody, diploid and with fleshy, multiseeded fruits (Table 10.5, p. 404), and thus are quite different from aposporous and diplosporous species.

Geographical distribution

It has frequently been pointed out (e.g. Stebbins, 1950), that gametophytic agamospermy becomes more frequent as one proceeds further north or further up mountains, just as does polyploidy and the frequency of perennial herbs. As far as I am aware, this relationship has not been investigated with increasing latitude southwards. Various explanations for increasing levels of agamospermy with increased latitude northwards can be presented.

1. Arctic, boreal and northern temperate floras have been more thoroughly examined for agamospermy than have tropical floras, and thus the relationship with latitude and altitude is an artefact.
2. For non-pseudogamous agamosperms, the lack of a requirement for pollination is an advantage in areas with cold summers.
3. The heterotic vigour of agamospermous hybrids is an advantage, and the lack of precise adaptation to a complex habitat is not a disadvantage in open arctic, montane or boreal habitats.
4. The onset of glacial epochs encouraged migration, resulting in hybridization; and encouraged rapid evolution, optimized by the genetic isolation and variability of polyploids. Both polyploidy and hybridization are associated with agamospermy, as discussed above, thus agamospermy in high latitudes and altitudes is an accidental correlate of hybridization and polyploidy.
5. Agamosperms can establish from single seeds, so that they could efficiently colonize arctic and alpine regions postglacially ('Baker's Rule' p. 304). In the *Hieracium pilosella* group, Gadella (1991) has emphasized that remote or disjunct populations are always apomictic.
6. Agamospermy is chiefly found in a few families which show a preadaptation towards it; these families predominate at high latitudes and altitudes for reasons which do not, or only partially, coincide with the preadaptive attributes.

It is likely that all these explanations contribute to the correlation between agamospermy and latitude/altitude, and most are to some extent interdependent or interrelated.

It has frequently been noted that agamospermous complexes have a wider distribution geographically and ecologically than their sexual counterparts and progenitors, and this has been explained by the greater levels of heterozygosity, capacity for plasticity, reproductive efficiency, and short-term fitness of agamospermous genotypes (e.g. Babcock and Stebbins, 1938; Stebbins, 1950 for North American *Crepis*). However, this apparent relationship appears to be the product of some confused and possibly wishful thinking, and requires a closer examination. For instance, agamospermous dandelions (*Taraxacum*) undoubtedly cover a wider area of the earth's surface than do sexual dandelions; probably about twice the area. Yet, there are about 10 times as many agamospermous taxa as sexual taxa in *Taraxacum*. Very many of the agamospecies are extremely localized, whereas sexual species such as *T. serotinum*, *T. brevirostre* or *T. bicolor* range for thousands of kilometres (van Soest, 1963; Richards, 1973). Typically, sexual species are palaeoendemic and may have wide, interrupted distributions, whereas agamospecies are neoendemic with narrow, but expanding, distributions, particularly as a result of introduction by man.

There is also a danger of comparing sexual species, as a group or individually, with agamospecies, as a group or individually. The philosophy surrounding specific limits is unavoidably very different for diverse breeding systems (p. 402), and it is difficult to make direct comparisons between the two. However, it is possible to point out that, for *Crepis*, *Taraxacum*, or other partially agamospermous genera, the agamospermous mechanism appears to be currently more successful, than sexuality for it is found in more taxa and over a wider area in these genera. Such a statement infers nothing about the ecological versatility of the agamospermous genotype, but does highlight its greater fitness in some genera. In others (for instance *Sorbus*), the agamospecies are numerically and distributionally much more limited than the sexual species, and some are verging on extinction in the British Isles (e.g. Clapham, Tutin and Warburg, 1962; Perring and Sell, 1968).

GENETIC DIVERSITY IN AGAMOSPERMS

The literature for agamospermy will shortly reach its centenary. Typically, *éminences grises* stressed the invariability of obligately agamospermous families and populations, prophesying a poor evolutionary prognosis for such asexual lines (Muller, 1964; Mayr, 1970). 'With the loss of sexual recombination the apomict . . . is cut off from ultimate survival . . . after a brief prosperity succumb(ing) to a changing environment. Apomixis is an escape from sterility . . . but it is an escape which leads only to extinction' (Darlington, 1939). '. . . all agamic complexes are closed systems and evolutionary "blind alleys" . . . the life expectancy of agamic complexes is shorter than that of sexual groups . . . agamic complexes are destined to produce only new variations on an old theme' (Stebbins, 1950). Stebbins and Darlington both noted that apomixis shows no evidence for long-term evolutionary success, for there is no taxon at generic rank or above which has become exclusively apomictic. However, this evidence is better cited in favour of the ongoing success of sexuality (Chapter 2), which is not the same thing!

To be fair to Darlington and Stebbins, they both noted the pervasiveness among apomicts of what Darlington called 'subsexual' reproduction. In particular, Stebbins was heavily influenced by the works of Gustafsson (1946–7) who described in detail the opportunities for recombinational and segregational variability available to most apomicts. Clearly, Gustafsson regarded truly obligate agamospermy as a rarity, a theme later taken up by Asker (1979, 1980), Mogie (1992) and others. These authors stress both the evolutionary aggressiveness and the versatility of most asexual systems, for which the archesporium often still provides a sexual alternative.

It is certainly the case, however, that sexual populations are very much more variable than are apomictic populations, although they may be less heterozygous (p. 399). Table 10.1, which compares levels of genetic variability, and of heterozygosity for typical populations of apomictic and sexual *Taxaxacum* (Hughes and Richards, 1988) makes this point very clearly.

Nevertheless, genetic variability often occurs within predominantly asexual plant lines (e.g. reviews in Gustafsson, 1946–7; Asker, 1979, 1980). At least some of the variation encountered is due to residual sexuality (e.g. Dickinson and Phipps, 1986; Nybom, 1988; Proctor, Proctor and Groenhof, 1989; Gadella, 1991). Where apomictic *Taraxacum* occur together with sexuals, gene exchange can occur freely between diploid sexuals and polyploid apomicts (Richards, 1970b; Muller, 1972; Jenniskens, den Nijs and Huizing, 1984; Sterk, 1987) to the extent that coexisting apomicts can show panmictic genotype distributions (den Nijs, Menken and Vlot, 1987). Where apomictic *Taraxacum* bear pollen, as most do, they can only be regarded as being obligate apomicts when no sexuals occur in the locality.

Variability in apomictic populations

In recent years, many papers have stressed that populations of obligately agamospermous *Taraxacum* are normally genetically diverse (e.g. Mogie and Richards, 1983; Lyman and Ellstrand, 1984; Ford and Richards, 1985; Mogie, 1985; van Oostrum, Sterk and Wijsman, 1985; Hughes and Richards, 1988). As Kirschner and Stepanek (1993) have pointed out, there are both semantic and evolutionary difficulties in the safe interpretation of this information. Nevertheless, it is interesting that Ford and Richards (1985), Battjes, Menken and den Nijs (1992) and Kirschner and Stepanek (1993) all show that some agamospecies (e.g. *T. subnaevosum* Richards, *T. hollandicum* v. Soest and *T. alpestre* (Tausch.) are highly invariable enzymatically, whereas others, such as *T. vindobonense* v. Soest show high levels of zymatic diversity (Table 10.6). In contrast, Proctor *et al.* (1989) find little enzymic variation within and between populations of any endemic British agamospecies of *Sorbus*, although sexual species were highly variable.

It is important that such populational studies on agamosperms have this kind of taxonomic input. When Solbrig and Simpson (1974) or Lyman and Ellstrand (1984) show that lawn populations of north American *Taraxacum* are genetically variable, without discriminating agamospecies, one might equally well indiscriminately sample the genotypes of grasses growing in that lawn, whatever their species.

Variability among apomictic offspring

There is no way that we can discover what kind of origin, or how old, genetic variability among agamospermous populations is. Consequently, an investigation of any genetic variability arising from obligate agamospermy must concentrate on asexually generated offspring families from single mothers. In these mothers there should be no sexual fusion, and meiosis is normally asynaptic and fully restitutional, so that recombination and segregation should not occur (Richards, 1973).

In such plants, theoretically, offspring variability could nevertheless arise through one or more of the following mechanisms:

1. Accumulation of mutational changes to the DNA, which cannot be shed latterly by recombination, through 'Müller's ratchet' (e.g. Maynard Smith, 1978);
2. Accumulation of gross changes to the cytology through disjunctional accidents, resulting in polyploid, haploid, oligosomic and polysomic aberrants amongst the offspring;
3. Somatic recombination resulting from chromosomal translocations;
4. Mutational or chromosomally based changes to those genes in the maternal genome concerned with the control of agamospermous behaviour.

The first reports of offspring variability among apomictic *Taraxacum* stem from the work of the young Icelander Gudjonsson whose work was

Table 10.6 Number of individuals per clone for *Taraxacum* section *Palustria* species from Czechoslovakia. No: number of unique clones in sample. Nc: total number of clones in sample. Ni: number of individuals in sample. Hg: clonal diversity (After Battjes, Menken and den Nijs, 1992)

		Clones																													
Locality	Species	119	106	126	120	127	125	1	2	17	13	16	14	24	53	54	56	61	66	71	75	84	88	90	100	99	97	No	Nc	Ni	Hg
1	*T. subalpinum*	5																										0	1	5	0
	T. paucilobum		1																									0	1	1	0
	T. ancoriferum			1																								0	1	1	0
	T. irrigatum																											1	1	1	0
	T. hollandicum				105																							0	1	105	0
2	*T. hollandicum*				43																							0	1	43	0
	indeterminate specimens					8																						0	1	8	0
	T. ancoriferum			2																								0	1	2	0
3	*T. hollandicum*				8																							0	1	8	0
4	*T. hollandicum*				11																							0	1	11	0
5	*T. hollandicum*				14																							0	1	14	0
	indeterminate specimens						8																					0	1	8	0
6	*T. hollandicum*				5																							0	1	5	0
7	*T. hollandicum*				42																							3	4	45	0.13
8	*T.* type 1							17																				0	1	17	0
9	*T. uliginosum*								4																			0	1	4	0
10	*T. uliginosum*								24																			0	1	24	0
11	*T. vindobonense*									2	2	2	2	2														23	28	33	0.96
	indeterminate specimens																											22	22	22	1.00
12	*T. olivaceum*														5	15	5											5	8	30	0.69
13	*T. vindobonense*																	2	2									8	10	12	0.89
14	*T. vindobonense*																			5	4							11	13	20	0.87
15	*T. vindobonense*																					5	5	2				10	13	22	0.87
	indeterminate specimens		1																						4	2	2	16	20	25	0.93

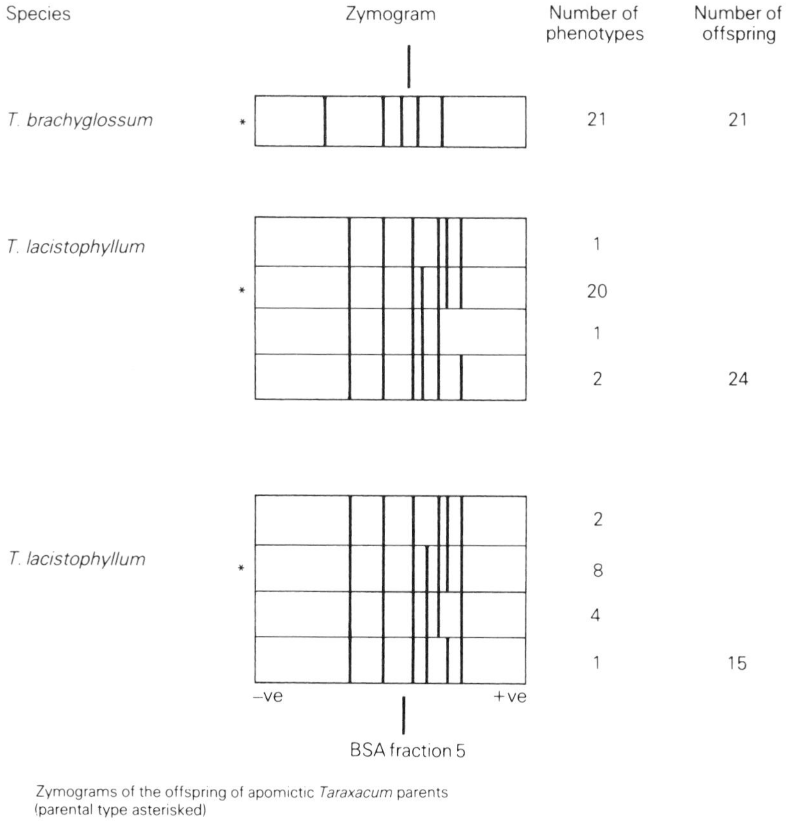

Species	Zymogram	Number of phenotypes	Number of offspring
T. brachyglossum	*	21	21
T. lacistophyllum		1	
	*	20	
		1	
		2	24
T. lacistophyllum		2	
	*	8	
		4	
		1	15

–ve +ve

BSA fraction 5

Zymograms of the offspring of apomictic *Taraxacum* parents (parental type asterisked)

Fig. 10.12 Representation of esterase gel-electrophoretic band phenotypes in 60 offspring of *Taraxacum* from Northumberland, UK. Three offspring families were examined, from two parents (*). BSA is the bovine serum albumin standard run with each gel.

tragically cut short by an accident. Working with large families from single mothers of well-understood triploid agamospecies in *Taraxacum* section Ruderalia, considerable phenotypic variability was found among families which could mostly be assigned to disjunctional accidents resulting in polyploidy, pentasomy, tetrasomy, disomy and monosomy (Sorensen and Gudjonsson, 1946). Such was the variety of phenotypes so produced that Gudjonsson even claimed to be able to predict the identity of the missing chromosome from the appearance of the plant. With the benefit of hindsight, we now know that this was based on faulty karyological interpretations (e.g. Mogie and Richards, 1983). Nevertheless, it is possible to find 'Gudjonsson-mutants' today, and to show that they are chromosomally based, fertile and heritable, if usually somewhat inviable.

Undoubtedly, many cytological 'offtypes' produced mitotically or meiotically will suffer from the adverse effects of imbalances in gene dosage, being inviable or weak, as Sorensen and Gudjonsson (1946) discovered. However, Mogie (1986) points out that nearly all apomicts are polyploid, for which gene dosage effects and the accumulation of disadvantageous recessive mutants by 'Müller's ratchet' are likely to prove less severe. In fact he argues that this should select for polyploidy among apomicts, as disadvantageous mutants can be accommodated within the apomictic line during which time advantageous dominant mutants may be acquired. It is certainly the case that some successful *Taraxacum* apomicts can be stable hypertriploids (Malecka, 1967).

Mogie (1982) initiated a detailed study of karyological and zymatic variation within and between the closely related members of one section of *Taraxacum* (Hamata). He noted very considerable karyological differentiation within families, although this was less than that found between families. Data from the frequency of anaphase bridges (Mogie, 1992) (Table 10.7) strongly suggested that some of this *de novo* karyological variation resulted from somatic recombination among apomictic lineages, as bridge formation proved to be rare in sexuals. Due to the technical difficulties of karyological analysis for small chromosomes, Richards (1989a) concentrated on karyological variability within asexual lineages of the readily recognizable NOR chromosome. This work clearly demonstrates that remarkably high levels of chromosome breakage and somatic recombination occur within some *Taraxacum* families, and that marked variation for this trait apparently occurs between lines, even within an agamospecies (Table 10.8, Figs. 10.13 and 10.14).

Table 10.7 Percentage mean similarity of karyotypes (KSI) between cells within individuals and percentage of anaphase cells showing bridges, in *Taraxacum* of different breeding systems (after Mogie, 1982)

Species	Breeding system	Mean KSI (%)	Cells with bridges (%)
T. pseudohamatum	obligate agamosperm	87.5	10.5
T. lacistophyllum	obligate agamosperm	89.6	3.8
T. unguilobum	obligate agamosperm	90.9	6.6
T. brevifloroides × *oliganthum*	wide sexual hybrid	85.6	8.1
T. alacre	sexual outbreeder	97.5	0
T. brevifloroides	sexual outbreeder	99.4	–
T. bessarabicum	sexual inbreeder	100	1.0

Table 10.8 Means and standard deviations (in μm) and coefficients of variability in lengths and ratio of the body, satellite and body/satellite ratio in sexual and agamospermous *Taraxacum* spp. studied

	Body/satellite ratio			
Taxon	mean	sd	cv	*n*
'French sexual'				
F	1.878	0.238	0.126	45
H	1.960	0.292	0.149	21
J	1.800	0.181	0.101	32
K	2.180	0.277	0.127	15
overall	1.905	0.274	0.143	113
T. pseudohamatum, agamosperm				
A	1.966	0.474	0.241	50
C	2.060	0.263	0.128	20
D	2.430	0.978	0.402	20
overall	2.096	0.624	0.298	90
T. brachyglossum, agamosperm				
E	2.210	0.570	0.258	16
G	2.020	0.300	0.149	21
overall	2.103	0.442	0.210	37

So far, four investigations have specifically sought for electrophoretic (isozymal) variation among seedlings from a single family. Mogie (1985), working solely with esterase, found that only one seedling of the agamospecies *T. unguilobum* Dahlst out of 232 seedlings (0.3%) of the agamospecies *T. pseudohamatum* Dahlst. and *T. unguilobum* displayed a non-parental zymogram with a missing band. The abnormal phenotype was repeatedly found to be present for this individual, but its heritability was not tested. A second non-parental phenotype could not be repeated in subsequent tests.

The scrupulous work of Hughes (1987) shows how important it is for the heritability of *de novo* variants to be tested. For 56 sister seedlings of *T. pseudohamatum* tested for 15 loci, no non-parental genotypes were found at all. However, for esterase, which, due to its complex and unanalysable band phenotype, was treated separately, one out of 31 sister seedlings displayed a non-parental phenotype with a single extra band. Although this phenotype was consistently displayed by this individual, all its offspring showed the original, grandparental phenotype lacking the extra esterase band (this was discovered after the report in Richards (1986)). Hughes also found a small amount of non-heritable esterase variation in *T. brachyglossum* (Dahlst.).

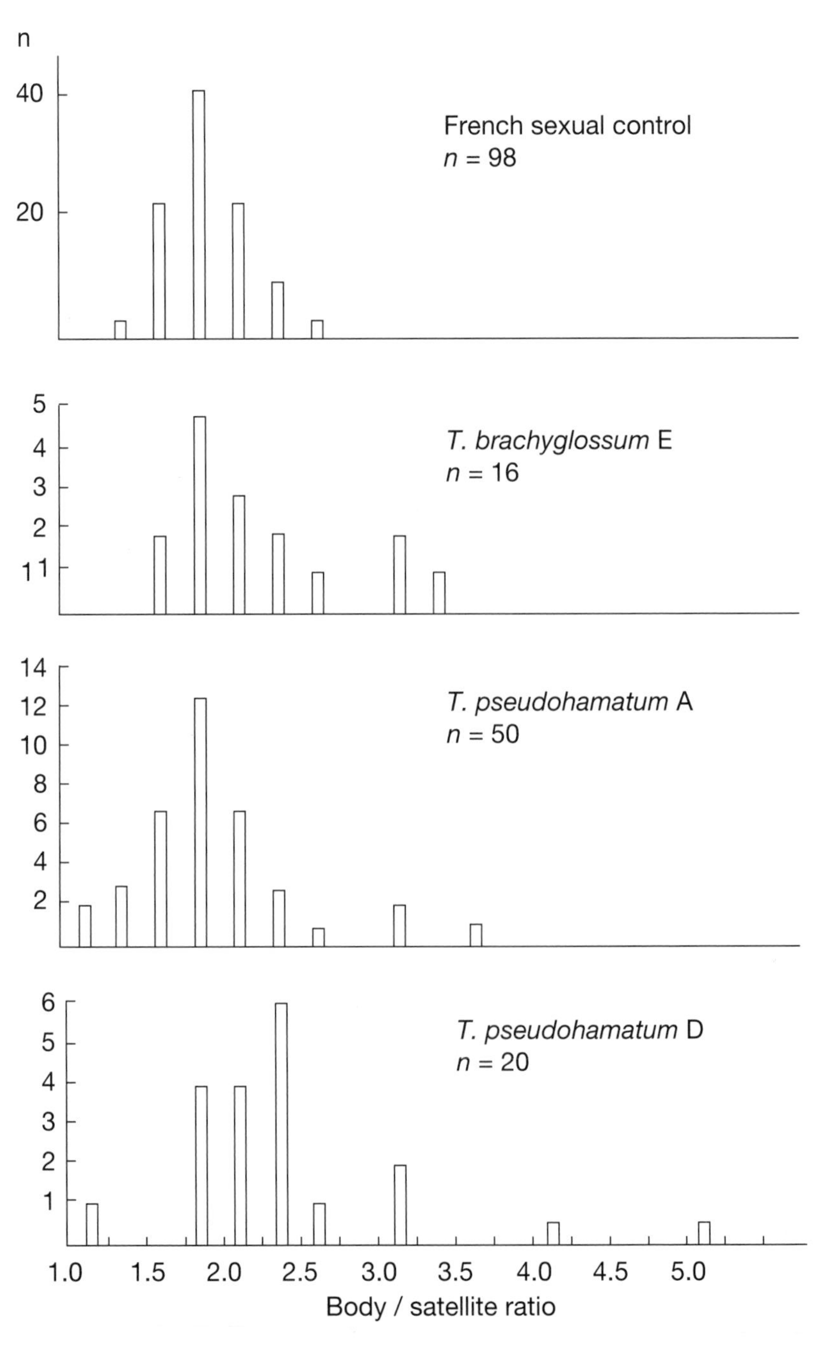

Fig. 10.13 Distribution of the body/satellite ratio for the NOR-chromosome for the 'French sexual' individuals combined (above), and three agamospermous individuals of *Taraxacum*. Readings are summed for units of 0.25, e.g. 1.00–1.24, 1.25–1.49, etc.

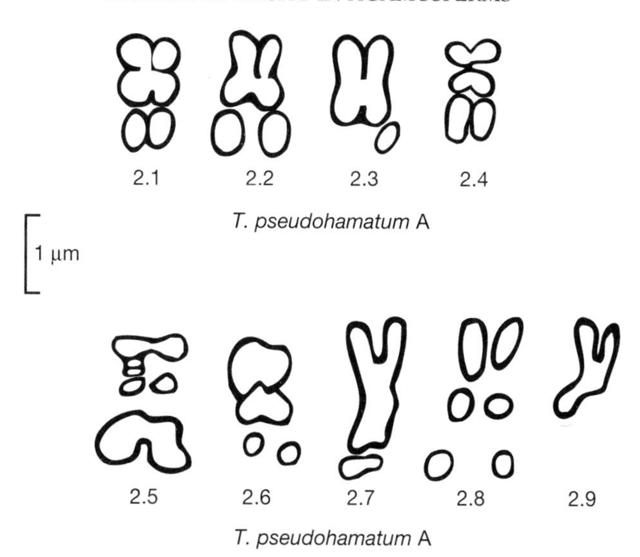

Fig. 10.14 Tracings of photographs of NOR-chromosomes in *Taraxacum pseudohamatum*. Chromosomes 2.1 and 2.2 are considered not to have undergone rearrangement. Chromosomes 2.3 to 2.9 have undergone rearrangement. Refer to text for further details.

Most recently, agamospermous *Taraxacum* has been subject to molecular investigations (King and Schaal, 1990). These authors probed variation in three classes of DNA, ribosomal DNA, chloroplast DNA, and the alcohol dehydrogenase (ADH) gene by using restriction site specific enzymes. They surveyed 714 offspring from 31 lineages. The only variation they found was for the offspring of two mothers who were identical sisters. In one case, all 35 offspring were non-parental for a rDNA restriction fragment, whereas in the other only one offspring out of a family of 26 carried the same non-parental rDNA genotype. Interestingly, non-parental genotypes in both families also carried non-parental genotypes for ADH, although these new ADH fragments differed between the two families. Tests showed that non-parental genotypes did not differ from parents cytologically, or in detectable quantities of DNA, and electrophoretic phenotypes for ADH were the same for parental and non-parental genotypes.

Somatic recombination

In recent years, it has become clear that certain codons, known as transposons, have the capability of silencing neighbouring genes and breaking the DNA and the chromosome proximally, so that these genes 'roam round the genome' changing phenotypic expression and causing somatic recombination, wherever they occur (Calos and Miller, 1980; Finnegan, 1981). Somatic recombination, resulting from such translocations, differs in its effect from meiotic recombination in that recombination occurs not only between homologous chromosomes, but also throughout the genome.

Obligate parthenogenesis is frequently encountered in animals as well as in plants. Earthworms (Muldal, 1952; Omodeo, 1952), weevils (Suomalainen, 1969), moths (Lokki *et al.*, 1976a,b), aphids and others (Suomalainen, Saura and Lokki, 1976) can all show genetic variation within single asexual families. Saura *et al.* (1976) suggest that transposon-induced somatic recombination may be a frequent cause of such variability.

In higher plants, transposon-induced behaviour seems to be particularly associated with hybrids (Woodruff, Thompson and Lyman, 1979; Woodruff and Thompson, 1980). It seems likely that species-specific quiescent transposons will be activated when exposed hemizygously in hybrids. High levels of chromosome breakage/refusion, often detected by anaphase bridge formation, seem also to be typical of many apomicts (e.g. *Hieracium*, *Calamagrostis*, *Dichanthium*, *Poa*, *Ranunculus* and *Hierachloë*, as well as in *Taraxacum* (reported in Mogie, 1992). With the benefit of hindsight, this activity can also be explained by transposons, particularly as most if not all apomicts are formed from relatively wide hybrids. Mogie (1992) has indeed suggested that by releasing transposon activity and stimulating somatic recombination, hybridity may have particularly favoured the creation of successful apomictic lines.

Asexual evolution

In this context, it is interesting that somatic recombination appears to play an important role among many asexuals. Such recombination is not comparable to meiotic recombination, where daughter cells differ genetically. Barring chromosomal accidents, mitotic daughter cells are identical, so that disadvantageous alleles cannot be 'shed' in the same way. Nevertheless, somatic recombination may cause such alleles to be removed to a chromosomal environment in which they are disabled or silenced. By breaking Müller's ratchet, transposon-induced somatic recombination may prove to be an essential component in the long-term survival of any asexual line.

Such a line can apparently diversify and evolve without sexual intervention to a point at which taxonomists are encouraged to describe a number of agamospecies, as has been shown for the 24 species classified within *Taraxacum* section Hamata (Mogie and Richards, 1983). Modern thinking is now leaning towards a concept, at least for *Taraxacum*, whereby most 'speciation' within the genus may have occurred asexually (Kirschner and Stepanek, 1993). It remains to be seen whether this randomly fixed variation can ever allow the line which possesses it to evolve into a distinctively new ecological niche. Having acquired a line of *Taraxacum hamiferum* Dahlst. which had apparently evolved copper tolerance (Table 10.9) (Richards, 1989b) and then having carelessly lost it again, I find this question more tantalizing than most!

Selection for male fertility in obligate apomicts

The fertility, regularity, and the more or less reductional nature of the pollen of many agamospermous dandelions may be adaptive (Richards, 1986; Mogie, 1992). Mutants that led to asynapsis in the female meiosis alone may have been more successful than those that also gave asynapsis in male meiosis. Agamosperms with reductional pollen can not only act as female parents with a full maternal genetic control, but can also act as male parents to coexisting sexuals. In many parts of Europe and Asia, such as Japan (Morita, 1976, 1980), sexuals and agamosperms regularly occur together. Even in areas such as the UK and the USA, where sexuals are rare or absent, extant agamosperms must have originally evolved agamospermy sympatrically with sexuals. Thus, obligate agamosperms may also be able to generate genetic variability (and new agamospermous

Table 10.9 Dry weight of seedlings of *Taraxacum hamiferum* harvested at 6 weeks after 22 days' growth in water culture containing various concentrations of copper salts. 'Parys' plants originated from a plant found growing in copper waste at Parys Mountain, Wales; control plants from an unpolluted roadside. Both parents were completely apomictic

Cu^{2+} (μg/g)	Parys	Control
0	32 ± 19	27 ± 13
	40 ± 9	25 ± 10
1	–	20 ± 9
	47 ± 8	19 ± 7
5	56 ± 16	8 ± 5
	47 ± 8	8 ± 4
10	–	10 ± 8
	38 ± 16	10 ± 6
20	24 ± 11	5 ± 9
	24 ± 11	3 ± 4
50	14 ± 7	6 ± 2
	15 ± 7	3 ± 2

biotypes) by hybridization with sexuals (Fig. 10.15). Sexual/agamosperm hybrid swarms of this kind have been described by Fürnkranz (1961, 1965) and Richards (1970c).

Variability in agamospermous behaviour

There are several reported cases in which the reproductive behaviour of facultative agamosperms (usually pseudogamous and aposporous) varies with environmental conditions. Perhaps the best known is in the grass *Bothriochloa* (Saran and de Wet, 1970) which behaves sexually during the summer, but as a facultative agamosperm during the Oklahoma winter. Embryological studies of plants kept experimentally under different day-length regimes demonstrated that the relevant environmental trigger was only two hours' difference in photoperiod (Table 10.10).

There are several other examples of related Andropogonoid grasses, in which apospory is related to photoperiod, and in all cases, short days favour high levels of agamospermy. Thus, Knox (1967) and Knox and Heslop-Harrison (1963) show that in northern Australian populations (9° S) of *Dichanthium aristatum*, with less than 14 hour days at flowering, 91% of embryo-sacs are aposporous. Further south (to 36° S), the longer day-lengths result in only 60% of the embryo-sacs being aposporous. These

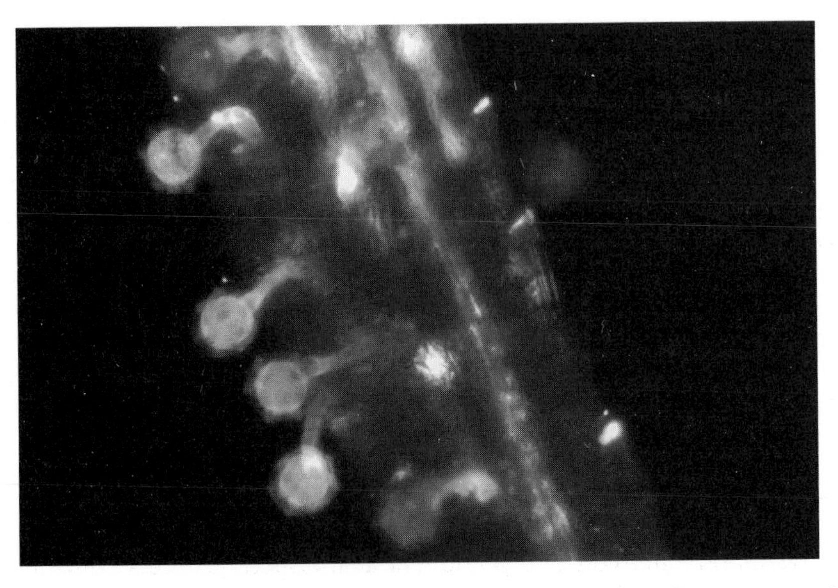

Fig. 10.15 Pollen germinating and penetrating stigmatic papillae in a sexual *Taraxacum*. Microphotograph under ultraviolet illumination (×400).

Table 10.10 The effect of day length (photoperiod) on the production of sexual (S) or aposporous (A) ovules in *Bothriochloa* spp. (after Saran and de Wet, 1970)

	Ovules			
Day length (photoperiod) (hours)	% S embryo-sacs only	% S embryo-sacs with non-functional A sacs	% with functional S and A embryo sacs	% with functional A sacs only
14	78.4	21.6		
12	36.6		20.0	43.4

responses are plastic, without genetic differentiation between areas. As the summer progresses and day-lengths decline, the proportion of apospory increases, and this effect can be repeated in the photoperiodically controlled experimental chamber. Somewhat unexpectedly, short days also resulted in high levels of pollen sterility. It is difficult to see how the pseudogamous apospory can operate in such conditions, without fertile pollen to fertilize the endosperm nucleus.

Nevertheless, we may presume that these plastic responses to photoperiod are adaptive. Short days relate to very different adverse conditions, both of which might favour agamospermy, in the two areas. Short days in Oklahoma presage the cold plains winter. Short days in Australia are associated with the searing summer heat of the northern tropics. However, in pseudogamous species such as these, there is no advantage in having the reproductive efficiency imposed by agamospermy in hostile conditions, as pollination and endosperm fertilization is required in any case. Conceivably, harsh conditions for seedling establishment favour low genetic variability, as is associated with the agamospermous mode of reproduction, among the offspring. If the mother is adapted to resist harsh conditions, seedlings of a maternal type may also be better suited to become established in these conditions. Seedlings establishing at a more favourable time of year, or a more favourable locality, might be able to colonize a wider range of niches in the environment if they displayed the genetic variability expected to result from sexual reproduction.

ADVENTITIOUS EMBRYONY

Adventitious embryony, or sporophytic agamospermy, differs from other apomictic mechanisms so much that I am dealing with it separately, although it has frequently been referred to in earlier sections of this book.

Adventitious embryony is very untypical of other apomictic mechanisms, in that it predominates in diploid, tropical, woody plants with large fleshy fruits (Table 10.5), and tends to be controlled by a single, dominant gene, as in *Citrus*. Also, the control of adventitious embryony can be at least partially hormonal, and labile, so that the predominance of asexual or sexual embryony can vary with the environment and can be modified experimentally, as can some types of apospory (p. 441).

In adventitious embryony, a 'proembryo' is formed vegetatively by the sporophytic tissues (usually the nucellus, sometimes the integument) of the ovule. The initial development of this proembryo is somewhat disorganized, appearing to be quasi-tumerous (p. 444). Only at a later stage does the proembryonic tissue organize itself into a cylindrical or tuberoid proembryo resembling the embryo formed by sexual reproduction. I have argued (p. 404) that plants forming this type of relatively undifferentiated embryo may be preadapted to adventitious embryony. As no embryo-sac is involved, the mechanism is entirely sporophytic. As the archesporium is not involved in the agamospermy, most sporophytic apomicts can also undertake perfectly regular sexual reproduction.

Normally, adventitious embryony shows no systematic or developmental relationships with gametophytic apomixis. Thus, the report that both adventive apomictic embryos, and those arising from mitotic diplospory occur together in the whitebeam *Sorbus intermedia* is remarkable (Jankun, 1994).

Garcinia

Apart from *Citrus*, sporophytic agamospermy has been studied relatively sparingly, perhaps because most examples grow well away from the homes of northern embryologists. *Garcinia* seems to form a typical example, and deserves a full discussion.

Garcinia is a large genus of small evergreen understorey trees from wet non-seasonal forest from around the Indian Ocean. There may be 400 species. All have fleshy fruits with few, large seeds, and most, at least, are dioecious.

Apomixis was first suspected to occur in *Garcinia* at an early date, because the prized Malay fruit tree the mangosteen, *G. mangostana,* occurs only as a female, but produces fertile seeds (Richards, 1990c dismisses the report of a single male mangosteen tree in Idris and Rukayah, 1987 as almost certainly referring to a hybrid). Richards (1990c) shows that the female mangosteen is of hybrid, allotetraploid origin, the parents (*G. hombroniana* and *G. malaccensis*) both being facultative apomicts which show sporophytic agamospermy in the females. I argued that the female mangosteen only arose once, from a single cross, inheriting apomictic behaviour from both parents. Thus, there is no male mangosteen, and mangosteens can only reproduce asexually. All mangosteens should be genetically identical.

Richards (1990a) studied a number of other *Garcinia* species in Malaysia, concentrating on the mangosteen parent *G. hombroniana*. In this species on average, 62% of ovules produced adventitious proembryos in the absence of sexual embryos (Figs. 10.16 and 10.17). There was a pronounced tendency for large ovules in small ovaries to produce a much higher proportion of proembryos than did small ovules in large ovaries, suggesting that there is some developmental control on proembryo initiation.

Where sexual fertilization also occurred in an ovule, this tended to inhibit the subsequent formation of proembryos. However, some proembryos had already formed precociously before sexual fertilization could occur, and the presence of these did not apparently inhibit the later development of sexual embryos. About 40% of ovules developed sexual embryos, and this figure did not differ significantly between controlled pollinations and open pollinations where male and female trees grew together. Despite the competition between sexual embryos and asexual proembryos, 8% of ovules contained both sexual and asexual embryos, and about the same proportion of polyembryonous seeds was later found.

Fig. 10.16 The female flowers, young and mature fruits of the Malaysian fruit tree *Garcinia hombroniana*. This species is facultatively apomictic so that an isolated female tree will set fruit with fertile seeds autonomously.

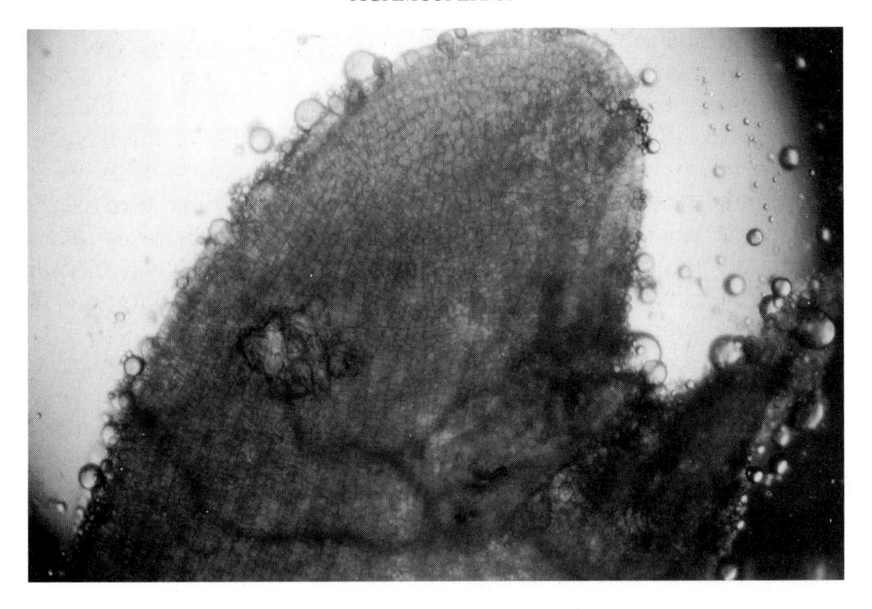

Fig. 10.17 A dissected-out ovule of *Garcinia hombroniana* about two days after flowering. A tumour-like proembryo initial has formed asexually in the nucellus; later, this will organize itself into a tuberoid proembryo. Interestingly, the proembryo is already vascularized.

Garcinia species are mostly pollinated by sweat bees (*Trigona* spp.) (Richards, 1990b, Fig. 4.33). Dioecious species such as these, occurring at very low density as understorey species in exceptionally diverse wet tropical forest, will almost certainly suffer from poor reproductive assurance (p. 303), and this may favour the evolution of apomixis.

Apomictically produced embryos will give rise to female offspring only, whereas I am assuming that sexually produced embryos give rise to male and female offspring in equal ratio. Thus, it should be possible to estimate the proportion of offspring produced by sexual and by apomictic reproduction from the gender ratio of trees (other things, for instance the survivorship of males and females, being equal, which may not be the case, p. 346). Two estimates of gender ratio in *G. hombroniana* gave 13% males and 21% males. From this we might expect about 60–70% of offspring to arise by apomictic reproduction, which fits in well with rough estimates (above) of about 62% of embryos being formed apomictically and 38% sexually.

Apomixis in a dioecious species renders the males relatively rare. Thus, the majority of individuals are productive, the proportion of 'wasteful males' being low. Levels of cross-pollination between trees will be low, so that the other main advantage of sporophytic apomixis to tropical trees may accrue from the autonomous setting of seed. These points have also been made for the other dioecious genus with apomixis, *Antennaria* (Bierzychudek, 1987).

In this case, we would expect successful examples of adventitious embryony to have become non-pseudogamous, i.e. seed development should be independent of the fertilization of the PEN (polar nucleus). In this, *G. hombroniana* and *G. mangostana* apparently differ in their solution to this problem. In *G. hombroniana*, endosperm only develops after fertilization of the PEN, but successful proembryo development occurs equally well in the absence of an endosperm. This species controls its female reproductive load by a massive early abortion of ovaries. In contrast, mangosteen proembryos tend to develop faster and more reliably when an endosperm is present, but this is formed autonomously, in the absence of pollination. Ovules lacking an endosperm tend to abort, which seems to be the mechanism by which this species controls its female load.

Where a dioecious forest fruit tree shows a massive predominance of female plants, it is reasonable to suspect that apomixis may be occurring. Using this and other types of evidence, Richards (1990a) suggests that at least ten Malaysian *Garcinia* species may be apomictic. In fact, there is only one species so far investigated, *G. cantleyana*, in which apomixis is probably absent. Thus, further work may show that literally hundreds of *Garcinia* species may be apomictic. Remarkably, not all *Garcinia* apomicts have adventitious embryony. In *G. parvifolia* and relatives, aposporous parthenogenesis occurs instead (Ha *et al.*, 1988). Occasional haploid parthenogenesis has been observed in the mangosteen as well, suggesting that elements of this mechanism may have survived in the species with sporophytic apomixis.

Other cases of sporophytic agamospermy

Adventitious embryony had also been recorded in a number of unrelated tropical fruit trees such as many *Citrus* (summarized by Cameron and Soost in Simmonds (1976). In this genus, proportions of apomictically produced embryos vary markedly between species and varieties. In some varieties of the orange (*C. sinensis*), mandarin (*C. reticulatus*) and grapefruit (*C. paradisi*), almost all embryos are produced asexually, whereas in the pumelo (*C. grandis*) and citron (*C. medica*) only sexual reproduction is known.

Other examples of sporophytic agamospermy among economically important fruits include mango (*Mangifera indica*), *Ochna*, *Pachira*, *Opuntia* and langsat (*Lansium domesticum*) (Prakash, Lim and Manurung, 1977).

Studies on the giant emergent dipterocarp trees of the south-east Asian forests have revealed triploidy and polyembryony in a number of species of *Shorea*, *Hopea* and *Dipterocarpus*, and adventitious embryony has been conclusively proved for two *Shorea* species (Kaur *et al.*, 1986). Kaur *et al.* (1978) suggest that the sparse distribution and outbreeding mating system of many tropical forest trees may have encouraged the evolution of sporophytic apomixis among many such species. Few of the thousands of species have yet been investigated, and this is certainly a fruitful field for further investigation. Oliveira *et al.* (1992) produced the first report of sporophytic agamospermy for a neotropical tree (*Eriotheca pubescens*).

AGAMOSPERMY AND PLANT BREEDING

For a plant breeder, the 'ideal' plant is highly heterozygous, with hybrid vigour and the potential for future variability, but is true breeding. Plant breeders have to face the basic problem that heterozygosity and true breeding are antagonistic for sexual plants, as heterozygotes release variability. In contrast, agamospermy is potentially an ideal breeding system from a crop breeding viewpoint, for an agamosperm is both heterozygous and true breeding.

Unfortunately, agamospermy is very rare in crop plants, the best known cases being in *Citrus* (oranges, lemons, grapefruits, etc.), and *Mangifera* (mango), although it also occurs in some minority crops such as prickly pear, caper, mangosteen, brambles, species apples, and in some bluegrasses (*Poa* spp.).

A great deal of money and effort has been expended in trying to induce agamospermy among sexual crops, but as yet without much success. However, Dujardin and Hanna (1989) report that an apospory gene has been transferred from the grass *Pennisetum squamulatum* to its sexual relative pearl millet, *P. americanum*. Currently, molecular techniques involving the location, sequencing, cloning and transduction of genes (so-called 'genetic engineering') are widely considered to have the potential for apomixis to be introduced into any crop plant, so that the topic has a new popularity.

Tacitly, these research directions make two important assumptions:

1. Apomixis is entirely controlled by one, or at the most two, localized genes.
2. Genes which promote apomixis in one species will have a similar effect in another, unrelated species.

Mogie (1986, 1988, 1992) has argued the case that agamospermy in many plants is indeed controlled by a single, localized recessive gene, but that this gene is only likely to become established and expressed in polyploids, as evidenced by the extremely strong relationship between gametophytic apomixis and polyploidy (e.g. Asker and Jerling, 1992).

In fact, as yet, there seems to be no single case of a gametophytic apomixis for which the genetic control is fully understood, where apomixis is expressed only by a single gene, or where the genetic control is completely recessive (Table 10.3). Undoubtedly, important elements of the apomictic process, such as the ability to produce aposporous embryo-sacs, have been shown to be controlled essentially by a single gene. Conceivably, the parthenogenetic development of aposporous egg-cells in such cases could be related to the precocious development of the aposporous embryo-sac, as Mogie has suggested. Nogler (1984a) suggests rather that genes for parthenogenesis may be linked to those for apospory, so that the two functions may at first appear to be controlled by a single gene.

There is now widespread recognition that apomixis has evolved polyphyletically in so many different genera, and in so many different ways, that it is unlikely that elements of gametophytic apomixis, cloned from one species, would be likely to have similar effects when introduced into the genome of completely unrelated species. Efforts are now concentrated on the identification of mutants within crop plants, or their relatives, which might contain the elements of apomixis. Ideally, different mutants could be cloned and then combined by transduction within a single host, thus mimicking the process by which apomixis evolved.

Unfortunately, some key mutant phenotypes, as for instance apospory, seem to mutate spontaneously only on very rare occasions if at all (Nogler, 1984a), although they have been reported in the intergeneric hybrid × *Raphanobrassica* (Ellerstrom and Zagorcheva, 1977) (Table 10.2).

More promising, perhaps, might be a search for parthenogenetic mutants, which are known in many sexual species (e.g. Asker and Jerling, 1992), and for 'apomeiotic' mutants, i.e. those for which the meiosis is bypassed or results in unreduced products. Such mutants are also not very uncommon (Table 10.2). Barcaccia *et al.* (1995) report a good example in alfalfa, *Medicago sativa*. Once RAPD or RFLP markers are established for such genes, it is but a short step for them to be cloned and the possibility then exists for these genes to be transduced into sexual relatives.

I emphasize that such a programme is only likely to succeed for gametophytic apomixis if the mutants are discovered within the gene pool of the crop plant, or of close relatives, in which it is hoped that apomixis will be developed. Also, host targets for cloned genes for apomixis will have to be polyploid, thus severely limiting possible candidates. If apomictic genes are introduced into diploids, these genes would be unlikely to persist through a sexual cycle, due to the probability of their linkage to lethals (p. 417).

The potential for sporophytic agamospermy

As yet, nobody seems to have investigated the possibility of isolating and cloning a gene for adventitious embryony (p. 442). However, it seems that such genes should have none of the problems associated with those for gametophytic apomixis. Because the evolution of sporophytic agamospermy involves only one new function, it may usually be found to be under the control of a single dominant gene, and sporophytic agamospermy is commonly found in diploids (p. 404), whereas it is not often associated with hybridity. Because sporophytic agamospermy does not involve the archesporium, it is commonly found that the sexual system persists alongside the apomictic production of proembryos. Thus the possibility to breed new apomictic genotypes often persists in such plants. Occurring mostly among tropical trees, the study of adventitious embryony is at present unfashionable. Until it attracts more interest, I suspect that the problems associated with the genetic engineering of apomixis will persist.

CONCLUSION

For agamospermy to succeed, it is necessary for the independent functions of avoidance of meiosis, and failure of fertilization/autonomous embryony to occur together. Except in the case of adventitious embryony, at least two separate mutants need to be combined. Although such mutations can arise in sexual populations, they are not widespread, and the reduced sexual function that they engender singly reduces the likelihood of their co-occurrence in the same genet. As a result, agamospermy is an uncommon phenomenon, which is most frequent in a few families of plants that may show preadaptation to the origin of certain types of agamospermy.

Obligate agamospermy is rare, and is probably limited to a few diplosporous genera in which pollen is missing (unusual as male sterile mutants cannot be recombined), or sexuals are absent. Most agamosperms retain some sexuality (facultative) which allows them to maintain some genetic variability. Nearly all agamosperms with apospory or adventitious embryony can be sexual, and retain the requirement for fertilization of the endosperm nucleus (pseudogamy). They thus often retain good pollen function, which can also be used in sexual reproduction. At least some obligate agamosperms are also able to generate some genetic variability by somatic recombination, and/or various forms of autosegregation. Thus, it is probably untrue that agamosperms generate no genetical variability, and have no evolutionary potential. In any case, it seems that some agamospecies are able to persist for thousands of years.

Nevertheless, agamosperms may suffer reduced fitness in comparison with sexuals by generating very much less variability, which reduces the number of niches in the environment that they are able to fill. They are also likely to suffer from the accumulation of disadvantageous mutants in the agamospermous linkage group, and from poor recombination and migration of advantageous mutants as a result of reduced sexual function.

However, agamosperms possess a number of theoretical advantages over sexuals which should result in their automatic success in competition with sexuals, at least in the unreal condition of a single environmental niche. These are as follows.

1. As agamospermous offspring are of a maternal genotype, the agamospermous mother wastes no energy succouring offspring (seeds) which through the variability engendered by sexuality are relatively maladapted and unfit; this advantage is difficult to quantify.
2. Agamosperms donate genes to sexuals as male parents, but receive no genes from sexuals; this advantage is frequency dependent (Lloyd, 1980b), and decreases with the frequency of agamosperms in the population.
3. Agamosperms have an advantage over dioecious and androdioecious sexuals, in that they produce no males; this advantage is independent of the sex ratio of the diclinous population (Charlesworth, 1980), and should be 2:1.
4. Male-sterile agamosperms, which are uncommon, have an advantage over hermaphrodite sexuals; this is not a function of the relative resource used in male and female reproduction, but depends on the fraction of the total energy output of the plant used for male reproduction (Chapter 2); however, male steriles can suffer a lowering of male fitness by not being able to donate apomictic genes to sexuals.

Table 10.11 Energetic advantages of agamosperms over coexisting sexuals

	Male function of agamosperm with equal male and female expenditure		
	No pollen	Non-functional pollen (e.g. sexuals absent)	Functional pollen
Dioecious sexual with equal sex ratios and equal male and female expenditure	2:1	1><2:1	2>:1 (but unlikely)
Hermaphrodite sexual with equal male and female expenditure	1>≪2:1	1:1	3:2 to 2:1 (depending on agamosperm frequency)

5. Non-pseudogamous agamosperms have a reproductive advantage over outbreeding sexuals in conditions of limited cross-pollination.

It should also be noted that male-fertile agamosperms, with pseudogamy, possess no automatic advantage over sexual hermaphrodites when they are unable to act as male parents (for instance when sexuals are absent), except that of avoiding the 'cost of meiosis'. Some of these points are expressed in Table 10.11. It is difficult to quantify advantages of different types together.

One of the features of agamosperms is that nearly all have close sexual relatives, so that it is possible to suggest what breeding system plants that have given rise to agamosperms are likely to have had.

Nearly all such probable sexual progenitors are perennial hermaphrodite outbreeders with limited powers of vegetative spread. Nearly all diplosporous agamosperms, and most aposporous agamosperms, differ from their sexual progenitors in being polyploid (often of uneven 'ploidy') and hybrid. In contrast, this is not true of agamosperms with adventitious embryony which can arise by a single mutation.

It is supposed that polyploidy and hybridity encourage the evolution of agamospermy by leading to the failure of meiosis, by recombining the separate functions of agamospermy from different parents, and by being seed sterile, thus giving an immediate advantage to agamospermous seeds. Perenniality and vegetative persistence should allow sterile clones, which possess one of the functions of agamospermy, to persist until a second function is incorporated into that clone (linkage group) by mutation or hybridization.

We can conclude that such features as perenniality, persistence, polyploidy and hybridity predispose certain plants to the evolution of gametophytic apomixis. This may explain at least in part why apomixis has a limited and uneven geographical and systematic distribution. Agamosperms have both advantages and disadvantages in comparison with sexuals. At times, it is clear that agamospermy has proved to be a successful form of reproduction for a number of genera in which new variation can be generated and old variation fixed.

CHAPTER 11

Conclusion

This book has attempted to show the following points:

1. Seed plants have a wide variety of breeding systems. Mate choice is controlled by a diversity of mechanisms. Many of these mechanisms are interrelated. They occur together within one plant or population, and influence one another.
2. Plant breeding systems are not static, or fixed, but are variable and flexible. Genetically controlled and heritable variation for all breeding-system attributes occurs between plants in many populations. Further variation, or novel attributes can also arise by mutation. As a result, breeding systems are subject to natural selection and adaptive evolution in nearly all plants in every generation.
3. Breeding-system attributes have profound effects on the reproductive fitness of individual genets in all plants.
4. Breeding-system attributes influence the genetic structure of individuals. They determine patterns of gene linkage and levels of heterozygosity in vitally important ways. Thus, they also influence the genetic structure of populations – the ways that plants within a population differ from one another genetically.
5. As a result, breeding systems control the nature of evolution in seed plants. Because breeding systems themselves evolve constantly, they are an integral part of this evolution, and are subject to feedback effects from the results of the evolution they control.
6. The taxonomic and classificatory systems that we impose on plant variation are influenced by the pattern of variation that we perceive. No two plant populations have exactly similar breeding systems and exactly similar patterns of variation. Our taxonomic philosophy should take account of this. At present it conspicuously fails to do so, resulting in tedious and unnecessary controversies of the 'splitter'/'lumper' type.

In this brief concluding chapter I shall examine some of the effects that different plant breeding systems have on the genetic variability of plant populations. This account owes much to the reviews by Hamrick *et al.* (1979, 1991). Despite much current work with DNA, most of our knowledge of genetic variation in plants has relied on the use of gel electrophoresis to examine the distribution of isozymes, the direct products of genes.

Hamrick, Linhart and Mitton (1979) compared isozyme attributes with types of breeding system and other life-history characters for 113 taxa of seed plants. They examined three features of genetic variability as revealed by isozyme electrophoresis (Chapter 9): PLP (P), the percentage of loci polymorphic per population (or species); *k* (*A*) the mean number of alleles per locus per population (or species); PI the mean proportion of loci expected to be heterozygous if the population is panmictic (Hardy–Weinberg frequencies, Chapter 2). (Symbols in parentheses are those used by the authors, which differ from those used in this book; note that these figures relate to variation in populations, rather than *H* figures which refer to heterozygosity in individuals.)

If mean values are calculated for all 113 taxa, an estimate of values typical of seed plants can be obtained. Interestingly, seed plants have average levels of genetic variability considerably higher than those estimated for vertebrate animals, which have fewer reproductive options (Chapter 2), but are roughly similar to estimates for invertebrates which have more reproductive options (Table 11.1).

In a comparison of genetic variability between major plant taxa, and different plant life strategies, some interesting features emerge (Table 11.2). We can expect Gymnosperms on average to be more variable than Angiosperms, for most Gymnosperms are long-lived woody perennials with efficient seed and pollen dispersal (usually by wind), and they are either monoecious or dioecious. When monoecious, they are often markedly protandrous, so they are usually outcrossed. It is more surprising that monocotyledons tend to be slightly more genetically variable than are dicotyledons. However, there are fewer monocotyledonous than dicotyledonous annuals, and many of the monocotyledons in the sample are grasses that have wind pollination and fairly efficient seed dispersal.

As expected, long-lived perennials are genetically far more variable than plants with shorter life histories. In long-lived perennials, reproductive fitness is most often optimized by vegetative rather than sexual reproduction (‘*K*’ rather than ‘*r*’ strategy), and so selfing and agamospermous

Table 11.1 Mean values for estimates of genetic variation in seed plants, invertebrates and vertebrates (see the text for explanation of symbols) (after Hamrick *et al.*, 1979)

Group	PLP (%)	*k*	PI
Seed plants	36.8	1.69	0.141
Invertebrates	46.9	–	0.135
	39.7	–	–(different estimates)
Vertebrates	24.7	–	0.061
	17.3	–	–(different estimates)

Table 11.2 Mean values and standard errors for estimates of genetic variation in populations of seed plants with different life strategies, and belonging to different major taxa (see text for explanation of the symbols) (after Hamrick, Linhart and Mitton, 1979)

Group	PLP (%)	*k*	PI	Mean number of loci examined	Number of taxa examined
Gymnosperms	67.0 ± 8.0	2.1 ± 0.2	0.27 ± 0.04	9.2	11
Dicotyledons	31.3 ± 3.3	1.5 ± 0.1	0.11 ± 0.01	11.4	74
Monoctyledons	39.7 ± 6.0	2.1 ± 0.2	0.16 ± 0.3	11.6	28
annual	39.5 ± 4.3	1.7 ± 0.1	0.13 ± 0.02	11.2	42
biennial	15.8 ± 5.1	1.3 ± 0.1	0.06 ± 0.02	17.2	13
short-lived perennial	28.1 ± 5.1	1.5 ± 0.1	0.12 ± 0.02	12.0	31
long-lived perennial	65.7 ± 5.1	2.0 ± 0.1	0.27 ± 0.03	7.6	27
colonizer	29.7 ± 3.8	1.6 ± 0.1	0.12 ± 0.01	12.5	54
mid-successional	37.9 ± 4.4	1.6 ± 0.1	0.14 ± 0.02	9.7	49
climax vegetation	62.8 ± 5.3	2.1 ± 0.2	0.27 ± 0.04	12.0	10
xerophyte	15.4 ± 8.2	1.1 ± 0.1	0.05 ± 0.04	8.8	4
hydrophyte	27.7 ± 10.3	1.6 ± 0.2	0.14 ± 0.05	13.0	8
mesophyte	36.0 ± 3.6	1.6 ± 0.1	0.15 ± 0.02	11.4	82
haploid chromosome number					
$n = 5–10$	35.5 ± 3.6	1.5 ± 0.1	0.11 ± 0.01	13.1	50
$n = 11–15$	37.4 ± 5.3	1.7 ± 0.1	0.17 ± 0.02	10.0	44
$n = 16 +$	41.6 ± 7.2	2.1 ± 0.1	0.22 ± 0.03	8.9	16

mechanisms that respond to pressures for the optimization of seed-set and reduce genetic variability are relatively unusual. It is interesting that annuals are more variable than biennials or short-lived perennials, at least for the percentage of loci that are polymorphic per population (PLP). This attribute examines variability at the populational rather than individual level, so that it cannot distinguish between homozygotes and heterozygotes. Thus, where an inbred population has fixed both of the genotypes *AA* and *aa* in a population polymorphic for the alleles *A* and *a*, this will register as polymorphic for that locus.

The correspondence of patterns of genetic variation with ecological attributes are very much as one might predict. Colonizing species, and species occurring in very dry habitats and in water, are less variable than plants that occur in stable climax and mesic habitats. Extreme habitats

Table 11.3 Mean values for estimates of genetic variation in populations of seed plants with different breeding system attributes (see the text for explanation of symbols) (after Hamrick, Linhart and Mitton, 1979)

Breeding system	PLP (%)	*k*	PI	Mean number of loci examined	Mean number of taxa examined
autogamous	19.0 ± 3.5	1.3 ± 0.1	0.06 ± 0.02	14.2	33
zoophilous	38.8 ± 3.9	1.5 ± 0.1	0.13 ± 0.01	9.5	55
anemophilous	57.4 ± 6.3	2.3 ± 0.2	0.26 ± 0.03	10.7	23
fruit dispersal					
epizoochorous	28.8 ± 5.5	1.5 ± 0.1	0.09 ± 0.02	11.1	16
endozoochorous	33.0 ± 8.2	1.4 ± 0.1	0.13 ± 0.04	7.0	20
anemochorous	44.9 ± 7.2	1.9 ± 0.1	0.19 ± 0.03	12.2	21
self-fertilizing	17.9 ± 3.2	1.3 ± 0.1	0.06 ± 0.01	14.2	33
mixed selfing and crossing	14.2 ± 4.9	1.8 ± 0.1	0.18 ± 0.02	8.6	42
cross fertilizing	51.1 ± 4.9	1.8 ± 0.1	0.18 ± 0.02	11.3	36

impose severe stabilizing selection, and encourage the evolution of breeding systems through which the release of genetic variation is minimized. Colonizing species can best initiate a new population, and optimize reproduction by seed, when they are self-fertilizing, and self-fertilization will reduce genetic variability. Mesic and stable habitats will provide more niches and will encourage the evolution of breeding systems that maximize the release of genetic variability.

The role played by the chromosome number of the plant in controlling the release of genetic variation is well illustrated. Chromosomes are linkage groups, and the more chromosomes that are present, the greater the potential for segregational variability (Chapter 2). Polyploids, with high haploid numbers, also buffer the effects of selfing with respect to loss of variability, and do not expose lethals to selection in the gametophyte (Chapter 9). High chromosome numbers will be advantageous for plants with outbreeding strategies, whereas plants that are favoured by a low release of genetic variability may select for low chromosome numbers through translocations.

Hamrick, Linhart and Mitton (1979) also compared the genetic structure of seed plant populations with different breeding system attributes, and these results are central to the theme of this book (Table 11.3). Once again, expectations are fulfilled well. Results from self-pollinating and self-fertilizing species are very similar, probably because they refer to essentially the same samples. They show very much less genetic variability than do insect- and wind-pollinated species, and species that are predominantly cross-fertilized.

There is an interesting contrast between animal- and wind-dispersed pollination. As discussed in Chapter 5, wind-dispersed pollen travels much further than does animal-dispersed pollen, although as a pollination mechanism it is much less efficient, and thus more expensive. We must attribute the notably higher genetic variation of populations that are wind pollinated to this difference. Wind-pollinated plants will tend to have much larger neighbourhood areas A and sizes N_e than do insect-pollinated plants. As plant populations are frequently very small (Chapters 2 and 9), wind pollination may often serve to overcome the inbreeding effects engendered by small populations. We can conclude that animal-pollinated plants may concede benefits of outcrossing in favour of benefits of reproductive assurance and low male costs when contrasted with wind-pollinated plants.

The effectiveness of wind as a dispersal agent is also illustrated by the data for fruit dispersal. The genetic variability of populations that are dispersed on the outside of animals is less than that of populations for which the fruits are eaten by animals. Populations with wind-dispersed fruit show most variation. These differences in genetic variability are probably attributable to differences in seed dispersal variance, and hence in neighbourhood size (Chapter 5). Fruits that are dispersed the furthest give rise to the largest, and hence the most variable, populations. There is no significant difference in the genetic variability of populations with fruits of different sizes (results not given).

The comparison between the three sets of data for types of fertilization is particularly interesting. For the percentage of all loci that vary (PLP), mixed strategy plants do not differ from selfers; relatively few loci are polymorphic within a population in comparison with outcrossers. However, the mean number of alleles per locus (k) in mixed strategy plants does not differ from outcrossers, and is higher than in selfers. This can only mean that mixed strategy plants have more than two alleles at a locus in a population more frequently than do selfers. A combination of genetic fixation through selfing, and gene migration through bouts of outcrossing, may allow this to happen. Also, mixed strategy plants do not differ from outcrossers in the proportion of loci that are heterozygous (PI), and this proportion is much higher than in selfers. We can conclude that a mixture of outcrossing and selfing tends to promote gene migration, recombination and heterozygosity, and is as effective at this as is outcrossing. However, it does not allow so many disadvantageous recessives to be maintained, as these will be exposed to selection by bouts of selfing. Mixed strategy mechanisms are revealed to be evolutionarily rather efficient, which may help to explain why so many species of plants possess them.

Appendix A: Glossary

Abiotic Not involving living organisms.
Abscission Dropping off of organs, e.g. leaves, flowers, fruits.
Actinomorphic Radially symmetrical (rotate), of a flower.
Adventitious embryony A form of agamospermy in which embryos are budded directly from the nucellus without an intervening embryo-sac.
Agamic Without sex, asexual.
Agamospecies 'Microspecies' or 'segregate species' of low taxonomic amplitude used in **agamic** complexes.
Agamospermy The formation of seeds without sexual reproduction.
Agglutination The conglomeration and precipitation of molecules of a protein.
Allele One of two or more forms that a gene may take.
Allogamy Fertilization between pollen and ovules of different flowers.
Allopolyploid A plant with more than two sets of chromosomes which originate from two or more parents, at least some chromosomes of which are non-homologous with respect to each other.
Allozyme A form of an enzyme which behaves allelically with respect to other allozymes of that enzyme.
Anaphase The stage in chromosome division **meiosis** or **mitosis** at which the chromosomes or chromatids pull apart.
Androdioecy Where **male** and **hermaphrodite** genets coexist.
Androecy Maleness.
Andromonoecy Where a **hermaphrodite** bears both **male** and **hermaphrodite** flowers.
Anemochory Dispersal by wind.
Anemophily Pollination by wind.
Aneuploid A plant with a chromosome number which is not a direct multiple of the base chromosome number of the group.
Angiosperm A flowering plant, i.e. belonging to the class Angiospermae.
Anisogamy Sexual fusion between gametes of unequal size.
Anisoplethy The departure of ratios of genet morphs in population from the expectation of unity (one to one, e.g. of males and females, pins and thrums).
Annulus Constricting ring at the top of a floral tube.
Anther The part of the stamen that contains pollen (a microsporangium).
Antherozoid A motile male gamete, as in the Pteridophyta and Cycadopsida.
Anthesis The opening of a flower; usually used to denote the stage when the flower first donates pollen, or is receptive to pollen, whichever is the earlier.
Anthocyanin Group of pigments that most commonly give flowers pinkish, purplish or blueish tints.
Antigen Substance that stimulates the formation of an antibody, or seroprotein.
Antipodal Nucleus (usually one of three) at the chalazal end of the embryo-sac.

Apogamety Autonomous development of a nucleus apart from the egg nucleus into an embryo in an **agamosperm**.

Apomixis Asexual reproduction; includes both **agamospermy** and vegetative reproduction.

Apospory Development of the embryo-sac from a tissue apart from the **archesporium**, usually the nucellus, in **agamospermy**.

Archegonium Female sex organ, containing the egg, in the gametophyte generation of the Bryophytes, Pteridophytes and Gymnosperms.

Archesporium The tissue within the nucellus of a young ovule which usually gives rise to the embryo-sac mother cell, female **meiosis** and the embryo-sac.

Assortative mating Where mating occurs between gametes of the same morph more frequently than would be expected by a random mating system (**panmixis**).

Asynapsis The failure of **homologous** or **homoeologous** chromosomes to form chromosome associations and thus chiasmata at **meiosis**.

Autogamy Within-flower fertilization.

Automixis Fusion of embryo-sac nuclei.

Auto-orientation The positioning of multivalents at diakenesis and metaphase I of **meiosis** in a polyploid in such a way that a regular segregation of centromeres is achieved and fertile spores result.

Autopolyploid A plant with more than two sets of chromosomes which originate from a single parent, or from different parents which are **homologous** in their chromosomes (*see* **allopolyploid**).

Autosegregation The regular, or somewhat regular, disjunction of chromosomes at anaphase I of **meiosis** between which there has been poor synapsis or asynapsis, as in agamosperms or hybrids.

Autosome A chromosome that is not a sex chromosome.

Axile (placentation) Arrangement of the ovules down the central walls of septae in a fused ovary.

Batesian mimicry The adaptive mimicry of an organism or an organ to a model (*see* **Müllerian mimicry**).

Bract Leafy organ subtending an inflorescence.

Bryophyte Member of the subphylum Bryophyta, a moss (Musci) or liverwort (Hepaticae).

Bulbil Vegetative organ of dispersal or perennation, usually formed from an axillary bud.

Callose A complex carbohydrate β1,3,glucan, usually formed as a wounding reaction.

Calyx The sepals of a flower.

Campylotropous Form of an ovule in which the micropile and **chalaza** are placed laterally to the placenta (Fig. 3.6).

Cantharophily Pollination by beetles.

Capsule A dry, dehiscent fruit.

Carpel The segment of an ovary.

Caudicle The sticky, elastic base to an orchid pollinium, formed from the modified connective.

Centric Containing the centromere (of a chromosome).

Centromere Part of the chromosome which attaches by microfibrils to the spindle; composed of repetitive sequence DNA, recombination is restricted in its vicinity.

Certation Differential growth of pollen grains of different *S* allele constitutions when grown on a legitimate stigma; can cause cross-compatible genotypes to appear cross-incompatible.

Chalaza The opposite end of the ovule to the **micropile**.

Chiasma Where chromatids of different **homologous** chromosomes pairing at meiosis I break and reciprocally join with each other, resulting in genetic crossing over, and hence in recombination within linkage groups.

Chimaera The coexistence of cells of more than one **genotype** within a **genet**.

Chirepterophily Pollination by bats.

Chorology The study of the geographical distribution of organisms.

Chromatid The product of longitudinal division of a chromosome; at **meiosis** in a diploid, two homologous chromosomes produce four chromatids which may form chiasmata with each other resulting in recombination; each spore resulting from a meiosis will receive only one **chromatid** from each pair of **homologous** chromosomes (bivalent).

Chromosome A structural unit, or 'package' of genetic information contained within the nucleus and consisting of DNA and histones; chromosomes segregate independently from one another at **meiosis**, and so each chromosome forms a linkage group, which may be broken by chiasmata (**recombination**). The number of chromosomes is usually characteristic for a species.

Cistron A unit of genetic information coding for a polypeptide which cannot usually be broken by **recombination**.

Cleistogamy Where flowers do not open, and are thus inevitably autogamous.

Clone A number of ramets which belong to the same **genet**.

Cob A heteromorphic condition in the family Plumbaginaceae characterized by short stigmatic papillae.

Cohort 1. Individuals of similar age in a population. 2. Pollinators of different species exhibiting similar flower choice.

Column Part of the **gynoecium** of orchids, homologous with the style, which is a **gynostegium**, carrying the pollinia as well as a stigmatic cavity.

Complementation Interaction of two or more genetic loci in the expression of a single phenotypic character.

Corolla The petals of a flower.

Crypsis Where an organism or organ is difficult to perceive or see.

Cuticle The superficial proteinaceous and waxy layer secreted by the epidermis which covers aerial plant organs (*see* **pellicle**).

Cutinase Esterase enzymes that break down the waxy cuticle, especially of stigmatic papillae.

Cypsela The one-seeded fruit of Compositae (Asteraceae) which commonly bears a feathery pappus which is derived from the **calyx** and promotes dispersal by wind.

Deletion The loss of a gene or part of a chromosome.

Deme A population of potentially interbreeding individuals.

Demography The study of populations.

Diakenesis The stage during **meiosis** I at which chromosomes become fully contracted, and centromeres start to disjoin.

Di-allelic An incompatibility system containing only two *S* alleles, *S* and *s*, usually as the diploid genotypes *Ss* and *ss* which are between-morph compatible and within-morph incompatible; usually associated with **heteromorphy**.

Dialysis Technique for separating molecules of different sizes through a membrane of known pore diameter, usually across an electropotential gradient.

Dichogamy Separation of anther dehiscence and stigma receptivity within a flower in time so that **autogamy** cannot occur (*see* **protandry** and **protogyny**).

Dicliny Where not all genets in a population are regularly **hermaphrodite**; i.e. males, or females, or both occur.

Dicotyledon Member of the class Dicotyledones which includes the majority of angiosperms, and is characterized by such features as two cotyledons in a seedling, vascular bundles in a ring, secondary thickening, broad, net-veined leaves and floral parts in fours and fives.

Dihaploid Where the gametophyte generation, or the parthenogenetic development of it, has two sets of chromosomes.

Dikaryon A cell containing two nuclei.

Dimer A protein (e.g. an enzyme) containing two subunits.

Dimorphism (y) The coexistence of two genetically controlled floral types in a population, e.g. **pins** and **thrums** (**heterostyly**), **cobs** and **papillates**, or males and females.

Dioecy Where all **genets** in a population are either male or female.

Diploid A **genet** with two homologous sets of chromosomes (**genomes**).

Diplospory Development of the embryo-sac from the **archesporium**, often by way of an irregular **meiosis**, in **agamosperms**.

Disjunction The pulling apart of the chromosomes at **anaphase** in **meiosis** or **mitosis**.

Disomic Where a chromosome or gene is present twice in a **genet**.

Disseminule An organ of reproductive dispersal (*see also* **propagule**).

Disassortative mating Where mating occurs between gametes of different morphs more frequently than would be expected by a random mating system (**panmixis**).

Duftmale An area of a petal that secretes scent.

Duplex *AAaa* heterozygote in **tetrasomic** inheritance (*see* **simplex**, **triplex**).

Duplication Where a gene, supergene or part of a chromosome occurs twice within a **genome**.

EMC (the embryo-sac mother cell) The archesporial cell which undergoes female **meiosis**.

Ecodeme A population or gene pool adapted to a particular environment.

Elaiophore An oil-secreting gland on a flower.

Elaiosome A gelatinous projection of the testa of the seed, attractive to ants.

Electrophoresis Technique for separating molecules of different sizes and charges (especially **isozymes**) by diffusing eluates through gels of acrylamide or starch along an electropotential gradient, and staining the products produced by suitable substrates to give rise to visible bands (*see* **zymogram**).

Embryo The young stage of a new **genet** within the seed, usually developed from an egg cell, which on germination will give rise to a seedling.

Embryony The development of an embryo.

Embryo-sac The female gametophyte of flowering plants (**angiosperms**), contained within the ovule, developing from the surviving megaspore after female meiosis, and containing eight nuclei (Fig. 3.5).

Endemic A plant **taxon** (usually species) which is confined to a small geographical area.

Endoduplication The multiplication of **genomes** within cells of certain tissues.

Endosperm The nurse tissue of the embryo within the seed, formed from fusion of a sperm cell with the fused polar nucleus, and usually triploid.

Endozoochorous Dispersed within animals, i.e. eaten (of a seed or fruit).

Entomophily Pollination by insects.

Epidermis The outermost layer of cells of plant organs.

Epiphyty Of plants which live rooted to the aerial parts of other plants, e.g. on tree trunks.

Epizoochorous Dispersed on the outside of animals (of a seed or fruit).

ESS (Evolutionarily stable strategy) Where an equilibrium results from divergent but balanced selection pressures.

Esterase Enzymes that lyse lipids and waxes.

Ethology The study of animal behaviour.

Eukaryote A higher organism with clearly differentiated cell organelles such as nucleus, plastids, mitochondria (*see* **mitochondrion**) etc.; includes algae (except Cyanophyta), Bryophyta, Pteridophyta and Spermatophyta (Gymnospermae and Angiospermae).

Eutriploid With exactly three sets of **homologous** or **homoeologous** chromosomes (**genomes**).

Exine The outer wall of an Angiosperm pollen grain, columnar in structure, perforated with baculae, with an outer tectum and an inner nexine, and constructed from **lipoprotein** (**sporopollenin**) (Fig. 3.3).

Extrorse Of an anther which dehisces away from the centre of the flower.

Facultative Partial, not complete, as in facultative agamospermy where some sexuality is also found.

Fecundity Ability to reproduce, particularly as the production of seeds per **genet**.

Filiform apparatus Outgrowths of one or both synergid nuclei, which engulf the pollen tube apex just prior to double fertilization in the Angiosperms.

Fixation index (F_0) A measure of the genetic effects of inbreeding, based on the deficiency of the frequency of heterozygotes compared with those predicted by the Hardy–Weinberg Law.

Frameshift A form of DNA mutation in which deletion of one or two base-pairs of the DNA next to a 'stop' code causes a whole sequence of DNA triplets to be translated differently.

Fused polar nucleus (*see also* **primary endosperm nucleus**) The **dikaryon** in the **angiosperm** embryo-sac which after fusion with a sperm cell forms the **endosperm**.

Gamete Specialized reproductive cells (usually haploid) which fuse in sexual reproduction to give the **zygote**, and thus the embryo; in seed plants the male gametes are sperm cells or **antherozoids**, and the female gamete is the egg cell nucleus.

Gametophytic Of the gametophyte (usually **haploid**) generation; in the seed plants the contents of the pollen grain and the pollen tube, and the embryo-sac; in **gametophytic** incompatibility it is the **genotype** of the pollen grain contents which is significant.

Geitonogamy Fertilization between different flowers on the same **genet**.

Gene pool Concerning **genets** which are in genetic contact through sexual reproduction in time and space.

Genet A genetical individual, resulting from a single sexual fusion (**zygote**), consisting of one to many **ramets**, and usually genetically distinct from all other genets.

Genetic drift Non-adaptive change in gene frequencies in a population caused by chance (stochastic) fluctuations in small populations.

Genome A set of **homologous** chromosomes; a **diploid** has two **genomes**.

Genotype The genetical constitution of a **genet**; its phenotype will be the result of interaction between the **genet** and the environment.

Gibberellin A series of plant growth substances (gibberellic acids) which influence seed germination, growth, flowering and seed development.

Glycoprotein Complex moieties of protein and carbohydrate.

Grex A group of hybrids of different origins but similar parentage.

Gymnosperm A seed plant (Spermatophyta) belonging to the class Gymnospermae, characterized by naked ovules and the absence of vessels in xylem among many other features; includes conifers, cycads, yews, ginkgos and gnetums.

Gynodioecy Where female and **hermaphrodite** genets coexist.

Gynoecium (pistil) The female parts of the flower, including **stigma**, style and ovary.

Gynoecy Femaleness.

Gynomonoecy Where a **hermaphrodite** bears both female and hermaphrodite flowers.

Gynostegium A complex **gynoecium** on which are borne the anthers (as in Orchidaceae and Asclepiadaceae).

Half-siblings Offspring possessing one parent (usually the mother) in common.

Haploid A **genet** with one set of **homologous** chromosomes (**genome**).

Haustoria Organs which filamentously penetrate other organs or **genets** for nourishment.

Hemigamy Where gametes fuse, but form a **dikaryon** without the nuclei fusing.

Hemizygous The condition of X-linked genes in the **heterogametic** (XY) sex, in which recessive X-linked alleles may be expressed.

Herkogamy Separation of **anthers** and **stigma** in space within a flower in such a way that **autogamy** cannot occur in the absence of an insect visit (see **dichogamy**).

Hermaphrodite A **genet** with both male and female function; it may have either monoecious (single sex) or hermaphrodite (both sexes) flowers.

Heterochromatin Parts of chromosomes which remain more condensed at interphase and less condensed at metaphase and which tend to have replicated sequences of DNA which are non-functional, or service the functioning of the chromosome (centromere, telomeres, nucleolar organizer regions).

Heterogametic The sex of a unisexual organism which has the number of X chromosomes half the number of genomes, e.g. XY, XO; it is usually, but not always, male.

Heteromorphy The coexistence of two or three genetically controlled **hermaphrodite** floral types in a population, e.g. pins and thrums (**heterostyly**), cobs and papillates, and tristylous conditions (*see* **di-allelic**, **dimorphism**).

Heterosis Vigour associated with outbreeding, hybridity, or levels of heterozygosity; the corollary of inbreeding depression.

Heterosporous The production of two types of spore, **microspores** which give rise to male **gametophytes**, and **megaspores** which give rise to female gametophytes.

Heterostyly The coexistence of genetically controlled **hermaphrodite** floral types with different style lengths, and usually with reciprocal anther positions, e.g. **pins** and **thrums** (**distyly**), tristylous conditions; here usually used in place of **distyly**, with **tristyly** treated separately.

Heterozygous With more than one **allele** at a locus in a **genet**.

Hexaploid A **genet** with six sets of homologous or **homoeologous** chromosomes (**genomes**).

Hitch-hike Where genes may become linked to an advantageous gene or supergene and segregate with it disadvantageously in linkage disequilibrium.

Homoeologous (Of chromosomes) from different parents, having partial, but not complete homology (ability to pair and recombine in a hybrid).

Homogametic The sex of a unisexual organism which has the number of X chromosomes equal to the number of genomes, e.g. XX (see heterogametic); it is usually, but not always, female.

Homogamy Coincidence of anther dehiscence and stigma receptivity within a flower, so that autogamy is possible in the absence of herkogamy (*see* dichogamy).

Homoiothermic (Of animals) 'warm-blooded', i.e. with constant body temperature, as in birds and mammals.

Homologous (Of chromosomes) having the ability to pair and recombine freely at meiosis.

Homomorphy With only one floral type in the population; used in contrast with **heteromorphy** with respect to derivatives of usually heteromorphic species.

Homostyly **Homomorphy** specifically derived from **heterostyly** (as in *Primula*).

Homozygous With only one allele at a locus (*see* **heterozygous**).

Hydrophily Pollination by water.

Hypogynous Flower with a superior ovary, i.e. with ovary borne distal to the **receptacle**.

Illegitimate Pollination of a flower of the same self-incompatibility **genotype** as the pollen parent (*see* **legitimate**).

Inbreeding A breeding system which is non-panmictic to the extent that the frequency of heterozygotes falls below that expected by the Hardy–Weinberg equilibrium; may be caused by selfing, assortative mating or small populations.

Inbreeding depression Lack of vigour associated with inbreeding and low levels of heterozygosity; the corollary of **heterosis**.

Inflorescence A group of flowers borne on a stem in proximity to one another.

Integuments The outer tissues of the ovule which form the **testa** in the seed.

Interchange Where two non-homologous chromosomes break and exchange segments.

Intine The inner cellulose wall of an **angiosperm** pollen grain (*see* **exine**) (Fig. 3.3).

Introgression Where one species gains some genes from another species as a result of occasional hybridization across a partial breeding barrier.

Introrse Of an anther which dehisces towards the centre of a flower (see **extrorse**).

Inversion Where a chromosome breaks twice, and the intervening segment turns round.

Isozyme A form of an enzyme with a particular electrophoretic mobility (compare **allozyme**).

'K' **strategy ramets** or **genets** which assign low energy allocations to reproduction, having approached the biomass carrying capacity of the environment.

Karyology The study of chromosome morphology.

Karyotype The morphology and number of chromosomes commonly found within a plant.

Labellum The modified, usually lower, petal of the orchids.

Lamella The central layer of the cell-wall, made of pectate and pectin.

Lectin A chemical that binds specifically to another chemical.

Legitimate Pollination of a flower of a different self-incompatibility **genotype** to the pollen parent, likely to result in fertile seed-set (*see* **illegitimate**).

Leptokurtosis Shape of a graphical curve such that the greater the value on the x-axis, the lesser the reduction of the value of the y-axis becomes per x-axis unit.

Liane A woody climbing plant.

Linkage disequilibrium Where the frequency of chromosome genotypes for two or more linked polymorphic genes differs from that predicted by **allele** frequencies.

Linkage group Genes borne on the same chromosome, or which for other reasons do not segregate from each other at meiosis; the closer together genes occur on a chromosome, the less often they will be recombined, and the 'tighter' the linkage group will be.

Lipoprotein An intimate association between a lipid and a protein.

Locus The position of a gene on a chromosome; each gene consists of an **allele** at a locus.

Majoring The 'learning' of a rewarding floral patch by a bee; to 'major' is the North American word for to 'graduate'.

Megaspore Gives rise to a female **gametophyte**.

Megasporophyll Spore bearing leaf which bears **megaspores**; the ovary or carpel in flowering plants.

Meiosis Reduction division of chromosomes, involving recombination and segregation of the chromosomes, and resulting in (usually **haploid**) spores.

Meiotic sieve The tendency of **meiosis** to only function regularly in tissues without major chromosomal abnormalities, hence meiotic products tend to be chromosomally normal.

Melittophily Pollination by bees.

Micropile The pore by which the pollen tube gains access to the interior of the ovule.

Microspore Gives rise to a male **gametophyte** (includes pollen).

Microsporophyll Spore-bearing leaf which produces microspores in microsporangia (e.g. pollen in anthers).

Mis-sense A form of DNA mutation (frequently a deletion) in which the resulting coded polypeptide is non-functional.

Mitochondrion The cell organelle within which the terminal oxidase system of respiration resides.

Monocarpic Of a **genet** (or more rarely ramet) which only flowers once.

Monocotyledon Member of the class Monocotyledones which includes a minority of Angiosperms (*see* **dicotyledon**) and is characterized by such features as one cotyledon in a seedling, scattered vascular bundles, an absence of secondary thickening, narrow, parallel-veined leaves and floral parts in threes and sixes. Probably derived from the dicotyledons.

Monoecy Hermaphrodite genets in which anthers and gynoecia occur in different flowers (male and female function are separated).

Monomer A protein (e.g. an enzyme) with only one subunit (*see* **dimer**).

Monomorphic Of only one form or **genotype**, as in a population that is not polymorphic for a gene.

Monophylesis Having a single origin in evolutionary history.

Müllerian mimicry The sharing of a feature or signal by a number of different species to the mutual benefit of all (*see* **Batesian mimicry**).

Multi-allelic A gene with more than two alleles at a locus.

Multilocular Of a carpel with more than one ovule.

Mutagenesis Increase in the rate of mutation through the use of chemicals, radiation etc.

Myophily Pollination by flies.

Neighbourhood Theoretical concept of an area containing a number of genets between which there is **panmixis** (the effective population number, or neighbourhood size); see Chapter 5 for further explanation.

Neoendemic A species whose restricted geographical distribution is a function of its recent origin.

Neoteny The persistence of a juvenile condition into an adult stage.

Nexine The inner layer of the **exine**.

Non-disjunction The failure of homologous chromosomes to pull apart at anaphase in mitosis or **meiosis**.

Nonsense A form of DNA mutation in which one or more base-pairs of the DNA code change to read as the 'stop' code (ATG) thus causing a polypeptide translation to end prematurely.

Nucellus Tissue of the ovule, inside the integuments, which acts as a nurse to the **archesporium** and embryo-sac (or **prothallus**).

Nucleolus Body within the nucleus containing RNA; the number of nucleoli per nucleus is usually the same as the number of nuclear organizer regions in the **karyotype**, which is often the same as the number of genomes present in the **karyotype** (i.e. a diploid has two).

Nucleotide The nucleotide bases adenine, guanine, cytosine and thymidine are the material from which the triplet code of the DNA is formed.

Oligophilic A flower that is visited by few species of pollinator (*see* **polyphilic**).

Oligotropic A pollinator that visits few species of flower (*see* **polytropic**).

Ornithophily Pollination by birds.

Orthotropous Form of an ovule in which the **chalaza** is next to the placenta and the micropile is distal to the placenta.

Osmophore Floral organ, often formed from a petal, which is long, narrow, drooping, often dark, and usually foul-smelling, usually as part of a myophilous syndrome.

Outbreeding A breeding system which is panmictic, giving rise to heterozygote frequencies as predicted by the Hardy–Weinberg equilibrium (*see* **inbreeding**).

Ovary Part of the **gynoecium**, containing one or more carpels, which houses the ovules and forms the fruit.

Overdominance Condition in which a heterozygote is superior to either homozygote.

Oviposition Egg laying (of insects).

Ovule The female sporangium of seed plants, containing the integuments, nucellus and embryo-sac or **prothallus** when mature, and after fertilization forming the seed.

PMC (the pollen mother cell) The cell which undergoes male **meiosis** to give rise to microspores (pollen).

Palaeoendemic A species whose restricted geographical distribution is considered to be a relic from an earlier wider distribution.

Panicle An inflorescence which is a determinate branched **raceme**.

Panmixis A theoretical concept of a breeding system which has an infinitely large number of genets which are equally likely to mate with each other; *see* **outbreeding, inbreeding, neighbourhood**.

Papilionoid The typical flower form of the family Leguminosae (Fabaceae) subfamily Papilionaceae, with a dorsal standard, lateral wings, and two petals fused to form a keel, within which the stamens and style are held under tension (a 'pea flower').

Papillae The specialized often extruded epidermal cells of the stigma which receive the pollen grains, and which the pollen tubes penetrate.

Papillate A heteromorphic condition in the family Plumbaginaceae characterized by long stigma papillae (*see* **cob**).

Parietal (placentation) Arrangement of the ovules around the outer walls of a carpel.

Parthenogenesis The autonomous development of an egg into an embryo in the absence of fertilization.

Patch A group of flowers, often of the same species, presenting a similar syndrome to pollinators, which 'major' on it.

Pectins Complex polysaccharides forming the middle lamella of the cell wall; also occur as salts or pectates.

Pectinases Enzymes that lyse pectin and pectinates.

Pellicle The proteinaceous outer layer of the cuticle of stigma papillae.

PEN (primary endosperm nucleus) *see* **fused polar nucleus**.

Pentaploid With five sets of **homologous** or homoeologous chromosomes (**genomes**).

Perianth The petals and sepals, or tepals, of a flower.

Phalaenophily Pollination by moths.

Phenotype The expression of a **genotype**, i.e. the result of interaction between the **genotype** and the environment.

Phenotypic plasticity The capability of a **genotype** to assume different phenotypes.

Pheromone Volatile hormone used by insects as a mating attractant.

Phytosociology The classification of plant ecological communities.

Pin The long-styled morph of a distylous heteromorph.

Placenta The organ that attaches an ovule to the carpel wall.

Plasmagene A hereditary factor that is borne on an organelle outside of the nucleus and chromosomes, is (usually) maternally inherited, and is not subject to recombination and segregation.

Plasmalemma The reticulum of membranes that permeates the contents of a cell.

Plastids Cell organelles that contain (for instance) photosynthetic pigments.

Pleiotropy The ability of a gene to influence more than one phenotypic attribute.

Pollen Microspores, which germinate to give pollen tubes (male gametophytes).

Pollenkitt Coating of lipoprotein on the outside of pollen grains, derived from the **tapetum**, often coloured orange or yellow by caroteins, and involved in sporophytic incompatibility recognition mechanisms (*see also* **tryphine**).

Pollen tube The male gametophyte produced by the pollen grain which has three nuclei (in **Angiosperms**), penetrates the stigma papilla, and grows down the style to the ovule where it releases two sperm cells.

Pollinium The male dispersal unit in the orchids, derived from an anther locule, containing a sticky mass of pollen tetrads, and with a flexible sticky stalk (**caudicle**).

Polyembryony The occurrence of more than one embryo in an ovule, usually caused by the coincidence of **apomixis** and sexuality.

Polygamy Where genets with male only, female only, and gynomonoecious and/or andromonoecious and/or hermaphrodite flowers coexist.

Polymer A molecule made of repetitive subunits.

Polymorphism The occurrence of more than one allele at a locus in a population at a frequency greater than that assignable to mutation or migration.

Polyphilic A flower that is visited by many species of pollinator (*see* **oligophilic**).

Polyphylesis Having more than one origin in evolutionary history (*see* **monophylesis**).
Polyploid A **genet** with more than two **homologous** or **homoeologous** sets of chromosomes.
Polysaccharide Polymer of sugar molecules.
Polysomic Where a chromosome or gene is present more than twice in a **genet**.
Polytropic A pollinator that visits many species of flower (*see* **oligotropic**).
Poricidal Anther dehiscence through apical pores (as in Ericaceae).
Precocious embryony Development of the embryo asexually before the flower opens and anther dehiscence occurs.
Primary endosperm nucleus (PEN) *see* **fused polar nucleus**
Proboscis The tubular sucking mouthparts of insects such as lepidoptera.
Propagule An organ of reproduction (*see* **disseminule**), e.g. seed, bulbil, stolon.
Protandrous Dichogamous condition in which anther dehiscence precedes stigma receptivity within a flower (*see* **protogynous**).
Prothallus The **gametophyte** generation of Pteridophytes, and the female gametophyte of Gymnosperms (equivalent to the Angiosperm embryo-sac).
Protogynous Dichogamous condition in which stigma receptivity precedes anther dihiscence within a flower (*see* **protandrous**).
Pseudocompatibility Loss of self-incompatibility due to environmental stimuli such as heat, electrical stimulation, physical damage to the stigma, or bud pollination.
Pseudogamy Agamospermy in which pollination and fertilization of the primary endosperm nucleus is required for the successful development of the asexual embryo.
Pseudoviviparous Vegetative proliferation of plantlets in the inflorescence axes of (usually) grasses.
Psychophily Pollination by butterflies.
Pteridophyte Member of the subphylum Pteridophyta, a vascular plant without flowers or seeds, and with a free-living **gametophyte** generation, including ferns, horsetails, quillworts and clubmosses.

Quadrivalent An association caused by the chiasmatic pairing of four chromosomes at **meiosis**.

Raceme An unbranched inflorescence.
Ramet A physiologically independent individual; from one to many may make up a **genet** (*see also* **clone**).
Receptacle Part of a flower on which the **perianth**, stamens, and in **hypogynous** flowers, the **gynoecium** are borne.
Reciprocal Of experimental crosses between the same two genets in different directions, i.e. *A* male × *B* and *B* male × *A*.
Recombination The reassociation of genes and parts of chromosomes between **homologous** or **homoeologous** chromosomes after **meiosis** by the action of chiasmata, thus breaking up linkage groups.

Restitution Where most or all the chromosomes of a meiotic or mitotic division remain together in a single daughter nucleus, usually as the result of non-disjunction.
Rugose Roughly sculptured, as of the **exine** of a pollen grain.

***S* alleles** **Alleles** that control the self-incompatibility response.
Sapromyophily Pollination of flowers by flies, where the flower mimics rotting meat.
Saprophyty Where plants lack chlorophyll, and use decomposing vegetation as an energy source.
Satellite The euchromatic region of a chromosome distal to the heterochromatic nucleolar organizer region.
Section Taxonomic rank at a level between species and genus.
Self-compatible A **genet** capable of self-fertilization if self-pollinated.
Self-incompatible A **genet** incapable of self-fertilization, but capable of cross-fertilization with another **genet**.
Selfing Fertilization of an ovule by a pollen grain of the same **genet**.
Semi-compatibility Where two genets share some but not all gametophytic *S* alleles, and thus in crosses between them some pollen grains can effect fertilization and others cannot.
Septae Walls between carpels in a fused ovary.
Sere A succession of plant communities in a given habitat leading to a climax association.
Seroprotein Protein produced in response to an antigen in an immunological reaction.
Siblings Offspring of the same parents.
Simplex *Aaaa* heterozygote in tetrasomic inheritance (*see* **duplex**, **triplex**).
Solar furnace Flower that focuses solar radiation so that the temperature at the centre of the flower is above the ambient temperature.
Somatic Non-meiotic or non-sexual; thus **somatic** recombination occurs by chromosome breakage/refusion events in non-sexual tissues.
Sphingid Hawk-moth (family Sphingidae).
Sporangium Organ that contains spores.
Sporophyll Organ that bears sporangia.
Sporophytic Of the sporophyte (usually diploid) generation; in the seed plants the dominant generation, the **genet** that develops from a zygote, is the sporophyte; in sporophytic incompatibility it is the genotype of the anther that produces the pollen grain which is significant, the anther being of the sporophyte generation.
Sporopollenin The lipoprotein which is the main constituent of the exine of the Angiosperm pollen grain.
Stamen The pollen-bearing organ in spermatophytes, derived from the microsporophyll.
Staminode Floral organ derived from a stamen, but no longer bearing an anther.
Stigma Part of the gynoecium which receives pollen grains on stigma papillae, often borne on the end of a style.
Strategy A teleological but useful word used to describe a group of attributes which result in a particular life style or behaviour pattern; compare syndrome.

Style Part of the gynoecium which links the stigma to the ovary or carpel, down which the pollen tubes grow; usually slender.

Subandroecy Where wholly female genets coexist with andromonoecious genets.

Subgynoecious Where wholly male genets coexist with gynomonoecious genets.

Supergene Linkage group of coadapted loci.

Sympatric Having overlapping geographical distributions.

Synapsis The close association of regions of **homologous** or **homoeologous** chromosomes during zygotene of **meiosis** preparatory to the formation of chiasmata.

Syndrome A group of coadapted attributes, or attributes which occur together in a particular condition; a syndrome can contribute to a strategy.

Synergid Nuclei (usually two) positioned at the micropilar end of the **angiosperm** embryo-sac, and involved in the process of double fertilization (*see* **filiform apparatus**); sometimes known as the egg apparatus.

Tapetum The nurse tissue to the developing pollen grains, within the anther.

Taxon A taxonomic category, e.g. species, section, genus, family.

Tectum The outer layer of the pollen grain **exine** in Angiosperms.

Telomere Repetitive sequence DNA forming a protective cap to the end of a chromosome. Recombination is discouraged in its vicinity.

Tepal A perianth organ not differentiated into a petal or sepal.

Testa The seed coat, derived from the ovule integuments.

Tetrads Groups of spore nuclei in fours, the immediate product of meiosis.

Tetrasomic Where a chromosome or gene is present four times in a **genet**.

Tetrasomic inheritance Involves three types of heterozygote (*see* **simplex**, **duplex**, **triplex**).

Thrum The short-styled morph of a distylous heteromorph.

Transcription The formation of messenger RNA from the coded sequences of DNA genes on chromosomes prior to the formation of polypeptides at ribosomes (translation).

Translocation Where a chromosome breaks, and part joins to a non-homologous chromosome (*see* **interchange**).

Transposons Short sequences of DNA which promote chromosome breakage probably by the formation of endonucleases near to the transposon, which is this enabled to 'roam' around the **karyotype** in association with linked genes by a succession of breakage/refusion events.

Trap-lining Where certain tropical bees are oligotropic with respect to distantly spaced individuals of flowering plant, and thus travel long distances between bouts of flower visiting.

Tricolporate A pollen grain with three pores, each contained within a groove (sulca).

Trioecy Where genets with male only, female only, and **hermaphrodite** only flowers coexist; compare **polygamy**.

Triplex *AAAa* heterozygote in tetrasomic inheritance (*see* **simplex**, **duplex**).

Tripping The explosive release of the stamens and style held in tension in the keel of a papilionoid flower, by a pollinator.

Tristyly A heteromorphic heterostylous condition in which short-styled, mid-styled and long-styled morphs with reciprocal anther positions coexist.

Trivalent An association caused by the chiasmatic pairing of three chromosomes at **meiosis**.

Tryphine A coating of lipoprotein on the outside of pollen grains derived from the **tapetum**; more or less identical with **pollenkitt**, but not necessarily pigmented.

Turgor Where a cell has absorbed water through osmosis until the cell wall is rigid.

Unilocular Of a carpel with only one ovule.

Unisexual Of a **genet** which is either male or female, equivalent to dioecy in seed plants.

Vascular plant Plant containing vascular tissue, xylem and phloem, i.e. Pteridophytes, Gymnosperms, Angiosperms.

Viscidium The sticky cap to the rostellum on the end of the column to which the pollinia adhere in some species of orchid.

Xenogamy Fertilization between pollen and ovules of different genets.

Xeromorphic With phenotypic attributes enabling a plant to survive dry conditions.

Zoophily Pollination by animals.

Zygomorphic Bilaterally symmetrical (two-lipped), of a flower.

Zygote The result of sexual fusion between two gametes.

Zymogram Configuration of isozyme bands in the electrophoretic analysis of an enzyme system, typical of a genet.

References

Abbott, R. J. (1985) Maintenance of a polymorphism for outcrossing frequency in a predominantly selfing plant, in *Structuring and Functioning of Plant Populations*, II. *Phenotypic and Genotypic Variation in Plant Populations* (eds J. Haeck and Woldendorp J.), North Holland, Amsterdam, pp. 277–86.

Abbott, R. J. and Forbes, D. G. (1993) Outcrossing rate and self-incompatibility in the colonizing species *Senecio squalidus*. *Heredity*, **71**, 155–9.

Abbott, R. J. and Gomes, M. F. (1989) Population genetic structure and outcrossing rate of *Arabidopsis thaliana*. *Heredity*, **62**, 411–18.

Abbott, R. J., Ashton, P. A. and Forbes, D. G. (1992) Introgressive origin of the radiate groundsel, *Senecio vulgaris* L. var. *hibernicus* Syme.; *Aat*-3 evidence. *Heredity*, **68**, 425–35.

Adams, W. T. and Allard, R. W. (1982) Mating system variation in *Festuca microstachys*. *Evolution*, **36**, 591–5.

Adams, W. T., Birkes, D. S. and Erikson, V. J. (1992) Using genetic markers to measure gene flow and pollen dispersal in forest tree seed orchards, in *Ecology and Evolution of Plant Reproduction* (ed. R. Wyatt), Chapman & Hall, London, pp. 37–61.

Adams, W. T., Griffin, A. R. and Moran, G. F. (1992) Using paternity analysis to measure effective pollen dispersal in natural populations. *Am. Nat.*, **140**, 762–80.

Adey, M. E. (1982) Taxonomic aspects of plant-pollinator relationships in the Genistinae (Leguminosae). PhD thesis, University of Southampton, UK.

Alexander, H. M. and Antonovics, J. (1993) Spread of anther-smut disease (*Ustilago violacea*) and character correlation in a genetically variable experimental population of *Silene alba*. *J. Ecol.*, **83**, 783–94.

Allard, R. W. (1960) *Principles of Plant Breeding*. Wiley, New York.

Allard, R. W., Jain, S. K. and Workman, P. L. (1968) The genetics of inbreeding populations. *Adv. Genet.*, **14**, 55–131.

Altman, P. L. and Dittmer, D. S. (1964) *Biology Data Book*, Federation of American Societies for Experimental Biology, Washington, DC, pp. 428–31.

Al Wadi, H. and Richards, A. J. (1993) Primary homostyly in *Primula* L. subgenus *Sphondylia* (Duby) Rupr. and the evolution of distyly in *Primula*. *New Phytol.*, **124**, 329–38.

Anderson, E. (1924) Studies on self-sterility. VI. The genetic basis of cross-sterility in *Nicotiana*. *Genetics*, **9**, 13–40.

Anderson, G. J. and Stebbins, G. L. (1984) Dioecy versus gametophytic self-incompatibility: a test. *Am. Nat.*, **124**, 423–8.

Anderson, G. J. and Symon, D. E. (1985) Extrafloral nectaries in *Solanum*. *Biotropica*, **17**, 40–5.

Anderson, G. J. and Symon, D. E. (1988) Insect foragers on *Solanum* flowers in Australia. *Ann. Miss. Bot. Gard.*, **75**, 842–52.

Anderson, G. J. and Symon, D. E. (1989) Functional dioecy and andromonoecy in *Solanum*. *Evolution* **43**, 204–19.

Anderson, M. A. *et al.* (17 authors) (1986) Cloning of cDNA for a stylar glycoprotein associated with expression of self-incompatibility in *Nicotiana alata*. *Nature*, **321**, 38–44.

Anderson, M. K., Taylor, N. L. and Duncan, J. F. (1974) Self-incompatibility, genotype identification and stability as influenced by inbreeding in red clover (*Trifolium pratense* L.). *Euphytica*, **23**, 140–8.

Annerstedt, I. and Lundqvist, A. (1967) Genetics of self-incompatibility in *Tradescantia paludosa* (Commelinaceae). *Hereditas*, **58**, 13.

Antonovics, J. and Ellstrand, N. C. (1984) Experimental studies on the evolutionary significance of sex. I. A test of the frequency-dependent selection hypothesis. *Evolution*, **38**, 103–15.

Arasu, N. N. (1968) Self-incompatibility in angiosperms: a review. *Genetica*, **39**, 1–24.

Arroyo, M. T. K. de (1975) Electrophoretic studies of genetic variation in natural populations of allogamous *Limnanthes alba* and autogamous *Limnanthes floccosa* (Limnanthaceae). *Heredity*, **35**, 153–64.

Arroyo, M. T. K. de and Raven, P. H. (1975) The evolution of subdioecy in morphologically gynodioecious species of *Fuchsia* sect. *Encliandra* (Onagraceae). *Evolution*, **29**, 500–11.

Arroyo, M. T. K. and Uslar, P. (1990) Breeding systems in a temperate mediterranean-type climate montane sclerophyllous forest in central Chile. *Bot. J. Linn. Soc.*, **111**, 83–102.

Arroyo, M. T. K., Primack, R. and Armesto, J. (1982) Community studies in pollination ecology in the high temperate Andes of central Chile. 1. Pollination mechanisms and altitudinal variation. *Am. J. Bot.*, **69**, 82–97.

Ascher, P. D. and Peloquin, S. J. (1966) Effect of floral ageing on the growth of compatible and incompatible pollen tubes in *Lilium longiflorum*. *Am. J. Bot.*, **53**, 99–102.

Ashman, T. -L. and Schoen, D. J. (1996) Floral longevity: fitness consequences and resource costs, in *Floral Biology* (eds D. G. Lloyd and S. C. H. Barrett), Chapman & Hall, London, pp. 112–39.

Ashton, P. S. (1969) Speciation among tropical forest trees: some deductions in the light of recent evidence. *Biol. J. Linn. Soc.*, **1**, 155–96.

Asker, S. (1979) Progress in apomixis research. *Hereditas*, **91**, 231–40.

Asker, S. (1980) Gametophytic apomixis: elements and genetic regulation. *Hereditas*, **93**, 277–93.

Asker, S. and Jerling, L. (1992) *Apomixis in Plants*. CTC Press, Boca Raton.

Assouad, M. W., Dommée, B., Lumaret, R. and Valdeyron, G. (1978) Reproductive capacities in the sexual forms of the gynodioecious species *Thymus vulgaris L. Bot. J. Linn. Soc.*, **77**, 29–39.

Attia, M. S. (1950) The nature of incompatibility in cabbage. *Proc. Am. Soc. Hort. Sci.*, **56**, 369–71.

Babcock, E. B. and Stebbins, G. L. (1938) The American species of *Crepis*. *Publ. Carnegie Inst., Wash.*, **504**, 1–199.

Baker, H. G. (1948) Corolla-size in gynodioecious and gynomonoecious species of flowering plants. *Proc. Leeds Phil. Soc.* (Scientific section) for 1948, 136–9.

Baker, H. G. (1955) Self-compatibility and establishment after 'long-distance' dispersal. *Evolution*, **9**, 347–9.

Baker, H. G. (1966) The evolution, functioning and breakdown of heteromorphic incompatibility systems. 1. The Plumbaginaceae. *Evolution*, **20**, 349–68.

Baker, H. G. (1967) Support for Baker's Law – as a rule. *Evolution*, **21**, 853–6.

Baker, H. G. and Baker, I. (1973a) Amino acids in nectar and their evolutionary significance. *Nature*, **241**, 543–5.

Baker, H. G. and Baker, I. (1973b) Some anthecological aspects of the evolution of nectar-producing flowers, particularly amino acid production in nectar, in *Taxonomy and Ecology* (ed. V. H. Heywood), Academic Press, London, pp. 243–64.

Baker, H. G. and Baker, I. (1975) Studies of nectar-constitution and pollinator–plant coevolution, in *Coevolution of Animals and Plants* (eds L. E. Gilbert and P. H. Raven), University of Texas Press, Austin.

Baker, H. G. and Baker, I. (1977) Intraspecific constancy of floral nectar amino acid complements. *Bot. Gaz.*, **138**, 183–91.

Baker, H. G. and Baker, I. (1983) Floral nectar sugar constituents in relation to pollinator type, in *Handbook of Experimental Pollination* (eds C. E. Jones and I. Little), Von Nostrand, New York.

Baker, K. (1995) The Ecological Genetics of *Armeria maritima* (Mill.) Willd. PhD Thesis, University of Newcastle upon Tyne, UK.

Baker, K., Richards, A. J. and Tremayne, M. (1994) Fitness constraints on flower number, seed number and seed size in the dimorphic species *Primula farinosa* L. and *Armeria maritima* (Miller) Willd. New Phytol., **128**, 563–70.

Baker, R. R. (1969) The evolution of the migratory habit in butterflies. *J. Anim. Ecol.*, **38**, 703–46.

Barcaccia, G., Mazzucato, A., Flacinelli, M. and Veronesi, F. (1995) Callose localization in cell walls in meiotic and apomeiotic megasporogenesis in diploid alfalfa. *Apomixis Newsletter*, **8**, 34–5.

Barker, P., Freeman, D. C. and Harper, K. T. (1982) Variation in the breeding system of *Acer grandidentatum*. *For. Sci.*, **28**, 563–72.

Barlow, B. A. and Forrester, J. (1984) Pollen tube interactions in *Melaleuca*, in *Pollination 84* (eds E. G. Williams and R. B. Knox), University of Melbourne, Melbourne, pp. 154–60.

Barrett, S. C. H. (1984) Variation in floral sexuality of diclinous *Aralia*. *Ann. Miss. Bot. Gdn*, **71**, 278–88.

Barrett, S. C. H. (1988) Evolution of breeding systems in *Eichhornia* (Pontederiaceae), a review. *Ann. Miss. Bot. Gdn*, **75**, 741–60.

Barrett, S. C. H. (1992) Gender variation and the evolution of dioecy in *Wurmbea dioica* (Liliaceae). *J. Evol. Biol.*, **5**, 423–44.

Barrett, S. C. H. (1993) Heterostylous genetic polymorphisms: model systems for evolutionary analysis, in *Evolution and Function of Heterostyly* (ed. S. C. H. Barrett), Springer-Verlag, Berlin.

Barrett, S. C. H. and Anderson, J. M. (1985) Variation in expression of trimorphic incompatibility in *Pontederia cordata* L. (Pontederiaceae). *Theor. Appl. Genet.*, **70**, 355–62.

Barrett, S. C. H. and Charlesworth, D. (1991) Effects of a change in the level of inbreeding on the genetic load. *Nature*, **352**, 522–4.

Barrett, S. C. H. and Cruzan, M. B. (1994) Incompatibility in heterostylous plants, in *Genetic Control of Self-incompatibility and Reproductive Development in Flowering Plants* (ed. E. G. Williams), Kluwer, Dordrecht, Netherlands, pp. 189–219.

Barrett, S. C. M. and Eckert, C. G. (1990) Variation and evolution of mating systems in seed plants, in *Biological Approaches and Evolutionary Trends in Plants* (ed. S. Kawano), Academic Press, London, pp. 229–54.

Barrett, S. C. H. and Glover, D. E. (1985) On the Darwinian hypothesis of the adaptive significance of tristyly. *Evolution*, **39**, 766–74.

Barrett, S. C. H. and Harder, L. D. (1992) Floral variation in *Eichhornia paniculata* (Spreng.) Solms (Pontederiaceae) II. Effects of development and environment on the formation of flowers. *J. Evol. Biol.*, **5**, 83–107.

Barrett, S. C. H. and Helenurm, K. (1987) The reproductive biology of boreal forest herbs. I. Breeding systems and pollination. *Can. J. Bot.*, **65**, 2036–46.

Barrett, S. C. H. and Kohn, J. S. (1991) Genetic and evolutionary consequences of small population size in plants, in *Genetics and Conservation of Rare Plants* (eds D. A. Falk and K. E. Holsinger), Oxford University Press, New York, pp. 3–30.

Barrett, S. C. H., Morgan, M. T. and Husband, B. C. (1989) The dissolution of a complex genetic polymorphism: the evolution of self-fertilization in tristylous *Eichhornia paniculata* (Pontederiaceae). *Evolution*, **43**, 1398–416.

Barrett, S. C. H., Kohn, J. R. and Cruzan, M. B. (1992) Experimental studies of mating-system evolution: the marriage of marker genes and floral biology, in *Ecology and Evolution of Plant Reproduction* (ed. R. Wyatt), Chapman & Hall, London, pp. 192–230.

Barrett, S. C. H., Lloyd, D. G. and Arroyo, J. (1996) Stylar polymorphisms and the evolution of heterostyly in *Narcissus*, in *Floral Biology* (eds D. G. Lloyd and S. C. H. Barrett), Chapman & Hall, London, pp. 339–76.

Bateman, A. J. (1947) Contamination in seed crops. III. Relation with isolation distance. *Heredity*, **1**, 303–36.

Bateman, A. J. (1952) Self-incompatibility systems in angiosperms. I. Theory. *Heredity*, **6**, 285–310.

Bateman, A. J. (1954) Self-incompatibility systems in angiosperms. II. *Iberis amara*. *Heredity*, **8**, 305–32.

Battaglia, E. (1963) Apomixis, in *Recent Advances in the Embryology of Angiosperms* (ed. P. Maheshwari), University of Delhi, Delhi, pp. 221–64.

Battjes, J., Menken, S. B. J. and den Nijs, J. C. M. (1992) Clonal diversity in some microspecies of *Taraxacum* sect. *Palustria* (Lindb. fil.) Dahlst. from Czechoslovakia. *Bot. Jahr. Syst.*, **114**, 315–28.

Batygina, T. B. (1974) Fertilization process of cereals, in *Fertilization in Higher Plants* (ed. H. F. Linskens), North Holland, Amsterdam. pp. 205–20.

Baur, E. (1919) Uber Selbststerilität und über Kreuzungsversuche einer selbstfertilen und einer selbststerilen Art in der Gattung *Antirrhinum*. *Z. Indukt. Abst.*, **21**, 48–52.

Bawa, K. S. (1979) Breeding systems of trees in a tropical wet forest. *N. Z. J. Bot.*, **17**, 521–4.

Bawa, K. S. (1980) Evolution of dioecy in flowering plants. *Ann. Rev. Ecol. Syst.*, **11**, 15–39.

Bawa, K. S. (1981) Modes of pollination, sexual systems and community structure in a tropical lowland rainforest. *Abstracts, XIII International Botanical Congress, Sydney*, p. 103.

Bawa, K. S. and Beach, I. H. (1983) Self-incompatibility systems in the Rubiaceae of a tropical lowland wet forest. *Am. J. Bot.*, **70**, 1281–8.

Bawa, K. S. and Opler, P. A. (1975) Dioecism in tropical forest trees. *Evolution*, **29**, 167–79.

Beadle, G. W. (1930) Genetical and cytological studies of Mendelian asynapsis in *Zea mays*. *Cornell Univ. Agric. Expt. Sta. Mem.* **129**.

Beattie, A. J. (1976) Plant dispersion, pollination and gene flow in *Viola*. *Oecologia*, **25**, 291–300.

Beattie, A. J. (1978) Plant–animal interactions affecting gene flow in *Viola*, in *The*

Pollination of Flowers by Insects (ed. A. J. Richards), Academic Press, London, pp. 151–64.

Beattie, A. J. (1982) Ants and gene dispersal in flowering plants, in *Pollination and Evolution* (eds J. A. Armstrong, J. M. Powell and A. J. Richards), Royal Botanic Gardens, Sydney, pp. 1–8.

Beattie, A. J. (1985) *The Evolutionary Ecology of Ant–Plant Mutualisms.* Cambridge University Press, New York.

Beattie, A. J. (1991) Problems outstanding in ant–plant interaction research, in *Ant–Plant Interactions* (eds C. R. Huxley and D. F. Cutler), Oxford University Press, Oxford, pp. 559–76.

Beattie, A. J. and Culver, D. C. (1979) Neighbourhood size in *Viola. Evolution*, **33**, 1226–9.

Bell, G. (1982) *The Masterpiece of Nature. The Evolution and Genetics of Sexuality.* Croom Helm, London.

Bergstrom, G. (1978) Role of volatile chemicals in *Ophrys*–pollinator interactions, in *Biochemical Aspects of Plant and Animal Coevolution* (ed. J. B. Harborne), Academic Press, London, pp. 207–32.

Bertin, R. I. (1986) Consequences of mixed pollinations in *Campsis radicans. Oecologia*, **70**, 1–5.

Bertin, R. I. (1988) Paternity in plants, in *Plant Reproductive Biology* (eds J. Lovett Doust and L. Lovett Doust), Oxford University Press, Oxford, pp. 30–59.

Bertin, R. I., Barnes, C. and Guttman, S. I. (1989) Self-sterility and cryptic self-fertility in *Campsis radicans* (Bignoniaceae). *Bot. Gaz.*, **150**, 397–403.

Bertsch, A. (1984) Foraging in male bumblebees (*Bombus lucorum* L.): maximizing energy or minimizing water load? *Oecologia*, **62**, 325–36.

Best, L. S. and Bierzychudek, P. (1982) Pollinator foraging on foxglove (*Digitalis purpurea*): a test of a new model. *Evolution*, **36**, 70–9.

Bierzychudek, P. (1987) Pollinators increase the cost of sex by avoiding female flowers. *Ecology*, **68**, 444–7.

Bierzychudek, P. and Eckhart, V. (1988) Spatial separation of the sexes of dioecious plants. *Am. Nat.*, **132**, 34–43.

Bino, R. J., Devente, N. and Meeuse, A. D. J. (1984) Entomophily in the dioecious gymnosperm *Ephedra aphylla* Forsk. with some notes on *E. campylopoda*. II. Pollination droplets, nectaries and nectarial secretion in *Ephedra. Proc. Kon. Nederl. Akad. C*, **87**, 15–24.

Böcher, T. W. (1951) Cytological and embryological studies in the amphiapomictic *Arabis holboelii* complex. K. Dan. *Vid. Selsk Biol. Skr. VI*, **7**, 1–59.

Bookman, S. S. (1984) Evidence for selective fruit production in *Asclepias. Evolution*, **38**, 72–86.

Booth, T. A. and Richards, A. J. (1978) Studies in the *Hordeum murinum* aggregate: disc electrophoresis of seed proteins. *Bot. J. Linn. Soc.*, **76**, 115–25.

Bos, M., Harmens, H. and Vrieling, K. (1986) Gene flow in *Plantago* 1. Gene flow and neighbourhood size in *P. lanceolata. Heredity*, **56**, 43–54.

Bosch, J. (1991) Floral biology and pollination of three co-occurring *Cistus* species (Cistaceae). *Bot. J. Linn. Soc.*, **109**, 39–55.

Bowey, K. (1995) Moorhens feeding on pollen. *British Birds*, **88**, 111.

Boyd, M., Silvertown, J. and Tucker, C. (1990) Population ecology of heterostyle and homostyle *Primula vulgaris*: growth, survival and reproduction in field populations. *J. Ecol.*, **78**, 799–813.

Brantjes, N. B. M. (1976a) Senses involved in the visiting of flowers by *Cucullia umbratica* (Noctuidae, Lepidoptera). *Ent. Exp. Appl.*, **20**, 1–7.

Brantjes, N. B. M. (1976b) Riddles around the pollination of *Melandrium album* (Mill.) Garcke (Caryophyllaceae) during the oviposition by *Hadena bicruris* (Noctuidae, Lepidoptera). *Proc. K. Akad. Wet. C*, **79**, 125–41.

Brantjes, N. B. M. (1978) Sensory responses to flowers in night-flying moths, in *The Pollination of Flowers by Insects* (ed. A. J. Richards), Academic Press, London, pp. 13–19.

Breukelen, E. W. van, Ramanna, M. S. and Hermsen, J. G. T (1975) Monohaploids ($n + X + 12$) from autotetraploid *Solanum tuberosum* ($2n = 4X = 48$) through two successive cycles of female parthenogenesis. *Euphytica*, **24**, 567–74.

Brewbaker, J. L. (1957) Pollen cytology and incompatibility systems in plants. *J. Hered.*, **48**, 217–77.

Brewbaker, J. L. (1959) Biology of the angiosperm pollen grain. *Ind. J. Genet. Pl. Breed.*, **19**, 121–33.

Brewbaker, J. L. and Kwack, B. H. (1963) The essential role of calcium ions in pollen germination and pollen tube growth. *Am. J. Bot.*, **50**, 859–65.

Brochman, C. (1993) Reproductive strategies of diploid and polyploid populations of arctic *Draba* (Brassicaeae). *Pl. Syst. Evol.*, **185**, 55–83.

Brock, R. D. (1954) Fertility in *Lilium* hybrids. *Heredity*, **9**, 409–20.

Broker, W. (1963) Genetischphysiologische Untersuchungen über die Zinkverträglichkeit van *Silene inflata* Sm. *Flora, Jena*, B **153**, 122–56.

Brown, A. H. D. (1978) Isozymes, plant population genetic structure and genetic conservation. *Theor. Appl. Genet.*, **52**, 145–57.

Brown, A. H. D. (1979) Enzyme polymorphism in plant populations. *Theor. Popul. Biol.*, **15**, 1–42.

Brown, A. H. D. (1990) Genetic characterization of plant mating systems, in *Plant Population Genetics, Breeding and Genetic Resources* (eds A. H. D. Brown, M. T. Clegg, A. L. Kahler and B. S. Weir), Sinauer, Sunderland, MA., pp. 145–62.

Brown, A. H. D. and Allard, R. W. (1970) Estimation of the mating system in open-pollinated maize populations using isozyme polymorphisms. *Genetics*, **66**, 133–45.

Broyles, S. B. and Wyatt, R. (1990) Paternity analysis in a natural population of *Asclepias exaltata*: multiple paternity, functional gender, and the 'pollen donation hypothesis'. *Evolution*, **44**, 1454–68.

Broyles, S. B. and Wyatt, R. (1991) Effective pollen dispersal in a natural population of *Asclepias exaltata*: the influence of pollinator behavior, genetic similarity and mating success. *Am. Nat.*, **138**, 1234–49.

Bruun, H. G. (1932) Cytological studies in *Primula*. *Symb. Bot. Uppsala* **1**, 1–239.

Bullard, E. R., Shearer, H. D. H., Day, J. D. and Crawford R. M. M. (1987) Survival and flowering of *Primula scotica* Hook. *J. Ecol.*, **75**, 589–602.

Byers, D. L. and Meagher, T. R. (1992) Mate availability in small populations of plant species with homomorphic sporophytic self-incompatibility. *Heredity*, **68**, 353–9.

Cahalan, C. and Gliddon, C. (1985) Genetic neighbourhood sizes in *Primula vulgaris*. *Heredity*, **54**, 65–70.

Calos, M. P. and Miller, J. H. (1980) Transposable elements. *Cell*, **20**, 579–95.

Carlquist, S. (1974) *Island Biology*. Columbia University Press, New York.

Carman, J. (1995) Gametophytic angiosperm apomicts and the occurrence of polyspory and polyembryony among their relatives. *Apomixis Newsletter*, **8**, 39–53.

Carman, J. G., Crane, C. F. and Riera-Lizarazu, O. (1991) Comparative histology of cell walls during meiotic and apomeiotic megasporogenesis in two hexaploid Australasian *Elymus* species. *Crop Sci.*, **31**, 1527–32.

Catcheside, D. G. (1977) The genetics of recombination. *Genetics – Principles and Perspectives*, 2. Edward Arnold, London.

Charlesworth, B. (1980) The cost of sex in relation to the mating system. *J. Theor. Biol.*, **84**, 655–71.

Charlesworth, B. and Charlesworth, D. (1978) A model for the evolution of dioecy and gynodioecy. *Am. Nat.*, **112**, 975–97.

Charlesworth, D. (1979) The evolution and breakdown of tristyly. *Evolution*, **33**, 489–98.

Charlesworth, D. and Charlesworth B. (1979) A model for the evolution of distyly. *Am. Nat.*, **114**, 467–98.

Charlesworth, D. and Charlesworth, B. (1987) Inbreeding depression and its evolutionary consequences. *Ann. Rev. Ecol. Syst.*, **18**, 237–68.

Charlesworth, D. and Charlesworth, B. (1990) Inbreeding depression with heterozygote advantage and its effect on selection for modifiers changing the outcrossing rate. *Evolution*, **44**, 870–88.

Charlesworth, D. and Ganders, F. R. (1979) The population genetics of gynodioecy with cytoplasmic-genic male sterility. *Heredity*, **43**, 213–18.

Charlesworth, D., Morgan, M. T. and Charlesworth, B. (1990) Inbreeding depression, genetic load, and the evolution of outcrossing rates in a multilocus system with no linkage. *Evolution*, **44**, 1469–89.

Charnov, E. L. (1982) *The Theory of Sex Allocation*. Princeton University Press, Princeton.

Charnov, E. L. (1984) Behavioural ecology of plants, in *Behavioural Ecology, an Evolutionary Approach* (eds J. J. Krebs and N. B. Davies), Blackwell Scientific, Oxford, pp. 362–79.

Charnov, E. L., Maynard Smith, J. and Bull, J. J. (1976) Why be hermaphrodite? *Nature*, **263**, 125–6.

Chase, S. (1969) Monoploids, and monoploid-derivatives of maize (*Zea mays* L.). *Biol. Rev.*, **35**, 117–67.

Chu, C. and Hu, S. Y. (1981) The development and ultrastructure of wheat sperm cell. Abstracts, *XIII International Botanical Congress, Sydney*, p. 61.

Clapham, A. R., Tutin, T. G. and Warburg, E. F. (1962) *Flora of the British Isles*, 2nd edn. Cambridge University Press, Cambridge.

Clay, K. and Antonovics, J. (1985) Quantitative variation of progeny from chasmogamous and cleistogamous flowers in the grass *Danthonia spicata*. *Evolution*, **39**, 335–48.

Clayberg, C. D., Butler, L., Kerr, E. A. Rick, C. M. and Robinson, R. W. (1966) Third list of known genes in the tomato: with revised linkage map and additional rules. *J. Hered.*, **57**, 188–96.

Clegg, M. T. (1980) Measuring plant mating systems. *Bioscience*, **30**, 814–18.

Colin, L. J. and Jones, C. E. (1980) Pollen energetics and pollination modes. *Am. J. Bot.*, **67**, 210–15.

Connor, H. E. (1973) Breeding systems in *Cortaderia* (Gramineae). *Evolution*, **27**, 663–78.

Connor, H. E. (1979) Breeding systems in the grasses: a survey. *N. Z. J. Bot.*, **17**, 547–74.

Cook, L. M. (1971) *Coefficients of Natural Selection*. Hutchinson, London.

Cope, F. W. (1962) The mechanism of pollen incompatibility in *Theobroma cacao*. *Heredity*, **17**, 157–82.

Corbett, S. A. (1978) Bees and the nectar of *Echium vulgare*, in *The Pollination of Flowers by Insects* (ed. A. J. Richards), Academic Press, London, pp. 21–30.

Correns, C. (1912) Selbststerilität und Individualstoffe. *Festschr. d. mat. nat. Gesell. zur.* 84. Versamml. Deut. Naturforsch. Artze, Munster i W., 1–32.
Correns, C. (1913) Selbststerilität und Individualstoffe. *Biol. Centr.*, **33**, 389–423.
Correns, C. (1916a) Individuen und Individualstoffe. *Die Naturwissensch.*, **4**, 183–7, 193–8, 210–13.
Correns, C. (1916b) Untersuchungen über Geschlechtsbestimmung bei Distelarten. *Sitz. Konigl. Preuss. Akad. Wiss.*, **20**, 448–77.
Correns, C. (1928) *Bestimmung, Vererbung und Verteilung des Geschlechtes bei den hoheren Pflanzen.* Borntraeger, Berlin.
Cox, P. A. (1982) Vertebrate pollination and the maintenance of dioecism in *Freycinetia. Am. Nat.*, **120**, 65–80.
Cox, P. A. and Knox, R. B. (1986) Two dimensional pollination in hydrophytic plants. *Am. J. Bot.*, **76**, 164–75.
Cramer, J. M., Meeuse, A. D. J. and Teunissen, P. A. (1975) A note on the pollination of nocturnally flowering species of *Nymphaea. Acta Bot. Neerl.*, **24**, 489–90.
Crane, M. B. and Lawrence, W. J. C. (1929) Genetical and cytological aspects of incompatibility and sterility in cultivated fruits. *J. Pomol. Hort. Sci.*, **7**, 276–301.
Crane, M. B. and Lewis, D. (1942) Genetical studies in pears. III. Incompatibility and sterility. *J. Genet.*, **43**, 31.
Crawford, R. M. M. and Balfour, J. (1983) Female predominant sex-ratios and physiological differentiation in arctic willows. *J. Ecol.*, **71**, 145–60.
Crawford, T. J. (1984a) The estimation of neighbourhood parameters for plant populations. *Heredity*, **52**, 275–83.
Crawford, T. J. (1984b) What is a population? in *Evolutionary Ecology* (ed. B. Shorrocks), Blackwell Scientific, Oxford, pp. 135–73.
Crepet, W. C. and Friis, E. M. (1987) The evolution of insect pollination in Angiosperms, in *The Origin of Angiosperms and Their Biological Consequences* (eds E. M. Friis, W. G. Chaloner and P. R. Crane), Cambridge University Press, Cambridge, pp. 145–80.
Croat, T. B. (1979) The sexuality of Barro Colorado Island (Panama) flora. *Phytologia*, **42**, 319–48.
Crosby, J. L. (1949) Selection of an unfavourable gene-complex. *Evolution*, **3**, 212–30.
Crow, J. F. and Kimura, M. (1970) *An Introduction to Population Genetics Theory.* Harper and Row, New York.
Crowe, L. K. (1954) Incompatibility in *Cosmos bipinnatus. Heredity*, **8**, 1–11.
Crowe, L. K. (1971) The polygenic control of outbreeding in *Borago officinalis. Heredity*, **27**, 111–18.
Cruden, R. W. (1972) Pollinators in high elevation ecosystems: relative effectiveness of birds and bees. *Science*, **176**, 1439–40.
Cruden, R. W. (1988) Temporal dioecism: systematic breadth, associated traits, and temporal patterns. *Bot. Gaz.*, **149**, 1–15.
Cruden, R. W. and Hermann-Parker, S. (1977) Temporal dioecism: an alternative to dioecism. *Evolution*, **31**, 863–6.
Cruden, R. W. and Miller-Ward, S. (1981) Pollen–ovule ratio, pollen size, and the ratio of stigmatic area to the pollen-bearing area of the pollinator: an hypothesis. *Evolution*, **35**, 964–74.
Cruden, R. W., Kinsman, S., Stockhouse, R. E. and Linhart, Y. B. (1976) Pollination, fecundity, and the distribution of moth-flowered plants. *Biotropica*, **8**, 204–10.
Culwick, E. G. (1982) The biology of *Acaena novae-zelandiae* Kirk on Lindisfarne. PhD thesis, University of Newcastle upon Tyne.

Curtis, J. and Curtis, C. F. (1985) Homostyle primroses revisited. I. Variation in time and space. *Heredity*, **54**, 227–34.

Czapik, R. (1994) How to detect apomixis in Angiospermae. *Pol. Bot. St.*, **8**, 13–21.

Dafni, A. (1984) Mimicry and deception in pollination. *Ann. Rev. Ecol. Syst.*, **15**, 259–78.

Dafni, A. (1986) Pollination in *Orchis* and related genera: evolution from reward to deception, in *Orchid Biology, Reviews and Perspectives* (ed. J. Arditti), Comstock, Ithaca, pp. 81–103.

Dafni, A. and Dukas, R. (1986) Insect and wind pollination in *Urginea maritima*. *Pl. Syst. Evol.*, **154**, 1–10.

Dafni, A. and Werker, E. (1982) Pollination ecology of *Sternbergia clusiana* (Ker-Gawler) Spreng. (Amaryllidaceae). *New Phytol.*, **91**, 571–7.

Dafni, A. and Woodell, S. R. J. (1986) Stigmatic exudate and the pollination of *Dactylorhiza fuchsii*. *Flora*, **178**, 343–50.

Damme, J. M. M. van (1984) Gynodioecy in *Plantago lanceolata*. III. Sexual reproduction and maintenance of male steriles. *Heredity*, **52**, 77–94.

Damme, J. M. M. van and van Delden, W. (1984) Gynodioecy in *Plantago lanceolata*. IV. Fitness components of sex types in different life cycle stages. *Evolution*, **38**, 1326–36.

Darlington, C. D. (1939) *The Evolution of Genetic Systems*, Cambridge University Press, Cambridge.

Darlington, C. D. and Mather, K. (1949) *The Elements of Genetics*, Allen and Unwin, London.

Darwin, C. (1862) *The Various Contrivances by which Orchids are Fertilised*, Murray, London.

Darwin, C. (1871) *The Descent of Man and Selection in Relation to Sex*, 2 vols, Murray, London.

Darwin, C. (1876) *The Effects of Cross and Self Fertilisation in the Vegetable Kingdom*, Murray, London.

Darwin, C. (1877) *The Different Forms of Flowers on Plants of the Same Species*, Murray, London.

Dauphin-Guerin, B., Teller, G. and Durand, B. (1980) Different endogenous cytokinins between male and female *Mercurialis annua* L. *Planta*, **148**, 124–9.

Davey, A. J. C. and Gibson, C. M. (1917) Note on the distribution of the sexes in *Myrica gale*. *New Phytol.*, **16**, 147–51.

Dawkins, R. (1976) *The Selfish Gene*, Oxford University Press, Oxford.

Delannay, X. (1978) La gynodioecie chez les angiospermes. *Natur. Belges*, **59**, 223–37.

Delph, L. F. (1996) Flower size dimorphism in plants with unisexual flowers, in *Floral Biology* (eds D. G. Lloyd and S. C. H. Barrett), Chapman & Hall, London, pp. 217–37.

Denward, T. (1963) The function of the incompatibility alleles in red clover (*Trifolium pratense* L.). *Hereditas*, **49**, 289–334.

Devlin, B. (1989) Components of seed and pollen yield in *Lobelia cardinalis*. *Am. J. Bot.*, **76**, 204–14.

Devlin, B. and Ellstrand, N. C. (1990) Male and female fertility variation in wild radish, a hermaphrodite. *Am. Nat.*, **136**, 86–107.

Devlin, B., Clegg, J. and Ellstrand, N. C. (1992) The effect of flower production on male reproductive success in wild radish populations. *Evolution*, **46**, 1030–42.

Dickinson, H. G. (1994) Self pollination. Simply a social disease? *Nature*, **367**, 517–18.

Dickinson, H. G., Moriarty, J. and Lawson, J. (1982) Pollen–pistil interaction in *Lilium longiflorum:* the role of the pistil in controlling pollen tube growth following cross and self-pollination. *Proc. R. Soc. Lond. B*, **215**, 45–62.

Dickinson, H. G., Crabbe, M. J. C. and Gaude, T. (1993) Sporophytic self-incompatibility systems – *S*-gene products. *Int. Rev. Cytol.*, **140**, 525–61.

Dickinson, T. A. and Phipps, J. B. (1986) Studies in *Crataegus* (Rosaceae: Maloideae) XIV. The breeding system of *Crataegus crus-gallii* s.l. in Ontario. *Am. J. Bot.*, **73**, 116–30.

Dobrofsky, S. and Grant, F. (1980) An investigation into the mechanism for reduced seed yield in *Lotus corniculatus. Theor. Appl. Gen.*, **57**, 157–60.

Dodson, C. H. (1962) Pollination and variation in the subtribe Catasetinae (Orchidaceae). *Ann. Miss. Bot. Gard.*, **49**, 35–56.

Dodson, C. H. and Frymire, G. P. (1961) Natural pollination of orchids. *Bull. Miss. Bot. Gard.*, **49**, 133–52.

Dommée, B. (1976) La sterilité male chez *Thymus vulgaris* L.: repartition écologique dans la region mediterranéene française. *Compt. Rend. Hebd. Seances Acad. Sci.*, **282** D, 65–8.

Donk, J. A. W. van der (1974) Synthesis of RNA and protein as a function of time and type of pollen tube-style interaction in *Petunia hybrida* L. *Mol. Gen. Genet.*, **134**, 93–8.

Dowrick, V. P. J. (1956) Heterostyly and homostyly in *Primula obconica. Heredity*, **10**, 219–36.

Dressler, R. L. (1968) Pollination by euglossine bees. *Evolution*, **22**, 202–10.

Dressler, R. L. (1980) *The Orchids, Natural History and Classification*. Harvard University Press, Cambridge, MA.

Dujardin, M. and Hanna, W. W. (1989) Developing apomictic pearl millet-characterization of a BC3 plant. *J. Gen. Pl. Breed.*, **43**, 145.

Dulberger, R. (1970) Floral dimorphism in *Anchusa hybrida* Ten. *Israel J. Bot.*, **19**, 37–41.

Dulberger, R. (1975) *S* gene action and the significance of characters in the heterostylous syndrome. *Heredity*, **35**, 407–15.

Dulberger, R. (1993) Floral polymorphisms and their functional significance in the heterostylous syndrome, in *Evolution and Function of Heterostyly* (ed. S. C. Barrett), Springer-Verlag, Berlin, pp. 41–84.

Dulberger, R., Smith, M. B. and Bawa, K. S. (1994) The stigmatic orifice in *Cassia*, *Senna* and *Chamaecrista* (Caesalpinaceae): morphological variation, function during pollination, and possible adaptive significance. *Am. J. Bot.*, **81**, 1390–6.

Durand, B. (1963) Le complex *Mercurialis annua* L. s.l. Une étude biosystematique. *Ann. Sci. Nat., Bot. IV*, **4**, 627–36.

Durand, B. and Durand, R. (1991) Sex determination and reproductive organ differentiation in *Mercurialis. Plant Sci.* (*Limerick*), **80**, 49–66.

East, E. M. (1915a) An interpretation of self-sterility. *Proc. Natl. Acad. Sci.*, **1**, 95–100.

East, E. M. (1915b) The phenomenon of self-sterility. *Am. Nat.*, **49**, 77–88.

East, E. M. (1917a) The behaviour of self-sterile plants. *Science*, **46**, 221–2.

East, E. M. (1917b) The explanation of self-sterility. *J. Hered.*, **8**, 382–3.

East, E. M. (1918) Intercrosses between self-sterile plants. *Mem. Brooklyn Bot. Gdn.*, **1**, 141–53.

East, E. M. (1919a) Studies on self-sterility. III. The relation between self-fertile and self-sterile plants. *Genetics*, **4**, 341–5.

East, E. M. (1919b) Studies on self-sterility. IV. Selective fertilization. *Genetics*, **4**, 34–5.

East, E. M. (1919c) Studies on self-sterility. V. A family of self-sterile plants, wholly cross-sterile inter se. *Genetics*, **4**, 355–63.

East, E. M. (1940) The distribution of self-sterility in flowering plants. *Proc. Am. Phil. Soc.*, **82**, 449–518.

East, E. M. and Mangelsdorf, A. J. (1925) A new interpretation of the hereditary behaviour of self-sterile plants. *Proc. Natl. Acad. Sci. (Wash)*, **11**, 166–71.

East, E. M. and Park, J. B. (1917) Studies on self-sterility. 1. The behaviour of self-sterile plants. *Genetics*, **2**, 505–609.

Eenink, A. H. (1974) Matromorphy in *Brassica oleracea* L. II. Differences in parthenogenetic ability and parthenogenesis inducing ability. *Euphytica*, **23**, 435–45.

Eisikowitch, D. (1978) Insect visiting of two subspecies of *Nigella arvensis* under adverse seaside conditions, in *Pollination of Flowers by Insects* (ed. A. J. Richards), Academic Press, London, pp. 125–32.

Eisikowitch, D. and Woodell, S. R. J. (1975) The effect of water on pollen germination in two species of *Primula*. *Evolution*, **28**, 692–94.

El-Keblawy, A. A., Lovett Doust, J., Lovett Doust, L. and Shaltout, K. H. (1995) Labile sex expression and dynamics of gender in *Thymelaea hirsuta*. *Ecoscience*, **2**, 55–66.

El-Keblawy, A. A., Shaltout, K. H., Doust, J. L. and Doust, L. L. (1996) Maternal effects on progeny in *Thymelaea hirsuta*. *New Phytol.*, **132**, 77–85.

Elkington, T. T. (1969) Cytotaxonomic variation in *Potentilla fruticosa*. *New Phytol.*, **68**, 151.

Elkington, T. T. and Woodell, S. R. J. (1963) *Potentilla fruticosa* L. Biological flora of the British Isles. *J. Ecol.*, **51**, 769.

Ellerstrom, S. and Zagorcheva, L. (1977) Sterility and apomictic embryo-sac formation in *Raphanobrassica*. *Hereditas*, **87**, 107–20.

Ellstrand, N. C. (1984) Multiple paternity within the fruits of the wild radish, *Raphanus sativus*. *Am. Nat.*, **123**, 819–28.

Ellstrand, N. C. and Antonovics, J. (1985) Experimental studies on the evolutionary significance of sexual reproduction. II. A test of the density dependent selection hypothesis. *Evolution*, **39**, 657–66.

Ellstrand, N. C. and Elam, D. R. (1993) Population genetic consequences of small population size: implications for plant conservation. *Ann. Rev. Ecol. Syst.*, **24**, 217–42.

Emerson, S. (1939) A preliminary survey of the *Oenothera organensis* population. *Genetics*, **24**, 524–7.

Ennos, R. A. and Dodson, R. K. (1987) Pollen success, functional gender and assortative mating in an experimental plant population. *Heredity*, **58**, 119–26.

Epling, C. and Dobzhansky, Th. (1942) Genetics of natural populations. VI. Microgeographic races in *Linanthus parryae*. *Genetics*, **27**, 317–32.

Epperson, B. K. and Clegg, M. T. (1987) First-pollination primacy and pollen selection in the morning glory *Ipomoea purpurea*. *Heredity*, **58**, 5–14.

Ernst, A. (1933) Weitere Untersuchungen zur Phananalyse, zum Fertilitatsproblem and zur Genetik heterostyler Primeln. 1. *Primula viscosa*. *Arch. J. K.-Stift. Ver., Soc. Rass.*, **8**, 1–215.

Ernst, A. (1936) Weitere Untersuchungen zur Phananalyse, zum Fertilitatsproblem and zur Genetik heterostyler Primeln. II. *Primula hortensis* Wettst. *Arch. J. K.-Stift. Ver., Soc. Rass.*, **11**, 1–280.

Ernst, A. (1950) Resultate aus Kreuzungen zwischen der tetraploiden, monomorphen *Pr. japonica* und diploiden, mono- und dimorphen Arten der Sektion *Candelabra*. *Arch. J. K.-Stift. Ver., Soc. Rass.*, **25**, 135–236.

Ernst, A. (1955) Self-fertility in monomorphic primulas. *Genetica*, **27**, 91–148.

Ernst, A. (1957) Austausch und Mutation im Komplex-gen für Blutenplastik und Inkompatibilität bei *Primula. Z. Indukt. Abst. Vererb.*, **88**, 517–99.

Ernst, A. (1958) Untersuchungen zur Phananalyse, zum Fertilitatsproblem und zur Genetik heterostyler Primeln 4. Die F2-F5-Nachkommenschaften der Bastarde *Pr.* (*hortensis* × *viscosa*). *Arch. J. K.-Stift. Ver., Soc. Rass.*, **33**, 103–251.

Faegri, K. and van der Pijl, L. (1979) *The Principles of Pollination Ecology*, 3rd edn, Pergamon, Oxford.

Fagerlind, F. (1945) Die Bastarde der Canina-Rosen, ihre Syndese und Formbildungsverhaltnisse. *Acta Hort. Berg.*, **14**, 9–37.

Falk, D. A. and Holsinger, K. E. (1991) *Genetics and Conservation of Rare Plants.* Oxford University Press, New York.

Favre-Duchatre, M. (1974) Phylogenetic aspects of the spermatophytes double fertilization, in *Fertilization in Higher Plants* (ed. H. F. Linskens), North Holland, Amsterdam, pp. 243–52.

Feinsinger, P. and Swarm, L. A. (1982) 'Ecological release', seasonal variation in food supply, and the hummingbird *Amazilia tobaci* on Trinidad and Tobago. *Ecology*, **63**, 1574–87.

Feinsinger, P., Wolfe, J. A. and Swarm, L. A. (1982) Island ecology: reduced hummingbird diversity and the pollination biology of plants, Trinidad and Tobago, West Indies. *Ecology*, **63**, 494–506.

Felsenstein, J. (1974) The evolutionary advantage of recombination. *Genetics*, **78**, 737–56.

Felsenstein, J. (1988) Sex and the evolution of recombination, in *The Evolution of Sex* (eds R. E. Michod and B. R. Levin), Sinauer, Sunderland, MA, pp. 74–86.

Fenster, C. B. (1991a) Gene flow in *Chamaecrista fasciculata* (Leguminosae). 1. Gene dispersal. *Evolution*, **45**, 398–409.

Fenster, C. B. (1991b) Gene flow in *Chamaecrista fasciculata* (Leguminosae). II. Gene establishment. *Evolution*, **45**, 410–22.

Fernandes, A. (1935) Remarque sur l'heterostylie de *Narcissus triandrus* et de *N. reflexus*. *Brot. Bot. Soc. Broteriana ser. 2*, **10**, 278–88.

Ferrari, T. E., Lee, S. S. and Wallace, D. H. (1981) Biochemistry and physiology of recognition in pollen–stigma interactions. *Phytopathology*, **71**, 752–5.

Filzer, P. (1926) Selbststerilitat von *Veronica syriaca*. *Z. Indukt. Abstamm. Vererbl.*, **41**, 137–97.

Fincham, J. R. S. and Sastry, G. R. K. (1974) Controlling elements in maize. *Annu. Rev. Genet.*, **8**, 15–50.

Finnegan, D. J. (1981) Transposable elements and proviruses. *Nature*, **292**, 800–1.

Fisher, R. A. (1961) A model for the generation of self-sterility alleles. *J. Theor. Biol.*, **1**, 411–14.

Flanagan, L. B. and Moser, W. (1985) Flowering phenology, floral display and reproductive success in dioecious *Aralia nudicaulis* L. (Araliaceae). *Oecologia*, **68**, 23–8.

Flores, S. and Schemske, D. W. (1984) Dioecy and monoecy in the flora of Puerto Rico and the Virgin Islands: ecological correlates. *Biotropica*, **16**, 132–9.

Ford, H. and Richards, A. J. (1985) Isozyme variation within and between *Taraxacum* agamospecies in a single locality. *Heredity*, **55**, 289–91.

Ford, M. A. and Kay, Q. O. N. (1985) The genetics of self-incompatibility in *Sinapis arvensis* L. *Heredity*, **54**, 99–102.

Fowler, N. L. and Levin, D. A. (1984) Ecological constraints on the establishment of a novel polyploid in competition with its diploid progenitor. *Am. Nat.*, **124**, 705–11.

Fox, J. F. (1985) Incidence of dioecy in relation to growth form, pollination and dispersal. *Oecologia*, **67**, 244–9.
Frankie, G. W. (1976) Pollination of widely dispersed trees by animals in Central America, with emphasis on bee pollination systems, in *Tropical Trees: Variation, Breeding and Conservation* (eds J. Burley and B. T. Styles), Academic Press, New York, pp. 151–9.
Frankie, G. W., Opler, P. A. and Bawa, K. S. (1976) Foraging behaviour of solitary bees: implications for outcrossing of a neotropical forest tree species. *J. Ecol.*, **64**, 1049–58.
Freeman, D. C., Harper, K. T. and Ostler, K. (1980). Ecology of plant dioecy in the intermountain region of western north America and California. *Oecologia*, **44**, 410–17.
Freeman, D. C., Harper, K. T. and Charnov, E. L. (1980) Sex change in plants: old and new observations and new hypotheses. *Oecologia*, **47**, 222–32.
Freeman, D. C., McArthur, E. D., Harper, K. T. and Blauer, A. C. (1981) Influence of environment on the floral sex ratio of monoecious plants. *Evolution*, **35**, 194–7.
Fritsch, P. and Riesenberg, L. H. (1992) High outcrossing rates maintain male and hermaphrodite individuals in populations of the flowering plant *Datisca glomerata*. *Nature*, **359**, 633–6.
Fritz, A.-L. and Nilsson, L. A. (1996) Reproductive success and gender variation in deceit-pollinated orchids, in *Floral Biology* (eds D. G. Lloyd and S. C. H. Barrett), London, Chapman & Hall, pp. 319–38.
Fryxell, P. A. (1957) Mode of reproduction in higher plants. *Bot. Rev.*, **23**, 135–233.
Furnkranz, D. (1960) Cytogenetische Untersuchungen an *Taraxacum* im Raume von Wien. *Ost. Bot. Z.*, **107**, 310–50.
Furnkranz, D. (1961) Cytogenetische Untersuchungen an *Taraxacum* im Raume von Wien. II. Hybriden zwischen *T. officinale* und *T. palustre*. *Ost. Bot. Z.*, **108**, 408–15.
Furnkranz, D. (1965) Untersuchungen an Populationen des *Taraxacum officinale* – Komplexes im Kontaktgebiet der diploiden und polyploiden Biotypen. *Ost. Bot. Z.*, **113**, 427–47.

Gabe, D. R. (1939) Inheritance of sex in *Mercurialis annua*. *Compt. Rend. Acad. Sci. URSS*, **23**, 478–81.
Gadella, T. W. J. (1991) Variation, hybridization and reproductive biology of *Hieracium pilosella* L. *Proc. K. Ned. Akad. Wet.*, **94**, 455–88.
Gale, J. S. (1980) *Population Genetics*, L. Blackie, London.
Galen, C. (1992) Pollen dispersal dynamics in an alpine wildflower *Polemonium viscosum*. *Evolution*, **46**, 1043–51.
Galen, C. and Kevan, P. G. (1980) Scent and color, floral polymorphisms and pollination biology in *Polemonium viscosum* Nutt. *Am. Midl. Nat.*, **104**, 281–9.
Galen, C. and Kevan, P. G. (1983) Bumblebee foraging and floral scent dimorphisms: *Bombus kirbyellis* Curtis (Hymenoptera: Apidae) and *Polemonium viscosum* Nutt. (Polemoniaceae). *Can. J. Zool.*, **61**, 1207–13.
Galil, J. and Eisikowitch, D. (1969) Further studies on the pollination ecology of *Ficus sycamorus* L. *Tijdschr. Ent.*, **112**, 1–13.
Ganders, F. R. (1974) Disassortative pollination in the distylous plant *Jepsonia heterandra*. *Can. J. Bot.*, **52**, 2401–6.
Ganders, F. R. (1975) Mating patterns in self-incompatible distylous populations of *Amsinckia* (Boraginaceae). *Can. J. Bot.*, **53**, 773–9.
Ganders, F. R. (1979) The biology of heterostyly. *N.Z. J. Bot.*, **17**, 607–35.

Gastel, A. J. G. van (1972) Spontaneous stylar part mutations in *Nicotiana alata* Link and Otto. *Incomp. Newslett. Assoc. EURATOM-ITAL, Wageningen*, **1**, 12–13.
Gastel, A. J. G. van (1974) Radiogenetics of self-incompatibility. *Ann. Rep. Comm. Europ. Comm., Progr. Biol. Health Protection* (1974).
Gastel, A. I. G. van and de Nettancourt, D. (1974) The effects of different mutagens on self-incompatibility in *Nicotiana alata* Link and Otto. 1. Chronic gamma radiation. *Radiat. Bot.*, **14**, 43–50.
Gastel, A. J. G. van and de Nettancourt, D. (1975) The effects of different mutagens on self-incompatibility in *Nicotiana alata* Link and Otto. II. Acute irradiations with X-rays and fast neutrons. *Heredity*, **34**, 381–92.
Geber, M. A. and Charnov, E. L. (1986) Sex allocation in hermaphrodites with partial overlap in male/female resource inputs. *J. Theor. Biol.*, **118**, 33–43.
Gentry, A. H. (1974) Flowering phenology and diversity in tropical Bignoniaceae. *Biotropica*, **6**, 64–8.
Gentry, A. H. (1976) Bignoniaceae of southern Central America: distribution and ecological specificity. *Biotropica*, **8**, 117–31.
Gerstel, D. U. (1950) Self-incompatibility studies in Guayule. II. Inheritance. *Genetics*, **35**, 482–506.
Gerwitz, A. and Faulkner, G. J. (1972) *National Vegetable Research Station 22nd Annual Report 1971*, 32. Wellesbourne, Warwick.
Gibbs, P. E. and Bianchi, M. (1993) Post-pollination events in species of *Chorisia* (Bombacaceae) and *Tabebuia* (Bignoniaceae) with late-acting self-incompatibility. *Bot. Acta*, **106**, 64–71.
Givnish, T. J. (1980) Ecological constraints on the evolution of breeding systems in seed plants: dioecy and dispersal in gymnosperms. *Evolution*, **34**, 959–72.
Gleaves, J. T. (1973) Gene flow mediated by wind borne pollen. *Heredity*, **31**, 355–66.
Glover, B. J. and Abbott, R. J. (1995) Low genetic diversity in the Scottish endemic *Primula scotica* Hook. *New Phytol.*, **129**, 147–53.
Godley, E. J. (1964) Breeding systems in New Zealand plants. 3. Sex ratios in some natural populations. *N.Z. J. Bot.*, **2**, 205–12.
Godley, E. J. (1975) Flora and vegetation, in *Biogeography and Ecology in New Zealand* (ed. G. Kuschel), W. Junk, The Hague, pp. 177–229.
Godley, E. J. (1979) Flower biology in New Zealand. *N.Z. J. Bot.*, **17**, 441–66.
Goldman, D. A. and Willson, M. F. (1986) Sex allocation in functionally hermaphroditic plants. A review and a critique. *Bot. Rev.*, **52**, 157–94.
Golenberg, E. M. (1987) Estimation by gene flow and genetic neighbourhood sie by indirect methods in a selfing annual *Triticum dicoccoides*. *Evolution*, **41**, 1326–34.
Golynskaya, E. L., Bashkirova, N. V. and Tomchuk, N. N. (1976) Phytohaemagglutenins of the pistil in *Primula* as possible proteins of generative incompatibility. *Sov. Pl. Physiol.*, **23**, 69–77.
Grant, V. (1952) Isolation and hybridisation between *Aquilegia formosa* and *A. pubescens*. *Aliso*, **2**, 341–60.
Grant, V. (1963) *The Origin of Adaptations*, Columbia University Press, New York.
Grant, V. (1981) *Plant Speciation*, 2nd edn. Columbia University Press, New York.
Grant, V. and Grant, K. A. (1965) *Flower Pollination in the Phlox Family*. Columbia University Press, New York.
Grant, V. and Grant, K. A. (1983) Behaviour of hawkmoths on flowers of *Datura meteloides*. *Bot. Gaz.*, **144**, 280–4.
Gray, A. J. (1987) Genetic change during succession, in *Colonization, Succession and Stability* (eds A. J. Gray, M. J. Crawley and P. J. Edwards), Blackwell, Oxford, pp. 274–93.

Grewal, M. S. and Ellis, J. R. (1972) Sex determination in *Potentilla fruticosa*. *Heredity*, **29**, 359–62.

Grimanelli, D., Leblanc, O., Gonzalez de Leon, D. and Savidan, Y. (1995) Mapping apomixis in tetraploid *Tripsacum*, preliminary results. *Apomixis Newsletter*, **8**, 37–9.

Grime, J. P. (1973) Competition and diversity in herbaceous vegetation. *Nature*, **244**, 311.

Gustafsson, A. (1944) The constitution of the *Rosa canina* complex. *Hereditas*, **30**, 408.

Gustafsson, A. (1946–7) Apomixis in higher plants. I–III. *Lunds Univ. Arsskr.*, **42**, 1–67, **43**, 69–179, 183–370.

Gustafsson, A. and Hakansson, A. (1942) Meiosis in some rose hybrids. *Bot. Notiser*. (1942), 331–42.

Ha, C. O., Sands, V. E., Soepadmo, E. and Jong, K. (1988) Reproductive patterns of selected understorey trees in the Malaysian rain forest: the apomictic species. *Bot. J. Linn. Soc.*, **97**, 317–31.

Hagerup, O. (1944) On fertilization, polyploidy, and haploidy in *Orchis maculatus*. *Dansk. Bot. Ark.*, **11**, 1–26.

Hagerup, O. (1945) Facultative parthenogenesis and haploidy in *Epipactis latifolia*. *Kl. Danske Vidensk Selsk.*, **19**, 1–13.

Hagerup, O. (1947) The spontaneous formation of haploid, polyploid and aneuploid embryos in some orchids. *Kl. Danske Vidensk Selsk.*, **20**, 1–22.

Haig, D. (1990) New perspectives on the angiosperm female gametophyte. *Bot. Rev.*, **56**, 236–74.

Haig, D. and Westoby, M. (1988) Inclusive fitness, seed resources and maternal care, in *Plant Reproductive Biology; Patterns and Strategies* (eds J. Lovett Doust and L. Lovett Doust), Oxford University Press, Oxford, pp. 60–79.

Haig, D. and Westoby, M. (1989) Parent-specific gene expression and the triploid endosperm. *Am. Nat.*, **134**, 147–55.

Hainsworth, F. R. and Wolf, L. L. (1972) Energetics of nectar extraction in a small, high altitude, tropical hummingbird, *Selasphorus flammula*. *J. Comp. Physiol.*, **80**, 377–87.

Haldane, J. B. S. (1922) Sex-ratio and unisexual sterility in hybrid animals. *J. Genet.*, **12**, 101–9.

Hamrick, J. L., Linhart, Y. B. and Mitton, J. B. (1979) Relationships between life history characteristics and electrophoretically detectable genetic variation in plants. *Annu. Rev. Ecol. Syst.*, **10**, 175–200.

Hamrick, J. L., Godt, M. J. W., Murawski, D. A. and Loveless, M. D. (1991) Correlations between species traits and allozyme diversity, in *Genetics and Conservation of Rare Plants*, (eds D. A. Falk and K. E. Holsinger), Oxford University Press, New York, pp. 75–86.

Handel, S. N. (1983) Pollination ecology, plant population stucture, and gene flow, in *Pollination Biology* (ed. L. Real), Academic Press, New York, pp. 165–211.

Handel, S. N. and Le Vie Mishkin, J. (1984) Temporal shifts in gene flow and seed set: evidence from an experimental population of *Cucumis sativus*. *Evolution*, **38**, 1350–7.

Hanna, W., Powell, J., Millot, J. and Burton, G. (1973) Cytology of obligate sexual plants in *Panicum maximum* Jacq. and their use in controlled hybrids. *Crop. Sci.*, **13**, 695–7.

Harberd, D. J. (1961) Observations on population structure and longevity of *Festuca rubra* L. *New Phytol.*, **60**, 184–206.

Harder, L. D. (1986) Effects of nectar concentration and flower depth on flower handling efficiency of bumble bees. *Oecologia*, **69**, 309–15.

Harder, L. D. and Barrett, S. C. H. (1995) Mating cost of large floral displays in hermaphrodite plants. *Nature*, **373**, 512–15.

Harder, L. D. and Barrett, S. C. H. (1996) Pollen dispersal and mating patterns in animal-pollinated plants, in *Floral Biology* (eds D. G. Lloyd and S. C. H. Barrett), Chapman & Hall, London, pp. 140–90.

Harder, L. D. and Thomson, J. D. (1989) Evolutionary options for maximizing pollen dispersal of animal-pollinated plants. *Am. Nat.*, **133**, 323–44.

Harding, J., Allard, R. W. and Smeltzer, D. G. (1966) Population studies in predominantly self-pollinated species. IX. Frequency-dependent selection in *Phaseolus lunatus*. *Proc. Natl. Acad. Sci. USA*, **56**, 99–104.

Harper, J. L. (1977) *The Population Biology of Plants*. Academic Press, London.

Hawkins, R. P. (1971) Selection for height of nectar in the corolla tube of English singlecut red clover. *J. Agric. Sci.*, **77**, 348–50.

Heinrich, B. (1972a) Temperature regulation in the bumblebee *Bombus vagans*: a field study. *Science*, **175**, 185–7.

Heinrich, B. (1972b) Energetics of temperature regulation and foraging in a bumblebee, *Bombus terricola* Kirby. *J. Comp. Physiol.*, **77**, 49–64.

Heinrich, B. (1975) Energetics of pollination. *Annu. Rev. Ecol. Syst.*, **6**, 139–70.

Heinrich, B. (1976) The foraging specializations of individual bumblebees. *Ecol. Monogr.*, **46**, 105–18.

Heinrich, B. (1979a) Resource heterogeneity and patterns of movement in foraging bumblebees. *Oecologia*, **40**, 235–45.

Heinrich, B. (1979b) *Bumblebee Economics*. Harvard University Press, Cambridge, MA.

Heinrich, B. (1979c) 'Majoring' and 'minoring' by foraging bumblebees, *Bombus vagans*: an experimental analysis. *Ecology*, **60**, 245–55.

Heinrich, B. and Raven, P. H. (1972) Energetics and pollination ecology. *Science*, **176**, 597–602.

Heitz, B. (1973) Heterostylie et speciation dans le groupe *Linum perenne*. *Ann. Sci. Nat. Bot. Biol. Veg.*, **14**, 385–405.

Henderson, A. (1986) A review of pollination studies in the Palmae. *Bot. Rev.*, **52**, 221–59.

Hendrix, S. D. (1988) Herbivory and its impact on plant reproduction, in *Plant Reproductive Ecology* (eds J. Lovett Doust and L. Lovett Doust), Oxford University Press, Oxford, pp. 246–66.

Herrera, C. H. (1993) Selection on complexity of corolla outline in a hawkmoth-pollinated violet. *Evol. Tr. Pl.*, **7**, 9–13.

Herrera, C. H. (1996) Floral traits and plant adaptation to insect pollinators: a devil's advocate approach, in *Floral Biology* (eds D. G. Lloyd and S. C. H. Barrett), Chapman & Hall, London, pp. 65–87.

Heslop-Harrison, J. (1957) The experimental modification of sex expression in flowering plants. *Biol. Rev. Camb. Phil. Soc.*, **32**, 38–90.

Heslop-Harrison, J. (1975a) The physiology of the pollen grain surface. *Proc. R. Soc. Lond. B*, **190**, 275–99.

Heslop-Harrison, J. (1975b) Incompatibility and the pollen stigma interaction. *Annu. Rev. Pl. Physiol.*, **26**, 403–25.

Heslop-Harrison, J. (1978) Genetics and physiology of angiosperm incompatibility systems. *Proc. R. Soc. Lond. B*, **202**, 73–92.

Heslop-Harrison, J. (1979a) Aspects of the structure, cytochemistry and germination of the pollen of rye (*Secale cereale* L.). *Ann. Bot.*, **44** (Suppl.), 1–7.

Heslop-Harrison, J. (1979b) An interpretation of the hydrodynamics of pollen. *Am. J. Bot.*, **66**, 737–43.

Heslop-Harrison, J. (1982) Pollen–stigma interaction and cross-incompatibility in the grasses. *Science*, **215**, 1358–64.

Heslop-Harrison, J. (1983) Self-incompatibility: phenonemology and physiology. *Proc. R. Soc. Lond. B*, **218**, 371–95.

Heslop-Harrison, J. and Heslop-Harrison, Y. (1982) The pollen–stigma interaction in the grasses. 4. An interpretation of the self-incompatibility response. *Acta Bot. Neerl.*, **31**, 429–39.

Heslop-Harrison, J., Knox, R. B. and Heslop-Harrison, Y. (1974) Pollen-wall proteins: exine-held fractions associated with the incompatibility response in Cruciferae. *Theor. Appl. Genet.*, **44**, 133–7.

Heslop-Harrison, J. W. (1924) Sex in the Salicaceae and its modification by eriophyid mites and other influences. *Br. J. Exp. Biol.*, **1**, 445–72.

Heslop-Harrison, Y., Heslop-Harrison, J. and Shivanna, K. R. (1981) Heterostyly in *Primula*. 1. Fine-structural and cytochemical features of the stigma and style in *Primula vulgaris* Huds. *Protoplasma*, **107**, 171–87.

Hickey, L. J. and Doyle, J. A. (1977) Early Cretaceous fossil evidence for Angiosperm evolution. *Bot. Rev.*, **43**, 3–93.

Hildebrand, F. (1863) *De la variation des animaux et des plantes a l'état domestique*, C. Reinwald, Paris.

Hodgkin, T. and Lyon, G. D. (1984) Pollen germination inhibition in extracts of *Brassica oleracea* stigmas. *New Phytol.*, **96**, 293–8.

Hogenboom, N. G. (1972) Breaking breeding barriers in *Lycopersicon*. 1, 2, 3, 4, 5. *Euphytica*, **21**, 221–7, 228–43, 244–56, 397–404, 405–14.

Holsinger, K. E. (1992) Ecological models of plant mating systems and the evolutionary stability of mixed mating systems, in *Ecology and Evolution of Plant Reproduction* (ed. R. Wyatt), Chapman & Hall, London, pp. 169–91.

Holsinger, K. E. and Gottlieb, L. D. (1991) Conservation of rare and endangered plants: principles and prospects, in *Genetics and Conservation of Rare Plants*, (eds D. A. Falk and K. E. Holsinger), Oxford University Press, New York, pp. 195–208.

Hooper, J. E. and Peloquin, S. J. (1968) X-ray inactivation of the stylar component of the self-incompatibility reaction in *Lilium longiflorum*. *Can. J. Gen. Cytol.*, **10**, 941–4.

Horowitz, A. and Beiles, A. (1980) Gynodioecy as a possible population strategy for increasing reproductive output. *Theor. Appl. Genet.*, **57**, 11–15.

Horowitz, A. and Harding, J. (1972) The concept of male outcrossing in hermaphrodite higher plants. *Heredity*, **29**, 223–36.

Howlett, B. M., Knox, R. B., Paxton, J. H. and Heslop-Harrison, J. (1975) Pollen-wall proteins: physicochemical characterisation and role in self-incompatibility in *Cosmos bipinnatus*. *Proc. R. Soc. Lond. B*, **188**, 167–82.

Hughes, J. (1987) Variability in Sexual and Apomictic *Taraxacum* Weber. PhD. Thesis, University of Newcastle upon Tyne, UK.

Hughes, J. and Richards, A. J. (1988) The genetic structure of populations of sexual and asexual *Taraxacum* (dandelions). *Heredity*, **60**, 161–71.

Hughes, J. and Richards, A. J. (1989) Isozymes and the status of *Taraxacum* agamospecies. *Bot. J. Linn. Soc.*, **99**, 365–76.

Hughes, M. B. and Babcock E. B. (1950) Self-incompatibility in *Crepis foetida* L. subsp. *rhoedaifolia*. *Genetics*, **35**, 570–88.

Ibrahim, H. (1979) Population studies in *Primula veris* L. and *P. vulgaris* Huds. PhD thesis, University of Newcastle upon Tyne.

Idris, S. and Rukayah, A. (1987) Description of the male mangosteen (*Garcinia mangostana* L.) discovered in peninsular Malaya. *MARDI Res. Bull.*, **15**, 63–6.

Ingram, R., Weir, J. and Abbott, R. J. (1980) New evidence concerning the origin of inland radiate groundsel, *S. vulgaris* L. var. *hibernicus* Syme. *New Phytol.*, **84**, 543–6.

Inouye, D. W. (1980) The effect of proboscis and corolla tube lengths on patterns and rates of flower visitation by bumblebees. *Oecologia*, **45**, 197–201.

Irish, E. E. and Nelson, T. (1989) Sex determination in monoecious and dioecious plants. *Plant Cell*, **1**, 737–44.

Irwin, J. A. and Abbott, R. J. (1992) Morphometric and isozyme evidence for the hybrid origin of a new tetraploid radiate groundsel in York, England. *Heredity*, **69**, 431–9.

Jacob, F. and Monod, J. (1961) Genetic regulatory mechanisms in the synthesis of proteins. *J. Mol. Biol.*, **3**, 318–56.

Janick, J. and Stevenson, E. C. (1955) Genetics of the monoecious character in spinach. *Genetics*, **40**, 429–37.

Jankun, A. (1994) Embryological studies on *Sorbus intermedia* (Rosaceae). *Pol. Bot. St.*, **8**, 69–74.

Janzen, D. H. (1971) Euglossine bees as long-distance pollinators of tropical plants. *Science*, **171**, 203–5.

Janzen, D. H. (1977) What are dandelions and aphids? *Am. Nat.*, **111**, 586–9.

Jarne, P. and Charlesworth, D. (1993) The evolution of the selfing rate in functionally hermaphrodite plants and animals. *Annu. Rev. Ecol. Syst.*, **24**, 441–66.

Jenniskens, M.-J., den Nijs, J. C. M. and Huizing, B. A. (1984) Karyogeography of *Taraxacum* sect. *Taraxacum* and the possible occurrence of facultative agamospermy in Bavaria and north-west Austria. *Phyton*, **24**, 11–34.

Jewell, J., McKee, J. and Richards, A. J. (1994) The keel colour polymorphism in *Lotus corniculatus* L: differences in internal flower temperature. *New Phytol.*, **128**, 363–8.

Johnson, S. D. (1994) Evidence for Batesian mimicry in a butterfly-pollinated orchid. *Biol. J. Linn. Soc.*, **53**, 91–104.

Johri, B. M. (1981) Transfer cells: their role in reproductive structures of Angiosperms. *Abstracts, XIII International Botanical Congress, Sydney*, **61**.

Jones, D. A., Compton, S. G., Crawford, T. J., Ellis, W. M. and Taylor, I. M. (1986) Variation in the colour of the keel petals in *Lotus corniculatus* L. 3. Pollination, herbivory and seed production. *Heredity*, **57**, 101–12.

Kakizaki, Y. (1930) Studies on the genetics and physiology of self- and cross-incompatibility in the common cabbage. *Jap. J. Bot.*, **5**, 135–208.

Kandelaki, G. V. (1976) Remote hybridization and the phenomenon of pseudogamy, in *Apomixis and Breeding* (ed. S. S. Khokhlov), Amerind, New Delhi, pp. 179–89.

Kannenberg, L. W. and Allard, R. W. (1967) Population studies in predominantly self-pollinated species. VIII. Genetic variability in the *Festuca microstachys* complex. *Evolution*, **21**, 227–40.

Karron, J. D. (1987) A comparison of levels of genetic polymorphism and self-compatibility in geographically restricted and widespread plant congeners. *Evol. Ecol.*, **1**, 47–58.

Karron, J. D. (1989) Breeding systems and levels of inbreeding depression in geographically restricted and widespread species of *Astragalus* (Fabaceae). *Am. J. Bot.*, **76**, 331–40.

Karron, J. D. (1991) Patterns of genetic variation and breeding systems in rare plant species, in *Genetics and Conservation of Rare Plants* (eds D. A. Falk and K. E. Holsinger), Oxford University Press, New York, pp. 87–98.

Kato, M. (1995) The aspidistra and the amphipod. *Nature*, **377**, 293.

Kaur, A., Ha, C. D., Jong, K., Sands, V. E., Chan, H., Soepadmo, E. and Ashton, P. S. (1978) Apomixis may be widespread among trees of the climax rain forest. *Nature*, **271**, 440–1.

Kaur, A., Jong, K., Sands, V. E. and Soepadmo, E. (1986) Cytoembryology of some Malaysian Dipterocarps, with some evidence of apomixis. *Bot. J. Linn. Soc.*, **92**, 75–88.

Kay, Q. O. N. (1978) The role of preferential and assortative pollination in the maintenance of flower colour polymorphisms, in *The Pollination of Flowers by Insects* (ed. A. J. Richards), Academic Press, London, pp. 175–90.

Kay, Q. O. N. (1982) Intraspecific discrimination by pollinators and its role in evolution, in *Pollination and Evolution* (eds J. A. Armstrong, J. M. Powell and A. J. Richards), Publ. Royal Botanic Gardens, Sydney, pp. 9–28.

Kay, Q. O. N. (1985a) Nectar from willow catkins as a food source for blue tits. *Bird Study*, **32**, 41–5.

Kay, Q. O. N. (1985b) Hermaphrodites and subhermaphrodites in a reputedly dioecious plant, *Cirsium arvense* (L.) Scop. *New Phytol.*, **100**, 457–72.

Kay, Q. O. N. (1987a) Ultraviolet patterning and ultraviolet-absorbing pigments in flowers of the Leguminosae, in *Advances in Legume Systematics* vol. 3, (ed. C. H. Stirton), Royal Botanic Gardens, Kew, pp. 817–53.

Kay, Q. O. N. (1987b) The comparative ecology of flowering. *New Phytol.*, **106** (suppl.), 265–81.

Kay, Q. O. N. and Stevens, D. P. (1986) The frequency, distribution and reproductive biology of dioecious species in the native flora of Britain and Ireland. *Bot. J. Linn. Soc.*, **92**, 39–64.

Kay, Q. O. N., Dauoud, H. S. and Stirton, C. H. (1981) Pigment distribution, light reflection and cell structure in petals. *Bot. J. Linn. Soc.*, **83**, 57–84.

Kay, Q. O. N., Jack, A. J., Bamber, F. C. and Davies, C. R. (1984) Differences in floral morphology, nectar production and insect visits in a dioecious species, *Silene dioica*. *New Phytol.*, **98**, 515–29.

Kenrick, J. and Knox, R. B. (1985) Self-incompatibility in the nitrogen-fixing tree *Acacia retinoides*: quantitative cytology of pollen tube growth. *Theor. Appl. Gen.*, **69**, 481–8.

Kerster, H. W. (1964) Neighbourhood size in the rusty lizard, *Sceloporus olivaceus*. *Evolution*, **18**, 445–57.

Kerster, H. W. and Levin, D. A. (1968) Neighbourhood size in *Lithospermum carolinense*. *Genetics*, **60**, 577–87.

Kevan, P. G. (1972a) Floral colors in the high arctic with reference to insect flower relation and pollination. *Can. J. Bot.*, **50**, 2289–316.

Kevan, P. G. (1972b) Insect pollination in high arctic flowers. *J. Ecol.*, **60**, 831–47.

Kevan, P. G. (1975) Sun-tracking solar furnaces in high arctic flowers: significance for pollination and insects. *Science*, **189**, 732–6.

Kevan, P. G. (1978) Floral coloration, its colorimetric analysis and significance in anthecology, in *The Pollination of Flowers by Insects* (ed. A. J. Richards), Academic Press, London, pp. 51–78.

Kevan, P. G. (1984) Pollination by animals and angiosperm biosystematics, in *Plant Biosystematics* (ed. W. F. Grant), Academic Press, London.

Kevan, P. G. (1989) Thermoregulation in arctic insects and flowers: adaptation and

counter-adaptation in behaviour, anatomy and physiology, in *Thermal Physiology 1989* (ed. J. B. Mercer), Elsevier, Amsterdam, pp. 747–53.
Kevan, P. G. (1990) Sexual differences in temperature of blossoms on a dioecious plant, *Salix arctica*: significance for life in the arctic. *Arctic Alpine Res.*, **22**, 283–9.
Kevan, P. G. and Baker, H. G. (1983) Insects as flower visitors and pollinators. *Ann. Rev. Entomol.*, **28**, 407–53.
Kevan, P. G. and Lack, A. J. (1986) Pollination in a cryptically dioecious plant *Decaspermum parvifolium* (Lam.) in Indonesia. *Biol. J. Linn. Soc.*, **25**, 319–30.
Kevan, P. G., Eisikowitch, D. and Rathwell, B. (1989) The role of nectar in the germination of pollen in *Asclepias syriaca*. *Bot. Gaz.*, **150**, 266–70.
Kevan, P. G., Eisikowitch, D., Amrose, J. D. and Kemp, J. R. (1990) Cryptic dioecy and insect pollination in *Rosa setigera* Michx. (Rosaceae), a rare plant in Carolinian Canada. *Biol. J. Linn. Soc.*, **40**, 229–43.
King, L. M. and Schaal, B. A. (1990) Genotype variation within asexual lineages of *Taraxacum officinale*. *Proc. Natl. Acad. Sci. USA*, **87**, 998–1002.
Kirschner, J. and Stepanek, J. (1993) Clonality as part of the evolutionary process in *Taraxacum*. *Folia Geobot. Phytotax.*, **29**, 265–75.
Knight, R. and Rogers, H. H. (1955) Incompatibility in *Theobroma cacao*. *Heredity*, **9**, 69–77.
Knox, R. B. (1967) Apomixis: seasonal and population differences in a grass. *Science*, **157**, 325–6.
Knox, R. B. and Heslop-Harrison, J. (1963) Experimental control of aposporous apomixis in a grass of the Andropogoneae. *Bot. Notiser*, **116**, 127–41.
Knox, R. B., Willing, R. and Ashford, A. E. (1972) Role of pollen-wall proteins as recognition substances in interspecific incompatilbility in poplars. *Nature*, **237**, 381–3.
Knox, R. B., Clarke, A. E., Harrison, S., Smith, P. and Marchalonis, J. J. (1976) Cell recognition in plants: determinants of the stigma surface and their pollen interactions. *Proc. Natl. Acad. Sci. USA*, **73**, 2788–92.
Knuth, P. (1906–9) *Handbook of Flower Pollination*. Transl. J. R. Ainsworth Davis (3 vols., I, 1906, II, 1908, III, 1909). Oxford University Press, Oxford.
Kochmer, J. P. and Handel, S. N. (1986) Constraints and competition in the evolution of flowering phenology. *Ecol. Monogr.*, **56**, 303–25.
Kolreuter, I. G. (1763) Vorlaufige Nachricht von einigen das Geschlecht der Pflanzen betreffenden Versuchen und Beobachtungen, nebst Fortsetzungen 1, 2 v. 3, 266. *Ostwald's Klassiker*, 41. Engelmann, Leipzig.
Krebs, J. R. (1978) Optimal foraging, in *Behavioural Ecology: an Evolutionary Approach* (eds J. R. Krebs and N. B. Davies), Blackwell Scientific, Oxford.
Krohne, D. T., Baker, I. and Baker, H. G. (1980) The maintenance of the gynodioecious breeding system in *Plantago lanceolata* L. *Am. Midl. Nat.*, **103**, 269–79.
Kuhn, E. (1939) Selbstbestaubungen subdioeischer Blutenpflanzen, ein neuer Beweis für die genetische Theorie der Geschlechtsbestimmung. *Planta*, **30**, 457–70.
Kullenberg, B. (1956) On the scents and colours of *Ophrys* flowers and their specific pollinators among the aculeate Hymenoptera. *Svensk Bot. Tidskr.*, **50**, 25–46.
Kurian, V. (1996) Investigations into the breeding system supergene in *Primula*. PhD Thesis, University of Newcastle upon Tyne, UK.
Kurian, V. and Richards, A. J. (1997) A new recombinant in the heteromorphy '*S*' supergene in *Primula*. *Heredity*, **78**, in press.

Lahav-Ginott, S. and Cronk, Q. C. B. (1993) The mating system of *Elatostema* (Urticaceae) in relation to morphology: a comparative study. *Pl. Syst. Evol.*, **186**, 135–45.

Lande, R. and Schemske, D. W. (1985) The evolution of self-fertilization and inbreeding depression in plants. I. Genetic models. *Evolution*, **39**, 24–40.
Lanza, J., Smith, G. C., Sack, S. and Cash, A. (1995) Variation in nectar volume and composition in *Impatiens capensis* at the individual, plant and populational levels. *Oecologia*, **102**, 113–19.
Law, R., Bradshaw, A. D. and Putwain, P. D. (1977) Life-history variation in *Poa annua*. *Evolution*, **31**, 233–46.
Lawrence, M. J., Fearon, C. H., Cornish, M. A. and Hayward, M. D. (1983) The genetical control of self-incompatibility in rye grasses. *Heredity*, **51**, 461–6.
Lawrence, M. J., Marshall, D. F., Curtis, V. E. and Fearon, C. H. (1985) Gametophytic self-incompatibilty re-examined: a reply. *Heredity*, **54**, 131–8.
Lawton, J. H. (1973) The energy cost of 'food-gathering', in *Resources and Population* (eds B. Benjamin, P. R. Cox and J. Peel), Academic Press, London, pp. 59–76.
Ledig, F. T. (1986) Heterozygosity, heterosis, and fitness in outbreeding plants, in *Conservation Biology, the Science of Scarcity and Diversity* (ed. F. T. Soulie), Sinauer, Sunderland, USA, pp. 77–95.
Lee, H.-S., Huang, S. and Kao, T.-H. (1994) S protein control rejection of incompatible pollen in *Petunia inflata*. *Nature*, **367**, 560–3.
Leereveld, H. (1984) Anthecological relations between reputedly anemophilous flowers and syrphid flies. VI. Aspects of the anthecology of Cyperaceae and *Sparangium erectum* L. *Acta Bot. Neerl.*, **33**, 475–82.
Lehmann, E. (1926) The heredity of self-sterility in *Veronica syriaca*. *Mem. Hort. Soc. NY*, **3**, 315–20.
Lepart, J. and Dommee, B. (1992) Is *Phillyrea angustifolia* L. (Oleaceae) an androdioecious species? *Bot. J. Linn. Soc.*, **108**, 375–87.
Levin, D. A. (1969) The effect of corolla color and outline on interspecific pollen flow in *Phlox*. *Evolution*, **23**, 444–55.
Levin, D. A. (1972) Low frequency disadvantage in the exploitation of pollinators by corolla variants in *Phlox*. *Am. Nat.*, **104**, 455–67.
Levin, D. A. (1978) Pollinator behaviour and the breeding structure of plant populations, in *The Pollination of Flowers by Insects* (ed. A. J. Richards), Academic Press, London, pp. 135–50.
Levin, D. A. (1979) Pollinator foraging behaviour: genetic implications for plants, in *Topics in Plant Population Biology* (eds O. T. Solbrig, S. Jain, G. B. Johnson and P. H. Raven), Columbia University Press, New York, pp. 131–53.
Levin, D. A. and Berube, D. E. (1972) *Phlox* and *Colias*: the efficiency of a pollination system. *Evolution*, **26**, 242–50.
Levin, D. A. and Kerster, H. W. (1968) Local gene dispersal in *Phlox*. *Evolution*, **22**, 130–9.
Levin, D. A. and Kerster, H. W. (1969a) The dependence of bee-mediated pollen and gene dispersal upon plant density. *Evolution*, **23**, 560–71.
Levin, D. A. and Kerster, H. W. (1969b) Density-dependent gene dispersal in *Liatris*. *Am. Nat.*, **103**, 61–74.
Levin, D. A. and Kerster, H. W. (1971) Neighbourhood structure in plants under diverse reproductive methods. *Am. Nat.*, **105**, 345–54.
Levin, D. A. and Kerster, H. W. (1973) Assortative pollination for stature in *Lythrum salicaria*. *Evolution*, **27**, 144–52.
Levin, D. A. and Kerster, H. W. (1974) Gene flow in seed plants. *Evol. Biol.*, **7**, 139–220.
Levin, D. A. and Watkins, L. (1984) Assortative mating in *Phlox*. *Heredity*, **53**, 595–602.
Levin, D. A. and Anderson, G. J. (1986) Evolution of dioecy in an American

Solanum, in *Solanaceae: Biology and Systematics* (ed. W. G. D'Arcy), Columbia University Press, New York, pp. 264–73.
Lewis, D. (1941) Male sterility in natural populations of hermaphrodite plants. *New Phytol.*, **40**, 56–63.
Lewis, D. (1942a) The evolution of sex in flowering plants. *Biol. Rev.*, **17**, 46–67.
Lewis, D. (1942b) The physiology of incompatibility in plants. I. The effect of temperature. *Proc. R. Soc. Lond. B*, **131**, 13–26.
Lewis, D. (1947) Competition and dominance of incompatibility alleles in diploid pollen. *Heredity*, **1**, 85–108.
Lewis, D. (1949a) Incompatibility in flowering plants. *Biol. Rev.*, **24**, 427–69.
Lewis, D. (1949b) Structure of the incompatibility gene. II. Induced mutation rate. *Heredity*, **3**, 339–55.
Lewis, D. (1952) Serological reactions of pollen incompatibility substances. *Proc. R. Soc. Lond. B*, **140**, 127–35.
Lewis, D. (1954) Comparative incompatibility in Angiosperms and Fungi. *Adv. Genet.* **6**, 235–45.
Lewis, D. (1955) Sexual incompatibility. *Sci. Prog.*, **172**, 593–605.
Lewis, D. (1960) Genetic control of specificity and activity of the S antigen in plants. *Proc. R. Soc. Lond. B*, **151**, 468–77.
Lewis, D. (1964) A protein dimer hypothesis on incompatibility. *Proc. 11th Int. Congr. Genet.*, The Hague 1963, in *Genetics Today* (ed. S. J. Geerts), **3**, 656–63.
Lewis, D. (1975) Heteromorphic incompatibility system under disruptive selection. *Proc. R. Soc. London* B **188**, 247–56.
Lewis, D. and Crowe, L. K. (1953) Theory of revertible mutation. *Nature*, **171**, 501.
Lewis, D. and Crowe, L. K. (1954) Structure of the incompatibility gene IV. Types of mutation in *Prunus avium* L. *Heredity*, **8**, 357–63.
Lewis, D. and Crowe, L. K. (1956) The genetics and evolution of gynodioecy. *Evolution*, **10**, 115–25.
Lewis, D. and Crowe, L. K. (1958) Unilateral interspecific incompatibility in flowering plants. *Heredity* **12**, 232–56.
Lewis, D. and Jones, D. A. (1993) The genetics of heterostyly, in *Evolution and Function of Heterostyly* (ed. S. C. H. Barrett), Springer-Verlag, Berlin, pp. 129–50.
Lewis, D., Verma, S. C. and Zuberi, M. I. (1988) Gametophytic–sporophytic incompatibility in the Cruciferae – *Raphanus sativus*. *Heredity*, **61**, 355–66.
Lewis, H. (1962) Catastrophic selection as a factor in speciation. *Evolution*, **16**, 257–71.
Liljefors, A. (1955) Cytological studies in *Sorbus*. *Acta Hort. Berg.*, **17**, 47–113.
Linder, R. and Linskens, H. F. (1972) Evolution des acides amines dans le style d'*Oenothera missouriensis* vierge, autopollinise et xenopollinise. *Theor. Appl. Genet.*, **42**, 125–9.
Linhart, Y. B. (1973) Ecological and behavioural determinants of pollen dispersal in hummingbird-pollinated *Heliconia*. *Am. Nat.*, **107**, 511–23.
Linskens, H. F. (1960) Zurfrage der Entstehung der Abwehrkorper bei der Inkompatibilitätsreaktion von *Petunia*. III. Mitteilung: Serologische Teste mit Leitgewebs – und der Pollenextrakten. *Zeit. Bot.*, **48**, 126–35.
Linskens, H. F., Schrauwen, J. A. M. and van der Donk, M. (1960) Uberwindung der Selbstinkompatibilität durch Rontgenbestrahlung des Griffels. *Naturwissenschaften*, **46**, 547.
Liston, A., Rieseberg, L. H. and Elias, T. S. (1990) Functional androdioecy in the flowering plant *Datisca glomerata*. *Nature*, **343**, 641–2.
Lloyd, D. G. (1969) Petal colour polymorphism in *Leavenworthia* (Cruciferae). *Contr. Gray Herbarium Harvard*, **198**, 9–40.

Lloyd, D. G. (1972a) Breeding systems in *Cotula* L. (Compositae, Anthemideae). I. The array of monoclinous and diclinous systems. *New Phytol.*, **71**, 1181–94.
Lloyd, D. G. (1972b) Breeding systems in *Cotula* L. (Compositae, Anthemideae). 2. Monoecious populations. *New Phytol.*, **71**, 1195–2002.
Lloyd, D. G. (1974a) Female-predominant sex ratios in angiosperms. *Heredity*, **32**, 35–44.
Lloyd, D. G. (1974b) Theoretical sex ratios of dioecious and gynodioecious angiosperms. *Heredity*, **32**, 11–34.
Lloyd, D. G. (1975a) The maintenance of gynodioecy and androdioecy in Angiosperms. *Genetica*, **45**, 325–39.
Lloyd, D. G. (1975b) Breeding systems in *Cotula*. III. Dioecious populations. *New Phytol.*, **74**, 109–23.
Lloyd, D. G. (1980a) Sexual strategies in plants. I. An hypothesis of serial adjustment of maternal investment during one reproductive session. *New Phytol.*, **85**, 265–73.
Lloyd, D. G. (1980b) Benefits and handicaps of sexual reproduction, in *Evolutionary Biology* (eds M. K. Hecht, W. C. Steere and B. Wallace), Plenum Press, New York, pp. 69–110.
Lloyd, D. G. (1984) Gender allocations in outcrossing cosexual plants, in *Perspectives on Plant Population Ecology* (eds R. Dirzo and J. Sarukhan), Sinauer, Sunderland, MA, pp. 277–300.
Lloyd, D. G. (1992) Evolutionarily stable strategies of reproduction in plants: who benefits and how? in *Ecology and Evolution of Plant Reproduction* (ed. R. Wyatt), Chapman & Hall, London, pp. 137–68.
Lloyd, D. G. and Webb, C. J. (1977) Secondary sex characters in plants. *Bot. Rev.*, **43**, 177–216.
Lloyd, D. G. and Barrett, S. C. H. (eds) (1996) *Floral Biology*. Chapman & Hall, London.
Lloyd, D. G. and Webb, C. J. (1993a) The evolution of heterostyly, in *Evolution and Function of Heterostyly* (ed. S. C. H. Barrett), Springer-Verlag, Berlin, pp. 151–78.
Lloyd, D. G. and Webb, C. J. (1993b) The selection of heterostyly, in *Evolution and Function of Heterostyly* (ed. S. C. H. Barrett), Springer-Verlag, Berlin, pp. 179–207.
Lloyd, D. G., Webb, C. J. and Dulberger, R. (1990) Heterostyly in species of *Narcissus* (Amaryllidaceae), *Hugonia* (Linaceae) and other disputed cases. *Pl. Syst. Evol.*, **172**, 215–27.
Lokki, J., Saura, A., Kankinen, P. and Suomalainen, E. (1976a) Genetic polymorphism and evolution in parthenogenetic animals. V. *Genet. Res.*, **28**, 27–36.
Lokki, J., Saura, A., Kankinen, P. and Suomalainen, E. (1976b) Genetic polymorphism and evolution in parthenogenetic animals. VI. *Hereditas*, **82**, 209–16.
Lopez-Portillo, J., Eguiarte, L. E. and Montana, C. (1993) Nectarless honey mesquites. *Function. Evol.*, **7**, 452–61.
Lord, E. M. (1981) Cleistogamy: a tool for the study of floral morphogenesis, function and evolution. *Bot. Rev.*, **47**, 421–49.
Love, A. (1944) Cytogenetic studies on *Rumex* subgenus *Acetosella*. *Hereditas*, **30**, 1–136.
Love, A. and Sarker, N. (1956) Cytotaxonomy and sex determination of *Rumex paucifolius*. *Can. J. Bot.*, **34**, 261–8.
Lundqvist, A. (1956) Self-incompatibility in rye. I. Genetic control in the diploid. *Hereditas*, **42**, 295–348.

Lundqvist, A. (1960) The origin of self-compatibility in rye. *Hereditas*, **46**, 1–19.

Lundqvist, A. (1961) A rapid method for the analysis of incompatibilities in grasses. *Hereditas*, **47**, 705–7.

Lundqvist, A. (1962) The nature of the two-loci incompatibility system in grasses. I. The hypothesis of a duplicative origin. *Hereditas*, **48**, 153–68.

Lundqvist, A. (1964) The nature of the two-loci incompatibility system in grasses. IV. Interaction between the loci in relation to pseudocompatibility in *Festuca pratensis* Huds. *Hereditas*, **52**, 221–34.

Lundqvist, A. (1968) The mode of origin of self-fertility in grasses. *Hereditas*, **59**, 415–26.

Lundqvist, A. (1994) The self-incompatibility system in *Ranunculus repens* (Ranunculaceae). *Hereditas*, **120**, 151–7.

Lundqvist, A., Osterbye, U., Larsen, K. and Linde-Laursen, I. (1973) Complex self-incompatibility systems in *Ranunculus acris* L. and *Beta vulgaris* L. *Hereditas*, **74**, 161–8.

Lyman, J. C. and Ellstrand, N. C. (1984) Clonal diversity in *Taraxacum officinale* (Compositae), an apomict. *Heredity*, **50**, 1–10.

McClure, B. A., Gray, J. E., Anderson, M. A. and Clarke, A. E. (1990) Self-incompatibility in *Nicotiana alata* involves degradation of pollen rRNA. *Nature*, **347**, 757–60.

McCraw, J. M. and Spoor, W. (1983a) Self-incompatibility in *Lolium* species. I. *Lolium rigidum* Gaud. and *L. multiflorum* L. *Heredity*, **50**, 21–7.

McCraw, J. M. and Spoor, W. (1983b) Self-incompatibility in *Lolium* species. 2. *Lolium perenne* L. *Heredity*, **50**, 29–33.

McCusker, A. (1962) Gynodioecism in *Leucopogon melaleucoides*. *Proc. Linn. Soc. NSW*, **87**, 286–9.

Macior, L. W. (1966) Foraging behaviour of *Bombus* (Hymenoptera: Apidae) in relation to *Aquilegia* pollination. *Am. J. Bot.*, **53**, 302–9.

Macior, L. W. (1982) Plant community and pollinator dynamics in the evolution of pollination mechanisms in *Pedicularis* (Scrophulariaceae), in *Pollination and Evolution* (eds J. A. Armstrong, J. M. Powell and A. J. Richards), Royal Botanic Gardens, Sydney, pp. 29–45.

Macior, L. W. (1983) The pollination dynamics of sympatric species of *Pedicularis* (Scrophulariaceae). *Am. J. Bot.*, **70**, 844–53.

Macior, L. W. (1986) Floral resource sharing by bumblebees and hummingbirds in *Pedicularis* (Scrophulariaceae) pollination. *Bull. Torrey Bot. Club*, **113**, 101–9.

McKee, J. and Richards, A. J. (1996) Variation in seed production and germinability in common reed (*Phragmites australis* (Cav.) Trin. ex Steud) in Britain and France with respect to climate. *New Phytol.*, **133**, 233–43.

McLean, R. C. and Ivimey-Cook, W. R. (1956) *Textbook of Theoretical Botany*, vol. 2. Longman, London.

McNeilly, T. (1968) Evolution in closely adjacent plant populations. III. *Agrostis tenuis* on a small copper mine. *Heredity*, **23**, 99–108.

McNeilly, T. and Antonovics, J. A. (1968) Evolution in closely adjacent plant populations. IV. Barriers to gene flow. *Heredity*, **23**, 205–18.

McNeilly, T. and Roose, M. L. (1984) The distribution of perennial ryegrass genotypes in swards. *New Phytol.*, **98**, 501–13.

Maheshwari, P. (1949) The male gametophyte of Angiosperms. *Bot. Rev.*, **15**, 1–75.

Maheshwari, P. and Rangaswamy, N. S. (1965) Embryology in relation to genetics, *Adv. Bot. Res.*, **2**, 219–321.

Malecka, J. (1965) Embryological studies in *Taraxacum palustre*. *Acta Biol. Crac.*, **8**, 223–235.

Malecka, J. (1967) Cytoembryological studies in *Taraxacum scanicum* Dt. *Acta Biol. Crac.*, **10**, 195–206.

Malecka, I. (1971) Cytotaxonomic and embryological investigations on a natural hybrid between *Taraxacum kok-saghyz* Rodin and *T. officinale* Web. and their putative parent species. *Acta Biol. Crac.*, **14**, 179–96.

Malecka, I. (1973) Problems in the mode of reproduction in microspecies of *Taraxacum* section *Palustria* Dahlstedt. *Acta Biol. Crac. ser. Bot.*, **16**, 37–84.

Malepszy, S. and Niemirowicz-Szczytt, K. (1991) Sex determination in the cucumber (*Cucumis sativa*) as a model system for molecular biology. *Plant Sci. (Limerick)*, **80**, 39–48.

Manicacci, D. and Barrett, S. C. H. (1995) Stamen elongation, pollen size, and siring ability in tristylous *Eichhornia paniculata* (Pontederiaceae). *Am. J. Bot.*, **82**, 1381–9.

Manton, I. (1950) *Problems of Cytology and Evolution in the Pteridophyta*, Cambridge University Press, Cambridge.

Marshall, D. F. and Abbott, R. J. (1984a) Polymorphism for outcrossing frequency at the ray floret locus in *Senecio vulgaris* L. II. Confirmation. *Heredity*, **52**, 331–6.

Marshall, D. F. and Abbott, R. I. (1984b) Polymorphism for outcrossing frequency at the ray floret locus in *Senecio vulgaris* L. III. Causes. *Heredity*, **53**, 145–50.

Marshall, D. L. and Ellstrand, N. C. (1986) Sexual selection in *Raphanus sativus*: experimental data on nonrandom fertilization, maternal choice and consequences of multiple paternity. *Am. Nat.*, **127**, 446–61.

Marshall, D. L. and Folsom, M. W. (1992) Mechanisms of non-random mating in wild radish, in *Ecology and Evolution of Plant Reproduction*, (ed. R. Wyatt), Chapman & Hall, London, pp. 91–118.

Marshall, D. R. and Allard, R. W. (1970) Maintenance of isozyme polymorphism in natural populations of *Avena barbata*. *Genetics*, **66**, 393–9.

Mather, K. (1949) *Biometrical Genetics*, Methuen, London.

Mather, K. (1950) The genetical architecture of heterostyly in *Primula sinensis*. *Evolution*, **4**, 340–52.

Mather, K. and De Winton, D. (1941) Adaptation and counteradaptation of the breeding system in *Primula*. *Ann. Bot.*, II **5**, 297–311.

Mattson, O. (1983) The significance of exine oils in the initial interaction between pollen and stigma in *Armeria maritima*, in *Pollen Biology and Applications for Plant Breeding* (eds D. L. Mulcahy and E. Ottaviano), Elsevier, New York, pp. 257–67.

Matzk, F. (1991) A novel approach to differentiated embryos in the absence of endosperm. *Sex. Pl. Repr.*, **4**, 88–94.

Maynard Smith, J. (1971) The origin and maintenance of sex, in *Group Selection* (ed. G. C. Williams), Aldine Atherton, Chicago, pp. 163–75.

Maynard Smith, J. (1978) *The Evolution of Sex*, Cambridge University Press, Cambridge.

Mayo, O. and Hayman, D. L. (1968) The maintenance of two loci systems of gametophytically determined self-incompatibility. *Proc. 12th Int. Congr. Genet.* (ed. C. Oshima), p. 331.

Mayr, E. (1970) *Populations, Species and Evolution*, Belknap, Cambridge, MA.

Mazer, S. J. (1992) Environmental and genetic sources of variation in floral traits and phenotypic gender in wild radish, in *Ecology and Evolution of Plant Reproduction* (ed. R. Wyatt), Chapman & Hall, London, pp. 281–325.

Mazer, S. J. and Hultgaard, U.-M. (1993) Variation and covariation among floral traits within and among four species of northern European *Primula* (Primulaceae). *Am. J. Bot.*, **80**, 474–85.

Meagher, T. R. (1980) Population biology of *Chamaelirium luteum*, a dioecious lily. I. Spatial distributions of males and females. *Evolution*, **34**, 1127–37.

Meagher, T. R. (1981) Population biology of *Chamaelirium luteum*, a dioecious lily. II. Mechanisms governing sex ratios. *Evolution*, **35**, 557–67.

Meagher, T. R. (1986) Analysis of paternity within a natural population of *Chamaelirium luteum*. I. Identification of most-likely male parents. *Am. Nat.*, **128**, 199–215.

Meagher, T. R. (1991) Analysis of paternity with a natural population of *Chamaelirium luteum*. II. Patterns of male reproductive success. *Am. Nat.*, **137**, 738–52.

Meeuse, A. D. J. (1973) Anthecology and Angiosperm evolution, in *Taxonomy and Ecology* (ed. V. H. Heywood), Academic Press, London, pp. 189–200.

Meeuse, A. D. J. (1978) Entomophily in *Salix*: theoretical considerations, in *The Pollination of Flowers by Insects* (ed. A. J. Richards), Academic Press, London, pp. 47–50.

Meeuse, A. D. J., de Meijer, A. H., Mohr, O. W. P. and Wellinga, S. M. (1990) Entomophily in the dioecious gymnosperm *Ephedra aphylla* Forsk. with some notes on *E. campylopoda*. III. Further anthecological studies and relative importance of entomophily. *Isr. J. Bot.*, **39**, 113–23.

Meeuse, B. D. J. (1978) The physiology of some sapromyophilous flowers, in *The Pollination of Flowers by Insects* (ed. A. J. Richards), Academic Press, London, pp. 97–104.

Menges, E. S. (1991) Seed germination percentage increases with population size in a fragmented prairie species. *Cons. Biol.*, **5**, 158–64.

Menzel, M. Y. (1964) Meiotic chromosomes of monoecious Kentucky hemp (*Cannabis sativa*). *Bull. Torrey Bot. Club*, **91**, 193–205.

Michaelis, P. (1954) Cytoplasmic inheritance in *Epilobium* and its theoretical significance. *Adv. Gen.*, **6**, 287–401.

Miri, R. K. and Bubar, J. B. (1966) Self-incompatibility as an outcrossing mechanism in birdsfoot trefoil (*Lotus corniculatus*). *Can. J. Pl. Sci.*, **46**, 411–18.

Mitton, J. B., Linhart, Y. B., Davis, M. L. and Sturgeon, K. B. (1981) Estimation of outcrossing in ponderosa pine, *Pinus ponderosa* Laws. from patterns of segregation of protein polymorphism and from frequencies of albino seedlings. *Silva Gen.*, **30**, 117–21.

Mogford, D. J. (1974) Flower colour polymorphism in *Cirsium palustre*. 2. Pollination. *Heredity*, **33**, 257–63.

Mogford, D. J. (1978) Pollination and flower colour polymorphism, with special reference to *Cirsium palustre*, in *The Pollination of Flowers by Insects* (ed. A. J. Richards), Academic Press, London, pp. 191–9.

Mogie, M. (1982) The status of *Taraxacum* agamospecies. PhD thesis, University of Newcastle upon Tyne.

Mogie, M. (1985) Morphological, developmental and electrophoretic variation within and between obligately apomictic *Taraxacum* species. *Biol. J. Linn. Soc.*, **24**, 207–16.

Mogie, M. (1986) On the relationship between asexual reproduction and polyploidy. *J. Theor. Biol.*, **122**, 493–8.

Mogie, M. (1988) A model for the evolution and control of generative apomixis. *Biol. J. Linn. Soc.*, **35**, 127–53.

Mogie, M. (1992) *The Evolution of Asexual Reproduction in Plants*, Chapman & Hall, London.

Mogie, M. and Richards, A. J. (1983) Satellited chromosomes, systematics and phylogeny in *Taraxacum*. *Pl. Syst. Evol.*, **141**, 219–29.

Mohl, H. von (1863) Einige Beobachtungen uber dimorphe Bluten. *Bot. Z. Berl.*, **21**, 309.

Molau, U. and Prentice, H. C. (1992) Reproductive system and population structure in three arctic *Saxifraga* species. *J. Ecol.*, **80**, 149–61.

Moldenke, A. R. (1975) Niche specialization and species diversity along a Californian transect. *Oecologia*, **21**, 215–42.

Moore, D. M. and Lewis, H. (1965) The evolution of self-pollination in *Clarkia xantiana*. *Evolution*, **19**, 104–14.

Moran, G. F. and Brown, A. H. D. (1980) Temporal heterogeneity in outcrossing rates in alpine ash (*Eucalyptus delegatensis* R. Bak.). *Theor. Appl. Genet.*, **57**, 101–5.

Moret, J., Bari, A., Lethomas, A. and Goldblatt, P. (1992) Gynodioecy, herkogamy and sex ratio in *Romulea bulbocodium* v. *dioica* (Iridaceae). *Evol. Tr. Pl.*, **6**, 99–109.

Morita, T. (1976) Geographical distribution of diploid and polyploid *Taraxacum* in Japan. *Bull. Nat. Sci. Mus. Tokyo, ser B*, **2**, 23–38.

Morita, T. (1980) A search for diploid *Taraxacum* in Korea and eastern China, by means of pollen observations on herbarium specimens. *J. Jap. Bot.*, **55**, 35–44.

Muenchow, G. E. (1987) Is dioecy associated with fleshy fruit? *Am. J. Bot.*, **74**, 287–93.

Mukerji, S. K. (1936) Contributions to the autecology of *Mercurialis perennis*. *J. Ecol.*, **24**, 38–91, 317–39.

Mulcahy, D. L. (1967) Optimal sex ratio in *Silene alba*. *Heredity*, **22**, 411–23.

Mulcahy, D. L. (1968) The significance of delayed pistillate anthesis in *Silene alba*. *Bull. Torrey Bot. Club*, **95**, 135–9.

Mulcahy, D. L. and Mulcahy, G. B. (1983) Gametophytic self-incompatibility reexamined. *Science*, **220**, 1247–51.

Mulcahy, G. B. and Mulcahy, D. L. (1987) The effects of pollen competition. *Am. Sci.*, **75**, 44–50.

Mulcahy, D. L., Mulcahy, G. B. and Searcy, K. B. (1992) Evolutionary genetics of pollen competition, in *Ecology and Evolution of Plant Reproduction* (ed. R. Wyatt), Chapman & Hall, London, pp. 25–36.

Muldal S. (1952) The chromosomes of the earthworms. I. The evolution of polyploidy. *Heredity*, **6**, 55–76.

Muller, H. (1883) *The Fertilisation of Flowers*. Transl. W. D'Arcy. Thompson London.

Muller, H. J. (1964) The relation of recombination to mutational advance. *Mutat. Res.*, **1**, 2–9.

Muller, U. (1972) Zytologisch-embryologische Beobachtungen an *Taraxacum* Arten aus der Sektion Vulgaria Dahlst. in der Schweiz. *Ber. Geobot. Inst. Eth. Stif. Rubel*, **41**, 48–55.

Muntzing, A. (1945) The mode of reproduction of hybrids between sexual and apomictic *Potentilla argentea*. *Bot. Not.*, **107**, 49–71.

Murbeck, S. (1904) Parthenogenese bei den Gattungen *Taraxacum* und *Hieracium*. *Bot. Notiser*, **57**, 285–96.

Murfett, J., Atherton, T. L., Moa, B., Gasser, C. S. and McClure, B. A. (1994) S-RNase expressed in transgenic *Nicotiana* causes S-allele-specific pollen rejection. *Nature*, **367**, 563–6.

Murfett, J., Strabala, T. J., Zurek, D. M., Mou, B. Q., Beecher, B. and McClure, B. A. (1996) S-RNase and interspecific pollen rejection in the genus *Nicotiana* – multiple pollen-rejection pathways contribute to unilateral incompatibility between self-incompatible and self-compatible species. *Plant Cell*, **8**, 943–58.

Murray, B. G. (1979) The genetics of self-incompatibility in *Briza spicata*. *Incomp. Newslett.*, **11**, 42–5.

Nakagawa, H. (1990) Embryo-sac analysis and crossing procedure for breeding apomictic guineagrass (*Panicum maximum* Jacq.). *Jap. Agric. Res. Q.*, **24**, 163–8.

Nakanishi, T. and Hinata, K. (1973) An effective time for CO_2 gas treatment in overcoming self-incompatibility in *Brassica*. *Plant Cell Physiol.*, **14**, 873–9.

Nasrallah, J. B. and Nasrallah, M. E. (1993) Pollen-stigma signaling in the sporophytic self-incompatibility response. *Plant Cell*, **5**, 1325–35.

Nasrallah, J. B., Kao, T.-H., Goldberg, M. L. and Nasrallah, M. E. (1985) A cDNA clone encoding a S-locus-specific glycoprotein from *Brassica oleracea*. *Nature*, **318**, 263–7.

Nasrallah, M. E., Burber, J. T. and Wallace, D. H. (1969) Self-incompatibility proteins in plants: detection, genetics and possible mode of action. *Heredity*, **24**, 23–7.

Naumova, T., den Nijs, A. P. M. and Willemse, M. T. M. (1993) Quantitiative analysis of aposporous parthenogenesis in *Poa pratensis* genotypes. *Acta. Bot. Neerl.*, **42**, 299–312.

Nei, M. (1973) Analysis of gene diversity in subdivided populations. *Proc. Natl. Acad. Sci. USA*, **70**, 3321–3.

Nei, M. (1975) *Molecular Population Genetics and Evolution*, North Holland, Amsterdam.

Nettancourt, D. de (1972) Self-incompatibility in basic and applied researches with higher plants. *Genet. Agraria*, **26**, 163–216.

Nettancourt, D. de (1975) Facts and hypotheses on the origin and on the function of the S gene in *N. alata* and *L. peruvianum*. *Proc. R. Soc. Lond. B*, **188**, 345–60.

Nettancourt, D. de (1977) *Incompatibility in Angiosperms*, Springer-Verlag, Berlin.

Nettancourt, D. de, Ecochard, R., Perquin, M. D. G., van der Drift, T. and Westerhof, M. (1971) The generation of new *S* alleles at the incompatibility locus of *L. peruvianum* Mill. *Theor. Appl. Genet.*, **41**, 1120–9.

Nettancourt, D. de, Devreux, M., Laneri, U., Cresti, M., Pacini, E. and Sarfatti, G. (1974) Genetical and ultrastructural aspects of self- and cross-incompatibility in interspecific hybrids between self-compatible *Lycopersicon esculentum* and self-incompatible *L. peruvianum*. *Theor. Appl. Genet.*, **44**, 278–88.

New, J. K. (1959) A population study of *Spergula arvensis*. II. Genetics and breeding behaviour. *Ann. Bot.* n.s. **23**, 23–33.

Nijs, J. C. M. den and Sterk, A. A. (1980) Cytogeographical studies of *Taraxacum* sect. *Taraxacum* (sect. Vulgaria) in central Europe. *Bot. Jahrb. Syst.*, **101**, 527–54.

Nijs, J. C. M. den and Sterk, A. A. (1984) Cytogeography of *Taraxacum* section Taraxacum and section Alpestria in France and adjacent parts of Italy and Switzerland, including some taxonomic remarks. *Acta. Bot. Neerl.*, **33**, 1–24.

Nijs, J. C. M. den, Menken, S. B. M. and Vlot, L. (1987) Gene flow in a di-triploid mixed stand of *Taraxacum* (Asteraceae) in the Odenwald, BRD, as measured by isozyme analysis. *Abstr. 14th Int. Bot. Congress, Berlin*, p. 307.

Nitsch, J., Kurtz, E. B., Livermann, J. L. and Went, F. W. (1952) The development of sex expression in Cucurbit flowers. *Am. J. Bot.*, **39**, 32–43.

Nogler, G. A. (1972) Genetik der aposporie bei *Ranunculus auricomus*. II. Endospermzytologie. *Ber. Schw. Bot. Ges.*, **82**, 54–63.

Nogler, G. A. (1984a) Gametophytic apomixis, in *Embryology of the Angiosperms* (ed. B. M. Johri), Springer-Verlag, Berlin, pp. 475–518.

Nogler, G. A. (1984b) Genetics of apospory in apomictic *Ranunculus auricomus*. V. Conclusion. *Bot. Helv.*, **94**, 411–22.

Nogler, G. A. (1994) Genetics of gametophytic apomixis – a historical sketch. *Pol. Bot. St.*, **8**, 5–12.

Nordborg, G. (1967) Embryologic studies in the *Sanguisorba minor* complex (Rosaceae). *Bot. Notiser*, **129**, 109–20.
Nybom, H. (1985) Active self-pollination in blackberries (*Rubus* subgen. *Rubus*, Rosaceae). *Nord. J. Bot.*, **5**, 521–5.
Nybom, H. (1988) Apomixis versus sexuality in blackberries (*Rubus* subgen. *Rubus*, Rosaceae). *Pl. Syst. Evol.*, **160**, 207–18.
Nybom, H. and Schaal, B. (1990) DNA 'fingerprints' reveal genotypic distributions in natural populations of blackberries and raspberries (*Rubus*, Rosaceae). *Am. J. Bot.*, **77**, 883–8.
Nygren, A. (1967) Apomixis in the angiosperms. *Handb. der Pflanzenphys.*, **18**, 551–6.
Nyman, Y. (1993) Pollen collecting hairs of *Campanula* L. (Campanulaceae). I. Morphological variation in the retractive mechanism. *Am. J. Bot.*, **80**, 1427–36.

Ockendon, D. J. (1974) Distribution of self-incompatibility alleles and breeding structure of open-pollinated cultivars of Brussels sprouts. *Heredity*, **33**, 159–71.
Ockendon, D. J. (1977) Rare self-incompatibility alleles in a purple cultivar of Brussels sprouts. *Heredity*, **39**, 149–52.
Ockendon, D. J. (1980) Distribution of S-alleles and breeding structure of cape broccoli (*Brassica oleracea* var. '*italica*'). *Theor. Appl. Genet.*, **58**, 11–15.
O'Donnell, S. and Lawrence, M. J. (1984) The population genetics of the self-incompatibility polymorphism in *Papaver rhoeas*. IV. The estimation of the number of alleles in a population. *Heredity*, **53**, 495–508.
Olesen, J. M. and Warncke, E. (1989) Temporal changes in pollen flow and neighbourhood structure in a population of *Saxifraga hirculus*. *Oecologia*, **79**, 205–11.
Olesen, J. M. and Warncke, E. (1990) Morphological, phenological and biochemical differentiation in relation to gene flow in a population of *Saxifraga hirculus*. *Sommerfeltia*, **11**, 159–72.
Olesen, T. and Warncke, E. (1992) Breeding system and seasonal variation in seed set in a population of *Potentilla palustris*. *Nord. J. Bot.*, **12**, 373–80.
Oliveira, P. E., Gibbs, P. E., Barbosa, A. A. and Talavera, S. (1992) Contrasting breeding systems in two *Eriotheca* (Bombacaceae) species of the Brazilian cerrados. *Pl. Syst. Evol.*, **179**, 207–19.
Omodeo, P. (1952) Cariologica dei Lumbricidae. *Caryologia*, **4**, 173–275.
Ono, T. (1935) Chromosomen und sexualitat von *Rumex acetosa*. *Tohoku Imp. Univ. Sci. Rep. ser* **4**, **10**, 41–210.
Oostrum, H. van, Sterk, A. A. and Wijsman, H. J. W. (1985) Genetic variation in agamospermous microspecies of *Taraxacum* sect. *Erythrosperma* and sect. *Obliqua*. *Heredity*, **55**, 223–8.
Opler, P. A. and Bawa, K. S. (1978) Sex ratios in tropical forest trees. *Evolution*, **32**, 812–21.
Ornduff, R. (1969) Reproductive biology in relation to systematics. *Taxon*, **18**, 121.
Ornduff, R. (1979) The genetics of heterostyly in *Hypericum aegypticum*. *Heredity*, **42**, 271–2.
Ottaviano, E. and Mulcahy, D. L. (1989) Genetics of angiosperm pollen. *Adv. Genet.*, **26**, 1–64.

Palmer, R. G. (1971) Cytological studies of ameiotic and normal maize with references to premeiotic pairing. *Chromosoma*, **35**, 233–46.
Pandey, K. K. (1956) Mutations of self-incompatibility alleles in *Trifolium pratense* and *T. repens*. *Genetics*, **41**, 353–66.

Pandey, K. K. (1957) Genetics of incompatibility of *Physalis ixocarpa* Brot. A new system. *Am. J. Bot.*, **44**, 879–87.
Pandey, K. K. (1959) Mutations of the self-incompatibility gene (*S*) and pseudocompatibility in angiosperms. *Lloydia*, **22**, 222–34.
Pandey, K. K. (1962) Interspecific incompatibility in *Solanum* species. *Am. J. Bot.*, **49**, 874–82.
Pandey, K. K. (1967) Elements of the *S*-gene complex. II. Mutation and complementation at the *S*1 locus in *Nicotiana alata*. *Heredity*, **22**, 255–83.
Pandey, K. K. (1970) Time and site of the *S*-gene action, breeding systems and relationships in incompatibility. *Euphytica*, **19**, 364–72.
Pandey, K. K. (1973) Phases in *S*-gene expression and *S*-allele interaction in control of interspecific incompatibility. *Heredity*, **31**, 381–400.
Pandey, K. K. (1979) Overcoming incompatibility and promoting genetic recombination in flowering plants. *N.Z.J. Bot.*, **17**, 645–64.
Paton, D. C. (1982) The influence of honeyeaters on flowering strategies of Australian plants, in *Pollination and Evolution* (eds J. A. Armstrong, J. M. Powell and A. J. Richards), Royal Botanic Garden, Sydney, pp. 95–108.
Percival, M. S. (1961) Types of nectar in angiosperms. *New Phytol.*, **60**, 235–81.
Percival, M. S. (1965) *Floral Biology*. Pergamon, Oxford.
Perring, F. H. and Sell, P. D. (1968) *Critical Supplement to the Atlas of the British Flora*, Nelson, London.
Petanidou, Th. and Ellis, W. (1993) Pollinating fauna of a phryganic ecosystem: composition and diversity. *Biodiversity Lett.*, **1**, 9–22.
Petanidou, Th. and Vokou, D. (1990) Pollination and pollen energetics in Mediterranean ecosystems. *Am. J. Bot.*, **77**, 986–92.
Petanidou, Th. and Vokou, D. (1993) Pollination ecology of labiates in a phryganic (east Mediterranean) ecosystem. *Am. J. Bot.*, **80**, 892–9.
Philipp, M. (1980) Reproductive biology of *Stellaria longipes* Goldie as revealed by a cultivation experiment. *New Phytol.*, **85**, 557–69.
Philipp, M. and Schou, O. (1981) An unusual heteromorphic incompatibility system. Distyly, self-incompatibility, pollen load and fecundity in *Anchusa officinalis* (Boraginaceae). *New Phytol.*, **89**, 693–703.
Philipson, M. N. (1978) Apomixis in *Cortaderia jubata* (Gramineae). *N.Z.J. Bot.*, **16**, 45–59.
Pigott, C. D. (1969) The status of *Tilia cordata* and *T. platyphyllos* on the Derbyshire limestone. *J. Ecol.*, **57**, 491–504.
Pijl, L. van der (1978) Reproductive integration and sexual disharmony in floral functions, in *The Pollination of Flowers by Insects* (ed. A. J. Richards), Academic Press, London, pp. 79–88.
Piper, J. G. and Charlesworth, B. (1986) The evolution of distyly in *Primula vulgaris*. *Bot. J. Linn. Soc.*, **29**, 123–37.
Piper, J. G., Charlesworth, B. and Charlesworth, D. (1986) Breeding system evolution in *Primula vulgaris* and the evolution of reproductive assurance. *Heredity*, **56**, 207–17.
Placke, A. (1958) Effect of gibberellic acid on corolla size. *Nature*, **182**, 610.
Pleasants, T. M. (1981) Bumblebee response to variation in nectar availability, *Ecology*, **62**, 1648–61.
Plitmann, U. and Levin, D. A. (1983) Pollen–pistil relationships in the Polemoniaceae. *Evolution*, **37**, 957–67.
Pollard, A. J. and Briggs, D. (1984) Genecological studies of *Urtica dioica*. I. The nature of intraspecific variation in *U. dioica*. *New Phytol.*, **92**, 453–70.

Prakash, N., Lim, A. L. and Manurung, R. (1977) Embryology of duku and langsat varieties of *Lansium domesticum*. *Phytomorphology*, **27**, 50–9.

Prance, G. T. and Arias, J. R. (1975) A study of the floral biology of *Victoria amazonica* (Poepp.) Sowerby (Nymphaeaceae). *Acta. Amazon.*, **5**, 109–39.

Prell, H. (1921) Das Problem der Unfruchtbarkeit. *Naturw. Wochschr. N.F.*, **20**, 440–6.

Prentice, H. C. (1984) The sex ratio in a dioecious endemic plant, *Silene diclinis*. *Genetica*, **64**, 129–33.

Price, S. D. and Barrett, S. C. H. (1984) The function and adaptive significance of tristyly in *Pontederia cordata* L. (Pontederiaceae). *Biol. J. Linn. Soc.*, **21**, 315–29.

Primack, R. B. and Lloyd, D. G. (1980) Andromonoecy in the New Zealand montane shrub Manuka, *Leptospermum scoparium* (Myrtaceae). *Am. J. Bot.*, **67**, 361–8.

Primack, R. B. and Silander, J. A. (1975) Measuring the relative importance of different pollinators to plants. *Nature*, **255**, 143–4.

Proctor, M. C. F. (1978) Insect pollination syndromes in an evolutionary and ecosystemic context, in *The Pollination of Flowers by Insects* (ed. A. J. Richards), Academic Press, London, pp. 105–16.

Proctor, M. C. F. and Yeo, P. F. (1973) *The Pollination of Flowers*, Collins New Naturalist, London, p. 54.

Proctor, M. C. F., Proctor, M. E. and Groenhof, A. C. (1989) Evidence from peroxidase polymorphism on the taxonomy and reproduction of some *Sorbus* populations in south-west England. *New Phytol.*, **112**, 569–75.

Punnett, R. C. (1950) The early days of genetics. *Heredity*, **4**, 1–10.

Purrington, C. B. (1993) Parental effects on progeny sex ratio, emergence, and flowering in *Silene latifolia* (Caryophyllaceae). *J. Ecol.*, **81**, 807–11.

Putwain, P. D. and Harper, J. L. (1972) Studies in the dynamics of plant populations. V. Mechanisms governing the sex ratio in *Rumex acetosa* and *R. acetosella*. *J. Ecol.* **60**, 113–29.

Pyke, G. H. (1978a) Optimal foraging: movement patterns of bumblebees between inflorescences. *Theor. Popul. Biol.*, **13**, 72–98.

Pyke, G. H. (1978b) Optimal foraging in bumblebees and coevolution with their plants. *Oecologia*, **36**, 281–96.

Pyke, G. H. (1978c) Optimal foraging in hummingbirds: testing the marginal value theorem. *Am. Zool.*, **18**, 627–40.

Pyke, G. H. (1979) Optimal foraging in bumblebees: rule of movement between flowers within inflorescences. *Anim. Behav.*, **27**, 1167–81.

Pyke, G. H. (1980a) Optimal foraging in bumblebees: calculation of net rate of energy intake and optimal patch choice. *Theor. Popul. Biol.*, **17**, 232–46.

Pyke, G. H. (1980b) Optimal foraging in nectar-feeding birds and coevolution with their plants, in *Foraging Behaviour* (eds A. C. Kamil and T. D. Sargent), Garland Press, New York.

Pyke, G. H. (1981) Optimal foraging in hummingbirds: a test of optimal foraging theory. *Anim. Behav.*, **29**, 889–96.

Pyke, G. H. (1982a) Animal movements: an optimal foraging approach, in *The Ecology of Animal Movement* (eds I. R. Swingland and R. J. Greenwood), Oxford University Press, Oxford.

Pyke, G. H. (1982b) Evolution of inflorescence size and height in Waratahs (*Telopea speciosissima*): the difficulties of interpreting correlations between plant traits and fruit set, in *Pollination and Evolution* (eds J. A. Armstrong, J. M. Powell and A. J. Richards), Royal Botanic Gardens, Sydney, p. 914.

Pyke, G. H., Pulliam, H. R. and Charnov, E. L. (1977) Optimal foraging: a selective review of theory and tests. *Q. Rev. Biol.*, **52**, 137–54.

Queller, D. C. (1987) Sexual selection in flowering plants, in *Sexual Selection: Testing the Alternatives* (eds J. W. Bradbury and M. B. Anderson), John Wiley, New York, pp. 165–81.

Ramirez, B. W. (1969) Fig wasps: mechanism of pollen transfer. *Science*, **163**, 580–1.

Rangaswamy, N. S. and Shivanna, K. R. (1972) Overcoming self-incompatibility in *Petunia axillaris*. III. Two-site pollinations *in vitro*. *Phytomorphology*, **21**, 284–9.

Real, L. (1983) Microbehaviour and macrostructure in pollinator–plant interactions, in *Pollination Biology* (ed. L. Real), Academic Press, London, pp. 287–304.

Regal, P. J. (1982) Pollination by wind and animals: ecology of geographic patterns. *Annu. Rev. Ecol. Syst.*, **13**, 497–524.

Rhoades, M. M. (1956) Genetic control of chromosome behaviour. *Maize Genet. Coop. News Lett.*, **30**, 38–42.

Richards, A. J. (1970a) Eutriploid facultative agamospermy in *Taraxacum*. *New Phytol.*, **69**, 761–74.

Richards, A. J. (1970b) Hybridisation in *Taraxacum*. *New Phytol.*, **69**, 1103–21.

Richards, A. J. (1970c) Observations on *Taraxacum* sect. *Erythrosperma* in Slovakia. *Act. F.A.N. Univ. Comen., Bot.* **18**, 81–120.

Richards, A. J. (1972) The *Taraxacum* flora of the British Isles. *Watsonia*, 9 (suppl.), 1–141.

Richards, A. J. (1973) The origin of *Taraxacum* agamospecies. *Bot. J. Linn. Soc.*, **66**, 189–211.

Richards, A. J. (1975) Notes on the sex and age of *Potentilla fruticosa* L. in Upper Teesdale. *Trans. Nat. Hist. Soc. Northumbria*, **42**, 85–97.

Richards, A. J. (1982) The influence of minor structural changes in the flower on breeding systems and speciation in *Epipactis* Zinn. (Orchidaceae), in *Pollination and Evolution* (eds J. A. Armstrong, J. M. Powell and A. J. Richards), Royal Botanic Gardens, Sydney, pp. 47–53.

Richards, A. J. (1986) *Plant Breeding Systems* (1st ed.), Allen & Unwin, London.

Richards, A. J. (1988) Male predominant sex ratios in holly (*Ilex aquifolium* L., Aquifoliaceae) and roseroot (*Rhodiola rosea*, Crassulaceae). *Watsonia*, **17**, 53–7.

Richards, A. J. (1989a) A comparison of within-plant karyological heterogeneity between agamospermous and sexual *Taraxacum*, as assessed by the nucleolar organizer chromosome. *Pl. Syst. Evol.*, **163**, 177–85.

Richards, A. J. (1989b) The evolution of copper tolerance in apomictic *Taraxacum*. *Apomixis Newsletter*, **1**, 37–40.

Richards, A. J. (1990a) Studies in *Garcinia*, tropical dioecious fruit trees: agamospermy. *Bot. J. Linn. Soc.*, **103**, 233–50.

Richards, A. J. (1990b) Studies in *Garcinia*, tropical dioecious fruit trees: the phenology, pollination biology and fertilization of *G. hombroniana*. *Bot. J. Linn. Soc.*, **103**, 251–61.

Richards, A. J. (1990c) Studies in *Garcinia*, tropical dioecious fruit trees: the origin of the mangosteen. *Bot. J. Linn. Soc.*, **103**, 301–8.

Richards, A. J. (1990d) The implications of reproductive versatility for the structure of grass populations, in *Reproductive Versatility in the Grasses* (ed. G. P. Chapman), Cambridge University Press, Cambridge, pp. 131–153.

Richards, A. J. (1993) *Primula*, Batsford, London; Timber, New York, pp. 1–299.

Richards, A. J. and Ibrahim, H. (1978) Estimation of neighbourhood size in two populations of *Primula veris*, in *The Pollination of Flowers by Insects* (ed. A. J. Richards), Academic Press, London, pp. 165–174.
Richards, A. J. and Ibrahim, H. (1982) The breeding system in *Primula veris* L. II. Pollen tube growth and seed set. *New Phytol.*, **90**, 305–14.
Richards, A. J. and Mitchell, J. (1990) The control of incompatibility in distylous *Pulmonaria affinis* Jordan (Boraginaceae). *Bot. J. Linn. Soc.*, **104**, 369–80.
Richards, A. J., Lefebvre, C., Macklin, M., Nicholson, A. and Vekemans, X. (1989) The population genetics of *Armeria maritima* (Mill.) Willd. on the River South Tyne. *New Phytol.*, **112**, 281–93.
Richards, R. A. and Thurling, N. (1973) The genetics of self-incompatibility in *Brassica campestris* L. ssp. *oleifera* Metzg. *Genetica*, **44**, 428–38, 439–53.
Rick, C. M. and Hanna, G. C. (1943) Determination of sex in *Asparagus officinalis* L. *Am. J. Bot.*, **30**, 711–14.
Riesenberg, L. H., Hanson, M. A. and Philbrick, C. T. (1992) Androdioecy is derived from dioecy in Datiscaceae: evidence from restriction-site mapping of PCR-amplified chloroplast DNA fragments. *Syst. Bot.*, **17**, 324–36.
Riesenberg, L. H., Philbrick, C. T., Pack, P. E., Hanson, M. A. and Fritsch, P. (1993) Inbreeding depression in androdioecious populations of *Datisca glomerata* (Datiscaceae). *Am. J. Bot.*, **80**, 757–62.
Riley, H. P. (1936) The genetics and physiology of self-sterility in the genus *Capsella*. *Genetics*, **21**, 24–39.
Ritland, K. (1984) The effective proportion of self-fertilization with consanguineous matings in inbred populations. *Genetics*, **106**, 139–52.
Ritland, K. and Jain, S. (1981) A model for the estimation of outcrossing rate and gene frequencies using independent loci. *Heredity*, **47**, 35–52.
Roach, D. A. and Wulff, R. D. (1987) Maternal effects in plants. *Annu. Rev. Ecol. Syst.*, **18**, 209–35.
Roberts, R. H. (1945) *Proc. Am. Soc. Hort. Sci.*, **46**, 87.
Roggen, H. P. and van Dijk, A. J. (1972) Breaking incompatibility in *Brassica oleracea* I. by steel-brush pollination. *Euphytica*, **21**, 48–51.
Roggen, H. P., van Dijk, A. J. and Dorsman, C. (1972) 'Electric aided' pollination: a method of breaking incompatibility in *Brassica oleracea*. *Euphytica*, **21**, 181–4.
Rosov, S. A. and Screbtsova, N. D. (1958) Honeybees and selective fertilization of plants. XVII. *Int. Beekeeping Congr.*, **2**, 494–501.
Ross, M. D. (1973) The inheritance of self-incompatibility in *Plantago lanceolata*. *Heredity*, **30**, 169–76.
Ross, M. D. (1978) The evolution of gynodioecy and subdioecy. *Evolution*, **32**, 174–88.
Ross, M. D. (1982) Five evolutionary pathways to subdioecy. *Am. Nat.*, **119**, 297–318.
Ross, M. D. and Abbott, R. J. (1987) Fitness, sexual asymmetry, functional sex and selfing in *Senecio vulgaris* L. *Evol. Tr. Pl.*, **1**, 21–8.
Royama, T. (1971) Evolutionary significance of predators' response to local differences in prey density: a theoretical study, in *Dynamics of Populations* (eds P. J. den Boer and G. R. Gradwell), PUDOC, Wageningen.
Russell, S. D. (1981) Structure and quantitative cytology of male gametes of *Plumbago zeylanica* L. *Abstracts, XIII Int. Bot. Congr., Sydney*, p. 61.

Salisbury, E. J. (1942) *The Reproductive Capacity of Plants*, Bell, London.
Sampson, D. R. (1967) Frequency and distribution of self-incompatibility alleles in *Raphanus raphanistrum*. *Genetics*, **56**, 241–51.

Saran, S. and Wet, J. M. J. de (1970) The mode of reproduction in *Dicanthium intermedium* (sic) (Gramineae). *Bull. Torrey Bot. Club*, **97**, 6–13.

Saura, A., Lokki, J., Lankinen, P. and Suomalainen, E. (1976) Genetic polymorphism and evolution in parthenogenetic animals. III. *Hereditas*, **82**, 79–100.

Savidan, Y. and Pernes, J. (1982) Diploid–tetraploid–dihaploid cycles and the evolution of *Panicum maximum*. Jacq. *Evolution*, **36**, 596–600.

Savidan, Y., Grimanelli, D. and Leblanc, O. (1995) Apomixis expression in maize–*Tripsacum* hybrid derivatives and the implications regarding its control and potential for manipulation. *Apomixis Newsletter*, **8**, 35–7.

Savina, G. I. (1974) Fertilization in Orchidaceae, in *Fertilization in Higher Plants* (ed. H. F. Linskens), North Holland, Amsterdam, pp. 197–204.

Schaal, B. A. (1975) Population structure and local differentiation in *Liatris cylindracea*. *Am. Nat.*, **109**, 511–28.

Schaal, B. A. (1980) Measurement of gene flow in *Lupinus texensis*. *Nature*, **284**, 450–1.

Schaal, B. A. (1984) Life history variation, natural selection and maternal effects in plant populations, in *Perspectives on Plant Population Ecology* (eds R. Dirzo and J. Sarukhan), Sinauer, Sunderland, MA, pp. 188–206.

Schaal, B. A. and Levin, D. A. (1976) The demographic genetics of *Liatris cylindracea* Michx. (Compositae). *Am. Nat.*, **110**, 191–206.

Schaal, B. A. and Levin, D. A. (1978) Morphological differentiation and neighbourhood size in *Liatris cylindracea*. *Am. J. Bot.*, **65**, 923–8.

Schaffner, J. H. (1921) Influence of environment on sexual expression in hemp. *Bot. Gaz.*, **71**, 197–218.

Schaffner, J. H. (1923) The influence of relative length of daylight on the reversal of sex in hemp. *Ecology*, **4**, 323–34.

Scheiner, S. M. (1993) Genetics and evolution of phenotypic plasticity. *Annu. Rev. Ecol. Syst.*, **24**, 35–69.

Schemske, D. W. and Lande, R. (1985) The evolution of self-fertilization and inbreeding depression in plants. II. Empirical observations. *Evolution*, **39**, 41–52.

Schemske, D. W., Agren, J. and Le Corff, J. (1996) Deceit pollination in the monoecious neotropical herb *Begonia oaxacana*, in *Floral Biology* (eds D. G. Lloyd and S. C. H. Barrett), Chapman & Hall, London, pp. 292–318.

Schlessman, M. A. (1988) Gender diphasy ('sex choice') in *Plant Reproductive Ecology* (eds J. Lovett Doust and L. Lovett Doust), Oxford University Press, Oxford, pp. 139–156.

Schmitt, J. (1980) Pollinator foraging behaviour and gene dispersal in *Senecio* (Compositae). *Evolution*, **34**, 934–42.

Schmitt, J. and Antonovics, J. (1986) Experimental studies on the evolutionary significance of sexual reproduction. IV. Effect of neighbourhood relatedness and aphid infestation on seedling performance. *Evolution*, **40**, 830–6.

Schoen, D. J. (1982a) The breeding system of *Gilia achilleifolia*: variation in floral characteristics and outcrossing rate. *Evolution*, **36**, 352–60.

Schoen, D. J. (1982b) Genetic variation and the breeding system of *Gilia achilleifolia*. *Evolution*, **36**, 361–70.

Schoener, T. W. (1969) Models of optimal size for solitary predators. *Am. Nat.*, **103**, 277–313.

Schou, O. and Philipp, M. (1983) An unusual heteromorphic incompatibility system. 3. On the genetic control of distyly and self-incompatibility in *Anchusa officinalis* L. (Boraginaceae). *Theor. Appl. Genet.*, **68**, 139–44.

Schuster, A., Noy-Meir, I., Heyn, C. and Dafni, A. (1993) Pollination dependent female reproductive success in a self-compatible outcrosser *Asphodelus aestivus* Brot. *New Phytol.*, **123**, 167–74.

Sears, E. R. (1937) Cytological phenomena connected with self-sterility in the flowering plants. *Genetics*, **22**, 130–81.

Seavey S. R. and Bawa, K. S. (1986) Late-acting self-incompatibility in the angiosperms. *Bot. Rev.*, **52**, 195–219.

Sharma, S. and Shivanna, K. R. (1982) Effects of pistil extracts on in vitro responses of compatible and incompatible pollen in *Petunia hybrida* Vilm. *Ind. J. Exp. Biol.*, **20**, 255–6.

Shen, H. H., Rudin, D. and Lindgren, D. (1981) Study of the pollination pattern in a scots pine seed orchard by means of isozyme analysis. *Silva Gen.*, **30**, 7–15.

Shivanna, K. R. and Heslop-Harrison, J. (1981) Membrane state and pollen viability. *Ann. Bot.*, **47**, 759–70.

Shivanna, K. R. and Rangaswamy, N. S. (1969) Overcoming self-incompatibility in *Petunia axillaris*. I. Delayed pollination, pollination with stored pollen, and bud pollination. *Phytomorphology*, **19**, 372–80.

Shivanna, K. R., Heslop-Harrison, J. and Heslop-Harrison, Y. (1981) Heterostyly in *Primula*. 2. Sites of pollen inhibition, and effects of pistil constituents on compatible and incompatible pollen-tube growth. *Protoplasma*, **107**, 319–37.

Shivanna, K. R., Heslop-Harrison, J. and Heslop-Harrison, Y. (1983) Heterostyly in *Primula*. 3. Pollen water economy: a factor in the intramorph incompatibility response. *Protoplasma*, **117**, 175–84.

Shore, J. S. and Barrett, S. C. H. (1985) The genetics of distyly and homostyly in *Turnera ulmifolia* L. (Turneraceae). *Heredity*, **55**, 167–74.

Siafaca, L., Adamandiadou, A. and Margaris, N. S. (1980) Caloric content in plants dominating maquis ecosystems in Greece. *Oecologia*, **44**, 276–80.

Simmonds, N. W. (1971) The breeding system of *Chenopodium quinoa*. I. Male sterility. *Heredity*, **27**, 73–82.

Simmonds, N. W. (1976) *Evolution of Crop Plants*, Longman, London.

Sirks, M. J. (1927) The genotypical problems of self and cross-incompatibility. *Mem. Hort. Soc. New York*, **3**, 325–43.

Slatkin, M. (1985) Gene flow in natural populations. *Annu. Rev. Ecol. Syst.*, **16**, 393–430.

Smith, B. W. (1963) The mechanism of sex determination in *Rumex hastatulus*. *Genetics*, **48**, 1265–88.

Smith, D. B. and Adams, W. T. (1983) Measuring pollen contamination in clonal seed orchards with the aid of genetic markers, in *Proc. 17th South. Forest Tree Improvement Conf.*, pp. 64–73.

Snaydon, R. W. (1970) Rapid population differentiation in a mosaic environment. I. The response of *Anthoxanthum odoratum* populations to soils. *Evolution*, **24**, 257–69.

Snow, A. A. (1982) Pollination intensity and potential seed set in *Passiflora vitifolia*. *Oecologia*, **55**, 231–7.

Snow, A. A. and Lewis, P. O. (1993) Reproductive traits and male fertility in plants. *Annu. Rev. Ecol. Syst.*, **24**, 331–51.

Snow, A. A., Spira, T. P. Simpson, R. and Klips, R. A. (1996) The ecology of geitonogamous pollination, in *Floral Biology* (eds D. G. Lloyd and S. C. H. Barrett, Chapman & Hall, London, pp. 192–216.

Soest, J. L. van (1963) *Taraxacum* species from India, Pakistan and neighbouring countries. *Wentia*, **10**, 1–91.

Solbrig, O. T. and Rollins, R. C. (1977) The evolution of autogamy in species of the mustard genus *Leavenworthia*. *Evolution*, **31**, 265–81.

Solbrig, O. T. and Simpson, B. B. (1974) Components of regulation of a population of dandelions in Michigan. *J. Ecol.*, **62**, 473–86.

Solnetzeva, M. P. (1974) Disturbances in the progress of fertilization in angiosperms under hemigamy, in *Fertilization in Higher Plants* (ed. H. F. Linskens), North Holland, Amsterdam, pp. 311–24.

Sorensen, T. (1958) Sexual chromosome-aberrants in triploid apomictic *Taraxaca*. *Bot. Tidskr.*, **54**, 1–22.

Sorensen, T. and Gudjonsson, G. (1946) Spontaneous chromosome-aberrants in apomictic *Taraxaca*. *K. Danske Vid. Selsk. Biol. Medd.*, **4**, 3–48.

Soule, M. E. (1976) Allozyme variation: its determinants in space and time, in *Molecular Evolution* (ed. F. J. Ayala), Sinauer, Sunderland, MA, pp. 60–77.

Sparrow, F. K. and Pearson, N. L. (1948) Pollen compatibility in *Asclepias syriaca*. *J. Agric. Res.*, **77**, 187–99.

Spoor, W. (1976) Self-incompatibility in *Lolium perenne* L. *Heredity*, **37**, 417–21.

Sprengel, C. K. (1793) *Das Entdecke Geheimnis der Natur im Bau und in der Befruchtung der Blumen*. Berlin.

Stace, C. A. (1975) *Hybridisation and the Flora of the British Isles*, Academic Press, London.

Stanley, R. G. and Loewus, F. A. (1964) Boron and myo-inositol in pollen pectin biosynthesis, in *Pollen Physiology and Germination* (ed. H. F. Linskens), North Holland, Amsterdam, pp. 128–36.

Stanton, M. L., Ashman, T.-L., Galloway, L. F. and Young H. J. (1992) Estimating male fitness in natural populations, in *Ecology and Evolution of Plant Reproduction* (ed. R. Wyatt), Chapman & Hall, London, pp. 62–90.

Staudt, G. (1952) Genetische Untersuchungen an *Fragaria orientalis* Los. und ihre Bedeutung fur Arbildung und Geschlechts-Differenzierung in der Gattung *Fragaria* L. *Z. Indukt. Abst. Ver.*, **84**, 361–416.

Stebbins, G. L. (1950) *Variation and Evolution in Plants*, Columbia University Press, New York.

Stebbins, G. L. and Ferlan, L. (1956) Population variability, hybridization and introgression in some species of *Ophrys*. *Evolution*, **10**, 32–46.

Stelleman, P. (1979) The significance of biotic pollination in a nominally anemophilous plant: *Plantago lanceolata*. *Proc. Kan. Nederl. Akad. Wet. ser. C*, **87**, 95–119.

Stelleman, P. (1984) Reflections on the transition from wind pollination to ambophily. *Acta. Bot. Neerl.*, **33**, 497–508.

Sterk, A. A. (1987) Aspects of the population biology of sexual dandelions in the Netherlands, in *Vegetation between Land and Sea* (eds A. H. L. Huiskes, C. W. P. M. Blom and J. Rozema), Dordrecht, pp. 284–90.

Stevens, D. P. (1985) Studies in *Saxifraga granulata* L. PhD thesis, University of Newcastle upon Tyne.

Stevens, D. P. (1988) On the gynodioecious polymorphism in *Saxifraga granulata*. *Biol. J. Linn. Soc.*, **35**, 15–28.

Stevens, D. P. and Richards, A. J. (1985) Gynodioecy in *Saxifraga granulata*. *Pl. Syst. Evol.*, **151**, 43–54.

Stevens, D. P. and van Damme, J. M. M. (1988) The evolution and maintenance of gynodioecy in sexually and vegetatively reproducing plants. *Heredity*, **61**, 329–37.

Stevens, V. A. M. and Murray, B. G. (1982) Studies on heteromorphic self-incompatibility systems: physiological aspects of the incompatibility system of *Primula obconica*. *Theor. Appl. Genet.*, **61**, 245–56.

Stiles, F. G. (1975) Ecology, flowering phenology and hummingbird pollination of some Costa Rican *Heliconia* species. *Ecology*, **56**, 285–301.

Storey, W. B. (1975) in *Advances in Fruit Breeding* (eds J. Janick and J. N. Moore), Purdue University Press, West Lafayette, pp. 568–89.

Stout, A. B. (1916) Self- and cross-pollinations in *Cichorium intybus* with reference to sterility. *Mem. New York Bot. Gdn*, **6**, 333–454.

Stout, A. B. (1917) Fertility in *Cichorium intybus*. The sporadic occurrence of self-fertile plants among the progeny of self-sterile plants. *Am. J. Bot.*, **4**, 375–95.

Stout, A. B. (1918a) Experimental studies of self-incompatibilities in fertilization. *Proc. Soc. Exp. Biol. Med.*, **15**, 51–4.

Stout, A. B. (1918b) Fertility in *Cichorium intybus*: self-compatibility and self-incompatibility among the offspring of self-fertile lines of descent. *J. Genet.*, **8**, 71–103.

Stout, A. B. and Chandler, C. (1942) Hereditary transmissions of induced tetraploidy and compatibility in fertilisation. *Science*, **96**, 257.

Stoutamire, W. P. (1983) Wasp-pollinated species of *Caladenia* (Orchidaceae) in south-western Australia. *Aust. J. Bot.*, **31**, 383–94.

Strenseth, N. C., Kirkendall, L. R. and Moran, N. (1985) On the evolution of pseudogamy. *Evolution*, **39**, 294–307.

Strid, A. (1970) Studies in the Aegean Flora. XVI. Biosystematics of the *Nigella arvensis* complex. *Opera Bot.*, **28**, 1–169.

Suomalainen, E. (1969) Evolution in parthenogenetic Curculionidae. *Evol. Biol.*, **3**, 261–96.

Suomalainen, E., Saura, A. and Lokki, J. (1976) Evolution of parthenogenetic insects. *Evol. Biol.*, **9**, 209–57.

Sutherland, S. (1986) Patterns of fruit-set: what controls fruit–flower ratios in plants? *Evolution*, **40**, 117–28.

Sutherland, S. and Delph, L. (1984) On the importance of male fitness in plants; patterns of fruit set. *Ecology*, **65**, 1093–104.

Taktajhan, A. (1969) *Flowering Plants, Origin and Dispersal* (Transl. C. Jeffrey), Oliver and Boyd, Edinburgh.

Taliaferro, C. M. and Bashaw, E. C. (1966) Inheritance and control of obligate apomixis in breeding buffelgrass, *Pennisetum ciliare*. *Crop Sci.*, **6**, 473–6.

Tauber, H. (1965) Differential pollen dispersal and the interpretation of pollen diagrams. *Dan. Geo. Unders.* (Afh.) Racke 2 **89**, 1–70.

Thompson, M. M. (1979) Incompatibility alleles in *Corylus avellana* L. cultivars. *Theor. Appl. Gen.*, **55**, 29–33.

Thomson, J. D. (1982) Patterns of visitation by animal pollinators. *Oikos*, **39**, 241–50.

Thomson, J. D. and Barrett, S. C. H. (1981) Selection for outcrossing, sexual selection and the evolution of dioecy in plants. *Am. Nat.*, **118**, 443–9.

Thomson, J. D. and Brunet, J. (1990) Hypotheses for the evolution of dioecy in seed plants. *Trends Ecol. Evol.*, **5**, 11–16.

Thomson, J. D. and Plowright, R. C. (1980) Pollen carryover, nectar rewards, and pollinator behaviour with special reference to *Diervilla lonicera*. *Oecologia*, **46**, 68–74.

Thomson, J. D. and Thomson, B. A. (1992) Pollen presentation and viability schedules in animal-pollinated plants: consequences for reproductive success, in *Ecology and Evolution of Plant Reproduction* (ed. R. Wyatt), Chapman & Hall, London, pp. 1–24.

Thomson, J. D., Andrews, B. J. and Plowright, R. C. (1981) The effect of foreign pollen on ovule development in *Diervilla lonicera* (Caprifoliaceae). *New Phytol.*, **90**, 777–83.

Trela, Z. (1963) Embryological studies in *Anemone nemorosa* L. *Acta. Biol. Crac.*, **6**, 1–14.

Tremayne, M. and Richards, A. J. (1993) Homostyly and herkogamous variation in *Primula* L. section Muscarioides Balf. f. *Evol. Trends Plants*, **7**(2), 15–20.

Tremayne, M. and Richards, A. J. (1997) The effects of breeding system and seed weight on the plant fitness in *Primula scotica* Hook. *BES Symposium on Small Populations* (ed. J. Warren), (in press).

Uno, G. E. (1982) Comparative reproductive biology of hermaphroditic and male-sterile *Iris douglasiana* Herb. (Iridaceae). *Am. J. Bot.*, **69**, 818–23.

Usberti, J. A. and Jain, S. K. (1978) Variation in *Panicum maximum*; a comparison of sexual and asexual populations. *Bot. Gaz.*, **139**, 112–16.

Vaarama, A. and Jaaskelainen, O. (1973) Studies on gynodioecism in the Finnish populations of *Geranium sylvaticum* L. *Ann. Acad. Sci Fenn.*, **A108**, 1–39.

Valdeyron, G., Dommee, B. and Vernet, P. (1977) Self-fertilisation in male-fertile plants of a gynodioecious species: *Thymus vulgaris* L. *Heredity*, **39**, 243–9.

Vansell, G. H., Watkins, W. G. and Bishop, R. K. (1942) Orange nectar and pollen in relation to bee activity. *J. Econ. Entomol.*, **35**, 321–3.

Vekemans, X., Lefebvre, C., Belalia, L. and Meerts, P. (1990) The evolution and breakdown of the heteromorphic incompatibility system in *Armeria maritima* revisited. *Evol. Trends Plants*, **4**, 15–23.

Vogel, S. (1962) Duftdrusen im Dienste der Bestaubung: uber Bau und Funktion der Osmophoren. *Abh. Math.-Naturw. Kl. Akad. Wiss. Mainz*, **10**, 599–763.

Vogel, S. (1966) Scent organs of orchid flowers and their relation to insect pollination. *Proc. Fifth World Orchid Conference*, Long Beach, California, pp. 253–9.

Vogel, S. (1978) Evolutionary shifts from reward to deception in pollen flowers, in *The Pollination of Flowers by Insects* (ed. A. J. Richards), Academic Press, London, pp. 89–96.

Voight, P. W. and Burson, B. L. (1981) Breeding of apomictic *Eragrostis curvula. Proc. XIV Int. Grassland Congr*. Lexington, KY, pp. 160–3.

Vokou, D., Petanidou, Th. and Bellos, D. (1990) Pollination ecology and reproductive potential of *Jankaea heldreichii* (Gesneriaceae); a tertiary relict on Mt. Olympus, Greece. *Biol. Cons.*, **52**, 125–33.

Vries, H. de (1907) *Plant Breeding*. Open Court, Chicago.

Waddington, K. D. (1979) Divergence in inflorescence height: an evolutionary response to pollinator fidelity. *Oecologia*, **40**, 43–50.

Waddington, K. D. (1981) Factors influencing pollen flow in bumblebee pollinated *Delphinium virescens. Oikos*, **37**, 153–9.

Waddington, K. D. (1983) Foraging behaviour of pollinators, in *Pollination Biology* (ed. L. Real), Academic Press, New York, pp. 213–39.

Waddington, K. D. (1985) Cost-intake information used in foraging. *J. Insect Physiol.*, **31**, 891–7.

Waddington, K. D. and Heinrich, B. (1981) Patterns of movement and floral choice by foraging bees, in *Foraging Behaviour: Ecological, Ethological and Psychological Approaches* (eds A. Kamil and T. Sargent), Garland STPM Press, New York, pp. 215–30.

Waddington, K. D., Allen, T. and Heinrich, B. (1981) Floral preferences of bumblebees (*Bombus edwardsii*) in relation to intermittent versus continuous rewards. *Anim. Behav.*, **29**, 779–84.

Wade, K. M. (1981a) Experimental studies on the distribution of the sexes of *Mercurialis perennis* L. II. Transplanted populations under different canopies in the field. *New Phytol.*, **87**, 439–46.

Wade, K. M. (1981b) Experimental studies on the distribution of the sexes of *Mercurialis perennis* L. III. Transplanted populations under light screens. *New Phytol.*, **87**, 447–55.

Wade, K. M., Armstrong, R. A. and Woodell, S. R. J. (1981) Experimental studies on the distribution of the sexes of *Mercurialis perennis* L. 1. Field observations and canopy removal experiments. *New Phytol.*, **87**, 431–8.

Wahlund, S. (1928) Zusammensetzung von Populationen und Korrelationserscheinungen vom Standpunkt der Vererbungslehre ausbetrachtet. *Hereditas*, **11**, 65–106.

Waller, D. W. (1984) Differences in fitness between seedlings derived from cleistogamous and chasmogamous flowers in *Impatiens capensis. Evolution*, **38**, 427–40.

Waller, D. M. (1988) Plant morphology and reproduction, in *Plant Reproductive Ecology* (eds J. Lovett Doust and L. Lovett Doust), Oxford University Press, Oxford, pp. 203–27.

Warmke, H. E. (1946) Sex determination and sex balance in *Melandrium album. Am. Bot.*, **33**, 648–60.

Warncke, E., Terndrup, U., Michelsen, V. and Erhardt, A. (1993) Flower visitors to *Saxifraga hirculus* in Switzerland and Denmark, a comparative study. *Bot. Helv.*, **103**, 141–7.

Waser, N. M. (1983) The adaptive nature of floral traits: ideas and evidence, in *Pollination Biology* (ed. L. Real), Academic Press, London, pp. 242–77.

Waser, N. M. (1988) Comparative pollen and dye transfer by pollinators of *Delphinium nelsonii. Funct. Ecol.*, **2**, 41–8.

Waser, N. M. (1993) Population structure, optimal outbreeding and assortative mating in angiosperms, in *The Natural History of Inbreeding and Outbreeding: Theoretical and Empirical Perspectives* (ed. N. W. Thornhill), University of Chicago Press, Chicago.

Waser, N. M. and Price, M. V. (1981) Pollinator choice, and stabilizing selection for flower color in *Delphinium nelsonii. Evolution*, **35**, 376–90.

Waser, N. M. and Price, M. V. (1983) Optimal and actual outcrossing in plants, and the nature of plant-pollinator interaction, in *Handbook of Experimental Pollination Biology* (eds C. E. Jones and R. J. Little), Van Nostrand, New York, pp. 341–59.

Washitani, I., Osawa, R., Namai, H. and Niwa, M. (1994) Patterns of female fertility in heterostylous *Primula sieboldii* under severe pollinator limitation. *J. Ecol.*, **82**, 571–9.

Webb, C. J. (1979a) Breeding systems and the evolution of dioecy in New Zealand apioid Umbelliferae. *Evolution*, **33**, 662–72.

Webb, C. J. (1979b) Breeding system and seed set in *Euonymus europaeus* (Celastraceae). *Pl. Syst. Evol.*, **132**, 299–303.

Webb, C. J. (1981) Andromonoecism, protandry, and sexual selection in Umbelliferae. *N.Z.J. Bot.*, **19**, 335–8.

Webb, C. J. and Lloyd, D. G. (1980) Sex ratios in New Zealand apioid Umbelliferae. *N.Z.J. Bot.*, **18**, 121–6.

Wedderburn, F. M. (1988) A Comparison of Heteromorphic Incompatibilities in *Primula*. PhD. Thesis, University of Newcastle upon Tyne, UK.

Wedderburn, F. M. and Richards, A. J. (1990) Variation in within-morph incompatibility inhibition sites in heteromorphic *Primula* L. *New Phytol.*, **116**, 149–62.

Wedderburn, F. M. and Richards, A. J. (1992) Secondary homostyly in *Primula* L.; evidence for the model of the 'S' supergene. *New Phytol.*, **121**, 649–55.

Weiner, J. (1988) The influence of competition on plant reproduction, in *Plant Reproductive Ecology* (eds J. Lovett Doust and L. Lovett Doust), Oxford University Press, Oxford, pp. 228–45.

Weller, S. G. (1993) Evolutionary modifications of tristylous breeding systems, in *Evolution and Function of Heterostyly* (ed. S. C. H. Barrett), Springer-Verlag, Berlin, pp. 247–71.

Wendelbo, P. (1959) *Taraxacum gotlandicum*, a pre-boreal relic in the Norwegian Flora? *Nytt. Mag. Bot.*, **7**, 161–7.

Wendelbo, P. (1961a) An account of *Primula* subgenus *Sphondylia* with a review of the subdivisions of the genus. *Arbok Univ. Bergen, Mat-Nat.*, **11**, 33–43.

Wendelbo, P. (1961b) Studies in Primulaceae. III. On the genera related to *Primula* with special reference to their pollen morphology. *Arbok Univ. Bergen, Mat-Nat.*, **19**, 1–31.

Westergaard, M. (1940) Studies on polyploidy and sex determination in polyploid forms of *Melandrium album*. *Dansk. Bot. Arkiv.*, **10**(5), 1–131.

Westergaard, M. (1946) Aberrant Y chromosomes and sex expression in *Melandrium album*. *Hereditas*, **32**, 419–43.

Westergaard, M. (1948) The relation between chromosome constitution and sex in the offspring of triploid *Melandrium*. *Hereditas*, **34**, 257–79.

Westergaard, M. (1953) Uber den Mechanismus der Geschlechtsbestimmung bei *Melandrium album*. *Die Naturwiss.*, **9**, 253–60.

Westergaard, M. (1958) The mechanism of sex determination in dioecious flowering plants. *Adv. Genet.*, **9**, 217–81.

Wet, J. M. de and Stalker, H. T. (1974) Gametophytic apomixis and evolution in plants. *Taxon*, **23**, 689–97.

Whitehead, D. R. (1983) Wind pollination: some ecological and evolutionary perspectives, in *Pollination Biology* (ed. L. Real), Academic Press, London, pp. 97–109.

Whitehouse, H. L. K. (1950) Multiple-allelomorph incompatibility of pollen and style in the evolution of the angiosperms. *Ann. Bot. n.s.* **14**, 198–216.

Whitmore, T. C. (1984) *Tropical Rain Forests of the Far East*, 2nd edn. Oxford University Press, Oxford.

Williams, E. G., Kaul, V., Rouse, J. L. and Knox, R. B. (1984) Apparent self-incompatibilty in *Rhododendron ellipticum*, *R. championae* and *R. amamiense*: a post-zygotic mechanism. *Incompatibility Newsletter*, **16**, 10–11.

Williams, G. C. (1975) *Sex and Evolution*. Princeton University Press, Princeton, NJ.

Williams, N. H. and Dodson, C. H. (1972) Selective attraction of male euglossid bees to orchid floral fragrances and its importance in long distance pollen flow. *Evolution*, **26**, 84–95.

Williams, W. (1964) *Genetic Principles and Plant Breeding*, Blackwell, Oxford.

Willing, P. R., Bashe, D. and Mascarenhas, J. P. (1988) Analysis of the quantity of diversity of messenger RNAs from the pollen and shoots of *Zea mays*. *Theor. Appl. Genet.*, **75**, 75–83.

Willson, M. F. (1979) Sexual selection in plants. *Am. Nat.*, **113**, 777–90.

Winge, O. (1931) X- and Y-linked inheritance in *Melandrium*. *Hereditas*, **15**, 127–65.

Wolf, L. L. (1975) Energy intake and expenditures in a nectar-feeding sunbird. *Ecology*, **56**, 92–104.

Wolf, L. L. and Hainsworth, F. R. (1971) Time and energy budgets of territorial hummingbirds. *Ecology*, **52**, 980–8.

Wolf, L. L., Hainsworth, F. R. and Stiles, F. G. (1972) Energetics of foraging: rate and efficiency of nectar extraction by hummingbirds. *Science*, **176**, 1351–2.

Woodell, S. R. J. (1960a) Studies in British *Primulas*. VII. Development of seed from reciprocal crosses between *P. vulgaris* Huds and *P. veris* L. *New Phytol.*, **59**, 302–13.

Woodell, S. R. J. (1960b) Studies in British *Primulas*. VIII. Development of seed from reciprocal crosses between *P. vulgaris* Huds and *P. elatior* (L.) Hill. and between *P. veris* L. and *P. elatior* (L.) Hill. *New Phytol.*, **59**, 314.

Woodell, S. R. J. (1978) Directionality in bumblebees in relation to environmental factors, in *The Pollination of Flowers by Insects* (ed. A. J. Richards), Academic Press, London, pp. 31–9.

Woodruff, R. C. and Thompson, J. N. (1980) Hybrid release of mutator activity and the genetic structure of natural populations. *Evol. Biol.*, **12**, 129–62.

Woodruff, R. C., Thompson, J. N. and Lyman, R. F. (1979) Intraspecific hybridization and the release of mutator activity. *Nature*, **278**, 277–9.

Wright, J. W. (1953) Pollen-dispersion studies: some practical applications. *Forestry*, **51**, 114–18.

Wright, S. (1921) Systems of mating. *Genetics*, **6**, 111–78.

Wright, S. (1931) Evolution in Mendelian populations. *Genetics*, **16**, 97–159.

Wright, S. (1938) Size of population and breeding structure in relation to evolution. *Science*, **87**, 430–1.

Wright, S. (1939) The distribution of self-sterility alleles in populations. *Genetics*, **24**, 538–52.

Wright, S. (1940) Breeding structure of populations in relation to speciation. *Am. Nat.*, **74**, 232–48.

Wright, S. (1943) Isolation by distance. *Genetics*, **28**, 114–38.

Wright, S. (1946) Isolation by distance under diverse systems of mating. *Genetics*, **31**, 39–59.

Wright, S. (1951) The genetical structure of populations. *Ann. Eugen.* **15**, 323–54.

Wright, S. (1978) *Evolution and the Genetics of Populations*, Vol. 4. *Variability within and among natural populations*, University of Chicago Press, Chicago.

Wright, S., Dobzhansky, Th. and Hovanitz, W. (1942) Genetics of natural populations. VII. The allelism of lethals in the third chromosome of *Drosophila pseudoobscura*. *Genetics*, **27**, 363–94.

Wyatt, R. (1981) Components of reproductive output in five tropical legumes. *Bull. Torrey Bot. Club*, **108**, 67–75.

Wyatt, R. (1984) The evolution of self-pollination in granite outcrop species of *Arenaria* (Caryophyllaceae). Morphological correlates. *Evolution*, **38**, 804–16.

Wyatt, R. (1990) The evolution of self-pollination in granite outcrop species of *Arenaria* (Caryophyllaceae). 5. Artificial crosses within and between populations. *Syst. Bot.*, **15**, 363–9.

Wyatt, R. (ed.) (1992) *Ecology and Evolution of Plant Reproduction: New approaches*. Chapman & Hall, London.

Yamamota, Y. (1938) Karyogenetische Untersuchungen der Gattung *Rumex*. VI. Geschechtsbestimmung bei eu- und aneuploiden Pflanzen von *Rumex acetosa* L. *Kyoto Imp. Univ., Mem. Coll. Agri.*, **43**, 1–59.

Yampolsky, E. and Yampolsky, H. (1922) Distribution of sex forms in the phanerogamic flora. *Bibl. Genet.*, **3**, 1–62.

Yeboah Gyan, K. and Woodell, S. R. J. (1987) Nectar production, sugar content, amino acids and potassium in *Prunus spinosa* L., *Crataegus monogyna* Jacq. and *Rubus fruticosus* at Wytham, Oxfordshire. *Funct. Ecol.*, **1**, 251–9.
Yeo, P. F. (1975) Some aspects of heterostyly. *New Phytol.*, **75**, 147–53.
Yudin, B. F. (1970) Capacity for parthenogenesis and effectiveness of selection on the basis of this character in diploid and autotetraploid maize. *Genetika*, **6**, 13–22.

Zimmerman, J. K. (1980) Ecological correlates of labile sex expression in the orchid *Catasetum viridiflavum*. *Ecology*, **72**, 597–608.
Zimmerman, M. (1979) Optimal foraging: a case for random movement. *Oecologia*, **43**, 261–7.
Zimmerman, M. (1982) The effect of nectar production on neighbourhood size. *Oecologia*, **52**, 104–8.
Zimmerman, M. (1983) Plant reproduction and optimal foraging: experimental nectar manipulations in *Delphinium nelsonii*. *Oikos*, **41**, 57–63.
Zuberi, M. I. and Dickinson, H. G. (1985a) Pollen–stigma interaction in *Brassica*. III. Hydration of the pollen grains. *J. Cell Sci.*, **76**, 321–36.
Zuberi, M. I. and Dickinson, H. G. (1985b) Modification of the pollen–stigma interaction in *Brassica oleracea* by water. *Ann. Bot.*, **56**, 443–52.
Zuberi, M. I. and Lewis, D. (1988) Gametophytic–sporophytic incompatibility in the Cruciferae – *Brassica campestris*. *Heredity*, **61**, 367–77.
Zuk, J. (1963) An investigation on polyploidy and sex determination within the genus *Rumex*. *Acta. Soc. Bot. Pol.*, **23**, 5–67.

Index

Page numbers appearing in *italics* refer to tables and page numbers appearing in **bold** refer to figures.